# Fabrication Methods
# for Precision Optics

T0178596

# Fabrication Methods for Precision Optics

Hank H. Karow

A JOHN WILEY & SONS, INC., PUBLICATION

*Library of Congress Cataloging-in-Publication Data is available.*

ISBN 0-471-70379-6

10 9 8 7 6 5 4 3 2 1

*Dedicated to the memory of my parents
Dr. Fritz and Lotte Karow*

# Contents

# Preface

This book is intended to provide a theoretical foundation for the practicing optician to complement the practical experience gained on the shop floor. It should also serve the master optician, the manufacturing engineer, and the shop manager as a useful reference.

This book is a collection of carefully structured facts that relate to optics fabrication and testing. The topics are approached from the perspective of the optician who asks: "What do I need to know to do my job?" A good understanding of these facts will provide a solid basis for meaningful exploration of new methods and approaches, rather than the often aimless experimentation that is so prevalent in many optical shops.

The book is the result of collaborative efforts of several former opticians. Consequently it is the only optics fabrication book written by opticians specifically with the practicing optician in mind. The basic layout of the chapters and most of the mathematical presentations can be traced back to my translation of an unpublished manuscript in German by my former teacher Harry Schade. This manuscript was based on his book with the German title *Arbeitsverfahren der Feinoptik* which was very popular with German opticians during the 1950s and 1960s. Many of the illustrations that have been so expertly produced with the latest in computer graphics by Gerhard Wolf can also be traced to Schade's work. My contribution was a complete revision of the translated text to purge from it most of the outdated methods and references and to greatly expand the scope of the book. Many sections were added to include most aspects of the contemporary way of fabricating and testing optics. Numerous tables and illustrations were added in support of this expanded coverage.

The references and examples reflect the recognition that optics fabrication has become a global industry that prefers metric measure over the English measure. As a result most dimensions are expressed in metric and are immediately followed by the approximate English equivalent in brackets. The approximate linear conversion factors of 0.040 inch per millimeter or 25 millimeters per inch have been used to keep the numbers simple. Only where precision is needed, as for specific examples, were the exact conversion factors applied.

Some topics appear to have received a needlessly exhaustive coverage, for instance, testplate testing in the last chapter. These are deliberate efforts to document as much as possible the important aspects of those topics that are in danger of sliding into oblivion as the industry converts to more modern methods. A clear understanding of the basic principles that govern these topics will be essential in understanding the more complicated aspects of the modern way of making optics.

Other topics are not discussed at all in this book, or they are only mentioned in passing. That has also been a deliberate decision. The intent of this book is to focus on the facts that are important for the average commercial optical shop. Modern manufacturing methods such as single-point diamond turning, ductile regime nanogrinding, computer-controlled optical surfacing, computer-integrated optical manufacturing systems, gradient index optics, diffractive optical elements and micro-optics for opto-electronic applications are exciting topics about which much has already been written and more will undoubtedly be written in the years to come. But as interesting as these methods are, they are not an accepted way of producing optics in the majority of optical shops. I hope that in the not so distant future some of these methods will revolutionize the way optics are produced. When the time comes, another book will have to be written. In the meantime, the vast majority of optical components will continue to be produced by contemporary versions of traditional methods. It is with this in mind that this book has been written.

A long-term project such as writing a book can never be a one-person show. The help of others is needed. The writing of this book was no exception. Special recognition must go to Harry Schade who laid the foundation to this work. Unfortunately, he did not live long enough to see the book completed. I am especially indebted to my old friend and colleague Gerhard Wolf who produced the many professionally executed illustrations with great skill and patience. They greatly enhance the usefulness of this book. I must also thank my colleague Dr. Philip W. Baumeister who helped me in many ways with the pitfalls of the word processor software. Special thanks must go to my wife Diane. Her untiring help proved invaluable in the timely preparation of the extensive index.

I sincerely hope that this book will fulfill its intended purpose as a useful source of facts and as a valuable reference for the practicing optician and for all professionals who are involved in the fabrication of precision optics.

HANK H. KAROW

*Auburn, California*
*January, 1993*

Fabrication Methods for Precision Optics

# 1

# Optical Materials

*Optical materials* are solid substances that are either transparent in the optical part of the spectrum or are in other ways useful in transforming, modifying, or redirecting light in the ultraviolet (UV), visible (VIS), or infrared (IR) regions. Only a little more than a quarter century ago the term "optical material" was essentially synonymous with optical glass. Although a variety of optically useful crystals were available, most opticians rarely had the opportunity to work these materials.

The invention of the laser in the early 1960s and the rapid development of associated technologies initiated significant changes in optical materials, and the opticians of that time were confronted with a rapidly growing list of new and often exotic materials of which little or nothing was known. This was particularly true of the workability of these new materials, and opticians had to quickly learn to adapt and develop imaginative methods to deal with these new requirements. Some old and many new optical materials were evaluated for use as host materials for solid state laser rods, and in terms of the extent to which they could modulate, deflect, attenuate, alter, or otherwise manipulate the laser light that was emitted by these modern scientific wonders. Although 25 years have since past, this process is still continuing today.

The late 1960s and early 1970s saw the emergence of infrared materials mainly for military applications, and IR components have since become a very important part of optics. There are a number of optical companies in this country and in other parts of the world that specialize in producing optical components for IR systems. The 1970s also witnessed the development of single-point diamond turning (SPDT), which added a variety of metals to the growing list of practical optical materials. Although some of these metals had been previously polished for optical purposes, the use of SPDT fabrication methods made it much easier to produce very complex surfaces on metal substrates that could not be made with traditional methods.

More efficient laser glasses were perfected during the late 1970s and early 1980s. The impetus for this development effort came from the laser fusion efforts at a number of laboratories, such as the Shiva-Nova laser fusion facility at Lawrence Livermore National Laboratory in the United States. The emphasis during the 1980s was on new materials for fiber optics and

1

opto-electronics, and on the development of new laser rod materials and frequency doubling crystals to increase the versatility of lasers for use in medicine and a wide range of other commercial applications. Another growing area of material development for optical applications was in very large lightweight mirror structures for use in space. Besides these newer optical materials, there are still several hundred different optical glass types, an ever-growing list of optical crystals, both water soluble and insoluble, fused silica and fused quartz, glass ceramics, low expansion materials, technical glasses, and optical plastics. Table 1.1 lists all the optical materials covered in this chapter in the order in which they are discussed.

This chapter cannot give an exhaustive account of the great variety of optical materials and their characteristics. It is also quite certain that no optician would be familiar with all of these materials, and it is certainly not possible for anyone to know how to shape and polish every type of optical material. However, this chapter provides the optician and the optical manufacturing engineer with useful information on materials and their characteristics, together with suitable references to additional information. The chapter gives a fairly comprehensive overview that is designed with the optician and optical manufacturing engineer in mind to give them some basic reference of materials and the parameters used in fabricating them.

**Table 1.1.   Optical Materials**

| |
|---|
| • Glass materials |
|  * Optical glass |
|  * Laser glass |
|  * Color filter glass |
|  * Technical glass |
| • Low expansion materials |
|  * Fused silica and fused quartz |
|  * Crystal quartz |
|  * Mirror substrate materials |
| • Optical crystals |
|  * UV crystals |
|  * IR materials |
|  * Laser crystals |
|  * Hygroscopic crystals |
| • Miscellaneous materials |
|  * Optical plastics |
|  * Ceramics |
|  * Metals |

## 1.1. OPTICAL GLASS

Glass in its most basic form is a melt of silicon dioxide, better known as quartz, certain alkali, and calcium. The melt is cooled in such a way that it solidifies into a solid material without forming a crystalline structure. In more technical terms, glass is a material that is held in a supercooled state at temperatures below its melting point so that it cannot crystallize [1]. Glass is an amorphous and vitreous material, and these terms more accurately describe the noncrystalline structure of glass.

Optical glass is a high-quality glass material that has been specifically formulated to possess certain desirable characteristics that affect the propagation of light. It is produced with the utmost care, often in small quantities to ensure that the necessarily strict quality standards have been met. Optical glass must meet a number of important criteria. It must be highly transparent over the visible spectrum. It must also be optically homogeneous, inclusion free, and free of striae and strain. In addition optical glass must have accurately defined optical constants that have only a small temperature dependence. The mechanical strength of optical glass must be sufficiently high so that optical components made from them can be easily handled during manufacture and assembly. Resistance to chemical attack is another important requirement for optical glass [1].

Optical glasses are fabricated into lenses, prisms, and other optical components. These are used to redirect and modify light in predetermined ways. To accomplish that, optical glasses must have special attributes. These attributes require that the basic ingredients be mixed in carefully controlled proportions with other chemicals that are added to the melt in order to influence the refractive index, dispersion, and transparency of the glass. The melt process is briefly described in Chapter 2. The physical and chemical properties of the numerous optical glasses are listed in the catalogs of glass manufacturers. The important attributes and quality demands on optical glasses have been standardized [2].

### 1.1.1. Refractive Index and Dispersion

The two primary parameters that define the basic nature of optical glass is its refractive index and its dispersion. The *refractive index n* is a measure of the refractive powers of the glass relative to air which has an index of 1. The *dispersion,* expressed by the Abbe number $\nu$, is a measure of the dispersive powers of the glass. It defines how the glass affects light at different wavelengths. Since both values vary for different wavelengths of light, it is necessary to measure them at some known wavelength standard. The long-practiced convention is that this wavelength reference standard is the yellow helium line at 587.56 nm, which is labeled *d*. Therefore the refractive index is labeled $n_d$ and the Abbe number $\nu_d$.

There are other indices at other wavelengths that are primarily of interest to the optical designer. Two of these are of interest here because they define the Abbe value for dispersion. The first, labeled $n_F$ is the index at the blue hydrogen line at a wavelength $\lambda_F = 486.13$ nm. The other is $n_C$ at the red hydrogen line at $\lambda_C = 656.27$ nm. The Abbe number is defined as

$$v_d = \frac{n_d - 1}{n_F - n_C} \tag{1.1}$$

The term $(n_F - n_C)$ is called the *mean* or *principal dispersion,* which shows how much greater the refractive effect is on the shorter (blue) wavelengths relative to the longer (red) wavelengths. A large value for $(n_F - n_C)$ yields a small Abbe number, which means that the dispersive power of the glass is large. Optical glasses are uniquely defined by their refractive index and their Abbe number. One always finds in the glass catalogs of optical glass manufacturers an $n_d/v_d$ diagram which shows the various glass types. One such diagram is shown in Fig. 1.1 [3].

### 1.1.2. Glass Types and Melt Numbers

Glass types with an $n_d > 1.60$ and an $v_d > 50$, as well as those with $n_d < 1.60$ and $v_d > 55$, are called *crown glass.* They have a letter designation K. Glass types with $n_d > 1.60$ and $v_d < 55$ are called *flint glasses,* with the letter designation F. This delineation is shown on the $n_d/v_d$ diagram (Fig. 1.1). The terms "crown" and "flint" have a historical origin, but their exact meanings are unclear.

The traditional way to differentiate the various glass types, in addition to the basic crowns and flints, is through the use of a letter and number designation. The letter code identifies the basic chemical composition of the glass and certain special qualities such as density. The number code further differentiates glass types within the letter group. It is essentially a recipe code where 1 is the first and 10 is the tenth glass type developed for a particular type of optical glass. For instance, light flint glasses are classified in group LF and borosilicate glasses in group BK. Within each group the individual glasses are further classified by a number to read LF 5 or BK 7. This original code was developed by Schott in Germany [4], and it uses abbreviations for the German glass designations. Glass manufacturers in the United States [5, 6, 7] and England [8] have produced their own code based on English terminology. A Japanese version of glass designation also exists [9]. In the recent past, one Japanese glass manufacturer [10] that used the Schott designation for many years has developed its own unique glass code.

The most widely used optical glass is BK 7. This is the Schott designation in which the K stands for *"Kron,"* which means crown, and the B identifies it as a borosilicate glass. The English equivalent is BSC 2, which stands for borosilicate crown. A common glass on the flint side is SF 2 which stands for

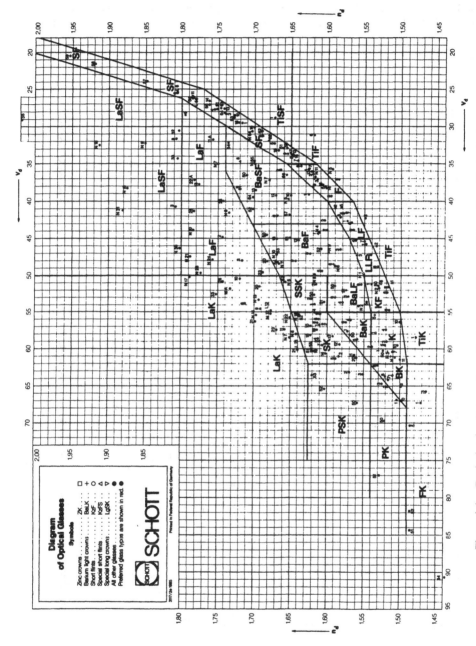

**Figure 1.1.** Typical $n_d/\nu_d$ diagram for optical glasses. (Source: Schott Glass Technologies, Inc.)

5

the German term *"Schwerflint,"* which means heavy or dense flint. The equivalent English designation is FD 2 which, not surprisingly, stands for dense flint. Table 1.2 lists the various glass groups by their designations. The table lists only the glass type designations used by the currently leading glass manufacturers. The contents of the table would even be more confused if the terminology of the American and French manufacturers were added as well.

**Table 1.2. Typical Glass Designation**

| Schott (4) | Ohara (10) | Hoya (9) | Chance Pilk. (8) | Glass type |
|---|---|---|---|---|
| FK | FPL, FSL | FC | | Fluor crown |
| PK | BSL | PC | | Phosphate crown |
| PSK | PHM, BAL | PCD | | Dense phosphate crown |
| BK | BSL | BSC | BSC | Borosilicate crown |
| BaLK | NSL | BaCL | | Light barium crown |
| K | NSL, FTL | C | | Crown |
| | | | HC | Hard crown |
| ZK | ZSL | ZnC | ZC | Zinc crown |
| BaK | BAL | BaC | | Barium crown |
| | | | MBC | Medium barium crown |
| SK | BAL, BSM | BaCD | DBC | Dense barium crown |
| KF | NSL, FTL | CF | | Crown flint |
| | | | TF | Telescope flint |
| BaLF | BAL | BaFL | | Light barium flint |
| SSK | BSM | BaCED | | Extra dense barium crown |
| LLF | PBL, FTM | FEL | ELF | Extra light flint |
| BaF | BAM, BAH | BaF | BF | Barium flint |
| LF | PBL | FL | LF | Light flint |
| F | PBM, FTM | F | | Flint |
| SF | PBM, PBH, TPH | FD | DF | Dense flint |
| BaSF | BAM, BAH | BaFD | | Dense barium flint |
| | | | EDF | Extra dense flint |
| | | | DEDF | Double extra dense flint |
| KzF | SSL | SbF | | Antimony flint |
| PKS | APL | PCS | | Special phosphate crown |
| KzSF | PBM, BPH, PBH | | BorF | Borate flint |
| SFS | | FDS | | Special dense flint |
| LaK | BSM, LAL | LaC | LAC | Lanthanum crown |
| LaF | LAM | LaF | LAF | Lanthanum flint |
| TiF | | FF | | Fluor flint |
| (FSK) | | FCD | | Dense fluor crown |
| (TaK) | | TaC | | Tantalum crown |
| (TaF) | | TaF | | Tantalum flint |
| (LaLK) | | LaCL | | Light lanthanum crown |
| (LaLF) | | LaFL | | Light lanthanum flint |
| (NbF) | | NbF | | Niobium flint |
| (NbSF) | | NbFD | | Dense niobium flint |

A more uniform method of identifying glass types was definitely needed. This was accomplished with the now universally accepted six-digit glass code. This simple code uniquely describes the basic optical characteristics of each glass type. The first three digits are derived from the refractive index which are the first three significant digits of the expression $(n_d - 1)$. The second set of three digits is derived from the Abbe number $v_d$ by multiplying it by 10 and using the first three significant digits of that product. This sounds a bit complicated, but it is really quite simple. It is best explained by an example.

The same glass types that were used to explain the letter and number designation will be used to explain the six-digit code. The refractive index of BK 7 is $n_d = 1.5168$. If 1 is subtracted from this value and the result rounded to three digits, it is 0.517. The first three significant digits are 517. The Abbe number for BK 7 is $v_d = 64.17$ which is first multiplied by 10 and then rounded to the nearest integer. This yields 642. The combination of these two numbers results in the glass code. The convention is not to hyphenate this code, so the code for BK 7 is 517 642. It uniquely identifies this particular type of optical glass. The code for SF 2 is derived in identical fashion from $n_d = 1.647689$ and $v_d = 33.85$. The resulting code for this type of glass is then 648 339. Because the six-digit code is universally applicable, it will eventually become the primary accepted designation for all optical glasses. In recognition of that fact, the major glass manufacturers list this code together with their own glass designations.

Although the production of optical glasses proceeds very carefully, the refractive index and dispersion of different melts of the same glass type often deviate a little from the catalog values. Therefore each glass shipment is accompanied by a test certificate with the melt number and the measured values of the melt. Since the optical designer makes use of this information and the design specifications for lenses often relate to a specific glass melt, it is essential that the workshop maintain proper traceability of these melt numbers.

### 1.1.3. Basic Glass Types and Their Compositions

Typical crown glasses, such as SK (dense barium crown), contain silicon dioxide ($SiO_2$) and barium oxide (BaO). These glasses are characterized by a high specific weight of about 3.5 $g/cm^2$, high refractive indices between 1.56 and 1.64, and low dispersion with Abbe numbers ranging from 55 to 61. Other crown glasses contain boron oxide ($B_2O_3$) and lanthanum oxide ($La_2O_3$) in addition to $SiO_2$.

Flint glasses contain the same basic chemical constituents as crown glasses, but they derive their particular characteristics, especially their high index and high dispersion, through the addition of lead oxide (PbO). While light flint (LF) contains only small amounts of PbO, dense flint (SF) can be composed of as much as 70% PbO. The specific weight for flints can range from a low of 2.8 $g/cm^3$ for extra light flint (LLF) to a high of about 6.3 $g/cm^3$

for dense flint. The refractive index can be as low as 1.54 for LLF or as high as 1.95 for SF. The typical high dispersion of flint glasses ranges from a high of $v_d = 50$ to a low of $v_d = 20$. See the $n_d/v_d$ diagram in Fig. 1.1 for clarification. Much more could be said about the composition of optical glass but that would go beyond the intent of this book. The interested reader should consult one of the excellent books written on this subject [1].

### 1.1.4.  Glass Parameters of Particular Interest to the Optician

Although the optical parameters must be known to the optician, they are primarily for the purpose of material identification. They do not help in deciding how to best handle and process the glass. It is quite reasonable to assume that different parts made from the same glass, say, BK 7, can be treated in similar fashion, but it is not necessarily true that the same assumption holds true for all other glasses of the BK group. More information is needed.

The glass catalogs of the optical glass manufacturers provide this added information in the form of data relating to the chemical, thermal, and mechanical behavior of the glasses. Data on chemical behavior are very important for polishing, handling, and storage of finished optics. Thermal data can provide clues on how carefully a particular glass must be treated during manufacture and data on mechanical characteristics of glass can be helpful during the shaping and grinding stages. Figures 1.2 through 1.4 show pages reproduced from glass catalogs for BK 7 (517 642) which give the data that will be discussed in this section.

### 1.1.5.  Chemical Behavior of Glass

The chemical behavior of glass is perhaps the most difficult set of data the optician has to deal with. The multitude of tests performed on a specific glass to determine its resistance to attack by water, moisture, acids and alkali is often confusing to the nonspecialist. This confusion is further complicated by different test approaches used by different optical glass manufacturers. This section attempts to reduce the confusion through brief descriptions of each test and the grading that results from that effort.

Simple glass is quite stable against chemical influences, and it is only noticeably attacked by hydrofluoric acid. The polished surfaces of many optical glasses, however, are easily affected by chemicals. Often rather mild chemical agents can readily dissolve from the glass matrix at the surface the ingredients that are added to the melt to obtain particular glass properties. The residue of the dissolved glass substance through such chemical attack often remains on the surface as a permanent stain. Since the stain has a refractive index that differs from that of the underlying bulk glass, it becomes visible as a thin layer because of the appearance of interference colors.

## BK 7 – 517 642

| | | |
|---|---|---|
| $n_d = 1.51680$ | $v_d = 64.17$ | $n_F - n_C = 0.008054$ |
| $n_e = 1.51872$ | $v_e = 63.96$ | $n_{F'} - n_{C'} = 0.008110$ |

### Refractive Indices

| | λ [nm] | |
|---|---|---|
| $n_{2325.4}$ | 2325.4 | 1.48929 |
| $n_{1970.1}$ | 1970.1 | 1.49500 |
| $n_{1529.6}$ | 1529.6 | 1.50094 |
| $n_{1060.0}$ | 1060.0 | 1.50669 |
| $n_t$ | 1014.0 | 1.50731 |
| $n_s$ | 852.1 | 1.50981 |
| $n_r$ | 706.5 | 1.51289 |
| $n_C$ | 656.3 | 1.51432 |
| $n_{C'}$ | 643.8 | 1.51472 |
| $n_{632.8}$ | 632.8 | 1.51509 |
| $n_D$ | 589.3 | 1.51673 |
| $n_d$ | 587.6 | 1.51680 |
| $n_e$ | 546.1 | 1.51872 |
| $n_F$ | 486.1 | 1.52238 |
| $n_{F'}$ | 480.0 | 1.52283 |
| $n_g$ | 435.8 | 1.52669 |
| $n_h$ | 404.7 | 1.53024 |
| $n_i$ | 365.0 | 1.53626 |

### Constants of Dispersion Formula

| | |
|---|---|
| $A_0$ | $2.2718929$ |
| $A_1$ | $-1.0108077 \cdot 10^{-2}$ |
| $A_2$ | $1.0592509 \cdot 10^{-2}$ |
| $A_3$ | $2.0816965 \cdot 10^{-4}$ |
| $A_4$ | $-7.6472538 \cdot 10^{-6}$ |
| $A_5$ | $4.9240991 \cdot 10^{-7}$ |

### Deviation of Relative Partial Dispersions ΔP from the "Normal Line"

| | |
|---|---|
| $\Delta P_{C,t}$ | 0.0210 |
| $\Delta P_{C,s}$ | 0.0083 |
| $\Delta P_{F,e}$ | -0.0009 |
| $\Delta P_{g,F}$ | -0.0008 |
| $\Delta P_{i,g}$ | 0.0029 |

### Relative Partial Dispersion

| | |
|---|---|
| $P_{s,t}$ | 0.3097 |
| $P_{C,s}$ | 0.5607 |
| $P_{d,C}$ | 0.3075 |
| $P_{e,d}$ | 0.2386 |
| $P_{g,F}$ | 0.5350 |
| $P_{t,h}$ | 0.7478 |
| $P'_{s,t}$ | 0.3075 |
| $P'_{C',s}$ | 0.6058 |
| $P'_{d,C'}$ | 0.2565 |
| $P'_{e,d}$ | 0.2370 |
| $P'_{g,F'}$ | 0.4755 |
| $P'_{t,h}$ | 0.7427 |

### Other Properties

| | |
|---|---|
| $\alpha_{-30/+70°C}$ [$10^{-6}$/K] | 7.1 |
| $\alpha_{20/300°C}$ [$10^{-6}$/K] | 8.3 |
| $T_g$ [°C] | 559 |
| $T_{10^{7.6}}$ [°C] | 719 |
| $c_p$ [J/g · K] | 0.858 |
| $\lambda$ [W/m · K] | 1.114 |
| $\rho$ [g/cm$^3$] | 2.51 |
| E [$10^3$ N/mm$^2$] | 81 |
| $\mu$ | 0.208 |
| HK | 520 |
| B | 0 |
| CR | 2 |
| FR | 0 |
| SR | 1 |
| AR | 2.0 |

### Internal Transmittance $\tau_i$

| λ [nm] | $\tau_i$ (5 mm) | $\tau_i$ (25 mm) |
|---|---|---|
| 2325.4 | 0.89 | 0.57 |
| 1970.1 | 0.968 | 0.85 |
| 1529.6 | 0.997 | 0.985 |
| 1060.0 | 0.999 | 0.998 |
| 700 | 0.999 | 0.998 |
| 660 | 0.999 | 0.997 |
| 620 | 0.999 | 0.997 |
| 580 | 0.999 | 0.996 |
| 546.1 | 0.999 | 0.996 |
| 500 | 0.999 | 0.996 |
| 460 | 0.999 | 0.994 |
| 435.8 | 0.999 | 0.994 |
| 420 | 0.998 | 0.993 |
| 404.7 | 0.998 | 0.993 |
| 400 | 0.998 | 0.991 |
| 390 | 0.998 | 0.989 |
| 380 | 0.996 | 0.980 |
| 370 | 0.995 | 0.974 |
| 365.0 | 0.994 | 0.969 |
| 350 | 0.986 | 0.93 |
| 334.1 | 0.950 | 0.77 |
| 320 | 0.81 | 0.35 |
| 310 | 0.59 | 0.07 |
| 300 | 0.26 | |
| 290 | | |
| 280 | | |

### Remarks

| | | |
|---|---|---|
| | | |

### Temperature Coefficients of Refractive Index

| | $\Delta n/\Delta T_{relative}$ [$10^{-6}$/K] | | | | | $\Delta n/\Delta T_{absolute}$ [$10^{-6}$/K] | | | | |
|---|---|---|---|---|---|---|---|---|---|---|
| [°C] | 1060.0 | s | C' | e | g | 1060.0 | s | C' | e | g |
| -40/-20 | 2.2 | 2.3 | 2.5 | 2.7 | 3.1 | 0.2 | 0.3 | 0.4 | 0.6 | 1.0 |
| -20/ 0 | 2.2 | 2.3 | 2.6 | 2.8 | 3.3 | 0.5 | 0.6 | 0.8 | 1.0 | 1.5 |
| 0/+20 | 2.3 | 2.4 | 2.7 | 2.8 | 3.4 | 0.9 | 1.0 | 1.2 | 1.3 | 1.9 |
| +20/+40 | 2.4 | 2.5 | 2.8 | 3.0 | 3.6 | 1.2 | 1.3 | 1.5 | 1.7 | 2.3 |
| +40/+60 | 2.5 | 2.6 | 2.9 | 3.1 | 3.8 | 1.3 | 1.4 | 1.7 | 1.9 | 2.6 |
| +60/+80 | 2.6 | 2.7 | 3.0 | 3.2 | 3.9 | 1.6 | 1.7 | 2.0 | 2.2 | 2.8 |

Schott Optical Glass 3111 e / USA

**Figure 1.2.** Data sheet for BK 7 (517 642) from Schott glass catalog.

# OHARA            (BK 7)

**516642**
**BSL 7**

| | | | | | | | |
|---|---|---|---|---|---|---|---|
| Refractive Index | $n_d$ | **1.51633** / 1.516330 | Abbe Number $\nu_d$ | **64.1** / 64.15 | Dispersion $n_F-n_C$ | **0.00805** / 0.008049 |
| Refractive Index | $n_e$ | 1.518251 | Abbe Number $\nu_e$ | 63.93 | Dispersion $n_{F'}-n_{C'}$ | 0.008106 |

### Refractive Indices

| | |
|---|---|
| $n_t$ (1.01398) | 1.50687 |
| $n_s$ (0.85211) | 1.50935 |
| $n_{A'}$ (0.76819) | 1.51097 |
| $n_r$ (0.70652) | 1.51243 |
| $n_C$ (0.65627) | **1.51385** |
| $n_{C'}$ (0.64385) | 1.51425 |
| $n_{He-Ne}$ (0.6328) | 1.51462 |
| $n_D$ (0.58929) | 1.51626 |
| $n_d$ (0.58756) | **1.51633** |
| $n_e$ (0.54607) | 1.51825 |
| $n_F$ (0.48613) | **1.52190** |
| $n_{F'}$ (0.47999) | 1.52236 |
| $n_{He-Cd}$ (0.44157) | 1.52564 |
| $n_g$ (0.43583) | **1.52621** |
| $n_h$ (0.40466) | 1.52976 |
| $n_i$ (0.36501) | 1.53577 |

### Constants of Dispersion Formula

| | |
|---|---|
| $A_0$ | 2.2697665 |
| $A_1$ | $-9.6395197\cdot10^{-3}$ |
| $A_2$ | $1.1025458\cdot10^{-2}$ |
| $A_3$ | $7.9465126\cdot10^{-5}$ |
| $A_4$ | $1.0120957\cdot10^{-5}$ |
| $A_5$ | $-4.4096694\cdot10^{-7}$ |

### Relative Partial Dispersions

| $n_C-n_t$ $\theta_{C,t}$ | $n_C-n_{A'}$ $\theta_{C,A'}$ | $n_d-n_C$ $\theta_{d,C}$ | $n_e-n_C$ $\theta_{e,C}$ |
|---|---|---|---|
| 0.006983 0.8676 | 0.002883 0.3582 | 0.002475 0.3075 | 0.004396 0.5462 |
| $n_g-n_d$ $\theta_{g,d}$ | $n_g-n_F$ $\theta_{g,F}$ | $n_h-n_g$ $\theta_{h,g}$ | $n_i-n_g$ $\theta_{i,g}$ |
| 0.009881 1.2276 | 0.004307 0.5351 | 0.003552 0.4413 | 0.009560 1.1877 |
| $n_{C'}-n_t$ $\theta_{C',t}$ | $n_e-n_{C'}$ $\theta_{e,C'}$ | $n_{F'}-n_e$ $\theta_{F',e}$ | $n_i-n_{F'}$ $\theta_{i,F'}$ |
| 0.007379 0.9103 | 0.004000 0.4935 | 0.004106 0.5065 | 0.013414 1.6548 |

### Deviation of Relative Partial Dispersions Δθ from "Normal"

| $\Delta\theta_{C,t}$ | $\Delta\theta_{C,A'}$ | $\Delta\theta_{g,d}$ | $\Delta\theta_{g,F}$ | $\Delta\theta_{i,g}$ |
|---|---|---|---|---|
| 0.0199 | 0.0046 | $-0.0039$ | $-0.0026$ | $-0.0002$ |

### Thermal Properties

| | | |
|---|---|---|
| Strain Point | StP(°C) | 511 |
| Annealing Point | AP(°C) | 547 |
| Transformation Temperature | Tg(°C) | 565 |
| Yield Point | At(°C) | 624 |
| Softening Point | SP(°C) | 715 |
| Expansion Coefficient ($\alpha\times10^7$) | (−30~ / +70°C) | 74 |
| | (+100~ / +300°C) | 86 |
| Thermal Conductivity | k(W/m·K) | 1.126 |

### Mechanical Properties

| | | |
|---|---|---|
| Young's Modulus | E($10^6$N/m²) | 802 |
| Rigidity Modulus | G($10^6$N/m²) | 332 |
| Poisson's Ratio | $\sigma$ | 0.207 |
| Knoop Hardness | Hk | 570 (6) |
| Abrasion | Aa | 95 |
| Photoelastic Constant | $\beta$ (nm/cm/$10^5$Pa) | 2.76 |

### Other Properties

| | | |
|---|---|---|
| Bubble Quality | Grp. B | 1 |
| Coloring | Coloring | 34/30 (W33/28) |
| Specific Gravity | d | 2.52 |
| Remarks | Remarks | |

### Chemical Properties

| | | |
|---|---|---|
| Water Resistance (Powder) Group | RW (P) | 3 |
| Acid Resistance (Powder) Group | RA (P) | 1 |
| Weathering Resistance (Surface) Group | W (S) | 1 |
| Acid Resistance (Surface) Group | SR | 1.0 |
| Phosphate Resistance | PR | 1.0 |

### Internal Transmittance

| $\lambda$(nm) | $\tau$ 10mm | W $\tau$10mm |
|---|---|---|
| 280 | | 0.03 |
| 290 | | 0.17 |
| 300 | 0.03 | 0.42 |
| 310 | 0.18 | 0.66 |
| 320 | 0.44 | 0.82 |
| 330 | 0.67 | 0.915 |
| 340 | 0.83 | 0.958 |
| 350 | 0.915 | 0.979 |
| 360 | 0.956 | 0.989 |
| 370 | 0.974 | 0.992 |
| 380 | 0.981 | 0.992 |
| 390 | 0.991 | 0.997 |
| 400 | 0.995 | 0.998 |
| 420 | 0.995 | 0.998 |
| 440 | 0.994 | 0.997 |
| 460 | 0.995 | 0.998 |
| 480 | 0.996 | 0.998 |
| 500 | 0.996 | 0.999 |
| 550 | 0.997 | 0.999 |
| 600 | 0.997 | 0.999 |
| 650 | 0.997 | 0.998 |
| 700 | 0.998 | 0.998 |
| 800 | 0.998 | 0.998 |
| 900 | 0.998 | 0.998 |
| 1060 | 0.997 | 0.998 |

### Temperature Coefficients of Refractive Index

| Range of Temperature (°C) | $dn/dt$ relative $\times10^6$/°C | | | | | | |
|---|---|---|---|---|---|---|---|
| | t | C' | He-Ne | D | e | F' | g |
| −40 ~ −20 | 2.1 | 2.4 | 2.4 | 2.5 | 2.6 | 2.9 | 3.1 |
| −20 ~ 0 | 2.2 | 2.5 | 2.5 | 2.6 | 2.7 | 3.0 | 3.3 |
| 0 ~ 20 | 2.3 | 2.7 | 2.7 | 2.8 | 2.9 | 3.1 | 3.4 |
| 20 ~ 40 | 2.4 | 2.8 | 2.8 | 2.9 | 3.0 | 3.3 | 3.5 |
| 40 ~ 60 | 2.5 | 2.9 | 2.9 | 3.0 | 3.1 | 3.5 | 3.7 |
| 60 ~ 80 | 2.7 | 2.9 | 2.9 | 3.0 | 3.2 | 3.5 | 3.8 |

**Figure 1.3.** Data sheet for BK 7 (516 642) from Ohara glass catalog.

$n_d = 1.51680$   $\nu_d = 64.20$   $n_F - n_C = 0.008050$

$n_e = 1.51872$   $\nu_e = 64.00$   $n_{F'} - n_{C'} = 0.008105$

## 517–642 BK7 BSC7

| Refractive Indices | | |
|---|---|---|
| | λ (nm) | |
| $n_t$ | 1014.0 | 1.50733 |
| $n_s$ | 852.1 | 1.50980 |
| $n_{A'}$ | 768.2 | 1.51143 |
| $n_r$ | 706.5 | 1.51289 |
| $n_C$ | 656.3 | 1.51432 |
| $n_{C'}$ | 643.8 | 1.51472 |
| $n_{632.8}$ | 632.8 | 1.51509 |
| $n_D$ | 589.3 | 1.51673 |
| $n_d$ | 587.6 | 1.51680 |
| $n_e$ | 546.1 | 1.51872 |
| $n_F$ | 486.1 | 1.52237 |
| $n_{F'}$ | 480.0 | 1.52282 |
| $n_g$ | 435.8 | 1.52667 |
| $n_h$ | 404.7 | 1.53022 |
| $n_i$ | 365.0 | 1.53622 |

| Constants of Dispersion Formula | |
|---|---|
| $A_0$ | 2.2702566 |
| $A_1$ | $-9.1988101 \times 10^{-3}$ |
| $A_2$ | $1.1609706 \times 10^{-2}$ |
| $A_3$ | $-7.6123911 \times 10^{-5}$ |
| $A_4$ | $2.8558727 \times 10^{-5}$ |
| $A_5$ | $-1.2566486 \times 10^{-6}$ |

| Relative Partial Dispersions | | | |
|---|---|---|---|
| $P_{A',t}$ | 0.5089 | $P'_{A',t}$ | 0.5055 |
| $P_{t,A'}$ | 0.1813 | $P'_{t,A'}$ | 0.1801 |
| $P_{C,t}$ | 0.1779 | $P'_{C',t}$ | 0.2257 |
| $P_{d,C}$ | 0.3080 | $P'_{d,C'}$ | 0.2569 |
| $P_{e,d}$ | 0.2387 | $P'_{e,d}$ | 0.2371 |
| $P_{F,e}$ | 0.4533 | $P'_{F',e}$ | 0.5060 |
| $P_{g,F}$ | 0.5342 | $P'_{g,F'}$ | 0.4748 |
| $P_{h,g}$ | 0.4408 | $P'_{h,g}$ | 0.4379 |
| $P_{i,h}$ | 0.7455 | $P'_{i,h}$ | 0.7404 |

| Deviation of Relative Partial Dispersions ΔP from the "Normal Line" | |
|---|---|
| $\Delta P_{C,t}$ | 0.0223 |
| $\Delta P_{C,A'}$ | 0.0044 |
| $\Delta P_{g,d}$ | 0.0007 |
| $\Delta P_{g,F}$ | 0.0015 |
| $\Delta P_{t,g}$ | 0.0132 |

| Chemical Properties | |
|---|---|
| $D_W$ (wt%) | 0.21 (3) |
| $D_A$ (wt%) | 0.09 (1) |
| $T_{Blue}$ (h) | >70 (1) |
| $D_{NaOH}$ [mg/(cm²·15h)] | 0.06 (2) |
| $D_{STPP}$ [mg/(cm²·h)] | < 0.01 (1) |
| $D_0$ [$10^{-3}$ mg/(cm²·h)] | < 0.3 (1) |

| Thermal Properties | |
|---|---|
| $T_g$ (°C) | 565 |
| $T_s$ (°C) | 630 |
| $T_{10^{14.5}}$ (°C) | 546 |
| $T_{10^{13}}$ (°C) | 571 |
| $T_{10^{7.6}}$ (°C) | 724 |
| $\alpha_{-30/+70°C}$ ($10^{-7}$/K) | 75 |
| $\alpha_{100/300°C}$ ($10^{-7}$/K) | 89 |
| $\lambda$ [W/(m·K)] | 1.21 |
| $c_p$ [kJ/(kg·K)] | 0.749 |

| Mechanical Properties | |
|---|---|
| $H_K$ | 595 (6) |
| $F_A$ | 100 |
| E (GPa) | 79.2 |
| G (GPa) | 32.7 |
| $\mu$ | 0.214 |
| $\sigma_b$ (MPa) | 106 |

| Electrical Properties | |
|---|---|
| $\epsilon_r$ | 5.8 |
| $\varrho_{V\,20°C}$ (Ω·cm) | $6.0 \times 10^{14}$ |
| $\varrho_{V\,200°C}$ (Ω·cm) | $2.5 \times 10^{11}$ |

| Temperature Coefficients of Refractive Index ($10^{-6}$/K at 632.8nm) | | |
|---|---|---|
| (°C) | (Δn/ΔT)rel. | (Δn/ΔT)abs. |
| −40/−20 | 2.1 | 0.1 |
| −20/ 0 | 2.1 | 0.4 |
| 0/+20 | 2.2 | 0.7 |
| +20/+40 | 2.3 | 0.9 |
| +40/+60 | 2.3 | 1.2 |
| +60/+80 | 2.4 | 1.4 |

| Stress-Optical Coefficient | |
|---|---|
| B ($10^{-12}$/Pa) | 2.82 |

| Other Properties | 2.82 |
|---|---|
| Specific Gravity | 2.54 |
| Bubble | 1 |

| Internal Transmittance | | |
|---|---|---|
| λ (nm) | $\tau_{5mm}$ | $\tau_{10mm}$ |
| 1550 | 0.994 | 0.989 |
| 1500 | 0.996 | 0.993 |
| 1400 | 0.987 | 0.974 |
| 1300 | 0.997 | 0.994 |
| 1200 | 0.998 | 0.995 |
| 1100 | 0.998 | 0.995 |
| 1060 | 0.998 | 0.996 |
| 1050 | 0.998 | 0.996 |
| 1000 | 0.999 | 0.997 |
| 950 | 0.999 | 0.997 |
| 900 | 0.999 | 0.998 |
| 850 | 0.999 | 0.998 |
| 830 | 0.999 | 0.999 |
| 800 | 0.999 | 0.999 |
| 780 | 0.999 | 0.999 |
| 750 | 0.999 | 0.999 |
| 700 | 0.999 | 0.999 |
| 650 | 0.999 | 0.999 |
| 600 | 0.999 | 0.999 |
| 550 | 0.999 | 0.999 |
| 500 | 0.999 | 0.999 |
| 480 | 0.999 | 0.999 |
| 460 | 0.999 | 0.999 |
| 440 | 0.999 | 0.999 |
| 420 | 0.999 | 0.999 |
| 400 | 0.999 | 0.999 |
| 390 | 0.999 | 0.997 |
| 380 | 0.996 | 0.992 |
| 370 | 0.994 | 0.989 |
| 360 | 0.990 | 0.981 |
| 350 | 0.978 | 0.957 |
| 340 | 0.952 | 0.906 |
| 330 | 0.904 | 0.817 |
| 320 | 0.81 | 0.65 |
| 310 | 0.64 | 0.40 |
| 300 | 0.39 | 0.15 |
| 290 | 0.15 | 0.02 |
| 280 | | |

| Coloration Code | |
|---|---|
| $\lambda_{80}/\lambda_5$ | 34/29 |

| Remarks |
|---|
| |

Figure 1.4. Data sheet for BK 7 (517 642) from Hoya glass catalog.

Optical glasses are not equally sensitive to chemical attack. Noted in the glass catalogs are chemical influences under which the glasses will react and how strong the reaction will be. Many different chemical durabilities are identified in the glass catalog, depending on glass manufacturer. Typically an assigned number indicates how strongly the glass reacts against the respective harmful agent. A higher number indicates a stronger sensitivity. Extremely sensitive glasses must be protected from disturbing influences after being worked on. Such glasses should not serve as exterior surfaces in optical instruments. Generally applicable handling procedures cannot be provided, however.

**Staining.** The chemical composition of glass determines to a large degree its resistance to attack by water, water vapors or water condensate, acids, alkali, and environmental gasses. For instance, aqueous solutions that are exposed to glass tend to become acidic (pH < 7) when the glass is attacked by hydrogen ions (H+) in the water but alkaline (pH > 7) in the presence of hydroxide ions (OH-) [3]. In general, glasses with good chemical resistance contain a higher percentage of silicon dioxide ($SiO_2$), aluminum oxide ($Al_2O_3$), titanium oxide ($TiO_2$), or rare earth oxides. The formation of stains can be expected for glasses that contain more soluble constituents such as alkali or earth alkali oxides and particularly boron oxide and oxides of phosphorous [3].

There are two basic reactions that occur when glass is immersed in water. Either the water can actually dissolve the glass network, or an ion exchange reaction takes place between certain alkaline constituents of the glass and hydroxyl ions ($H_3O+$) in the water. The latter case leads to the formation of a hydrated or silica gel layer on the surface of the glass, while other ions leach into the water [11]. Stains form on the glass surface when ion exchange reactions between the water and the glass cause the formation of a low index layer. As this layer increases in thickness, it can be seen as a brown to blue stain when its thickness is one quarter wave of visible light [3, 11]. This type of attack occurs most frequently in aqueous cleaning baths or when glass is permitted to stand submerged in static water bath for any length of time. The constant exposure of the surfaces of particularly sensitive glasses to water based polishing slurries can also lead to staining.

**Dimming Caused by Water Vapors.** Although some glassmakers do not appear to make a distinction between stains formed in the presence of water or those formed by water vapors, one Japanese glass manufacturer [9] makes a clear distinction in calling the latter process "dimming." Moisture condensing on glass in air can attack glass surfaces in the following manner: Alkali ions diffuse from the glass into small water droplets that have condensed on the surface. This turns them into concentrated alkaline solutions that erode the silica gel layer of the polished surface. When the condensation evaporates, silicon ions and alkali are deposited on the glass surface leaving a hazy film that cannot be removed by wiping [11].

**Other Sources of Surface Attack.** Certain gasses in the environment can also contribute to stain formation on glass when they react with water or condensed moisture. The pH of the water may become acidic when carbon dioxide ($CO_2$) or sulfur dioxide ($SO_2$) is absorbed by the water. Also improper handling of polished optics can cause the formation of permanent stains. Perspiration is acidic, which can cause fingerprints to be permanently etched into the surface of a chemically sensitive glass. Also saliva, which can be easily deposited on polished surfaces as so-called spittle marks when talking while handling parts, can cause permanent stains. The formation of such stains can be avoided by proper part handling practices [3].

### 1.1.6. Testing the Resistance of Optical Glass to Chemical Attack

All optical glass manufacturers devote a considerable part of their glass catalogs to describing the various tests to determine the resistance of optical glass to chemical attack and to grade them according to the test results. Although this fact points to the importance of the chemical behavior of optical glass, it is also a source of much confusion among opticians about what the grading of the glass actually signifies. The various test approaches will be examined in this section.

An understanding of the material covered in this subsection is not essential for those who are primarily interested in fundamental shop practice. This difficult topic may be hard to follow, and it can be reserved for later study. The reason for the in-depth discussion of the various tests performed by the glassmakers is to give the optical manufacturing or process engineer added information and to clarify the complexity of the chemical behavior of optical glasses. The author is indebted to the glass manufacturers referenced in this section for the excellent descriptions of their test methods which have been extensively used in this subsection.

There are basically only four tests performed. Each tests the resistivity of the glass to one of four chemical attack mechanisms, but the different test methods and the ways that resistance is measured can be confusing. Confusion is unavoidable when different tests are performed for the same parameter and lead to different results. Table 1.3 lists the chemical resistance tests performed by four of the leading optical glass manufacturers. If the tests performed by other glass manufacturers were added, it would complicate the task even more.

**Chemical Resistance of Optical Glasses to Attack by Water Vapors.** The mechanism for surface attack by water vapors was already discussed in the previous section. The condensation of water vapors is enhanced by high

**Table 1.3.  Test Methods for Determining the Resistance of Optical Glasses to Chemical Attack**

| 1) Chemical resistance to attack by water vapors | Class | Group | Ref. |
|---|---|---|---|
| ° Schott:  Climatic resistance          - | CR | 1 - 4 | (3) |
| ° Hoya:     Dimming resistivity           - | $D_W$ | 1 - 6 | (9) |
| ° Ohara:   Weathering resistance        - |  | 1 - 4 | (10) |
| **2) Chemical resistance to attack by water** | **Class** | **Group** |  |
| ° Schott:  Stain resistance             - | FR | 0 - 5 | (3) |
| ° Hoya:     Staining resistivity          - | $T_{BLUE}$ | 1 - 5 | (9) |
|     Chemical durability to water- | $D_O$ | 1 - 5 |  |
| ° Ohara:   Water resistance            - |  | 1 - 6 | (10) |

| 3) Chemical resistance to attack by acids | Class | Group | | Ref. |
|---|---|---|---|---|
| ° Schott:  Acid resistance           - | SR | 1 - 4, | 51 - 53 | (3) |
| ° Hoya:     Acid durability            - | $D_A$ | 1 - 6 | | (9) |
| ° Ohara:   Acid resistance  (bulk)    - |  | 1 - 6 | | (10) |
|     Acid resistance (surface)  - |  | 1 - 4, | 5a - 5c | |
| ° C - P:    Acid durability            - |  | 1 - 7, | .1  - .5 | (8) |

| 4) Chemical resistance to attack by alkalis | Class | Group |  |
|---|---|---|---|
| ° Schott:  Alkali resistance          - | AR | 1 - 4 | (3) |
| ° Hoya:     Latent scratch resistivity   - | $D_{NaOH}$ | 1 - 5 | (9) |
|     Latent scratch resistivity   - | $D_{STPP}$ | 1 - 5 |  |

relative humidity and temperature fluctuations. The methods used by two of the glassmakers [3, 10] are very similar in that freshly polished samples of the glass under test are exposed for many hours to a water vapor saturated atmosphere in an environmental test chamber at either a fixed temperature [10] or a fluctuating temperature [3]. The samples are then examined for surface scattering and compared to standard glasses.

Another manufacturer [9] opted for a different approach, which is called the "powdered glass method." A sample of the glass under test is crushed into about 0.5-mm (0.020-in.) diameter grains. A portion of this crushed glass is accurately weighed, put into a platinum basket, and immersed in a fused silica flask filled with pure water. The flask is then placed into boiling water for 1 hour. Then the crushed glass is removed and carefully dried. It is weighed again with the same accuracy as before to determine the weight loss. Based on the weight difference, the glass is classified for resistance to dimming.

Regardless which method is used, the glasses with the highest class or group ratings should be handled and stored with considerable care, and their

**Table 1.4. Optical Glass Types Susceptible to Dimming**

| Schott Optical Glass [3] | | Ohara Optical Glass [10] | | Hoya Optical Glass [9] | |
|---|---|---|---|---|---|
| CR class 4 | | Weathering group 4 | | $D_w$ class 5 | |
| PSK 2 | 569 631 | PSK 2 | 569 632 | FC 3 | 465 658 |
| PSK 50 | 558 673 | SK 18 | 639 554 | FC 5 | 487 704 |
| SK 14 | 603 606 | SSK 01 | 649 530 | PCD 2 | 569 631 |
| SK 16 | 620 603 | BaF 9 | 643 478 | LaCL 3 | 665 534 |
| LaKN 7 | 652 585 | BaF 13 | 669 450 | LaCL 4 | 670 517 |
| LaK 11 | 658 573 | BaF 01 | 683 447 | LaC 7 | 652 584 |
| LaKN 12 | 678 552 | BaSF 6 | 668 419 | SbF 2 | 529 516 |
| LaKN 13 | 694 533 | BaSF 8 | 723 379 | ADF 10 | 613 444 |
| LaK 21 | 641 601 | BaSF 12 | 670 393 | | |
| LaKL 21 | 640 597 | KzFS 5 | 654 397 | | |
| Lak 23 | 669 574 | KzFS 8 | 720 347 | | |
| BaSF 50 | 710 366 | LaK 7 | 652 587 | | |
| SF 59 | 953 204 | LaK 11 | 658 573 | | |
| TiK 1 | 479 587 | LaK 02 | 670 573 | $D_w$ class 6 | |
| KzFS 1 | 613 443 | LaK 04 | 651 562 | LaCL 1 | 641 568 |
| KzFS 6 | 592 485 | LaK 07 | 678 534 | SbF 5 | 521 528 |
| KzFS 8 | 720 346 | LaF 02 | 720 437 | SbF 6 | 527 511 |
| SK 55 | 620 601 | LaF 06 | 686 492 | ADF 1 | 565 530 |
| LaK 31 | 697 564 | LaF 011 | 720 460 | | |
| LaFN 23 | 689 497 | LaF 012 | 783 362 | | |

polished surfaces must be protected during processing and prolonged storage. Table 1.4 lists some of these glasses that require such special attention.

Glasses in CR class 4 [3] show light scatter after 5 hours of exposure. This condition is reached after 6 hours for weathering resistance group 4 [10]. The criterion for dimming resistivity [9] is a weight loss of 0.6% for class 5 and 1.1% for class 6.

*A word of caution:* Just because a glass type appears in Table 1.4 does not mean that the polished surface will definitely deteriorate from humidity, nor will a glass type not on the list not be affected by condensed water vapors. Much depends on the conditions during manufacture, handling, and storage. The glasses in Table 1.4 have shown a distinct tendency to be affected by humidity, so particular attention must be paid to this fact. This is true for all the glasses on the subsequent chemical durability tables.

**Chemical Resistance of Optical Glasses to Attack by Water.** Schott [3] calls this test "resistance to staining." The designation is FR with six classes from 0 to 5. The test is designed to determine how the polished surface of a glass is affected by small amounts of slightly acidic water. It attempts to simulate conditions that can occur during manufacture, handling, storage, and use from fingerprints and exposure to acidic condensation, which can occur when environmental gasses such as $CO_2$ and $SO_2$ are absorbed by atmospheric water vapors.

The test procedure is relatively simple. As mentioned in the glass catalog [3], it involves polished plano glass samples pressed against a small concave depression in a cuvette that contains a few drops of an acidic test solution. Test solution I has a pH of 4.6 (standard acetate), and test solution II has a pH of 5.6 (sodium acetate buffer). As the polished surface slowly decomposes, interference color stains develop. The glasses are graded according to the time of exposure to one of the two solutions that causes a brown-blue stain to form. The stain is approximately 0.1 $\mu$m thick or about $\lambda/4$ for visible light. The test solution II is used only on those glasses that have failed the test with test solution I. Glasses in stain class FR 0 show no interference colors even after 100 hours of exposure to test solution I. Those in stain class 3 show stains after 1 hour exposure to test solution I.

The most stain-sensitive glasses are those in classes 4 and 5. They are tested only with test solution II. Class 4 glasses show interference colors after 1 hour of exposure, while those in class 5 stain in less that 12 minutes. Glasses in stain class 4 and especially those in class 5 require special care during manufacture, handling, and storage. These glasses and those of two other manufacturers whose test methods are described in the following paragraphs are listed in Table 1.5.

Other glass manufacturers [9, 10] have chosen different approaches to test glasses for their resistance to water. One Japanese glassmaker uses the powdered glass method [10] which has already been described in a preceding paragraph. The reagent for this water resistance test is distilled water with a pH between 6.5 and 7.5. The criterion for grading glasses for water resistance is the percent weight loss of the dried sample. Glasses in group 1 have no or only negligible weight loss after the powder glass test, whereas those in group 6 have a weight loss in excess of 1.1%.

The other Japanese glassmaker [9] uses two separate tests to determine the chemical durability of glasses in the presence of water. One is a surface test which is designated $T_{BLUE}$; the other is a bulk test labeled $D_0$. The $T_{BLUE}$ test is also called "staining resistivity by the surface method." For this test a standard sample of the glass is polished on both sides. It is then immersed in pure water with a pH of 7.0. The water is stirred well and circulated through several layers of ion exchange resins. In predetermined intervals of time, the sample is removed from the water to check for interference colors on the surfaces under a 100-watt tungsten filament lamp. The time required to form a bluish layer, which occurs when the layer has reached a thickness of a quarter wave of visible light (about 0.13 $\mu$m), determines the staining resistivity class of the glass. Glasses in class 1 had to be immersed for over 45 hours before any evidence of interference colors was seen, whereas those in class 5 required only 5 hours. This test also distinguishes a separate class for

Table 1.5. Optical Glass Types Susceptible to Staining

**Schott Optical Glass**

FR class 5

| Type | Value |
|---|---|
| LaK 11 | 658 573 |
| BaFN 11 | 667 484 |
| BaF 13 | 669 450 |
| BaSF 12 | 670 392 |
| BaSF 52 | 702 410 |
| BaSF 54 | 736 322 |
| LaF 9 | 795 284 |
| SF 57 | 847 238 |
| SF 58 | 918 215 |
| SF 59 | 953 204 |

FR class 4-5

| Type | Value |
|---|---|
| SKN 18 | 639 554 |

FR class 4

| Type | Value |
|---|---|
| SK 16 | 620 603 |
| SK 52 | 639 555 |
| LaKN 6 | 643 580 |
| BaSF 50 | 710 366 |
| BaSF 51 | 724 381 |
| LaFN 8 | 735 416 |
| KzFS 1 | 613 443 |
| KzFSN 2 | 558 542 |
| KzFS 6 | 592 485 |

**Hoya Optical Glass**

$T_{Blue}$ class 5

| Type | Value |
|---|---|
| FC 5 | 487 704 |
| PCD 2 | 569 631 |
| BaCD 7 | 607 595 |
| BaCD 10 | 623 569 |
| BaCD 14 | 603 607 |
| LaCL 1 | 641 568 |
| LaCL 3 | 665 534 |
| LaCL 4 | 670 517 |
| LaCL 7 | 670 573 |
| LaCL 60 | 640 602 |
| LaC 6 | 643 580 |
| LaC 11 | 658 573 |
| LaC 12 | 678 555 |
| LaFL 2 | 697 485 |

There are 35 additional glass types which also fall into class 5. The surface attack is such that interference fringes are obscured by dissolution of the entire surface.

$D_o$ class 5

| Type | Value |
|---|---|
| PCD 2 | 569 631 |
| BaCD 14 | 603 607 |
| BaCD165 | 620 598 |
| LaCL 1 | 641 568 |
| LaCL 3 | 665 534 |
| LaCL 4 | 670 517 |
| LaCL 6 | 640 602 |
| LaCL 7 | 670 573 |
| LaCL 60 | 640 602 |
| LaC 6 | 643 580 |
| LaC 7 | 652 584 |
| LaC 11 | 658 573 |
| LaC 12 | 678 555 |
| BaF 21 | 664 489 |
| FDS 9 | 847 238 |
| LaFL 1 | 686 494 |
| LaFL 2 | 697 485 |
| LaF 7 | 750 350 |
| ADC 1 | 620 622 |
| ADF 1 | 565 530 |
| ADF 8 | 687 429 |
| ATC 1 | 620 622 |

**Ohara Optical Glass**

Water resistivity group 6

| Type | Value |
|---|---|
| FK 3 | 465 659 |
| BK 9 | 494 661 |
| KzF 2 | 529 517 |
| KzF 5 | 521 526 |
| KzF 6 | 527 511 |
| KzFS 5 | 654 397 |
| LaK 6 | 643 584 |
| LaK 7 | 652 587 |
| LaK 11 | 658 537 |
| LaK 01 | 640 601 |
| LaK 02 | 670 573 |
| LaF 012 | 783 362 |

Water resistivity group 5

| Type | Value |
|---|---|
| PSK 2 | 569 632 |
| BaLK 1 | 518 603, |
| SF 03 | 847 239 |
| LaK 04 | 651 562 |
| LaK 07 | 678 534 |
| LaK 013 | 641 569 |

which glasses show a significant surface attack beyond mere formation of interference stains.

The second test for resistance to water, the $D_0$ test, determines the intrinsic chemical durability to water. The test parameters are similar to those for $T_{BLUE}$, but the criterion that is used to classify the glasses is weight loss per unit time. This test attempts to account for the effect of water on immersed glass due to the leaching of soluble ions from the glass and the disintegration of the $SiO_2$ or $B_2O_3$ glass network through hydrolysis. The resistivity of glass to these effects is directly related to the intrinsic chemical durability $D_0$. The unit of measurement is weight loss which is measured in fractions of milligrams per cubic centimeter per hour. Glasses in class 1 experience virtually no weight loss. Those in class 5 have a weight loss of greater than 0.015 mg/$cm^3$/hr.

**Chemical Resistance of Optical Glasses to Attack by Acidic Solutions.** Various tests are performed by optical glass manufacturers to determine the resistance of glass surfaces or of bulk material to the effects of acids. The test results provide information on the decomposition of polished glass surfaces in the presence of acids or acidic solutions. It is more difficult to predict the resistance to acids of a polished glass surface on the basis of results from an acid durability test for bulk material, since these tests will typically predict a higher resistivity to staining than is experienced in practice. Therefore both surface and bulk acid resistivity tests are needed.

Schott [3] performs two separate acid resistance tests. For the first test, samples of the more acid resistant glasses are exposed to a $0.5N$ nitric acid solution with a pH of 0.3. The exposure time that it takes to dissolve a 0.1-mm layer of glass is the measure used to classify the glasses. Glasses in this category range from an acid resistance class SR 1 at more than 100 hours of exposure down to acid resistance class SR 4 for which the time to remove a 0.1-mm layer lies between 6 minutes and 1 hour. For glass types that are so sensitive to acid attack that the standard test would yield useless data, a much milder acetic acid solution with a pH of 4.6 must be used to be able to make a distinction between them. The acid resistance classification for these glasses is SR 51 to SR 53, with the first requiring between 1 and 10 hours to remove a 0.1-mm layer, whereas the same removal is already achieved after less than 6 minutes of exposure for the last class. Schott glasses in these low-acid-resistance classes are listed on Table 1.6 [3]. They require the utmost care during manufacture, handling, and storage, and they must be protected from exposure to any form of acidic solutions, such as can be found in perspiration, solvents for certain cements, and acids that can form in water vapors in the presence of atmospheric gasses.

The surface test method used by Ohara [10] to determine the acid resistance of their glasses is nearly identical to the one described in the preceding paragraph. The evaluation is also similar and involves determining at which time interval the characteristic purple interference color appears on the test

**Table 1.6. Optical Glass Types Susceptible to Attack by Acids**

| Schott Optical Glass | | Hoya Optical Glass | | Ohara Optical Glass | | | |
|---|---|---|---|---|---|---|---|
| Class SR 53 | | Class $D_A$ 6 | | Bulk resistivity group 6 | | Surface resistivity group 5c | |
| LaKN 6 | 643 580 | ADF 4 | 613 449 | FK 3 | 465 659 | SSK 01 | 649 530 |
| LaKN 7 | 652 585 | ADF 10 | 613 444 | FK 5 | 487 702 | BaF 9 | 643 478 |
| LaK 11 | 658 573 | Class $D_A$ 5 | | KzFS 5 | 654 397 | BaF 10 | 670 473 |
| LaK 12 | 678 552 | FC 1 | 471 673 | Bulk resistivity group 5 to 6 | | BaF 11 | 667 483 |
| LaKN 13 | 694 533 | FC 5 | 487 704 | KzFS 4 | 613 438 | BaF 13 | 669 450 |
| LaK 21 | 641 601 | PCD 2 | 569 631 | Hoya class $D_A$ 5 cont'd. | | SF 03 | 847 239 |
| LaKL 21 | 640 597 | PCD 4 | 618 634 | LaFL 1 | 686 494 | KzFS 5 | 654 397 |
| LaK 23 | 669 574 | PCD 5 | 617 628 | NbF 1 | 743 492 | KzFS 8 | 720 347 |
| BaSF 50 | 710 366 | PCD 53 | 620 635 | NbF 3 | 735 498 | SFS 1 | 923 209 |
| BaSF 52 | 702 410 | BaCD 16 | 620 603 | NbFD 3 | 805 396 | SFS 01 | 923 213 |
| LaFN 8 | 735 416 | BaCD 165 | 620 598 | NbFD 4 | 831 365 | LaK 01 | 640 601 |
| LaF 9 | 795 284 | LaCL 1 | 641 568 | NbFD 7 | 783 361 | LaK 02 | 670 573 |
| LaFN 23 | 689 497 | LaCL 6 | 640 602 | NbFD 8 | 807 355 | LaK 04 | 651 562 |
| SF 58 | 918 215 | LaCL 7 | 670 573 | NbFD 12 | 800 423 | LaK 06 | 678 507 |
| SF 59 | 953 204 | LaCL 30 | 658 534 | TaFD 9 | 850 322 | LaK 07 | 678 534 |
| KzFS 1 | 613 443 | LaC 6 | 643 580 | ADF 1 | 565 530 | LaK 08 | 694 508 |
| KzFS 6 | 592 485 | LaC 7 | 652 584 | ADF 8 | 687 429 | LaF 3 | 717 479 |
| 44 glass types in class SR 52 | | LaC 9 | 691 547 | ADF 40 | 613 443 | LaF 01 | 700 481 |
| 41 glass types in class SR 51 | | LaC 12 | 678 555 | ADF 50 | 654 396 | LaF 06 | 686 492 |
| | | TaC 2 | 741 526 | ADF 355 | 645 408 | LaF 012 | 783 362 |
| | | TaC 4 | 734 511 | ADF 405 | 677 375 | 44 glass types in group 5b | |
| | | SbF 5 | 521 528 | | | 32 glass types in group 5a | |
| | | SbF 6 | 527 511 | | | | |

19

samples. But Ohara adds a second test that uses the powdered glass method to determine the acid resistance of the bulk material. This method was described earlier. The solution used for this test is a $0.01N$ nitric acid solution. The validity of this test has been questioned because the acid resistivity determined by this method is typically higher for the glass type than any found in actual practice during polishing and handling. It is therefore necessary to also refer to the acid resistivity by the surface method. The Ohara classification uses the old Schott convention. Their acid resistance group 1 is equivalent to Schott's SR 1; the Ohara group 4 is equivalent to SR 4, and groups 5a, 5b, and 5c are equivalent to SR 51, SR 52, and SR 53, respectively. Ohara glasses in the latter group are listed on Table 1.6.

The Japanese optical glass manufacturer Hoya [9] uses the powdered glass method to determine the acid resistance of their glasses. Hoya calls this method the acid durability test $D_A$. It is identical to the staining resistivity described above, except that the reagent is a $0.01N$ nitric acid solution. The percent weight loss is measured, and the glasses are grouped into six classes, with class 1 showing a negligible weight loss of less than 0.2% for the 1 hour test, and class 6 an appreciable weight loss of greater than 2.2%. The Hoya glasses in class $D_A$ 6 are listed on Table 1.6.

Chance Pilkington [8], the English optical glass manufacturer, has developed a somewhat different approach to grading their glasses according to resistance to attack by acidic solutions. The test is similar to that described for Schott and Ohara in that polished samples of the glass under test are exposed to an $0.5N$ nitric acid ($HNO_3$) until interference colors are detected. The method of evaluation differs in that there are seven time interval codes with code 1 > 100 hours and code 7 < 0.001 hours or <3.6 seconds. In addition the severity of attack is subjectively judged from mild (code .1) to very heavy (code .5). The type of surface attack is judged as being acid stain (a) as evidenced by interference colors, acid polish (b) which shows signs of acid attack without the formation of interference colors, or acid etch (c) which is an uneven removal of the surface with the presence of an opaque residue on the surface. Thus the acid grade for a particular glass (BF 702 410, in this example) may read 7.3 a, which means that the acid attack is rapid (<0.001 hour), moderately severe, and in the form of interference colors. CP glasses of code 7 are also listed in Table 1.6.

**Chemical Resistance of Optical Glass to Attack by Alkaline Solutions.** During the manufacturing stages there is a greater chance for polished surfaces to come in contact with alkaline solutions than with any other chemically active solutions. Water-based slurries are required during the grinding and polishing stages, but studies have shown that the slurry can become increasingly alkaline as more glass is abraded and becomes a constituent of the slurry. The water reacts with the microscopically fine glass particles to leach out alkaline glass components; this raises the pH of the slurry. Although normally not a problem for brushed on or otherwise hand-fed slurries, this condition can develop into a serious problem for recirculating slurry systems unless its pH is regularly monitored and the slurry is replaced or diluted with fresh slurry when predetermined pH levels have been reached. Other obvious alkaline solutions that components made from optical glass are often exposed to are found in the detergent baths of aqueous cleaning lines. More will be said about this topic later in Chapter 5.

Hoya [9], in recognition of the importance of the effect of alkaline solutions on optical glass, has developed two tests to determine the alkali resistance of their glasses. These tests are both called "latent scratch resistivity," with the first (designated $D_{NaOH}$) testing the corrosion resistivity of the glass to hydroxyl ions and the second ($D_{STPP}$) testing the corrosion resistivity to polymerized phosphoric ions.

The term "latent scratch resistivity" resulted from the understanding that alkaline solutions tend to attack polished glass surfaces to clearly reveal subsurface scratches that otherwise could be seen only with considerable difficulty, if at all. This fairly typical surface attack is caused by corrosive ions that are formed from inorganic builders through the chemical process of hydrolysis. Such inorganic builders are certain constituents of detergents found in aqueous cleaning systems. One commonly used inorganic builder is sodium hydroxide (NaOH), another is sodium tripoly phosphate, or STPP, which is also known as trisodium phosphate (TSP). The process of hydrolysis creates either hydroxyl ions or polymerized phosphoric ions. Therefore two separate tests are required to define the resistivity of optical glasses to alkali attack.

For the $D_{NaOH}$ test a circular glass sample polished on both sides, is immersed in a well stirred $0.01 N$ NaOH solution at 50°C. The immersion time is 15 hours, after which time the weight loss of the sample is determined. Glasses in class 1 experience a negligible weight loss, while those in class 5 experience a weight loss that exceeds 0.31 mg/cm$^2$. Therefore the glasses in class 5 are most susceptible to attack by hydroxyl ions.

The $D_{STPP}$ test is performed in nearly the identical fashion, except that the solution is 0.01 M $Na_5P_3O_{10}$ (STPP) and that the immersion time is only 1 hour. There are also five classes with glasses in class 1 remaining nearly unaffected, while the weight loss of those glasses in class 5 exceed 0.61 mg/cm2. Again, glasses in $D_{STPP}$ class 5 are most susceptible to alkaline attack, this time from phosphoric ions.

Schott [3] elected to use a somewhat different approach for their alkali resistance (AR) test. Standard samples of the glasses are polished on both sides and immersed in an alkaline (pH 10) solution of sodium hydroxide at 90°C. The immersion time in minutes required to dissolve a 0.1 mm layer of the glass is used to classify the glasses. Glasses in class 1 require an immersion time of 120 minutes, whereas those in class 4 require less than 7.5 minutes to show effects.

Since the appearance of the attack varies from glass to glass, a further, though somewhat subjective, classification is required. This is similar to the method used by Chance Pilkington to classify glasses by their acid durability as has already been described earlier. Glasses in subclass .0 show no effect, in subclass .1 some effect but no interference colors, in subclass .2 interference colors are seen, in subclass .3 whitish stains have appeared, and finally in subclass .4 a sometimes thick white coating has formed on the surface.

These test results led Schott to express the alkali resistance of optical glasses as a two-digit number. The first is the AR class, and the second is the subjective classification number. For instance, SF 58 (918 215) has an alkali resistance value of 3.4 which means that alkali attack occurs in less than 30 minutes and that a whitish stain forms on the polished surface. This undesirable condition most definitely will lead to rejection of the optic. Polishing the glass in high-speed recirculating polishing systems and cleaning the finished parts in hot detergent baths will probably not yield acceptable results. The components must then be polished with a conventional process using hand fed slurry and then manually cleaned, using either mild neutral pH detergent solutions at room temperature or cleaning solvents such as acetone or methanol. The Hoya and Schott glasses that have been identified by various tests as being most susceptible to attack by alkaline solutions are listed on Table 1.7.

We close this lengthy section on the chemical behavior of optical glasses under various tests performed by the glass manufacturers with a caveat. The process engineer and the optician should always consult the glass catalog of the manufacturer who made the glass, whenever they are confronted with the task of processing a new, unfamiliar glass type. This readily available

**Table 1.7.  Optical Glass Types Susceptible to Alkali Attack**

| Schott Optical Glass [3] | | Hoya Optical Glass [9] | | | |
|:---:|:---:|:---:|:---:|:---:|:---:|
| AR class 4 | | $D_{NaOH}$ class 5 | | $D_{STPP}$ class 5 | |
| LaKN  7 | 652 585 | BaCD  10 | 623 569 | LaCL  1 | 641 568 |
| LaKN  11 | 658 573 | BaCD  14 | 603 607 | LaCL  3 | 665 534 |
| LaKN  12 | 678 552 | BaCD  165 | 620 598 | LaCL  4 | 670 517 |
| LaK  21 | 641 601 | LaCL  1 | 641 568 | LaCL  6 | 640 602 |
| LaKL  21 | 640 597 | LaCL  4 | 670 517 | LaCL  7 | 670 573 |
| LaK  23 | 669 574 | LaCL  6 | 640 602 | LaC  7 | 652 584 |
| KzF  6 | 527 511 | LaC  6 | 643 580 | LaC  9 | 691 547 |
| KzFS  1 | 613 443 | FDS  9 | 847 238 | LaC  11 | 658 573 |
| KzFSN  2 | 558 542 | LaFL  1 | 686 494 | LaC  12 | 678 555 |
| KzFSN  4 | 613 443 | LaFL  2 | 697 485 | LaFL  1 | 686 494 |
| KzFSN  5 | 654 396 | ADF  1 | 565 530 | ADC  1 | 620 622 |
| KzFS  6 | 592 485 | ADF  8 | 687 429 | ADF  1 | 565 530 |
| KzFSN  7 | 681 372 | | | ADF  8 | 687 429 |
| | | | | ATC  1 | 620 622 |

information has important implications for the success or failure of polishing, handling, and storing optics made from glasses that have been identified to be sensitive to the various chemical influences.

### 1.1.7. Mechanical and Thermal Properties of Optical Glass

The mechanical properties of optical glass should be of interest to the optician and the process engineer because they can provide valuable clues on the behavior of the glass during the shaping, grinding and polishing processes. The thermal properties provide an early indication how the material might behave when the operating temperature fluctuates over several hundred degrees. The main concern under these conditions which are found during the blocking operations is thermal fracturing of the glass. The coefficient of expansion of the glass can alert the optician that the parts must be handled with great care during the heating and cooling phases or that thermal gradients can set up considerable distortions in the interference fringes during testplate testing or testing in the interferometer. The *specific gravity* is a unique value for each glass that expresses how many grams a cube with 1-cm sides will weigh.

Unlike the chemical characteristics that were previously reviewed, a thorough knowledge of the mechanical properties is not essential or even directly applicable to the job. Nevertheless, an understanding of these properties can be very useful in predicting the behavior of glass types for which there is no prior experience. There are two characteristics under consideration. The first is the mechanical strength of the glass; this gives us some idea how much force can be applied to the glass or how easily the glass can chip or fracture. The second is the abrasion hardness of the glass; this is a relative value that represents the ability of the glass to resist abrasion. Another way to describe it is as a measure of the stock removal rate relative to a standard glass that can be expected during grinding operations.

The hardness of the glass is measured with a microhardness tester that uses a rhomboidal diamond point of exactly described dimensions that is forced to penetrate the glass surface by a precisely measured weight. The penetration leaves an indentation in the polished glass surface which is then measured with a microscope. From this information, the Knoop hardness is calculated. It is expressed in terms of kiloponds per square millimeters (kp/$mm^2$). The reported hardness values [12, 13] range from a low of 250 kp/$mm^2$ for some SF glasses to a high between 650 to 750 kp/$mm^2$ for a number of LaSF glasses. The *kilopond* is a measure of force in the mks system of measurement (mks = meter/kilo/second). That differentiates it from the measure of mass, which is expressed in kilograms (kg). The values vary somewhat between these two sources. That is due to the slightly different test approaches and the uncertainty in determining the exact indentation length. From these data the conclusion can be drawn that SF glasses are soft and should grind and polish faster than the LaSF glasses, which seem much harder. Although this is true in a general sense, there are numerous departures from this apparent relationship.

One way to get a better idea of how a glass will respond to grinding and polishing is to measure their abrasion hardness. This is done by grinding test samples of the optical glasses under clearly defined conditions and then comparing the amount of material removed after a specified time interval with that from a sample of BK 7 under identical conditions. In this way BK 7 has a value of 1.0, harder glasses have a value of less than 1, and softer glasses have a value greater than 1. To differentiate the abrasion values better, that of BK 7 is set at 100. The most abrasion resistant glasses have an abrasion hardness below 60 (LaSF); the abrasion hardness value will exceed 400 for the softest glasses (FK) [13]. In other words, compared to the BK 7 standard sample, it takes twice as long to remove a specific thickness from an LaSF glass, but the same amount of material can be removed from an FK glass in one-quarter of the time. The abrasion value can serve as a useful guide in predicting the rate at which material will be removed from the glass during grinding. It says nothing, however, how well the glass will polish.

Every material has a unique weight per unit volume. This weight is known as *specific weight* or also as its *density*, and it is expressed in grams per cubic centimeter ($g/cm^3$). Namely, it is the weight in grams of a cube of the material with one-centimeter sides. It is possible to determine the specific weight for nearly any substance by preparing such a cube and weighing it. The specific weight for optical glasses range from 2.25 $g/cm^3$ for some FK glasses to 6.25 $g/cm^3$ for a few SF glasses [12, 13]. BK 7 has a specific weight of 2.51 $g/cm^3$. This value is essential when the weights of lenses, prisms, or blanks must be calculated. The volume of the lens or prism must be calculated first from the part dimensions (in centimeters) which is then multiplied by the specific weight to obtain the weight of the part in grams. If the dimensions are in inches and the weight is to be expressed in pounds, then the specific weight in $g/cm^3$ must be first multiplied by 0.0361. The density of the material is then expressed as pounds per cubic inch ($lb/in.^3$).

The optician must be concerned about another property of optical glasses or of any other optical material that determines how the materials will respond to sudden temperature changes. Every material changes its volume as the temperature changes. Most expand as the temperature increases and contract when it decreases. A few odd materials behave in an opposite fashion, at least over a specific temperature range, but some specially formulated and treated materials do not change their volume over an appreciable range of temperatures. Such materials are known as *low-expansion materials,* and they will be discussed in Section 1.2.

This thermal behavior of materials is expressed in terms of their coefficient of thermal expansion (CTE). The CTE values range for optical glasses from a low of nearly $40 \times 10^{-7}/°C$ for a KzFS glass to about $150 \times 10^{-7}/°C$

for a few FK glasses [12, 13]. These values were measured over a temperature range that closely corresponds to the operating temperatures typically found in the optical shop. By definition, the coefficient of linear expansion expresses the increase (or decrease) in length per centimeter of ambient length per centigrade of temperature rise (or decrease). For instance, BK 7 has a CTE of $71 \times 10^{-7}$ cm/cm/°C. This means that for every centigrade increase in temperature, the original length of one centimeter will increase by 0.0000071 centimeter, or about 0.07 $\mu$m. The CTE values are between 10% and 20% higher at temperatures between 100°C to 300°C.

High CTE values indicate glasses that may be thermally sensitive. Sudden temperature changes, often called *thermal shock,* can result in thermal fracturing of the optical glass. Therefore considerable care must be exercised by the opticians when the parts made from such thermally sensitive glasses are heated up for blocking or cooled down after blocking. Any sudden cool drafts, such that emanate from air-conditioning vents, must be kept away from the heated parts. They must also not be placed on or touched by any cooler surface, for thermal fracturing will result. Even glasses with much lower CTE should be treated with care when they are hot, but the high CTE glasses warrant extraordinary attention.

### References for Section 1.1 (Optical Glass)

[1] Izumitani, T. S., *Optical Glass,* UCRL-Trans-12065, LLNL, 1985.
[2] per DIN 58 925.
[3] Schott Optical Glass, Inc., catalog 3060/72/USA IX/72 o.P.
[4] Schott Glass, Germany.
[5] Corning Glass Works, Corning, NY.
[6] Bausch & Lomb, Inc., Rochester, NY.
[7] Pittsburgh Plate Glass (PPG), Pittsburgh, PA.
[8] Chance Pilkington, St. Asaph, Clwyd, U.K., 1977 catalog.
[9] Hoya Corporation, Japan.
[10] Ohara Inc., Japan.
[11] Izumitani, T. S., *Optical Glass,* Chapter 2, see Ref. [1].
[12] *Optical Glass,* pocket catalog, edition 1986, Schott Glass Technologies, Inc., Duryea, PA.
[13] *Ohara Optical Glass,* pocket catalog, 1990, Ohara, Inc., Japan.

## 1.2.  SPECIAL PURPOSE GLASSES

A number of glasses are used for optical purposes, though they are not optical glasses in the strictest sense. Optical glass is specifically formulated for use in optical systems; the glasses described in this section are glass materials that are used for optical purposes even though they originally had other applications. These glasses are laser glass, color filter glass, and technical sheet glass.

### 1.2.1. Laser Glass

Laser glasses are optical quality glasses formulated for use in optically pumped high-power lasers. These glasses are typically neodymium doped. This gives them their characteristic light purple color. Laser rods made from these materials emit light in the near-IR region of the spectrum at about 1.06 $\mu$m.

Although most of the laser glass has been developed for use as laser rods and disk amplifiers in large laser fusion experiments, a number of commercially available laser systems use laser glass rods as the source element. A more recent development is the *zigzag* or *slab laser* where the lasing action takes place not in a rod but in a carefully fabricated slab of laser glass.

**Brief History.** Neodymium-doped laser glass has been available since the mid-1960s. A concentrated effort to improve these glasses and to develop new ones began in the early 1970s in support of several large-scale laser fusion efforts [14, 15, 16]. Several glass manufacturers [17, 18, 19, 20, 21] made significant contributions to the development and testing of a great variety of new laser glasses. They are listed in Table 1.8.

The development efforts took place in essentially three phases with silicate glasses in the early 1970s, phosphate and fluorophosphate glasses in the late 1970s into the early 1980s, and a concurrent separate effort [17] to develop fluoroberyllate glass. The development of platinum-free glass and high transmission (very low absorbing) glasses were additional significant results of these efforts.

As already indicated, there are five basic types of laser glass:

1. Silicate glass
2. Phosphate glass
3. Fluorosilicate glass
4. Fluorophosphate glass
5. Fluoroberyllate glass

These glasses are shown in the $n_d/v_d$ diagram in Fig. 1.5 [22]. They will be discussed separately in some detail in the following paragraphs.

Silicate glass is based on silicon dioxide ($SiO_2$), as the glass-forming oxide. It was the first laser glass produced, and it is the least expensive of all laser glasses. For instance Schott's LG-660 can be produced quite economically in a continuous melting process. Silicate laser glass has good thermal, chemical, and mechanical characteristics. From a lasing performance standpoint, however, it is the least efficient laser glass.

**Table 1.8. Laser Glass Summary**

| Silicates | | Phosphates | | Fluorophosphates | | Fluoroberyllates | | Faraday rotator glass | |
|---|---|---|---|---|---|---|---|---|---|
| Glass | Maker | Glass | Maker | Glass | Maker | Glass | Maker | Glass | Maker |
| ED-2 | OI [21] | EV-1 | OI | EVF-1 | OI | B-101 | Corning [17] | FR-4N | Hoya |
| ED-3 | OI | EV-2 | OI | | | | | FR-5 | Hoya |
| ED-8 | OI | EV-4 | OI | LG-10 | Hoya | | | FR-7N | Hoya |
| ED-9 | OI | | | LHG-104A | Hoya | | | FR-9 | Hoya |
| | | LGH-5 | Hoya | LHG-104B | Hoya | | | | |
| LSG-91h | Hoya [18] | LGH-6 | Hoya | LHG-105A | Hoya | | | | |
| | | LGH-7 | Hoya | | | | | | |
| Q-246 | Kigre [20] | LGH-8 | Hoya | FK-51 | Schott | | | | |
| | | LGH-91N | Hoya | FK-52 | Schott | | | | |
| LG-660 | Schott [19] | HAP-3 | Hoya | LHG-810 | Schott | | | | |
| LG-670 | Schott | | | LG-802 | Schott | | | | |
| LG-680 | Schott | Q-88 | Kigre | LG-810 | Schott | | | | |
| | | Q-98 | Kigre | LG-812 | Schott | | | | |
| | | Q-100 | Kigre | | | | | | |
| | | LG-700 | Schott | | | | | | |
| | | LG-703 | Schott | | | | | | |
| | | LG-706 | Schott | | | | | | |
| | | LG-750 | Schott | | | | | | |
| | | LG-760 | Schott | | | | | | |

27

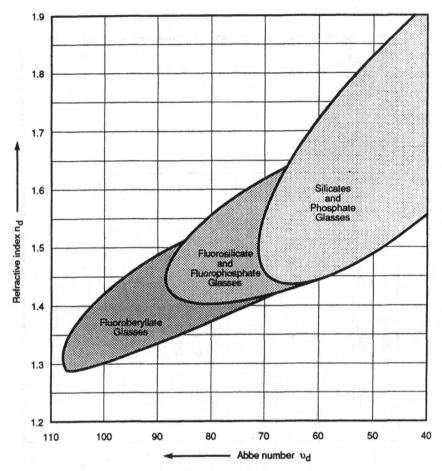

**Figure 1.5.** $n_d/\nu_d$ diagram for laser glass [22].

Silicate glasses are typically edge strengthened. This treatment is performed after the laser rods or disks are ground as nearly damage free as possible. This is an essential first step in this process to prevent thermal stress damage; thermal stress damage, which leads to eventual failure of the rod or disk, occurs at locations of subsurface flaws left over from the fine-grinding steps. The parts are then treated with acid to remove any residual flaws, and the surface is strengthened by an ion exchange process. This strengthening treatment has greatly increased the pump and laser power handling capabilities of the glass. It also increases the rupture strength of the glass by a factor of four, which greatly reduces the danger of damaging the rods or disks during handling and installation [23].

Phosphate glasses are more efficient as lasing medium than silicate glasses but they tend to be less stable in their thermal, chemical, and mechanical properties. Some phosphate glasses are considered *hygroscopic*. This means that they are affected by water or water vapors. As a result they have low chemical durability, which makes them harder to polish and handle than parts made from silicate glass. Moisture condensing on polished surfaces can create a permanent dull haze [24]. Good success has been achieved by polishing with alcohol-based polishing slurries. Other phosphate glasses, such as Hoya's LHG-6, have good chemical durability, and they are polishable with standard optical shop methods without staining.

The coefficient of thermal expansion (CTE) of phosphate glass is twice that of silicate glasses which greatly increases the danger of thermal fracture during pumping at high powers. This condition makes it essential that laser components made from phosphate glass be chemically strengthened. There has been some success to strengthen finished laser rods without affecting the quality of the optically polished surfaces [25]. Other phosphate glass variants were developed, such as Hoya's LGH-8 athermal glass which has a near-zero CTE of optical path length. This property is important in reducing thermal effects in the glass during the lasing action. The athermal glasses, however, are chemically and mechanically less durable than the other phosphate glasses [26].

Another improvement in phosphate laser glass is the development of platinum-free phosphate glass, such as Schott's LG-750, which was developed for the Nova Fusion Laser [14]. Microscopic traces of platinum in the glass can lead to laser damage because they absorb a sufficient amount of laser energy to vaporize which causes local microfractures in the glass and this leads eventually to the failure of the laser [27]. Schott improved its phosphate glass melting process to eliminate platinum contamination and the company earned an IR-100 award for this success [28].

In addition to silicate and phosphate laser glass, there are specially formulated laser glasses that are based on fluor, such as fluorophosphate, fluorosilicate, and fluoroberyllate glasses. The fluor component tends to decompose during the melting stage into fluorine gas, which is not only toxic but also introduces bubbles into the melt. This requires the addition of fining agents that control the formation of fluorine bubbles [29]. Fluorophosphate glass is particularly difficult to make because the fluor and phosphate batch ingredients do not mix well; this leads to the formation of numerous fine parallel striae in the glass. Fluoroberyllate glass, also known as beryllium fluoride ($BeF_2$) glass, has the lowest nonlinear index of all the laser glasses, so it has the greatest laser efficiency potential. However, the high cost of the raw materials, the limitations in size, and above all the beryllium toxicity problems make commercialization of this type of glass very difficult. Because of the toxicity concerns, the melting of $BeF_2$ glass and all subsequent cutting, shaping, grinding and polishing must be done in an OSHA approved controlled environment. All work must be done in airtight glove boxes, and all the air used for ventilation must be filtered [30].

Faraday rotator glass is mentioned here, although it is not a laser glass. Components made from this type of special glass are used in high-power fusion lasers as blocking filters for back-reflected laser energy. Faraday rotator glass is terbium-doped optical glass that utilizes the *Faraday effect*. That is, the plane of polarization of the light passing through it is rotated 45° when a strong magnetic field is applied. This allows the light to pass through in one direction unhindered; any of the light reflected off another surface in the optical train is blocked so that it cannot propagate in the opposite direction. This glass acts as one-way shutter that eliminates the potentially destructive behavior of self-focusing that can occur in high-power laser systems.

### 1.2.2. Color Filter Glass

Color filter glass is used in optical systems because it has specific spectral transmittance characteristics within the range of the optical spectrum that are useful to the optical systems designer. This glass is usually not high-quality optical glass, so it is important to know how and where in the system the filters are used. There are three major manufacturers of color filter glass: Schott [31], Hoya [32], and Corning [33]. Together, they produce more than two hundred different color glass filter types. Hoya lists nearly 100 in their catalog, Schott offers more than 80, and Corning more than 50.

Optical filters function in two basic ways in which ultraviolet (UV), visible (VIS), and infrared (IR) radiation are either uniformly attenuated (reduced in intensity) or selectively attenuated in specific spectral regions. Optical filters can act as neutral density filters, ND filters for short; these filters appear grey because they affect all wavelengths uniformly. Filters that take on many different colors are referred to as "color filters." Color filters are classified according to their effect on the optical spectrum. Bandpass filters selectively isolate specific spectral regions. Longpass filters reject shorter wavelengths, whereas shortpass filters reject longer wavelength regions.

The attenuation results from the absorption of certain wavelengths or wavelength regions. This gives the filters their characteristic color. Blue filters, for instance, absorb most of the yellow and red portion of the visible spectrum, whereas IR filters absorb all of the visible light and only pass the infrared portion of the optical spectrum. That is why the blue filter appears blue and the IR filter appears black.

**Coloration.** There are two basic processes that introduce coloration into glass. One uses ionic coloration. The color characteristic is achieved by specific ions that are in true solution in the glass. The second involves thermal treatment to segregate the colorants from the glass matrix as submicron particles. Table 1.9 lists some of the colorants used to achieve the various colors [34]. The spectral characteristics of color filter glass are determined by the type of colorant used, the density of coloration and the type of glass that serves as host. Processing conditions can also be manipulated; for instance, if desired, transmittance for a filter can be controlled by varying the thickness of the filter.

Table 1.9. Colorants for Specific Filter Colors [34]

| Ionic coloration | | Thermal coloration | |
|---|---|---|---|
| Colorant | Color | Colorant | Color |
| nickel oxide | purple | sulfur | lt. yellow |
| cobalt oxide | blue | cad. sulfide | yellow |
| chromium oxide | green | cad. selenide | orange-red |
| | | gold | rose to ruby |

The optician must work filter glass to shape it, and to reduce the thickness for controlling the transmission characteristics. Even though this material is not considered an optical quality material, the optician must have a basic knowledge of the thermal, chemical, and mechanical properties of color filter glass.

The thermal property of interest to the optician is the coefficient of thermal expansion because it is a measure of the susceptibility of the glass to thermal fracture. Filter glasses are difficult to work. They typically exhibit fairly high coefficients of thermal expansion ranging from a low of $40 \times 10^{-7}$ (cm/cm/°C) to $140 \times 10^{-7}$ (cm/cm/°C). In fact, about $\frac{2}{3}$ of all color filter glasses have a higher coefficient of thermal expansion than BK 7, which is about $70 \times 10^{-7}$ (cm/cm/°C). As a result filter glasses must be handled with great care when they are being heated up for blocking. The sensitivity to thermal breakage is further aggravated by unfavorable thickness-to-length aspect ratios, which for color filter glass can be as high as $40:1$.

The chemical characteristics of color filter glass provide clues to the problems that the optician can encounter when polishing these materials. The methods used by Hoya to classify these glasses are nearly identical to those described for optical glass in Section 1.1.5. Hoya uses the water durability ($D_W$) to identify the susceptibility to dimming and the acid durability ($D_A$) to classify the glasses according to their tendency to stain during polishing. Schott originally elected to provide a somewhat more subjective measure of the chemical stability [35], as shown in Table 1.10, but has since classified color filter glass according to the chemical resistance tests for FR, SR, and AR, as described in the Section 1.1 of this chapter.

All these methods, however, provide essentially the same information. They show that between 60% and 70% of the color filter glasses do not present any problems during polishing. Of the remainder most require only a little additional care; only a small percentage of the color glasses are likely to present significant problems during polishing. These glasses are listed in Table 1.11. In addition to special precaution during polishing, the optician has to apply a protective coating to the polished surfaces during storage.

Finally, it is important to know before shaping or grinding color filter glasses some of their mechanical properties, that is, whether a specific glass filter can be easily cut or ground or whether it may present some unforseen

**Table 1.10.  Chemical Stability of Color Filter Glasses [35]**

| Group | Weathering resistance | Polishing characteristics |
|-------|----------------------|---------------------------|
| 1 | very good | excellent |
| 2 | good | good |
| 3 | sufficient | requires care |
| 4 | poor | difficult to polish |

difficulties. One measure is the Knoop hardness which ranges from 400 to 600 kg/mm$^2$ for color filter glass, with about 66% between 500 and 600 kg/mm$^2$. In this sense color filter glasses are similar to most optical glasses.

A more useful measure of workability is the abrasion factor, denoted by Hoya as $F_A$. A factor greater than 100 indicates a glass that is softer (more rapidly abraded) than BK 7, whereas a harder glass has a factor less than 100. Of the 85 or so color glasses listed in the Hoya catalog, most have an abrasion factor between 100 and 150, which means that their abrasion factor is either similar to BK 7 or somewhat softer. Two glasses (B-370 and B-380) are slightly harder than BK 7, while six glasses (UV-22, UV-36, IR-80, IR-85, U-330, and U-340) are significantly softer than BK 7, with an abrasion factor exceeding 200 which is in the range of SF and SFS optical glasses. Almost all of these softer glasses are also listed in Table 1.11. Clearly considerable care must be exercised when working and handling these glasses.

Some color filters such as heat-absorbing filters are intended to be used near heat sources. They are either thermally or chemically strengthened to avoid thermally induced damage. But the treatment can lead not only to surface deformation and possible discoloration of the color filter but also to breakage. As a result these filters cannot be cut or ground without some

**Table 1.11.  Difficult to Polish Color Filter Glasses**

| Schott [36, 38] Long term attack group 5 | group 5 | $D_W$ group 4 | Hoya [37] $D_A$ group 4 | $D_A$ group 5 |
|---------|---------|---------|---------|---------|
| UG-5 | BG-39 | UV-22 | UV-22 | IR-80 |
| UG-11 | BG-40 | G-545 | RM-100 | IR-85 |
| BG-24 | WG-225 | G-550 | B-380 | ND-70 |
| KG-1 | KG-5 | N-70 | ND-50 | U-330 |
| KG-2 | | U-330 | U-340 | |
| KG-3 | | U-340 | | |
| KG-4 | | V-10 | | |

breakage. Before resurfacing or cutting a filter glass, the optician must make sure that the glass has not been "strengthened"; otherwise, the grinding and cutting operations may turn out to be a wasted effort.

### 1.2.3.  Technical Glass

The term *technical glass* encompasses all glass materials used for optical purposes that are not optical, laser, or filter glasses. Any sheet glass like plate glass, float glass, or drawn glass is considered technical glass. These glasses will have often undesirable quality attributes that are not acceptable for optical quality glass. They are typically used only as simple reflecting surfaces for illumination systems or as substrate for a wide variety of thin film filters.

Most flat glass is standard soda-lime glass that is produced by one of three methods: the Foucault method, Colburn method, or float glass method. In the Foucault method it is drawn vertically upward from a molten pool. In the Colburn method it is drawn horizontally through an annealing oven. But most flat glass manufacturers use the float process because it is the most economical way to produce flat glass in sufficient quantities. The quality of float glass is also superior to flat glass produced by the other methods. Only flat glass that is no more than 2 mm (0.080 in.) thick is still drawn today [39]. Soda-lime glass is composed of about 70% silica ($SiO_2$) and 15% sodia ($Na_2O$).

Float glass is produced by casting the molten glass in a continuous band onto a bed of molten tin. This produces sheets of glass that are surprisingly flat and also quite smooth. The natural thickness of float glass is 7 mm (0.280 in.), but it can be stretched down to 1.8 mm (0.070 in.) [40]. Although not specifically produced for optical purposes, the extraordinarily flat and smooth surfaces of float glass makes it an ideal choice for low end reflective optics and for filter substrate. A comparative study [41] of the quality of float glass from German, Belgian, and American sources showed that the thickness tolerances were between ±0.001 and ±0.005 in. and that the parallelism was between 1 and 3 arc minutes. Flatness in terms of spherical power was as good as 0.5 fringe and no worse than 3 fringes over any 0.5-in. square area. The irregularity in the interference fringes, however, varied over a wide range, up to 5 fringes over the same surface area. Most of the samples evaluated met or exceeded 80-50 surface quality requirements.

Thin glass sheets below 2.5 mm (0.100 in.) are often needed as thin film filter sustrate. Many filters must be much thinner. Since most flat glass is not easily ground and polished, it must be produced to the design thickness. Thin glass sheets can be produced by one of several methods. They can be drawn upward, drawn downward, or mechanically stretched to the desired thickness by a process called *fusion;* they can be produced by the float method to a thickness of no less than 2 mm (0.080 in.); or they can be rolled and extruded to a specified thickness. The fusion method has been used to produce thin glass sheets as thin as 0.25 mm (0.010 in.). Thin glass sheets with a thickness of 1 mm (0.040 in.) or less are usually referred to as micro-sheet.

Prominent among other technical glass besides the most common soda-lime glasses are borosilicate glass and alumino silicate glass. These glasses are either drawn or cast. Drawn sheets have fire-polished surfaces that are under significant surface stress. To improve the optical surface quality of these glass sheets, one might be tempted to grind and polish the side where the better finish is desired. Because of the high surface stresses, however, both sides need to be ground and polished to avoid significant surface distortions that would otherwise result. This product is commercially available as twin ground sheet, but it is significantly more expensive than the drawn sheet because of the additional costly double lapping and two-sided polishing steps [42].

Two technical grade borosilicate glasses must be mentioned here because they are important in low-end optics where high precision is not needed but where low cost is a consideration. One of these is B-270 [43]. This is a water white glass that provides good optical quality at low cost. It is available as drawn sheet, extruded rods, or cast blocks or sheets. Strip glass is produced by a continuous casting process in standard sizes of 300 to 450 mm (12 to 18 in.) by 800-mm (32-in.) long in eight standard thicknesses ranging from 17 mm (0.68 in.) to 60 mm (2.4 in.). This glass can also be pressed into lens and prism blanks. It is an economic alternative to BK 7 which makes it an ideal choice for noncritical transmissive or reflective optics. Its $n_d = 1.523$ and the dispersion value is $v = 58.8$. This gives it a six-digit glass type code of 523 588. Its chemical resistance is similar to that of BK 7, as is its Knoop hardness. The coefficient of thermal expansion, however, is about 33% higher than that of BK 7 glass. Another borosilicate glass is Code 7059 [44], which is available as drawn sheet. This glass has been specifically formulated for use as fotomask or as optical thin film filter substrate. Its borosilicate composition has a very low alkali content, and this reduces the danger of alkali ions leaching out of its surface. Such ions can cause pinholes in thin film coatings. Code 7059 as drawn sheet is not very flat, nor is it very smooth, especially when it is compared with float glass. Nevertheless, if optical quality is not a criterion, it is a cost-effective choice.

Some technical glasses are thermally strengthened (tempered) or chemically strengthened by an ion exchange process with alkali nitrate salts. Either treatment can set up highly compressive surface stresses, and when these stressed surfaces are scored or damaged, the glass can fail catastrophically and disintegrate through stress relief process called dicing. Strengthened glasses cannot be ground or polished [45].

### References for Section 1.2 (Special Purpose Glass)

[14] SHIVA and NOVA Laser Fusion Program at (LLNL) Lawrence Livermore National Lab, Livermore, CA.

[15] Omega Fusion Laser at the University of Rochester, Rochester, NY.

[16] Gekko Fusion Laser at the University of Osaka, Japan.

[17] Corning Glass Works, Corning, NY.

[18] Hoya Optics, Inc., Fremont, CA.

[19] Schott Glass Technologies, Inc., Duryea, PA.

[20] Kigre, Inc., Hilton Head, SC.

[21] Owens Illinois (laser glass technology sold to Schott).

[22] Weiss, J. A., "Optical materials and the great leap fusionward," *Optical Spectra,* May 1977.

[23] Schott catalog, *Laser Glass,* No. 7525 e USA, VI/83.

[24] News item, "Phosphate-glass report: Nonlinear index is low, but so is the glass' durability," *Laser Focus,* June 1975.

[25] *Photonics Spectra,* September 1986.

[26] Hoya Corporation, product report "New Hoya glasses for high-power laser systems," June 1977.

[27] News item, "LLNL, Schott, and Hoya develop platinum-free laser glass, *Laser Focus* E/O, November 1987.

[28] "Improved laser glass," *Photonics Spectra,* November 1987.

[29] Bailey, Donald L., "Future optical glass," *Optical Shop Notebook I,* Section II: *Optical Materials,* May 1976.

[30] "Corning develops special laser glass," *Optical Spectra* October 1977.

[31] Schott Optical Glass, Inc., Duryea, PA.

[32] Hoya Corporation, Tokyo, Japan.

[33] Corning Glass Works, Corning, NY.

[34] Corning product literature, "Color filter glass," CFG 5/80 MA.

[35] Schott product literature, "Schott color filter glass," CF 5/9.

[36] Schott product literature, "Color filter glass from Schott," CF 15-22.

[37] Hoya catalog, "Hoya color filter glass," No. 7601 E.

[38] Schott catalog, "Optical glass filters," No. 3555e/USA IX/84.

[39] Dart, William C., "Glass for industry," *Optical Spectra,* October 1977.

[40] Shoemaker, A. F., "Thin glass for solar applications," Corning Glass Works, 1978.

[41] Karow, Hank H., "Float glass study," OCLI, 1980.

[42] Lind, M. A., and Rusin, J. M., "Heliostat glass survey and evaluation," Batelle Pacific Northwest Laboratories, 1978.

[43] Schott product literature, "B-270," 1980.

[44] Corning product information, "Code 7059 superflat substrates," 1979.

[45] Corning Glass Works, "Characteristics of chemically strengthened glass substrates for magnetic memory disks (Glass: Code 0313), 1971.

## 1.3. LOW EXPANSION MATERIALS

Optical materials such as fused quartz, fused silica, crystal quartz, and low-expansion materials such as Pyrex [46], Cer-Vit [47], and Zerodur [48] are usually not discussed under one heading. However, their grinding and polishing characteristics are sufficiently similar for it to make sense, from a fabricator's point of view, to do that. These materials share one particular property. They are relatively easy to polish. They not only polish well; they hold their figure over a wide range of operating temperatures, thanks to their very low coefficient of thermal expansion (even though these materials are quite hard compared to many optical glasses). They are also chemically stable and mechanically strong (the same properties that contribute to the good polishing characteristics of these low-expansion materials).

### 1.3.1. Fused Quartz and Fused Silica

Fused quartz and fused silica are terms used to describe a family of very hard and transparent materials of excellent optical clarity. The raw material is nearly 100% pure silicon dioxide ($SiO_2$) which is a crystalline material found in great abundance on earth. It is fused at high temperatures into an amorphous or vitreous material. These terms mean that the material is not crystalline but more like a glass. Both fused quartz and fused silica are basically amorphous silicon dioxide. The high optical purity of these materials and their thermal, chemical and mechanical stability make fused quartz and fused silica excellent choices for many critical applications in high-quality optical systems.

Until the recent past no distinction was made between fused quartz and fused silica by manufacturers, distributors, or users. These terms were used interchangeably and further complicated by the fact that different manufacturers described these materials as silica glass, vitreous quartz, and vitreous silica. But as requirements became more stringent and applications more diverse, a distinction had to be made to avoid unnecessary confusion. The generally accepted terms are now "fused quartz" and "fused silica," and the distinction is primarily based on the raw material used.

Fused quartz is of mineralogical origin. It is produced either from mined natural quartz or from quartzite sand. The natural quartz is in most cases high-quality Brazilian rock crystal which is then converted at very high temperatures into transparent, noncrystalline fused quartz boules or ingots. Natural crystal quartz has the highest homogeneity, with a purity exceeding 99.97% with aluminum oxide ($Al_2O_3$) as primary impurity. A translucent form of fused quartz results from the fusion process when the raw material is a highly pure and specifically treated quartzite sand.

Fused silica is a synthetic amorphous silicon dioxide. It is produced by flame or vapor phase hydrolysis of a silicon halide [49]. The fusion process yields an amorphous optical material of excellent homogeneity and purity, although it is not quite as homogeneous as fused quartz. It is claimed, however, that fused silica with a purity of 99.8% is one of the most transparent optical materials made. Commercially available optical grade fused quartz and fused silica are listed by manufacturer and designation in Table 1.12.

**Table 1.12. Fused Quartz and Fused Silica**

| Manufacturer | Material designation | FQ | FS |
|---|---|---|---|
| Amersil Hereaus | Optosil 1 (T-12) | X | |
| | Optosil 2 (T-13) | X | |
| | Optosil 3 (T-14) | X | |
| | Homosil (T-15) | X | |
| | Ultrasil (T-16) | X | |
| | Infrasil 1 (T-17) | X | |
| | Infrasil 2 (T-18) | X | |
| | Suprasil 1 (T-19) | | X |
| | Suprasil 2 (T-20) | | X |
| | Suprasil W1 (T-22) | | X |
| | Suprasil W2 (T-23) | | X |
| | SR Optical (T-32) | X | |
| Corning | Code 7940 (standard) | | X |
| | Code 7940 (UV grade) | | X |
| | Code 7957 (IR grade) | | X |
| | Code 7958 (prism) | X | |
| Dynasil | UV No. 1000 | | X |
| | Optical No. 4000 | | X |
| | Optical No. 5000 | | X |
| General Electric | Type 104 | X | |
| | Type 105 | X | |
| | Type 106 | X | |
| | Type 124 | X | |
| | Type 125 | X | |
| | Type 151 | | X |
| Nippon Silica Glass | ES | | |
| | OV | | |
| | OX | | |
| | OX-1 | | |
| | OX-UV | | |
| | IR | | |
| Thermal American | Spectrosil A | X | X |
| | Spectrosil B | X | X |
| | Spectrosil WF | X | X |
| | Vitreosil 055 | X | |
| | Vitreosil 066 | X | |
| | Vitreosil IR | X | |

Fused quartz is produced by fusing powdered natural crystal quartz in an oxygen-hydrogen flame. Water vapors are a by-product of this process; some of the vapors are then bound in the amorphous structure of the fused quartz in the form of hydroxyls or OH radicals. This causes a deep water absorption band in the near-IR spectral region at 2.72 $\mu$m. Secondary absorption bands are found at 1.38 $\mu$m and 2.22 $\mu$m. The useful transmission range in the IR region does not go much beyond 3.0 $\mu$m.

Fused silica is a synthetic variant of this process. It is similarly produced by passing silicon halides, such as high purity silicon tetrachloride vapors (SiCl$_4$), through an oxygen-hydrogen flame. This vapor phase reaction process is called *flame hydrolosis*. Bound hydroxyls are a by-product of this process and produce the same absorption bands at the same infrared wavelengths as already described for fused quartz.

The water content is considered an undesirable impurity but it is an unavoidable by-product of the manufacturing processes. The hydroxyl content lies somewhere between 1200 ppm for standard grade material to less than 8 ppm for specially treated material designed to eliminate the absorption bands for improved IR transmission. Table 1.13 shows the hydroxyl content for three basic quality grades as reported by two manufacturers.

To improve the IR transmittance of fused quartz and fused silica, the flame hydrolosis is replaced by an electric arc melting process or a plasma

Table 1.13.  Typical Hydroxyl
Content of Fused Silica and Fused
Quartz

| Hydroxyl content | Manufacturer | Product |
|---|---|---|
| 1200 ppm | Amersil [56] | Suprasil |
|  |  | SR |
|  |  | SR |
|  | NSG [51] | ES |
| 180 ppm | Amersil | Ultrasil |
|  |  | Homosil |
|  |  | Optosil |
| 150 ppm | NSG | SG |
|  |  | OY |
|  |  | OX |
|  |  | OX-1 |
| 8 ppm | NSG | IR |
|  | Amersil | Infrasil |
| 5 ppm |  | Suprasil-W |

fusion method that is done in a vacuum furnace. Both approaches eliminate the need for hydrogen. In the absence of hydrogen, no water forms and there are no water absorption bands in the near-IR region.

Fused quartz and fused silica have excellent UV transmission with fused quartz transmitting down to 180 nm, while fused silica extends this down to about 160 nm. The lower values are achieved by subjecting the standard material to a heat treatment that drives off impurities that absorb UV energy.

As a result of these differing processes, there are three fundamental transmission curves for fused quartz and fused silica which are a function of the hydroxyl content and the heat treatment used to improve the UV transmission. Additional differences exist between the transmission curves for fused quartz and fused silica from manufacturer to manufacturers. The curves shown in Fig. 1.6 represent only the fundamental differences between the three basic types.

Fused quartz made from natural rock crystal contains about 20 to 30 ppm of metallic impurities [56] that can affect the optical quality of the material. The primary metallic impurities in descending order are aluminum (Al), lithium (Li), and sodium (Na). Synthetic fused silica is carefully controlled to hold metallic impurities to 1 ppm with iron (Fe), aluminum (Al), chromium (Cr), and magnesium (Mg) as primary impurities in descending order. Chlorine (Cl), at nearly 90 ppm is another primary impurity in fused silica, while calcium (Ca), potassium (K), and particularly sodium (Na) can also assume relatively significant levels. The latter impurities can be greatly reduced through special process techniques.

**Chemical Properties.** Both fused quartz and fused silica are chemically stable under nearly all conditions found during manufacture and typical use. They have also excellent chemical resistance to such common acids as sulfuric acid ($H_2SO_4$), nitric acid ($HNO_3$), and hydrochloric acid (HCl), although there can be a slight attack at elevated temperatures. Prolonged exposure to boiling water has a very small but detectable effect.

Fused quartz and fused silica are attacked by hydrofluoric acid (HF) and by phosphoric acid ($H_2PO_4$) but at a slower rate than for most optical glasses. The attack by these acids accelerates at higher temperatures. HF will attack these materials at room temperature at a fairly slow rate, but the rate of attack increases rapidly as the temperature rises. For instance, Corning [57] reports that a 1 $\mu$in. (0.000001-in.) of material is etched away after a 24-hour exposure to a 5% HF solution at 95°C. Phosphoric acid, on the other hand, has little or no effect below 150°C but becomes aggressive above that temperature.

These materials are also subject to appreciable attack by strong alkaline solutions. Sodium hydroxide (NaOH), ammonium hydroxide ($NH_4OH$), and potassium hydroxide (KOH) will attack fused quartz and fused silica at varying degrees, especially at elevated temperatures. For instance, Corning [57] notes that an appreciable 500 microinch (0.0005 in.) of material was removed during a 24-hour exposure to a 5% solution of NaOH at 95°C. That

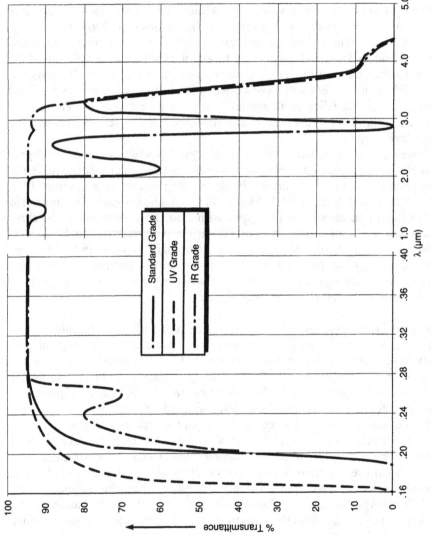

**Figure 1.6.** Typical transmission curves for fused silica.

Standard Grade
UV Grade
IR Grade

% Transmittance

λ (µm)

40

is 500 times the rate of attack reported for HF. To avoid unpleasant and expensive surprises, it therefore is very important that the chemistry of aqueous cleaning baths be properly understood.

**Optical Characteristics.** Idealized transmission curves for the three basic material types have already been discussed. Other optical characteristics of interest to the optician are the refractive indices at the primary wavelengths and the dispersion or Abbe values. The commonly accepted values are $n_d = 1.45845$ and $v_d = 67.8$. Therefore fused quartz and fused silica can be defined by the six-digit code [58] as a 458 678 glass. Amersil [56] reports slightly different values for fused quartz and fused silica. For fused quartz (natural), $n_d = 1.45857$ and $v_d = 67.6$, whereas for fused silica (synthetic) the values the $n_d = 1.45847$ and $v_d = 67.7$ Dynasil [59] lists the refractive index for three common laser wavelengths: (Argon) $n_{4880} = 1.46302$, (HeNe) $n_{6328} = 1.45702$, and (Nd:YAG) $n_{10650} = 1.44962$.

Another optical characteristic of fused quartz is fluorescence. Fluorescence can become a problem when the optic is used in the UV region below 290 nm. The material gives off a faint but discernable colored glow if exposed to UV radiation. This interferes with critical measurements and detracts from the performance of the system of which the optic is a part. Fluorescence is typically not a factor during fabrication. Clear fused quartz made from rock crystal will fluoresce, whereas synthetic fused silica is nearly fluorescence free.

**Thermal Characteristics.** Although the thermal behavior of optical glasses is a major concern for opticians, the low coefficient of thermal expansion for fused quartz and fused silica gives them considerable latitude in how to process these materials. The thermal expansion coefficient is $5.5 \times 10^{-7}$ cm/cm/°C over a temperature range from 20°C to 300°C which represents the normal thermal environment that the material is exposed to during manufacture and typical use. This low coefficient is about $\frac{1}{7}$ of the lowest value for optical glass and about $\frac{1}{8}$ that of Pyrex. This permits the quenching of very hot parts in ice water without any thermal fracturing. This is done for demonstrative purposes only to show the thermal stability of the material. Such quenching serves no useful purpose during the manufacture of optical components and the optician is expected to treat fused quartz or fused silica parts like any other optical quality product. Although it has no bearing on the normal processing of fused quartz and fused silica components, it may be of interest that their thermal stability permits their use at 900°C for a prolonged time and up to 1200°C for shorter exposures. The fusion temperatures lie between 1700°C and 1800°C.

**Mechanical Properties.** Although mechanical properties of optical materials are rarely, if ever, considered during the manufacture and use, it is important to have some understanding of the relative strengths of such materials.

**Table 1.14. Mechanical Properties of Fused Quartz and Fused Silica**

| Mechanical characteristic | Value | Dimension |
|---|---|---|
| Young's modulus (E) | $7.4 \times 10^5$ | $kg/cm^2$ |
|  | $10.4 \times 10^6$ | psi |
| Compressive strength | $1.75 \times 10^5$ | psi |
| Tensile strength | $7.0 \times 10^3$ | psi |
| Knoop hardness | 600 | $kg/mm^2$ |
| Density | 2.2 | $g/cm^3$ |

Therefore typical values are listed in Table 1.14. The values in the table are composites based on data from the various sources that were consulted. They are not absolute values, and they should be used for reference only. Some difference in values exist from source to source. Different systems of measurement were also used to express these values. If accurate values are required, it is best to contact the manufacturers directly.

A typical feature of amorphous optical materials is that their tensile strength is only a small fraction of their compressive strength. In the case of fused quartz and fused silica, the tensile strength is only $\frac{1}{25}$ of the compressive strength. This means that while these materials can support considerable loads in compression, they tend to fail under relatively light tensile loads. Tests with drawn glass fibers have shown, however, that the inherent tensile strength of glasses is much higher than has been measured. This discrepancy exists because microcracks in the surface resulting from the shaping and grinding of the glass test pieces weaken the material sufficiently to cause premature failure when under a tensile load. The reason for the higher tensile strength of drawn glass fibers is the total absence of any mechanical abrasion. The surfaces of the fibers are essentially fire-polished.

Since opticians don't have the luxury of fire-polished surfaces during the manufacture of precision optical components, this characteristic behavior of amorphous materials must be kept in mind, especially when clamping devices are used to hold parts during some operations. The problem may become more acute when parts are subjected to thermal fluctuations while clamped. If care in the design and use of the fixtures is not taken, breakage is certain to occur.

**Internal Material Quality.** As with the requirements for optical glass, not just the optical transmission characteristics of fused quartz and fused silica are important but also by how much, or rather how little, an image or a wavefront that is projected through the material is degraded by internal imperfections. These imperfections cannot be totally eliminated in most cases, so a careful selection process has been developed to grade fused quartz and fused silica according to the level of the imperfections listed below:

- Index inhomogeneity ($\Delta n$)
- Striae (rated per MIL-G-174A)
- Inclusions, bubbles, and seeds
- Stress birefringence (nm/cm)
- Granularity of the material

Index inhomogeneity have to be very low for any critical application for which distortion of transmitted wavefronts must be minimal. Typical values for $\Delta n$ are $5 \times 10^{-5}$, and for specially selected grades they can be as low as $\Delta n = 6 \times 10^{-6}$. NSG [54] reports $\Delta n$ values as low as $3 \times 10^{-6}$ for their best-grade fused quartz. Corning fused silica is available in several index homogeneity grades ranging from $\Delta n = 1 \times 10^{-6}$ to $5 \times 10^{-6}$. Corning code 7958 can meet these values in one, two, or three orthogonal axes. Amersil [56] will also select material that have very low index inhomogeneities in one, two, or all three orthogonal axes. The single-axis material is suitable for precision windows, the two-axis material is required for critical beam splitters, and certain prisms, and the three-axis material is used only for the most critical applications such as spectrographic prisms because it is by far the most expensive fused silica available.

Striae are local index variations that can result from the material manufacturing process. This topic has already been covered under optical glass. For per MIL-G-174A there are several grades of striae that are graded under prescribed viewing conditions. Grade A is the best, and grade D the worst of the striae grades. In order not to degrade a transmitted wavefront or a high resolution image, the striae grade must be A. Striae grade B can be an acceptable choice for less-stringent requirements, while grade C should be reserved for reflective optics only. Commercial grade fused quartz or fused silica have typically a striae rating D.

Inclusions such as bubbles and seeds are often an unavoidable by-product of the material manufacturing process. The selection process must reduce these imperfections to a minimum for precision applications. The method adopted by the industry is to limit the number and size of bubbles or seeds per cubic centimeter or cubic inch. Another way to express the same thing is by the average projected area of the inclusions per inch of path length. Less than 0.05% projected area constitutes good material. The term "bubble free" refers typically to a highly select material that contains no bubbles or seeds above a certain size. It does not mean, however, that absolutely no bubbles or seeds are present. Since this is a volume dependent measure, it becomes much more difficult, and consequently more expensive, to maintain the same bubble count and size per unit volume as the volume increases. Therefore, as parts get larger, the permissible bubble size and count must also increase proportionally. This is why the average projected area method per inch of path length is a better measure for the purpose of comparing the internal quality from an inclusion standpoint of two optical components of different dimensions.

Stress birefringence is also a material imperfection resulting from the material manufacturing process. It is caused by either mechanical stress or stress introduced into the material during locally uneven cooling. It can be partially reduced through a costly and time-consuming annealing process, but a residual stress will remain even with the best process. The residual stress causes light to be affected in slightly different ways depending on the direction the light takes through the material. This condition is called *birefringence* (or double refraction), hence stress birefringence. Stress birefringence measures what is called "retardation of the light wave" which is expressed in nanometers per centimeter (nm/cm) of pathlength. A stress birefringence value of 5 nm/cm is attainable for the best-fused quartz and fused silica.

Granularity is a material imperfection that is peculiar to fused quartz and fused silica. It also results from the manufacturing process and ranges from nearly invisible to quite noticeable under prescribed viewing conditions. The effect of this material property on the performance of the components made from it is not clear, but that must depend on the particular application. It is certain that the presence of noticeable granularity cannot help the performance of a critical optical or laser system.

### 1.3.2. Crystal Quartz

Crystalline quartz is a natural mineral that is found in many places in the world. The largest crystals with the highest optical quality are mined in the mountains of Brazil. These are called Brazilian rock crystal, and they are the raw material for synthetic crystal quartz and for fused quartz. These natural crystals are always unevenly hexagonal in cross section, and the optical axis runs along the length of the so-called stone [60]. Until the 1950s only the naturally occurring crystals were available for optical use. The strongly birefringent nature of natural quartz makes it an excellent choice for retarders, wave plates, and polarizing components. The main problem with natural quartz is the unpredictable quality of the raw material that tends to vary from stone to stone. In addition local imperfections in each crystal can make most of it unusable for precision optics. These problems require a time-consuming and very costly selection process to isolate and recover those regions in the crystal that will yield good parts. Such uncertainties are overcome by the newer synthetic crystal quartz that has been available for over 30 years.

The primary interest for developing synthetically produced crystal quartz came from the U.S. government and the consumer market. The need for large quantities of frequency control devices for military and commercial communications systems and for industrial piezoelectric applications could simply not be filled by the uncertain supply of natural quartz and the unpredictable yields from the crystals. Synthetic crystal quartz, also called *cultured quartz,* is grown in high-pressure reactors by a hydrothermal process, which is described in Chapter 2.

The optical, thermal, mechanical, and chemical properties of natural and cultured quartz are nearly identical. These two materials differ only in one significant respect. Micron-size inclusions are typically found in cultured quartz, whereas natural crystal quartz is free of this type of imperfection. The inclusions are particles of sodium iron silicate originating from the steel walls of the growth vessel. They appear either white when viewed by reflection, or dark when inspected with transmitted light. The density of these inclusions is on the order of three to four per cubic centimeter, and this becomes a problem only for the most critical applications [61]. The materials producers are continuing their efforts to reduce the inclusion concentration, and their efforts to date have resulted in considerable improvement of optical grade cultured quartz.

The optical transmission of cultured quartz is slightly better than for natural quartz. This is especially true in the UV region. The UV cutoff for cultured quartz is at 147 nm. This is not only a measure of the high purity of the material, but it is also useful for a variety of UV applications. Unlike fused quartz, both natural and cultured quartz do not fluoresce when exposed to high intensity UV radiation. This characteristic makes cultured quartz the material of choice for UV laser tube windows.

The refractive index of both natural and synthetic crystal quartz is essentially identical over the region from 0.19 to 1.5 $\mu$m. Another important material property of great interest to the optical user is internal homogeneity. Both natural and synthetic quartz can have local strain locked into the bulk during the growth process. To guarantee optical homogeneity, the material producer inspects each synthetically produced crystal scheduled for precision optics use with a conoscope. With this instrument, divergent light from a monochromatic source is passed through the crystal along the optical axis, the Z-axis. Concentric interference fringes are generated and projected on a screen or a film plane. These are then inspected for departure circularity which identifies the presence of internal strain. Fringe distortion exceeding 0.1 fringe per inch of optical pathlength causes rejection of the crystal for critical optical applications [61].

The growth process for cultured quartz results in a unique internal material structure that limits the choices the optician has on how to cut parts from a bar. As the quartz crystal grows around the seed bar, regions develop that have a crystal orientation that is not useful for optical applications. A more detailed description can be found in Chapter 2. These regions, typically the x-growth regions, must be trimmed away from the as-grown crystal. But careful control of the growth process results in sizable regions of high optical quality that is uniform from crystal to crystal, unlike natural crystals which often suffer from twinning, veils, and solid inclusions that are randomly distributed throughout the crystal [60].

Both natural and cultured quartz are *anisotropic*. This means that they exhibit slightly different optical properties along different crystal axes. This behavior is called "birefringence," which makes crystal quartz, either natural or synthetic, the preferred material for waveplates, Brewster laser windows, and birefringent prisms such as Wollaston prisms.

The thermal, mechanical, and chemical properties of crystal quartz are very similar to those of fused quartz. Both natural and synthetic quartz crystals are thermally stable, quite hard and mechanically strong, and chemically impervious to most chemical agents.

### 1.3.3. Mirror Materials

Low-expansion optical materials are characterized by their very low coefficient of thermal expansion over the range of normal operating temperatures. This characteristic makes them an ideal choice for large reflective optics, such as astronomical telescope mirrors, and any other optical application where the precision of the reflective surfaces must be maintained over a substantial range of temperatures. The thermal stability of these materials makes it also possible to more easily control an accurate surface figure since, unlike optical glass, local temperature variations on the surface during polishing have only a negligible effect on the figure.

It is possible to use some of these materials for transmissive optics as well, but the internal optical quality is not good enough for use in precision optical systems. Their use in transmission is limited to commercial optics applications where they can represent an economical alternative to some of the lower-end optical glasses.

Low-expansion mirror materials can be divided into three basic groups, namely, low expansion glasses, titanium silica glass, and glass ceramics. Table 1.15 lists the material producers and their commonly known low-expansion products.

**Low-Expansion Glasses.** The optical materials in this group have an amorphous structure like any other glass, but unlike most other glasses, they have a very low coefficient of expansion. This makes them useful for reflective optics and for noncritical transmissive optics which are used in close proximity to fluctuating heat sources.

The most commonly used low-expansion glass is Pyrex [65]. It has an equivalent in E6 which is produced in Japan [63]. Both are basically borosilicate glass that has been specially treated to achieve the characteristic low coefficient of expansion. These materials are primarily used for reflective optics such as small- to medium-size telescope mirrors. Larger mirrors are now made from more advanced materials that will be discussed later in this

Table 1.15. Low-Expansion Materials

| Manufacturer | Low expansion glass | Titanium silica glass | Glass ceramics |
|---|---|---|---|
| Corning [51] | Pyrex | | |
| Corning | Vycor | | |
| Schott [62] | Tempex | | |
| Ohara [63] | E 6 | | |
| Corning | | ULE | |
| Ol Kimble [64] | | | Cervit |
| Schott | | | Zerodur |

section. But in 1936 a special variant of Pyrex, Corning code 7160, was used for casting the blank for the 200-in. Hale Telescope on Mount Palomar, which is still the second largest telescope mirror in the world.

Pyrex and E6 play an important role in most optical shops, where they serve as optical flats and as testplate materials. They are also used quite frequently as precision plano blocking tools or as contact blocks for critical blocking requirements. Some opticians prefer to make blocking and grinding tools from Pyrex for prototype work. The thermal, optical, and mechanical properties of these and other low-expansion materials are listed in Table 1.16. The low coefficient of thermal expansion makes these glasses much less sensitive to thermal shock than is typical for optical glass such as BK 7. This feature permits their use in close proximity to fluctuating thermal sources, such as arc lamps, projector lamps, or other high-intensity light sources, where ordinary glass would quickly fail due to thermal fracturing. Therefore materials like Vycor [66] and Tempax [62] find use as windows, filters, and other commercial quality optics in such applications.

**Titanium Silica Glass.** There is currently only one material in the titanium silica glass category called ULE [67], which stands for ultra low expansion. This glass is primarily used for large astronomical mirrors, either as solid monoliths or as lightweight honeycomb structures. The blank for the 2.4-m (98-in.) diameter primary mirror for the Hubble Space Telescope is a ULE lightweight, fusion-bonded square cell structure, which was produced by Corning in 1978 [68]. It was designed to be the most precise and most powerful astronomical telescope ever used, theoretically extending our reach into the depth of the universe sevenfold to about 14 billion light-years. Unfortunately, a testing error has prevented this goal from being reached, and a special shuttle mission is scheduled for 1993 to correct the problem.

ULE is titanium silicate that is produced by flame hydrolosis of silicon and titanium tetrachloride gas. This process is similar to that used to pro-

**Table 1.16. Thermal, Optical, and Mechanical Properties of Low-Expansion Materials**

| Properties | Parameters | Values | Pyrex | E 6 | ULE | Cer-Vit | Zerodur |
|---|---|---|---|---|---|---|---|
| Thermal | Expansion coefficient | 10 E-7°C | 34 | 24 | 0.5 | < 1 | 0.1 |
| | Thermal conductivity @25°C | cal/cm/sec/°C | 0.0027 | 0.0023 | 0.0031 | 0.0040 | |
| | Specific heat @25°C | cal/g/°C | 0.18 | 0.17 | | 0.20 | |
| Optical | Refractive index | $n_d$ | | 1.4670 | 1.4828 | 1.5402 | 1.5430 |
| | Abbe value (dispersion) | $v_d$ | | | 53.1 | 57.3 | 56.1 |
| Mechanical | Density | gm/cm$^3$ | 2.23 | 2.18 | 2.21 | 2.50 | 2.53 |
| | Young's modulus | 10E6 psi | 9.5 | 8.3 | 9.8 | 13.4 | 13.1 |
| | Knoop hardness | Kg/mm$^2$, 100gr (200gr*) | 481 | 520 | 460* | 540* | 630 |
| | Poisson ratio | | 0.20 | 0.19 | 0.17 | 0.25 | |
| | Modulus of rupture (abrad.) | 10E3 psi | 6.1 | | | 8.3 | |

duce fused silica, but unlike fused silica which is nearly 100% silicon dioxide, ULE is only 92.5% $SiO_2$ while the remaining 7.5% is titanium oxide ($TiO_2$). The homogeneity and clarity of ULE is similar to that of fused silica. It is an isotropic, single-phase amorphous optical material. This makes it a true silica glass; ULE is not a glass ceramic [69].

The most important characteristic of ULE is its practically zero-expansion coefficient which greatly improves the thermal stability of the surface figure for large telescope mirrors. The linear coefficient of thermal expansion is $0.5 \times 10^{-7}$/°C over the normal range of operating temperatures. ULE is readily polishable, much like fused silica, but its low residual internal strain can distort an accurate surface figure if cutting or shaping must be done after polishing [70]. Although ULE is typically not used for transmissive elements, its internal quality is sufficient for most commercial optical applications. Typical thermal, optical, and mechanical parameters for ULE are listed in Table 1.16.

The chemical durability of ULE is quite similar to that of fused silica. It is nearly impervious to most chemical agents, except for exposure to hydrofluoric acid (HF) and concentrated alkaline solutions, such as sodium hydroxide (5% NaOH). The rate of chemical attack increases with increasing temperatures.

**Glass Ceramics.** Glass ceramics are represented by Cer-Vit C-101 [64] and Zerodur [62]. Both materials have played a major role in the successful development of the ring laser gyro. Glass ceramics have both amorphous and crystalline phases. Therefore they are neither a true glass nor are they a true crystalline material. They combine the properties of both. Glass ceramics contain certain impurities that give them their characteristic light amber coloration. The presence of these impurities leads to the absorption of some of the incident light. Internal light scattering is another inherent property of glass ceramics. This tendency to scatter light can be minimized through special thermal treatment during manufacture, but it cannot be totally eliminated. The amount of light scatter is a function of crystal size, typically about 50 nm, and the index differential between the glass and the crystal regions. Because of the inherent bulk absorption and bulk scatter, glass ceramics are rarely used for transmissive optics [71].

The most noteable application of Cer-Vit in recent times has been for cavity blocks and laser mirrors for ring laser gyros (RLG). It is now precision molded into RLG cavity blocks, with most of the bores and cavities molded in to minimize the need for extensive machining [72]. But the use of Cer-Vit is not limited to small precision jobs. It also serves as mirror blank material for some of the largest telescopes. For instance, the largest Cer-Vit piece ever cast was the 4.21-m (166-in.) diameter by 0.56-m (22-in.) thick mirror blank for the International Observatory in La Palma on the Canary Islands. At this size and a weight of more than 36,300 kg or 40 tons, it is the third largest mirror in the world after the 6.1-m (240-in.) diameter telescope in the Caucasus of Russia and the 5.1-m (200-in.) Hale Telescope at Mount Palomar in the United States.

Zerodur is a glass ceramic produced by Schott. It is an inorganic, pore-free material that has both vitreous (glasslike) and crystalline phases. About 75% of the material's volume is in the crystalline phase, and the remainder is amorphous. The crystalline phase has a negative coefficient of expansion, and that of the vitreous phase is positive. In this way a nearly zero expansion coefficient is achieved over the range of normal operating temperatures [73]. The actual value is $1 \times 10^{-7}/°C$ over a range of 20°C to 300°C. Like Cer-Vit, Zerodur is a poor choice for transmissive optics because of the inherent bulk absorption and internal scatter, although it is optically homogeneous and exhibits good transparency in the visible region. The totally nondirectional structure of Zerodur allows it to be ground and polished like fused silica or optical glass.

Some typical uses for Zerodur are for laser gyroscope bodies, astronomical telescope mirrors, reference bars for precision X–Y stages used in coordinate measuring machines (CMM), laser resonator spacer rods, and precision length gauges [74]. Although most of the applications listed here are for relatively small components, Zerodur has been used as primary mirror material for the 3.5-m (138-in.) diameter telescope for the German–Spanish Astronomical Center in Calar Alto, Spain. The weight of the primary mirror is about 11,800 kg or 13 tons [75]. Table 1.16 lists the thermal, optical and mechanical parameters for both Cer-Vit and Zerodur.

The chemical properties of glass ceramics are similar to those of fused silica, although there are some differences. Glass ceramics are not attacked by water or water vapors, even after prolonged exposure. They do not form stains during polishing and normal cleaning methods. They are attacked, however, by certain acids and concentrated alkaline solutions. For instance, exposure to hydrofluoric acid (HF) must be avoided. Also use of sulphuric acid ($H_2SO_4$) can lead to chemical attack, especially at higher temperatures. While Cer-Vit can be safely immersed in hot nitric acid ($HNO_3$), Zerodur will be rapidly etched. Strong alkaline solutions, especially at elevated temperatures, must also be avoided because of the danger of chemical attack [73].

In recent years two additional low-expansion materials have been announced. Hoya [76] makes a low thermal expansion glass with the tradename Cryston, and it has high chemical durability [77]. The other is a lithium-aluminum-silicate glass ceramic that is produced by the controlled crystallization of specific glass compositions. The optical transmission is similar to that of fused silica with an OH absorption band at 2.8 $\mu$m. The OH absorption can be eliminated through vacuum melting [78].

### References for Section 1.3 (Low-Expansion Materials)

[46] Corning Glass Works, Corning, NY.

[47] Owens-Illinois, Kimble Division, Toledo, OH.

[48] Schott Glass Technologies, Inc., Duryea, PA.

[49] Dynasil Corp. of America catalog No. 702A.

[50] Heraeus Amersil, Inc., Buford, GA; Heraeus Quartzschmelze, GmbH, Germany.

[51] Corning Glass Works, Corning, NY.

[52] Dynasil Corporation of America, Berlin, NJ.

[53] General Electric, Richmond Heights, OH.

[54] Nippon Silica Glass USA, Inc., Somerville, NJ; Nippon Silica Glass Co., Ltd., Japan.

[55] Thermal American Fused Quartz Co., Georgetown, DE.

[56] Amersil, Inc. catalog EM-9227, "Optical fused quartz and fused silica."

[57] Corning product literature, "Code 7940."

[58] per MIL-G-174A.

[59] Dynasil Corp. of America catalog 702B.

[60] Kinloch, Durand R., Brumbaugh, James D., "Inclusions in cultured quartz crystals—New developments and lower concentrations," *Technical Digest,* Workshop on Optical Fabrication and Testing, December 1982.

[61] Sawyer Research Products, Inc., Eastlake, OH.

[62] Schott Glass Technologies, Inc., Duryea, PA.

[63] Ohara, Inc., Japan.

[64] Owens-Illinois, Kimble Division, Toledo, OH.

[65] Corning code 7740.

[66] Corning code 7913.

[67] Corning code 7971.

[68] Corning product literature, ULE, 1988.

[69] News item, "Details on space telescope mirror substrate," *Electro-Optical Systems Design,* September 1979.

[70] Lobdell, A., Itek Advanced Technology, Data Sheet 67-1.

[71] Brock, T. W., "Optical properties of selected Cer-Vit materials," Owens-Illinois, 1982.

[72] Ownes-Illinois, Kimble Div., product literature, 1981.

[73] Schott Zerodur product literature, Z 10/7 and 3039e USA, 1977.

[74] Schott Zerodur product literature, 313/e USA 111/82.

[75] *Schott Information,* No. 1, English edition, 1985.

[76] Hoya Corporation, Japan.

[77] News item, "Optical manufacturers intensify efforts," *Laser Focus Electro-Optics,* July 1987.

[78] Dana, Brian, "Glass-ceramic hybrids transmit from UV to IR," *Laser Focus E/O,* August 1988.

## 1.4.  IR MATERIALS

The infrared region of the optical spectrum ranges from $\lambda$ 0.8 $\mu$m to more than $\lambda = 40$ $\mu$m. This range is not continuous because there are deep atmospheric absorption bands that limit the wavelength regions that can be used. There are three such regions or windows, with one in the near-IR from 0.8 to 2.4 $\mu$m, another in the mid-IR from 3 to 5 $\mu$m, and the most important one between 8 to 12 $\mu$m. A number of optical materials transmit in this spectral region, and they are referred to as IR materials. Each of the materials discussed in this section will transmit well over one or more of these atmospheric windows.

### 1.4.1. Germanium

Metallic germanium (Ge) is not found in nature. It is typically a trace element found in sulfide ores of copper, zinc, lead, and tin [79]. Primary sources are Zaire in Africa, Bolivia and Peru in South America, and Russia, central Europe, and the United States. Even the richest germanium-bearing ores, such as germanite, contain less than 9% Ge. Almost all commercially used germanium is recovered from the processing residue of other metals such as copper and zinc. There are a number of producers, reprocessors and distributors of germanium [80, 81, 82, 83, 84, 85, 86] which process it into forms usable for IR optics. The methods of production, which are described in Chapter 2 in greater detail, are Czochralski (CZ) growth of germanium boules for single-crystal germanium or casting of polycrystal germanium for larger parts.

The scarcity of the raw material and the complexity of its recovery and refinement makes germanium very expensive. The total annual worldwide production of germanium is estimated to be only about 150 tons. More specifically, the total consumption of germanium in the Free World market economies was 80 tons in 1986 [87]. Because of its extensive use in military IR systems, germanium is designated a strategic material.

Despite its cost the unique optical, mechanical, and chemical properties have made germanium the material of choice for many IR applications. Ge also plays a major role in many other industries. It was originally developed as a semiconductor material until it was replaced by the much less expensive silicon during the mid-1960s. Fortunately the useful optical properties of Ge were recognized at about the same time, and since the early 1970s, Ge has evolved as the foremost IR material for use in the 8- to 12-$\mu$m region of the spectrum.

The 8- to 12-$\mu$m window is of primary importance for forward-looking infrared (FLIR) system, thermal imaging, and night vision devices. The lasing wavelength of $CO_2$ lasers at 10.6 $\mu$m also happens to fall right near the center of this window as well. Nowadays, about half of all the Ge consumed in the western world, is used for IR optics, and this application is still a growing segment of Ge consumption. In 1986, the annual growth rate for Ge for IR optics stood at 15%. The rate is probably less today because of reduced world tensions. However, growth in the use of Ge in commercial IR systems will soon make up for a declining military market.

Germanium is either grown from the melt using the CZ method, or it is cast into larger shapes. Although single crystal material would be preferrable because of its high uniformity, and in the absence of impurities, lower absorption, it is more expensive than polycrystal Ge. For instance, for a typical series of CZ growth cycles, only half of them will yield single-crystal Ge while the rest goes either partially or totally polycrystal. Cast germanium is always polycrystalline.

It is not enough to classify Ge as either single or polycrystal. Considerable differences can exist from blank to blank within each group. Dislocations and slip planes can vary in single crystal, and the crystal size and distribution can vary in polycrystal material. These differences can affect optical quality. These realities require that the optical quality of each ingot be tested independently. The parameters of interest are resistivity and absorption. Germanium is a semiconductor material and like silicon, it can be either $p$-type or $n$-type. While either type silicon can be used for IR applications, only $n$-type Ge can be used for IR optics. The reason for this is that $p$-type Ge has an absorption coefficient that is 10 to 20 times greater than $n$-type at a bulk resistivity between 5 and 25 ohm-cm [88]. This is shown in Fig. 1.7.

Resistivity measurements provide an indirect means to determine the optical quality of semiconductor materials used for IR applications. The property that affects the optical performance of germanium the most is its

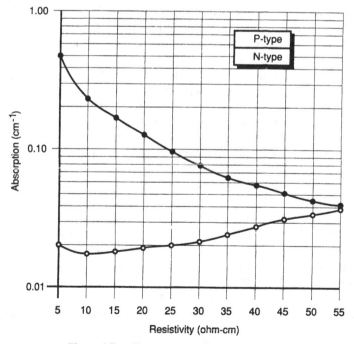

**Figure 1.7.** Absorption coefficient of Ge [88].

absorption coefficient. The presence of unintentional impurities in the material can raise its resistivity, which signals an increase in absorptivity and this reduces the optical quality. Deliberate impurities can lower resistivity and thus can improve optical quality. Resistivity measurements are made at room temperature with either a two-point, three-point, or four-point probe on an abraded surface. Several resistivity readings must be averaged for polycrystal material. The absorption coefficient of polycrystal Ge is usually higher than that of single-crystal material. In addition to low absorptivity, optical grade germanium must be inclusion free and fine-annealed to improve its homogeneity. The residual inhomogeneity for single-crystal Ge has been reported to be as low as <10 ppm [89].

Although germanium is opaque in the visible region, which makes it difficult to align Ge optics, it is highly transparent between 8 and 12 $\mu$m, especially when it is antireflection (AR) coated. Coatings are required because of Ge's unusually high refractive index of 4.0 at 10.6 $\mu$m; otherwise, the transmittance through Ge would be less than 50%. However, the same high index makes it possible to design optical lens systems with relatively short focal lengths using much longer lens radii than would otherwise be required for a lens system of equivalent focal length using optical glass. This advantage translates into reduced Ge consumption because longer radii have less sag depth and thus require less thickness. The longer radii are also easier to generate and polish, which also reduces fabrication cost.

Germanium has high mechanical strength, which makes it less susceptible to breakage than other IR materials. It has excellent hardness, which makes it fairly easy to achieve a quality polish [90]. It is, however, quite brittle, and it can fail by brittle fracture when subjected to high mechanical stress. This means that germanium can be easily chipped.

Although Ge has good thermal conductivity and fairly low absorption, it cannot be used in high power lasers or for optical components used at elevated temperatures, such as IR windows for high-performance aircraft, because of an undesirable property called *thermal runaway*. The change in refractive index as a function of change in temperature ($dn/dt$) is quite large in germanium, and this leads to a deterioration in optical performance as temperatures increase [91]. As the operating temperature increases, the IR transmission rapidly decreases to the point where the component becomes essentially opaque for thick parts at temperatures above 50°C.

If the heat cannot be conducted away at least as fast as it is introduced, the absorption will continue to increase and will heat up the part. This will further increase absorption, since it is temperature dependent. The process can rapidly escalate into thermal runaway. Since any absorbed energy reduces transmission, the optical performance continues to deteriorate to the point where no further transmission of IR energy takes place [92]. In $CO_2$ lasers with an output exceeding 100 watts, thermal runaway can lead to irreversible damage, although it is more likely that the laser will shut down due to thermally induced defocusing before it reaches this critical stage.

Germanium is chemically stable at room temperatures. It does not readily oxydize, nor is it attacked by hydrofluoric (HF) or hydrochloric (HCl) acids. Attack by room temperature nitric and sulfuric acids is quite slow. Ge is also not attacked by most dilute alkaline solutions, but it is vigorously attacked by alkaline hydrogen peroxide or sodium hypochlorite solutions. At elevated temperatures Ge oxidizes to form GeO or $GeO_2$. Germanium or germanium dioxide ($GeO_2$) has shown no evidence of toxicity [79].

### 1.4.2. Silicon

Next to oxygen, silicon (Si) is the most abundant element on earth. But despite this abundance it is not inexpensive because of the high cost of extraction and purification required to produce a material that is sufficiently pure for high-technology applications. Nevertheless, silicon is the most economical IR material available today. The vast majority of silicon grown in the world is for semiconductor applications. Silicon grown specifically for optical purposes represents only a small fraction of the total.

Silicon is produced as either single or polycrystal by the CZ crystal growth method or as single crystal by the float zone method. It can also be cast to produce large polycrystalline ingots for less demanding applications. These methods are described in Chapter 2. When specifying silicon for transparent optics, it should be stated whether it must be single or polycrystal material and also if it should be CZ grown, float zoned, or cast [93].

Unlike germanium, Si is acceptable as both $p$-type and $n$-type for IR optics applications, provided that the resistivity is high enough. The recommended resistivity specification for IR transmissive silicon material is >20 ohm-cm for $n$-type and >40 ohm-cm for $p$-type [93]. The resistivity is of little concern for Si used for reflective optics.

Resistivity measurements provide an indirect measure of the impurity level of silicon, since impurities provide an electrical path that lowers resistivity. High optical transparency requires ultrapure silicon, but even the most refined material still has some residual bulk absorption caused by impurities. This residual absorption lowers transparency [94]. The highest purity Si is float-zoned material. The process, however, limits the diameter of the ingot and thus the size of the parts.

Oxygen (O) and carbon (C) are two common unintentional impurities in silicon that result from the growth process [95]. In particular oxygen from the fused quartz crucible forms defect sites within the atomic structure of silicon. The presence of this impurity alters the resistivity of the material and lowers its optical transmission [96]. Of the three basic fabrication methods, CZ grown, float zoned, or cast, only float-zoned material has a low level of oxygen contamination because the molten Si is not in direct contact with the crucible.

Material defects in silicon are of great interest to the integrated circuit manufacturers, but since they also affect the optical quality, it is prudent to have some understanding of them. Numerous structural faults can be found in even the most carefully prepared single-crystal silicon. Such faults are stacking faults, point defects, and dislocations. In polycrystal Si, impurities such as oxygen, carbon, metallic compounds, and amorphous precipitates are not uncommon; all tend to congregate at the grain boundaries [96]. Amorphous precipitates can also form in single-crystal material. For instance, $SiO_x$ can form precipitate dislocations and large stacking faults that not only makes the material less desirable for semiconductor use but can also adversely affect its optical quality. The level of these impurities in Si both lowers the optical properties and interferes with polishing efforts to achieve a high degree of surface finish.

The semiconductor industry deals with oxygen and other impurities by subjecting polished wafers to a very high temperature treatment in a carefully controlled inert atmosphere. This process is called *gettering*. This treatment creates a nearly impurity-free zone to a depth of about 50 $\mu$m (0.002 in.). This zone is known as the *denuded zone* [97]. This process may have an application for low-scatter silicon laser mirrors but it would be of little use for transmissive silicon optics.

Because of its limited transmission range silicon is not considered a good candidate for transmissive IR optics, except in the 3- to 5-$\mu$m region [98]. For instance, with strong absorption bands at >6.5 $\mu$m, it cannot be used for transmissive $CO_2$ elements. These absorption bands are especially pronounced in the 8- to 12-$\mu$m region [99]. Silicon is, however, the material of choice for $CO_2$ laser mirrors because of its excellent thermal conductivity, hardness, chemical durability, and ready acceptance of exceptionally high-surface finishes using chemo-mechanical polishing methods.

### 1.4.3. Gallium Arsenide

Gallium arsenide is a binary compound resulting from the chemical combination of gallium (Ga) and arsenic (As). These elemental raw materials are by-products of aluminum and zinc smelting. GaAs is grown by several methods. The horizontal Bridgeman (HB) method is preferred for optical grade material, but the vertical Bridgeman (VB) and the liquid encapsulated Czochralski (LEC) methods yield material better suited for electronic applications. These methods are described in detail in Chapter 2.

GaAs is a semiconductor material with some properties superior to those of silicon. Most GaAs is produced for semiconductor applications such as substrates for VLSI (very large scale integration) circuits. Only a relatively small percentage is produced for opto-electronic devices such as light-emitting diodes (LED), diode lasers, space-based solar cells, and IR detectors. An even smaller percentage is specifically grown for use as IR materials for night vision devices and $CO_2$ optics. The use of GaAs for IR optics is the focus of this section.

Although single-crystal GaAs would be preferable because of its inherently lower impurity level, this approach is not only size limited but would also be much too costly for IR applications [100]. Semi-insulating chromium-doped polycrystal GaAs grown by the HB technique has been specifically developed for use as IR material. The chromium doping is necessary to overcome the detrimental effects of silicon from the fused quartz crucible. Fused quartz is $SiO_2$ and a small amount of the Si is absorbed by the molten GaAs, where it becomes an impurity that affects not only the electrical properties of GaAs but the optical properties as well. Since GaAs is primarily grown for use as semiconductor material, there is some concern that the availability of optical grade GaAs may be compromised by the demands for semiconductor grade material [101].

GaAs is a suitable alternative to zinc selenide (ZnSe) as a material for high-power $CO_2$ optics. It can handle high-power densities much better than ZnSe optics without thermally induced optical distortion [102]. Although the absorption coefficient for GaAs is about ten times higher than for ZnSe, its nearly three times greater thermal conductivity compensates sufficiently. Therefore GaAs can outperform ZnSe in high-power $CO_2$ systems [103]. For this reason GaAs has sometimes replaced ZnSe as $CO_2$ material. Nevertheless the debate over the merits of that alternative continues, especially since there is no cost advantage in choosing GaAs over ZnSe [101, 102]. Size limitation is possibly another argument against the use of GaAs, but the present limit of 75-mm (3.0-in.) diameter is not a factor for $CO_2$ optics since components are typically smaller than that.

Since the use of GaAs as IR material is a fairly new development, many problems with the optical quality of the material remain and must be resolved before it can be considered as reliable as the other primary IR materials that are discussed in this section. Residual impurities and bulk defects become apparent to the optical fabricator only after the optic is reported by customers to have failed in the field. Good optical grade GaAs must have low absorption and that implies the absence of absorbing impurities in the material. The growers of GaAs are typically not equipped to measure and verify the optical quality of the material with some degree of certainty. This forces the fabricator into a time-consuming and costly evaluation of the incoming material.

Because of these still very common problems, the rest of this section is devoted to a description of the tests that must be performed to ensure that the optical quality of GaAs is sufficient for the intended application. If the material quality cannot be improved, and expensive testing and selection must be performed, then the future of GaAs as an IR material may be in doubt. The description of these tests has a bearing on other IR materials, specifically Ge and Si, which can suffer from similar problems, though to a lesser degree [104].

The two most important material specifications for $CO_2$ grade GaAs are the resistivity and the absorption coefficient. These two properties are related in that a high resistivity implies a low-absorption coefficient. However, that is true in practice only in a general sense; exceptions caused by measurement difficulties exist. In theory a high-absorption coefficient and low resistivity in the GaAs indicate the presence of impurities which provide a conducting path. Such impurities lower the resistivity and absorb a greater percentage incident $CO_2$ energy. This absorbed energy causes the temperature of the part to rise to the point where damage to the part or to the laser can occur. Thermal runaway, a condition that is described in detail in the section on Ge, is typically not a problem with GaAs since it is initiated at above 250°C. The laser would have shut down long before damage would occur [105]. This means that optical grade GaAs should have an absorption coefficient that is as low as possible and consequently a resistivity that is as high as possible.

The accepted method used to measure bulk absorption is laser calorimetry for which the sample is irradiated with a $CO_2$ laser beam of a specified power level, typically 50 watt, and the temperature rise in the part is measured over a specified time interval, typically 30 seconds. Using this method, absorption coefficients in good grade GaAs have been measured from a low of 0.0025/cm to a high of 0.0075/cm, with the average about 0.005/cm. When parts fail the absorption test, it is typically by a wide margin. For instance, absorption coefficients as high as 0.06/cm and only as low as 0.02/cm have been measured on a one rejected lot, although the resistivity measurements made with a four-probe meter had indicated acceptable material with a resistivity of $>10^7$ ohm-cm.

Laser calorimetry measurements are a reliable indicator how the bulk material will perform in the field [106]. But the quality of surface finish of the test parts becomes a major factor in the measurement. For instance, absorption measurements on unpolished but fineground GaAs blanks yielded very high values regardless of whether the bulk material was absorbing or not. Obviously surface absorption overwhelmed bulk absorption. The same blanks were remeasured after an inspection polish. They yielded much lower readings, but they were still higher than expected (see Fig. 1.8). Only a high polish on both sides of the sample brought the readings into the anticipated range. What this means is that to yield reliable and accurate bulk absorption values, the blanks must be polished on both sides to a high optical finish. This fact alone makes laser calorimetry an expensive, slow, and thus unsuitable method for production. The manner in which the absorption measurements are taken affects the results as well. The highly polished samples were first tested after 10-second exposure to the laser beam. When the results were not only high but also scattered, the exposure time was increased to 30 seconds. This not only reduced the absorption coefficient but also the scatter of the data points. These conditions are shown in Fig. 1.8.

Indirect methods, particularly resistivity measurements, are typically used to evaluate GaAs and other semiconducting materials for optical quality. A high-resistivity value indicates low absorption. There are a number of resistivity meters on the market with just as many ways to evaluate the

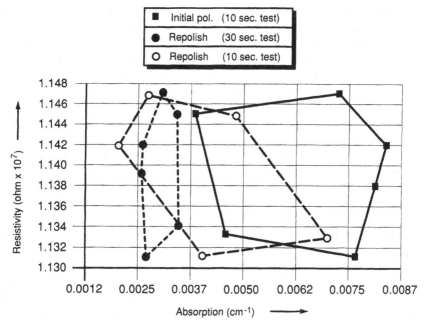

**Figure 1.8.** Absorption versus resistivity.

results. Some use two-point, some three-point, and still others use four-point probes. However, no two instruments, even of the same design, seem to exactly agree with each other. Although they tend to be quite accurate on single crystals, they have been shown to be less reliable on polycrystal materials. The primary reason for that is that the points will contact different parts of the polycrystal structure with each measurement and and thus give different reading each time. It is therefore recommended to take several readings on the same part and average the results. But that will increase the inspection time appreciably.

A more serious problem is the still tenuous correlation that exists between resistivity measurements and absorption measurements. Unless the resistivity is very high such as $>1.2 \times 10^7$ ohm-cm or very low such as $<10^7$ ohm-cm, there seems to be little or no correlation between the results from these two measurement techniques. Unfortunately, the majority of parts produced currently fall into this intermediate range, making resistivity measurements a questionable way to determine material quality. In response to customer complaints, at least one supplier of GaAs [107] has responded to bridge the gap by using an independent measurement technique that is called the *breakdown voltage measurement* (BVM). A voltage potential is placed across the test parts and increased until breakdown occurs. The voltage reading at which this discharge occurs is a comparative measure of the level of impurities in the material. The breakdown is thought to occur along a path through the bulk material that is defined by trace impurities. Therefore the higher the voltage, the fewer impurities are present and the lower the bulk absorption. BMV values above 200 volts have not resulted in absorption rejects, while some of the parts with values below 160 volts had unacceptable absorption. The exact acceptance level has not yet been determined.

On particularly poor GaAs, voids, and zonal inclusions have been found on freshly polished surfaces which were at first thought to be surface defects. The voids are caused by gasses trapped in the crystal during cool down, while the zonal inclusions are aggregates of excess arsenic compounds and other impurities. Most of the affected parts were also highly absorbing. A simple IR viewer [108] was successfully used to look through the parts against an incandescent light source. All parts that had the surface defects from the bad material showed dark regions, and many of them were nearly totally opaque. A number of those that appeared to be of visually acceptable quality had dark zones and spots in the material. Therefore the initial sorting of newly prepared blanks is done by visually inspecting them for dark regions. That inspection, however, only removes the most obviously defective parts. It will be found that many fairly clear blanks still exhibit high absorption. This inspection must be followed by additional tests, such as BVM, resistivity, and calorimetry.

### 1.4.4. Zinc Selenide

Zinc selenide (ZnSe) has been used as IR material for several decades. Until the early 1970s, ZnSe was known as Irtran 4, a polycrystalline hot-pressed material [109]. Since that time, chemical vapor deposited or CVD process ZnSe has become available, and it has now all but replaced Irtran 4. Most properties of Irtran 4 and ZnSe are nearly identical. The hot-pressing and CVD processes are described in Chapter 2.

CVD process ZnSe is a polycrystal zinc chalcogenide with a composition of about 45% selenium and 55% zinc. It has found many uses, particularly as FLIR windows and as $CO_2$ components. It is also used for chromatic correction in predominantly germanium FLIR lens systems. Since ZnSe has fairly good transmission in the visible part of the spectrum, it has found use for multispectral windows, which permits it to be used in any one of the three atmospheric IR windows as well as in the upper portion of the visible spectrum above 0.5 $\mu$m. This permits the testing and alignment of IR windows and lenses in transmission with HeNe lasers or with interferometers that use a HeNe laser as the light source.

Unlike germanium, ZnSe can be used as IR component at elevated temperatures. Because of its very low absorption (<0.0005/cm at 10.6 $\mu$m), thermal runaway can become a problem only at temperatures of 300°C or above. This good temperature performance makes ZnSe a good choice for

IR windows on high-performance aircraft, but its inherent low strength has caused problems with rain erosion. To overcome this problem, the primary producers of CVD process ZnSe [110, 111] have developed a process whereby a thin layer of ZnS is deposited on a ZnSe window. Although ZnS has inferior optical performance, it is much harder than ZnSe. The thin layer of ZnS is thought to give the ZnSe added strength without decreasing its optical performance. This sandwich process is still quite costly, and stress problems caused by the differential coefficient of thermal expansion (CTE) of the two materials have not yet been overcome [112]. It is also considered problematic from an optical design standpoint [113].

ZnSe has found use in high-power HF-DF lasers and particularly in high-power $CO_2$ lasers. Its low absorption coefficient makes thermal runaway a remote possibility. Even if there were a rise in temperature due to absorption, optical distortion caused by such temperature rise would shut the laser down before damage could occur. ZnSe is used for partial reflectors, output couplers, and focusing lenses in and for $CO_2$ systems. The relatively high refractive index of 2.403 at 10.6 $\mu$m requires effective AR coatings. ZnSe has fewer thermal focusing problems and is the material of choice for $CO_2$ output couplers. GaAs is considered a good alternative; it also can handle higher power densities better. The best choice depends on the specific application.

The internal quality of CVD process ZnSe must still be further improved. Nevertheless, high quality is not impossible to get from any of the current manufacturers if one can pick and choose. The problems that still must be resolved are internal haze, crystal structure, and solid inclusions. The haze that is often found in CVD process ZnSe can range from very light to be nearly undetectable to very heavy when the part becomes almost translucent. This cloudiness or haze is from light scattered by bulk impurities such as excess zinc or selenium atoms or from zinc oxide (ZnO) in the 1-ppm range [114, 115].

The crystal structure is a function of grain size. Different crystal grain sizes have been reported over time for ZnSe but much depends from which region in the slab the blank was cut. Grain sizes tend to be smaller near the graphite mandrel and become larger as more and more ZnSe is deposited. The growth of grain size seems to follow the same model of columnar growth that has been developed for multilayer thin film deposition. The outside skin of the CVD process ZnSe slab has a very bumpy appearance. When parts are cut from a region that is near the surface of the slab, a characteristic pattern of a very coarse structure will be exposed. This effect is so noticeable that the parts need not be polished to detect it. It has been called "alligator skin" by the opticians.

CVD process has an elongated microstructure with a length of the grain of about 46 $\mu$m and average grain diameter of about 50 $\mu$m [116]. Another report says that the average grain size is 70 $\mu$m [117]. Still another reports a typical grain size between 50 and 100 $\mu$m [115].

The thermal properties of ZnSe are quite good, and it resists thermal shock sufficiently so that standard optical fabrication methods, such as pitch and wax blocking, can be used successfully when adequate care is taken. ZnSe is chemically inert and insoluble in water. However, exposure to acids must be avoided because chemical reactions can liberate toxic hydrogen selenide ($H_2Se$) gas. ZnSe should also be worked wet since ZnSe dust, primarily selenium dioxide ($SeO_2$), can be toxic when inhaled. Therefore inhalation protection such as respirators should be worn when ZnSe dust is present. For the same reason ZnSe swarf must be collected and disposed of according to accepted safety rules. Swarf is finely abraded material from cutting and/or grinding operations.

ZnSe suffers from a problem that many opticians may have come across during their efforts to polish it. This problem is particularly noticeable on large surfaces such as FLIR windows and on multilens blocks for $CO_2$ focus lenses. Certain regions of the surface, typically the center of the blocks, appear to remain underpolished regardless how long the polishing cycle continues. This seems to be one manifestation of a problem that has been reported as being characteristic of ZnSe. It is called *environmentally assisted slow crack growth* with moisture being the corrosion medium [116].

Another way how this problem has been manifesting itself is through catastrophic failures of ZnSe windows after a limited number of pressure cycles from normal pressure to several thousand pounds per square inch [118]. The pressure window failures were most likely the result of stress corrosion or static fatigue caused by slow crack growth. The environmentally assisted crack growth occurs when the surface of the window is under tensile stress, as it would be during polishing. Liquid water or water vapors are the most active environmental constituents that assist in this process. Therefore the observed difficulties in polishing some ZnSe surfaces could be the result of this problem, aggravated by an unfavorable crystal structure. As the crack growth precedes the polish, it must appear to an observer that no progress is being made even as polishing times are greatly extended.

Because of slow crack growth failures, it is inadvisable to put scribe marks on any brittle crystalline materials like ZnSe since microscopic cracks in the scribe line can grow to critical size under mechanical and/or thermal loads, at which point the optical component will fail catastrophically. For the same reason the use of harsh stock removal processes and coarse abrasives must be avoided when working ZnSe or similar materials.

### 1.4.5. Zinc Sulfide

Zinc sulfide (ZnS) is also produced by the CVD process which is very similar to the one used to produce ZnSe. The CVD process for both ZnS and ZnSe is described in Chapter 2. Prior to the development of the CVD process, ZnS was commercially available Irtran 2 [119], a hot-pressed, polycrystalline material made from zinc sulfide powder.

Like zinc selenide, ZnS is also a polycrystal zinc chalcogenide. It differs from ZnSe in that it has better mechanical strength, but it suffers from higher absorption and thus from reduced optical transmission. ZnS is typically used for windows or domes in FLIR systems. Because of its greater strength relative to ZnSe, ZnS is better suited to serve as window material for IR windows in high-speed aircraft. It also provides adequate rain erosion resistance. The optical transmission of ZnS is, however, inadequate for broadband applications. Therefore a sandwich-type material has been developed for which a thin layer of ZnS is deposited on a previously polished ZnSe window via the CVD process. This way the useful properties of both materials are combined to yield components that exhibit both strength and optical performance.

ZnS has been successfully CVD processed into dome shapes using "muffin tin" mandrils. This was done to reduce the amount of material that had to be removed to produce the dome shape. Hundreds of these dome blanks have been produced for a military program [120].

CVD process ZnS has an elongated grain structure with an aspect ratio of 6 to 1. The diameter of that grain is about 6 to 8 $\mu$m. It is this small grain size that gives ZnS its mechanical strength [121]. Because of this elongated grain structure, ZnS must be cut only perpendicular to the long grain axis; otherwise, the material would have reduced strength and would be much more difficult to polish.

CVD process ZnS has a reddish orange color which results from the presence of zinc hydride ($ZnH_2$) impurities or from excess zinc or sulfide ions [122, 123]. Although ZnS is chemically quite stable, it does react with water to liberate small, but detectable, amounts of hydrogen sulfide ($H_2S$) gas during the polishing operations. The presence of this gas is easily detected by its strong "rotten egg" odor which is detectable at very low gas concentrations. Adequate ventilation must be provided when polishing ZnS with aqueous polishing slurries. The use of acids must be avoided because they can liberate toxic quantities of $H_2S$ gas [124]. Airborne ZnS dust can also be toxic; therefore adequate inhalation protection must be provided when such dust is suspected.

A variant of the CVD process for ZnS produces clear or multispectral ZnS. After the standard CVD process, the ZnS is subjected to a thermal treatment under significant pressure. This treatment causes the material to become quite transparent in the visible region while maintaining the IR transmission properties of standard CVD process ZnS. The absorption coefficients of multispectral ZnS are also much improved, ranging between 1/10 to 1/100 than those for standard CVD process ZnS.

The crystal grains of multispectral ZnS are about ten times larger than they are for the standard ZnS. This grain growth reduces the mechanical strength of the material. For instance, the Knoop hardness drops from 250 kg/mm$^2$ for standard ZnS to 160 kg/mm$^2$ for multispectral, which makes the polishing process more difficult. At the same time the flexural strength is cut in half to a mere 7500 psi, which makes multispectral ZnS much more fragile that standard ZnS.

### 1.4.6.  Infrared Glasses

Infrared glasses are amorphous (noncrystalline) optical materials that transmit in the IR region of the spectrum. Despite intensive efforts in developing a variety of such materials, IR glasses have, for the most part, not been commercially successful. They must be mentioned in this chapter for the sake of completeness and also because the optician may be asked on occasion to work these or similar materials. IR glasses can be categorized roughly into germanate glasses, chalcogenides, and chalcopyrites.

Germanate glasses are optical materials that have a composition similar to silicate glasses, except that the $SiO_2$ has been replaced by $GeO_2$. Compared to silicate glasses, germanate glasses have a higher refractive index, greater dispersion, and greater density. They transmit well into the IR, but numerous absorption bands limit their usefulness to about 5.0 $\mu$m. The transmission spectrum exhibits particularly deep hydroxyl (water) absorption bands at about 2.9 $\mu$m and between 10.5 and 13.5 $\mu$m [125]. Germanate glasses typically suffer from poor chemical durability, which makes them difficult to polish, clean, and handle. Some compositions, however, have shown improved resistance to chemical attack.

Chalcogenides are binary compounds of a chalcogen such as oxygen (O), sulphur (S), selenium (Se), or tellurium (Te) and a more electropositive element such as arsenic (As) [126]. One of the earliest of those was arsenic trisulfide ($As_2S_3$). It is an amorphous material, and thus a glass, which is useful to about 10 $\mu$m but with a deep water absorption band at 2.8 $\mu$m. Arsenic trisulfide is toxic and must be processed and handled with care [127]. Arsenic triselenide ($As_2Se_3$) is a closely related material to arsenic trisulfide with very similar characteristics and a useful transmission range from 1 to 18 $\mu$m [128]. One major problem with these materials is the high absorption that limits their use to only low power applications. Other nonoxide chalcogenides are germanium trisulfide ($Ge_2S_3$) and germanium triselenide ($Ge_2Se_3$) [129].

Chalcogenide IR-transmitting glasses are typically ternary compounds of Tl–As–Se, Tl–As–S, Ca–La–S, and Ge–As–Se. Thallium arsenide sulfide ($Tl_3AsS_4$) and thallium arsenide selenide ($Tl_3AsSe_3$) are chalcogenide glasses that are useful for acousto-optic applications with transparencies well into the IR [130, 131]. $CaLa_2S_4$ may become a substitute for more commonly used and more expensive IR materials such as germanium. It has good IR transmission from 2.5 $\mu$m to 16 $\mu$m and a much lower refractive index than Ge. Its hardness is similar to that of ZnS [132].

The best known of these ternary compounds are those composed of Ge–As–Se or Ge–Sb–Se. They are better known as TI 1173 and AMTIR 1, 2, and 3. For instance, TI 1173 is composed of 28% of 6N's Ge, 12% of 5N's Sb, and 60% of 5N's Se [133]. AMTIR 1, for example, is $Ge_{10}As_{20}Se_{70}$, which is the same as saying 10% Ge + 20% As + 70% Se [129]. Despite efforts to commercialize these materials, they have enjoyed only limited success. They may find use as IR fiber optic materials. These materials have several drawbacks that make them difficult to work with. They have low fracture strength, which makes them quite brittle, and they are also quite soft, which makes it difficult to control surface figure and center thickness. They further are sensitive to thermal shock, which is greatly aggravated by the existing surface condition [134]. These materials contain toxic ingredients in chemically unbound form, such as arsenic and selenium, which requires strict safety precautions when working with and handling these materials.

Finally, there are the chalcopyrites which are also ternary compounds composed of Ag–Ga–S or Ag–Ga–Se. Most of these materials are still in the development stage for use as IR second-harmonic generators (SHG) or other IR applications. Silver thiogallate ($AgGaS_2$) transmits from 0.5 to 13 $\mu$m, and silver selenogallate ($AgGaSe_2$) transmits from 0.7 to 17 $\mu$m [135].

### 1.4.7. Miscellaneous IR Materials

There are many other IR-transmitting crystals. Some are not of commercial value although of great interest to researchers. We will limit the number discussed in this section to a few of the more familiar crystals that are useful in the IR region.

Heavy metal salts such as silver chloride (AgCl), silver bromide (AgBr), and thallium bromide iodide (KRS-5) transmit well into the infrared. They are all quite soft and easily hot pressed. Their softness makes them very difficult to polish. Heavy metal salts have a tendency to darken upon exposure to light due to a photochemical reaction [136]. For these and other reasons heavy metal salts are not useful in laser or thermal imaging optics. Although more suitable IR materials are readily available, heavy metal salts have played an important role in IR spectrometry for many decades, and they may find use in IR optical fibers in the future.

In the 1950s and 1960s the group of IR-transmitting materials used for IR spectrometry were the hot pressed, polycrystal Irtran materials [137]. They were made from ultrapure powders of the base crystal. These powders were compacted under heat and pressure into a solid with randomly oriented microcrystals in intimate contact with one another. There are six Irtran materials which are listed in Table 1.17.

Hot-pressed materials, starting from powders, have a tendency to exhibit significant bulk scatter. This scatter is caused by submicron voids in the bulk material. It gives these materials, such as the Irtran group, a translucent quality. Their translucence prevents their use in the visible region despite the fact that almost all of them transmit well into the visible part of the

**Table 1.17.   List of Irtran IR Materials**

| Irtran 1 | Magnesium fluoride | $MgF_2$ |
|----------|--------------------|---------|
| Irtran 2 | Zinc sulfide | ZnS |
| Irtran 3 | Calcium fluoride | $CaF_2$ |
| Irtran 4 | Zinc selenide | ZnSe |
| Irtran 5 | Magnesium oxide | MgO |
| Irtran 6 | Cadmium telluride | CdTe |

spectrum. Another limitation for Irtran materials is that the hot-press method is limited to about a 7-in. diameter. This method is described in detail in Chapter 2.

Since the Irtran group has been all but replaced by the same materials produced by more advanced techniques, their properties are discussed under the heading for the appropriate crystal. For example, Irtran 2 and Irtran 4 have now been replaced by CVD-process ZnS and ZnSe, respectively. The respective properties for these CVD materials also apply for the most part to Irtran 2 and Irtran 4. If additional information is needed for these materials, there is an exhaustive product description available from the manufacturer [138].

Cadmium sulfide (CdS) and cadmium telluride (CdTe) are useful IR materials, although their use has somewhat declined over the years. CdTe was hot pressed as Irtran 6, but it did not develop into a commercial success. Both materials are now produced by the CVD process which yields polycrystal and by other crystal growth techniques which yield single crystal. Optical grade CdTe has been used as a window or output coupler for low-power $CO_2$ lasers.

Sapphire is aluminum oxide ($Al_2O_3$). Processes have been developed which produce single-crystal sapphire in appreciable sizes and in commercial quantities. Large sizes are now routinely grown using the HEM method which is described in Chapter 2. Sapphire is very hard, second only to diamond. It is also the only optical material that is practically free from internal scatter [136]. In fine powder form, aluminum oxide finds wide use as abrasive and as polishing powder for optical fabrication. Diamond powders must be used to achieve a good polish on sapphire.

Diamond is a material known to everyone. When we think of diamond, we immediately think of valuable gem stones. But by far the greatest number of diamonds find use in industry. It may be surprising that diamond is also an excellent IR material. Not only is it the hardest material in nature, which makes it very difficult and costly to polish, but it is still a very expensive crystal which is available only in small sizes. It has one of the highest thermal conductivities known at five times that of copper, which makes diamond a very efficient heat sink.

**References for Section 1.4 (Infrared Materials)**

[79] Metz, F. I., ed., *New Uses for Germanium,* Midwest Research Institute, 1974.

[80] MHO—Metallurgie Hoboken-Overpelt, Belgium.

[81] Otavi Minen, AG, Germany.

[82] Preussag Pure Metals, GmbH, Germany.

[83] Afrimet Indussa, Inc., New York, NY.

[84] Eagle Picher Industries, Inc., Quapaw, OK.

[85] Exotic Materials, Inc., Costa Mesa, CA.

[86] Vinten Penarroya, Inc., Fountain Valley, CA.

[87] Marc van Sande, et al., "Applications of germanium anno 1986," paper presented at the symposium, "100 Jahre Germanium," Freiburg, Germany, 1986.

[88] Adams, Jack H., "Specifications for optical grade germanium and silicon blanks," *SPIE Proceedings,* vol. 406, April 1983.

[89] News item, "What's new in IR", *Photonics Spectra,* June 1986.

[90] Miles, Perry A., "Crystalline infrared transmissive materials," *Optical Shop Notebook II.* Section XVI: *Infrared Optics,* November 1977.

[91] Hilton, A. Ray, "Production and fabrication of blanks from germanium and chalcogenide glasses," *Optical Shop Notebook* II. Section II: *Optical Materials,* November 1977.

[92] Sherman, Glen H., "$CO_2$ laser optics: Absorption's dominant role," *Electro-Optical Systems Design,* June 1982.

[93] Adams, Jack H., "Specifications for optical grade germanium and silicon blanks," *SPIE Proceedings,* vol. 406, April 1983.

[94] News item, "Studies of low loss window materials shifting to chemical laser wavelengths," *Laser Focus,* March 1977.

[95] Liaw, H. M., "Trends in semiconductor material technologies for VLSI and VHSIC applications," *Solid State Technology,* July 1982.

[96] Huff, Howard R., "Chemical impurities and structural imperfections in semiconductor silicon—Part II," *Solid State Technology,* April 1983.

[97] Swaroop, R. B., "Advances in silicon technology for the semiconductor industry—Part II," *Solid State Technology,* July 1983.

[98] News item, "What's new in IR," *Photonics Spectra,* June 1986.

[99] Truett, W. L., "IR optics revisited," *Photonics Spectra,* April 1984.

[100] Miles, Perry A., "Crystalline infrared transmissive materials," *Optical Shop Notebook* II. Section XVI: *Infrared Optics,* November 1977.

[101] Sherman, G. H., "Optics for industrial $CO_2$ lasers," *Lasers and Optronics,* September 1988.

[102] Sherman, G. H., "Latest optics for industrial $CO_2$ lasers," International Conference on Laser Material Processing—Science and Applications, Osaka, Japan, May 1987.

[103] Sherman, Glen H., "$CO_2$ laser optics: Absorption's dominant role," *Electro-Optical Systems Design,* June 1982.

[104] Karow, H. H., "Revised material specifications for gallium arsenide," CCG memo, December 1987.

[105] Truett, W. L., "IR optics revisited," *Photonics Spectra*, April 1984.

[106] Welch, M., "Techniques for testing infrared windows," *Lasers and Applications*, August 1984.

[107] Crystal Specialties, Inc., Colorado Springs, CO.

[108] FJW Optical Systems, Elgin, IL, IR-scope model: "Findr-Scope."

[109] Eastman Kodak Co., Rochester, NY.

[110] Raytheon Company, Lexington, MA.

[111] CVD, Inc., Woburn, MA.

[112] Cooper, Bill, "IR window advances: FLIR and beyond," *Photonic Spectra*, October 1988.

[113] Moore, D. T., "Infrared gradient index design," *Photonics Spectra*, July 1988.

[114] Truett, W. L., "Optical materials—The role of impurities," *Lasers and Applications*, August 1984.

[115] Miles, P., "High transparency infrared materials—A technology update," *Optical Engineering*, vol. 15, no. 5, September/October 1976.

[116] AFML-TR-77-23.

[117] Taylor, R. L., "An infrared alternative: Vapor-deposited materials," *Laser Focus*, July 1981.

[118] Annett, C., "The latest on the dangers of ZnSe," *Optical Spectra*, September 1979.

[119] Eastman Kodak Co., Rochester, NY.

[120] Maverick program.

[121] AFML-TR-77-23.

[122] Taylor, R., "CVD IR materials: New challenges," *Photonics Spectra*, July 1982.

[123] Truett, W. L., "Optical materials—The role of impurities," *Lasers and Applications*, August 1984.

[124] Taylor, R. L., "Infrared materials—CVD materials," *Technical Digest*, Workshop on Optical Fabrication and Testing, December 1982.

[125] Metz, F. I., ed., *New Uses for Germanium*, Midwest Research Institute, 1974.

[126] Tebo, A. R., "Infrared optical fibers—The promise of the future," *Electro-Optical Systems Design*, June 1983.

[127] McCarthy, D. E., "The reflection and transmission of infrared materials. Part I: Spectra from 2 to 50 microns," *Applied Optics*, vol. 2, no. 6, June 1963.

[128] Drexhage, M. G., "IR transmitting fluoride glasses," *Laser Focus*, October 1980.

[129] Miller, S. A., "Ultralow-loss communications in the mid-IR," *Photonics Spectra*, July 1986.

[130] News item, "Infrared a–o material family expands, but crystal growth remains a problem," *Laser Focus*, May 1977.

[131] Gottlieb, M., "Infrared acousto-optic materials: applications, requirements, and crystal development," *Optical Engineering*, vol. 19, no. 6, November/December 1980.

[132] Szappanos, J., "Infrared optics: The newest frontier," *Photonics Spectra*, July 1986.

[133] Klocek P., "Chalcogenide glass optical fibers and image bundles: Properties and applications," *Optical Engineering,* vol. 26, no. 2, February 1987.

[134] Hilton, A. R., "Production and fabrication of blanks from germanium and chalcogenide glasses," *Optical Shop Notebook* II. Section II: *Optical Materials,* presented at OF&T 77, November 1977.

[135] Feigelson, R. S., "Recent developments in the growth of chalcopyrite crystals for nonlinear infrared applications," *Optical Engineering,* vol. 26, no. 2, February 1987.

[136] Truett, W. L., "Optical materials—The role of impurities," *Lasers and Applications,* August 1984.

[137] Eastman Kodak Co., Rochester, NY.

[138] "Kodak IRTRAN Infrared Optical Materials," Eastman Kodak product publication U-72.

## 1.5.  OPTICAL CRYSTALS

Most optical materials are amorphous such as optical glass (or fused silica), but there are a wide variety of optical crystals that have unique properties that are useful for specific applications. Some of these have good transmission from the ultraviolet to the near-IR regions of the optical spectrum. Other crystals serve as host for a number of solid state laser rods. Water soluble crystals have excellent transparency over a wide transmission range. Some of these have electro-optic properties that makes them useful for laser beam modulation. A wide variety of new crystal materials has been developed in the recent past for converting laser energy from the characteristic wavelength to the second or even third harmonic. The realm of optical crystals is probably the most challenging for the optician because each crystalline material has its own unique properties that require special attention during the manufacture of optical components.

### 1.5.1.  Alkaline Earth Fluorides

Alkaline earth fluorides are a family of optical crystals that have excellent transmission from the vacuum ultraviolet (VUV) to the mid-IR region of the spectrum. These materials are further characterized by relatively low refractive indexes and low absorption coefficients. Table 1.18 lists the pertinent properties for these crystals.

The properties listed in the table make these materials useful as optical components for excimer lasers and HF-DF chemical lasers. Excimer lasers emit intense light pulses in the VUV and UV region at wavelengths between 158 nm ($F_2$) and 351 nm (XeF), while HF-DF lasers operate in the near-IR over the 2.6 to 4 $\mu$m range.

The most prominent members of this family of crystals are calcium fluoride ($CaF_2$), magnesium fluoride ($MgF_2$), barium fluoride ($BaF_2$), and strontium fluoride ($SrF_2$). These fluorides will be described in more detail. Other

**Table 1.18. Optical Properties of Alkaline Earth Fluorides**

| Property | Dimension | @ wvl (µm) | BaF$_2$ | CaF$_2$ | LiF | MgF$_2$ | SrF$_2$ | Ref. |
|---|---|---|---|---|---|---|---|---|
| Refr. index | | 0.2 | 1.52 | 1.50 | | | | (139) |
| | | 0.5 | 1.48 | 1.44 | 1.40 | 1.38 | 1.44 | (140) |
| | | 1.0 | 1.47 | 1.43 | | 1.38 | | (141) |
| | | 2.7 | 1.46 | 1.42 | 1.37 | | | (142) |
| | | 3.8 | 1.46 | 1.41 | 1.35 | 1.35 | 1.44 | |
| | | 5.3 | 1.45 | 1.40 | 1.32 | | | |
| | | 6.0 | | 1.37 | | 1.32 | | |
| | | 8.0 | | 1.35 | 1.20 | 1.26 | | |
| | | 10.0 | 1.40 | | | | | |
| Transmission | 10% T. | Min (µm) | 0.14 | 0.13 | 0.11 | 0.12 | 0.13 | (140) |
| | | Max (µm) | 15.0 | 12.0 | 9.0 | 9.7 | 14.0 | (141) |
| | | | | | | | | (143) |
| | Useful range | Min (µm) | 0.14 | 0.13 | 0.11 | 0.12 | 0.13 | |
| | | Max (µm) | 12.0 | 8.0 | 8.0 | 8.0 | 11.0 | |
| | VUV cut-off | nm | 135 | 123 | 105 | 114 | | (144) |
| | | | | | | | | (145) |
| Absorption | .01/cm | @ 2.7 µm | 0.18 | 0.08 | | 0.15 | 0.15 | |
| | | @ 3.8 µm | 0.20 | 0.03 | | 0.07 | 0.07 | (142) |
| | | @ 5.3 µm | 0.03 | 0.05 | | 1.40 | 0.01 | |

**Table 1.18a. Physical Properties of Alkaline Earth Fluorides**

| Property | Description | Dimension | BaF$_2$ | CaF$_2$ | LiF | MgF$_2$ | SrF$_2$ | Ref. |
|---|---|---|---|---|---|---|---|---|
| Therm. expan. | CTE | E10-6/C | | | | 13.7∥ to c | | |
| | | | | | | 8.5 ⏊ to c | | (141) |
| Therm. cond. | E10-4 | cal/cm/sec/C | 287 | 237 | 270 | 360 | 239 | (142) |
| Density | | g/cm$^3$ | | | | 3.18 | | (141) |
| Hardness | Knoop | kg/mm$^2$ | 65 | 120 | 102 | 415 | 130 | (141) |
| | Mohs | | 4.0 | 4.0 | 3.0 | 6.0 | 4.0 | (142) |
| Fract. streng | psi | pol. crystal | | 16 000 | | 7 600 | 12 100 | (141) |
| | | monocrystal | | 25 000 | | | 11 300 | (146) |
| Young's mod. | psi | E10+6 | | | | 20.1 | | (141) |
| Solubility | g/100g H20 | E10-3 | 140.0 | 1.5 | 270.0 | 7.6 | 11.7 | (140) |
| | | | | | | | | (142) |
| Chemical res. | | | | inert to H20 | | durable | | (141) |
| Polishability | | | | | difficult, soft | rel. easy | | (141) |

members of this family of crystals that do not play as significant a role in optics are lithium fluoride (LiF), lanthanum fluoride ($LaF_2$), cadmium fluoride ($CdF_2$), lead fluoride ($PbF_2$), and sodium fluoride (NaF).

A natural form of $CaF_2$ is optical quality fluorspar, also known as *fluorite*. The alkaline earth fluorides used in optics today, however, are synthetically produced either by the BSS growth method, which yields single-crystal boules, or by the fusion casting or hot isostatic pressing methods, which yield polycrystalline ingots. These methods are described in greater detail in Chapter 2.

The hardness ranges from the relatively hard $MgF_2$ to the rather soft LiF. Hot-pressed fine-grain polycrystalline materials have inherently greater mechanical strength than the corresponding single crystals. But since the optimum strength is a function of surface finish and a better polish is usually obtained on single-crystal material, it has been shown that single-crystal $CaF_2$ has a fracture strength that is 50% greater than polycrystal $CaF_2$ [146]. The thermal properties of these materials require careful handling during manufacture. Magnesium fluoride, the most resistant in the group, has a coefficient of thermal expansion that is nearly twice that of BK 7 glass and a thermal conductivity that is only a quarter of that of germanium.

The excellent UV transmission of $MgF_2$ makes it the most useful material for use in excimer lasers. Single-crystal $MgF_2$ is virtually insoluble, and its favorable hardness makes it relatively easy to polish. It is quite resistant to damage from mechanical or thermal shock. Because of these properties $MgF_2$ is the preferred material for excimer lasers, although it is more expensive than $CaF_2$ which has similar properties except that it is not as durable. $CaF_2$ has good UV transmission, which makes it a good choice for UV optics. It transmits well into the IR, which makes it also useful for mid-IR applications. $CaF_2$ is the least expensive of all the alkaline fluorides. Single-crystal $MgF_2$ is birefringent, which makes it a useful material for polarizers and waveplates. The refractive indices for $MgF_2$ at 0.5 $\mu$m are $n_o = 1.38$ and $n_e = 1.39$. $BaF_2$ and $CaF_2$ are not birefringent.

Alkaline earth fluoride crystals tend to discolor when exposed to intense UV radiation. Color centers in the crystal structure cause the crystal to noticeably darken. In extreme cases the crystals can become nearly opaque. This reduction in transmission is accompanied by a proportional increase in bulk absorption. With the exception of LiF, which discolors permanently, the discoloration can be bleached out through an appropriate heat treatment. Nondiscoloring crystals of $CaF_2$ can be selected from a group of boules, but that requires a time-consuming and expensive functional test. $MgF_2$ resists discoloration more than any of the other alkaline earth fluorides, but it is still possible for the crystal to discolor at very high UV intensities.

A group of materials that is closely related to alkaline earth fluorides are the noncrystalline heavy metal fluoride glasses. These glasses were developed in the early 1980s for use as laser window material or for mid-IR optical fibers. They are known as *fluorozirconates* and *fluorohafnates*. The fluorozirconate glass, which is identified as ZBT, is made up for the most part of zirconium fluoride ($ZrF_4$). The fluorohafnate glass, labeled HBT, contains hafnium fluoride ($HfF_4$) as the primary ingredient.

ZBT glasses have a low refractive index of 1.52 and a dispersion value of 80. They are transparent between 0.2 and 7.0 $\mu$m. Their densities range from 4.8 to 6.2 $g/cm^3$ depending on composition. The Knoop hardness of 250 kg/$mm^2$ is similar to that of the softer flint glasses such as Schott SF-59 (953 204). The properties of HBT are very similar [147].

Both glasses are chemically resistant to the effects of water. ZBT resists acid attack of even hydrofluoric acid (HF). Conventional polishing techniques can be used. They can also be molded by hot pressing. Although they have, supposedly, low toxicity, it is always prudent to exercise reasonable care when working or handling unfamiliar materials.

ZBT and HBT have a low bulk absorption coefficient of about 0.01/cm in the 3- to 5-$\mu$m region. This property makes theses glasses useful for mid-IR laser and optical fiber applications [148]. They have also low UV absorption, which makes them potential candidates for excimer lasers.

Heavy metal fluoride glasses are melted in an inert gas atmosphere in vitreous carbon or in platinum crucibles using standard optical glass-making processes. High purity starting materials and extreme cleanliness is required during the melting process in order to achieve the required low-absorption values.

A related glass is fluoroberyllate ($BeF_2$) glass, which was originally developed as a laser glass. It has an unusually low refractive index of 1.30 and an unusually high dispersion value of 100. It transmits from 0.15 to 5.0 $\mu$m. The high beryllium content makes $BeF_2$ glass potentially toxic. It is also hygroscopic. These two properties make $BeF_2$ glass difficult to work with. ZBT and HBT present themselves as better alternatives. $BeF_2$ glass is described in greater detail in Section 1.2.1 (Laser glass) in this chapter.

### 1.5.2.  Laser Crystals

Laser crystal rods and rods made from laser glass are the heart of solid state lasers. These crystals or glasses, also called *laser hosts,* are doped with ions of chromium (Cr+) or neodymium (Nd+) to provide the proper gain medium necessary for laser action. Laser crystals are examined in this section, and laser glasses are discussed separately in Section 1.2.1.

The first laser in 1960 was a small ruby rod that was surrounded by a helical flash lamp. It was a modest beginning to what has become one of the most pervasive inventions of the twentieth century. Ruby is chromium-doped sapphire (Cr+ : $Al_2O_3$). The chromium doping gives ruby its characteristic red color. The intensity of the color can be varied by varying the doping concentration. This is why the color of ruby can range from deep red to a bright pink color, with the lighter shades of red found in those crystals with the least doping.

Ruby is very hard, has high mechanical strength, high thermal conductivity, and is chemically stable. The high hardness of ruby and sapphire requires the use of diamond tools and diamond powders to shape, grind, and polish the laser rods. Ruby crystals are grown by the CZ method which is described in Chapter 2. By the 1970s ruby crystals had been grown to 4-in. diameters and 12-in. lengths [149, 150]. Although ruby was an important laser host in the early days, it has been all but replaced by a growing number of other laser crystals.

The second solid state laser host that has also played an important role in the development of the laser is neodymium-doped calcium tungstate $(Nd+ : CaWO_4)$. This crystal is also grown by the CZ method. The Nd-doping ions which are typically between 1 to 5 M% give calcium tungstate its characteristic purplish-blue color [151].

Calcium tungstate is much softer than ruby and thus much easier to fabricate into laser rods. The crystal will easily take on a high degree of polish using standard optical shop methods. Although chemically inert in most environments, calcium tungstate is not nearly as strong as ruby and can break when mishandled. It can also fail due to thermal shock, especially when fine surface scratches are present. Calcium tungstate has also been replaced by improved laser glasses and more efficient laser crystals. By the end of the 1960s, Nd:YAG and some Nd-doped laser glasses had become the dominant laser hosts.

Neodymium-doped yttrium aluminum garnet $(Nd : Y_3Al_5O_{12})$ or Nd:YAG for short has emerged as the leading solid state laser host material. Like ruby and calcium tungstate, it is also a CZ grown crystal. But unlike these, it is nearly colorless. It is quite hard, mechanically strong but brittle, and chemically stable in most acids, including nitric acid $(HNO_3)$ and hydrofluoric acid $(HF)$. It can be attacked, however, by phosphoric acid $(H_3PO_4)$, especially at elevated temperatures [152].

By the early 1980s, the demand for Nd:YAG laser rods had outstripped the available supply, and delivery times sometimes exceeding one year were not uncommon [153]. Part of the problem was the then-limited growth facilities, the relatively low yield, and the long growth cycles, which can exceed several weeks for boules of even modest size. This shortage encouraged the development of substitute crystals, but Nd:YAG still remains the most widely used solid state laser material at the writing of this chapter.

Other crystals developed for use as laser hosts are YALO or YAP, YLF, pentaphosphate, and emerald. YALO is neodymium-doped yttrium aluminum perovskite $(Nd : YAlO_3)$. This crystal material is also known as YAP. It has properties very similar to those of Nd:YAG, but it can be CZ grown at rates ten times faster [154]. Another crystal developed at about the same time is neodymium-doped yttrium lithium fluoride $(Nd : YLiF_4)$ or Nd:YLF for short. YLF is a laser host useful for UV lasers, and it is emerging as a serious competitor to Nd:YAG. Neodymium-doped pentaphosphate $(Nd : P_5O_{14})$ is another crystal mentioned in the literature, but like YALO and other crystal materials, it has also never rivaled the commercial importance of Nd : YAG.

Synthetic emerald emerged during the early 1980s as a new and promising laser crystal. Emerald is like ruby and alexandrite, a gem laser material. It is chromium-doped beryllium aluminum silicate [Cr : $Be_3Al_2(SiO_3)_6$]. Because of its inherently lower polarization effects as compared to alexandrite, emerald may be a better choice for slab lasers [155]. To become a practical reality, the optical quality of emerald still needs considerable improvement. Another suggested application for emerald is for ultrashort laser pulse generation [156] which has become an important field in the 1980s. The latest addition to the growing list of laser crystals is chromium-doped forsterite (Cr : $MgSiO_4$). It is quite similar to alexandrite. Depending on the axis through which the forsterite crystal is viewed, it appears blue, violet, or green [157].

The 1980s were also a period of development for a number of crystal materials that make tunable solid state lasers possible. The first of these was alexandrite (Cr : $BeAl_2O_4$). Alexandrite has optical, mechanical, thermal, and chemical characteristics quite similar to those of ruby [158]. This crystal offers continuous wavelength tunability in the near-IR region. Alexandrite is a synthetic form of chrysoberyl which is a rare natural crystal. Although the material itself is not toxic, one of its constituents (beryllium) is, and safety precautions should be applied accordingly when working with this material. Other tunable laser hosts are chromium-doped gadolinium scandium gallium garnet (Cr : GSGG), cobalt-doped magnesium fluoride (Co : $MgF_2$), scandium borate ($ScBO_3$), and titanium-doped sapphire (Ti : $Al_2O_3$). The last of these has become the focus of interest for laser researchers.

### 1.5.3. Alkali Halides

Alkali halides are salt crystals that are either grown by the solution growth method or by the Stockbarger method. These methods are described in Chapter 2. More specifically, alkali halides are binary (two-part) compounds of a halogen such as Br, Cl, F, or I and an alkali such as Cs, Li, Na, K, or Rb. Of the 20 possible alkali halides only 10 or so have found useful optical applications, and of the 10 only half are in common use. These 5 are cesium bromide (CsBr), cesium iodide (CsI), potassium bromide (KBr), potassium chloride (KCl), and the most common of all: sodium chloride (NaCl). Others used much less often are potassium iodide (KI), sodium bromide (NaBr), and cesium chloride (CsCl).

Because of the inherent tendency of alkali halides to form perfect crystals under properly controlled growth conditions, it is relatively easy to grow high-purity single crystals of appreciable size from a saturated solution (see Chapter 2). Sodium chloride has been grown by this method for over 60 years. The starting materials for crystal growth must be of the highest purity, $6N$'s or better, which means 99.9999% pure or an impurity level of 1 ppm or less.

During the 1970s, hot-forged polycrystal alkali halides were produced. These crystals are considerably stronger than the corresponding single crystals. The use of alkali halides in high-power $CO_2$ lasers required the development of low-absorbing materials by chemical purification during the growth cycle. A further refinement in the growth of alkali halides was the development of the reactive atmosphere process (RAP), which essentially eliminated oxygen as a major impurity. Also the hot-forging process was improved by introducing alloying agents for grain-size control. One such example is the addition of rubidium chloride to potassium chloride to form a polycrystalline alloy (KCl : RbCl). This method has increased the yield strength by a factor of ten over that of single crystal KCl [159].

Alkali halides have excellent optical transparency, and their optical characteristics are not much affected by changes in temperature. Since they are cubic crystals, they are optically isotropic, which is another way of saying that they are not birefringent [160]. All of them are, however, highly water soluble (or hygroscopic), very soft, and quite sensitive to thermal shock. This makes working and handling these crystals a difficult task.

The unusually broad transmission range from the UV to well into the IR region and the typically very low absorption coefficients have made alkali halides a good choice for multispectral applications. They are particularly useful as components for excimer lasers (UV) and for $CO_2$ lasers (IR). For instance, polycrystal potassium chloride has been successfully used as window material for excimer lasers [161] and for optics used in large high-power $CO_2$ lasers [162]. Large polycrystal sodium chloride windows have also been used for the Helios high-energy $CO_2$ laser at Los Alamos Lab (LASL) [163, 164].

The primary characteristic that limits the use of alkali halides is their very high water solubility. In more technical terms, alkalide halides are strongly hygroscopic, which means that they are readily attacked by hydrolysis. This process contaminates the material with oxygen and/or hydroxyls [165]. Even normal atmospheric moisture above 30% or 40% relative humidity (RH) can attack polished surfaces and destroy them. Fortunately the moisture absorption is reversible by placing the affected components in a vacuum oven for a period of time. The polished surfaces must be protected during storage to avoid atmospheric attack. Optical coatings for these materials can be applied, but no suitable coating has yet been found that will protect the crystals from water vapors for any length of time.

Most IR materials fail by brittle fracture, which in turn is a function of grain size. This means that polycrystal materials are more resistant to failure by brittle fracture than single-crystal material. For instance, single-crystal KCl can fail at loads as low as 200 psi, whereas hot-pressed polycrystal KCl can withstand loads that are at least 20 times greater [166]. Almost all alkali halides are low-strength materials with a rupture modulus in the 200- to 400-psi range. Similarly single-crystal alkali halides are much more susceptible to thermal fracture than hot-forged polycrystals. These facts and the greater mechanical strength of polycrystal material makes them much easier to work with.

The polishing of these difficult to work salt crystals must be done with nonaqueous polishing slurries, such as fine alumina in anhydrous ethylene glycol. All operations, such as fab, clean, and coat must be performed in either a dry box or a room in which the relative humidity can be lowered and maintained at 30% RH or less. All cleaning steps must be performed with "dry" solvents such as reagent or electronic grade. At the very least finger cots must be worn when handling these crystals. Surgical gloves are much better. Simple respirators must be worn when working in close proximity to these crystals to avoid breathing on the polished surfaces or to deposit destructive spittle marks on them when talking. Alkaline halides have been surfaced by SPDT in the fly-cutting mode. The results have not been convincing [167].

### 1.5.4.   KDP and Homologs

An important subgroup of alkali halides are the KDP group of homologs. Homologs are chemical variations of a base crystal, in this case KDP, that share the same crystal structure [168]. The KDP family of crystals is further subdivided into those grown from aqueous solutions ($H_2O$) and from deuterium ($D_2O$, heavy water) grown crystals. Each of these two groups can be divided again into phosphates and arsenates.

These crystals have been used for over 25 years as electro-optical polarization rotators to vary the intensity or modulate laser light that passes through them [169]. More recently KDP and its homologs, particularly the deuterated versions, have emerged as important frequency doubling crystals, also called *second harmonic generators* (SHG). Such a crystal is used, for instance, to convert the lasing frequency of an Nd:YAG laser at its characteristic near-IR wavelength of 1.06 $\mu$m to its second harmonic at a wavelength of 0.532 $\mu$m, which lies in the yellow region of the visible spectrum.

The transmission properties of these crystals range from 0.22 to 1.6 $\mu$m for phosphates and from 0.26 to 1.6 $\mu$m for arsenates. The transmission range of deuterated versions of these crystals is extended to 1.9 $\mu$m [170]. The refractive indices at the HeNe wavelength of 0.633 $\mu$m range from a low of 1.50 for D-KDP and RDP to a high of 1.57 for ADA [171]. These crystals are also birefringent. For instance, the indices for KDP at 0.5 $\mu$m are $n_o$ = 1.515 and $n_e$ = 1.470 [172].

Like the solution-grown alkali halides, all crystals in the KDP family are quite hygroscopic, very fragile, and sensitive to thermal shock. The solution growth process for KDP is described in Chapter 2 in some detail. The water solubility of these crystals is high with 33 g/100 g $H_2O$ for KDP and about 37 g/100 g $H_2O$ for ADP which is about the same as for sodium chloride (NaCl) [173]. Therefore the same processing, handling, and storage precautions apply that were discussed in the preceding section on alkali halides.

The KDP family of crystals also suffers from lack of mechanical strength, which causes it to chip and break easily. The thermal sensitivity of these crystals requires that any temperature change experienced by the crystals must be limited to 5°C/min [174]. This means that any drafts, sudden heating, or sudden cooling by evaporation of solvents must be avoided. Even touching a crystal at room temperature with unprotected fingers could easily result in thermal fracture.

Despite these difficult properties KDP-type crystals have been successfully grown, ground, and polished to a high surface finish and flatness ranging from slender modulator crystals with 2-mm (0.080-in.) cross sections and 50-mm (2.0-in.) lengths to SHG crystals of appreciable size. For instance, KDP frequency doubling arrays were produced for the LLNL Novette Laser which were 740 mm (29.1 in.) in diameter after assembly. These arrays were either in a 5 × 5 matrix of 150-mm (5.9-in.) crystals or a 3 × 3 matrix of 270-mm (10.6-in.) crystals [175]. They were produced in this fashion because large enough KDP crystals were not available at that time. The primary grower of alkali halides [176] managed to produce even larger monolithic KDP crystals up to 390 mm (15.35 in.) in diameter for Japan's GEKKO inertial confinement fusion laser.

### 1.5.5. Miscellaneous Optical Crystals

In addition to the more commonly used optical crystals, there are a number of crystal materials used for electro-optic, acousto-optic, frequency doubling, integrated optics, and other such specialized applications. Many of these crystals are also called nonlinear crystals because they exhibit nonlinear effects that permit them to become optically active to modulate, deflect, or otherwise modify laser light.

Electro-optical crystals are typically used as modulators of laser light. These crystals can vary (or modulate) the intensity of the light, and this allows the beam to carry a signal. Acousto-optic (A/O) devices act as either optical on–off switches for laser beams or beam deflectors to redirect the beams. Both functions can be performed with the same device, permitting the simultaneous modulation (on–off) and scanning (deflection) of the beam. Frequency doubling, also called *frequency down conversion* or *second-harmonic generation* (SHG), is an application of current interest in the laser community. Certain nonlinear crystals have the capability to convert laser light from its characteristic wavelength to its second harmonic (which is a wavelength half of the original). For instance, the second harmonic of a Nd:YAG laser is 0.532 $\mu$m, which is exactly half of the characteristic Nd:YAG wavelength of 1.064 $\mu$m. This conversion effectively turns a YAG laser that normally emits in the invisible near-IR into a visible laser in the green region of the spectrum. Certain nonlinear SHG crystals can also be used as optical parametric oscillators (OPO) that continuously vary the output of a laser source over a specific range of wavelengths by using frequency upconversion. Finally, one of the upcoming developments is the field of integrated optics which is the optical analog to integrated optical (I/O) circuits. These will be essential for the development of super-high-speed optical computers. Some of the optical crystals mentioned in this section and their applications are listed in Table 1.19.

Lithium niobate ($LiNbO_3$) is a very versatile nonlinear optical crystal that has proved useful for many applications. This CZ-grown birefringent crystal is relatively hard and chemically inert and thus takes on a good polish. This material transmits from 0.4 to 5.0 $\mu$m [177]. It is the preferred substrate material for integrated optics.

Lithium iodate ($LiIO_3$) and lithium tantalate ($LiTaO_3$) are two birefringent crystals that are used for electro-optic (E/O) applications. Lithium iodate is hygroscopic and should be processed and handled like the alkaline halides. The transparency of lithium iodate ranges from 0.3 to 4.0 $\mu$m, which makes it useful as SHG in the UV region. Lithium tantalate has similar characteristics and uses.

There has been much interest in certain efficient SHG crystals in the last few years. The most recent excitement has been over BBO or beta-barium borate ($\beta$-$BaB_2O_4$), which has proved to be a very efficient SHG crystal for YAG and excimer lasers. This material is also useful as OPO for continuous wavelength tuning from 0.4 to 2.5 $\mu$m [177]. BBO transmits from 0.2 to 3.5 $\mu$m, and its birefringent index values at 1.064 $\mu$m are $n_o = 1.657$ and $n_e = 1.539$ [178]. Most BBO commercially available is grown in the People's Republic of China by a high-temperature flux growth technique [179]. It is presently available in only very small sizes and it is very expensive. A "standard" $3 \times 3 \times 5$ mm ($0.12 \times 0.12 \times 0.20$ in.) crystal costs $4000, which roughly translates to a material cost of $1.5 million per cubic inch. At these

**Table 1.19. Typical Applications for Nonlinear Optical Crystals**

|  | E/O | A/O | SHG | OPO | IO |
|---|---|---|---|---|---|
| $LiNbO_3$ | • | • | • |  | • |
| $LiIO_3$ | • | • | • | • |  |
| $LiTaO_3$ | • |  |  |  |  |
| BBO |  |  | • | • |  |
| KTP |  |  | • |  |  |
| KTN |  |  | • |  |  |
| BSN |  |  | • |  |  |
| $KNbO_3$ |  | • |  |  |  |
| $BaTiO_3$ | • |  |  |  |  |
| $PbMoO_4$ |  | • |  |  |  |
| DAN |  |  | • |  |  |

prices it is unlikely that many opticians will have the opportunity to work this material.

Another popular SHG crystal is KTP or potassium titanyl phosphate ($KTiOPO_4$). It is preferred as a doubling crystal for YAG lasers, since it can convert the laser's fundamental output wavelength of 1.064 to 0.532 $\mu$m in the green region of the visible spectrum. KTP is grown either by the hydrothermal process in an autoclave similar to the method used to grow cultured quartz or by the solution or flux growth method, which is quite similar to the method for growing KDP. Both of these techniques are described in Chapter 2. At present crystal sizes are limited to a 10-mm (0.400-in.) cube by the autoclave method and a 13-mm (0.520-in.) cube with the flux growth method [180]. The length of the growth cycles for KTP are between five and six weeks [181]. Potassium tantalate niobate (KTN) is also a good SHG crystal, but it is no longer in vogue because of the availability of the more efficient and presumably more exciting KTP and BBO.

Barium sodium niobate ($Ba_2NaNb_5O_{15}$), or BSN, also known as "Banana" has promising SHG characteristics. But crystal growth from the melt suffers from a severe reproducibility problem that makes this crystal a poor economic choice [181]. Other optical crystals that play a role in electro-optics, acousto-optics, and second-harmonic generation are potassium niobate ($KNbO_3$), KB5 ($KB_5O_8$), barium titanate ($BaTiO_3$, also known as perovskite), lead molybdate ($PbMoO_4$), and photorefractive materials such as SBN and BSKNN. More recently organic crystals have been found useful for SHG, with identifiers such as DAN, mNA, MNA, MAP, POM, and COANP. These crystals are for the most part still in the R&D stage [182]. These numerous crystals and their often confusing identifiers are sometimes referred to as the "alphabet soup" of optical crystals.

The list of optical crystals is constantly expanding as new developments in crystal applications lead to increasing demands on the crystal grower. The purpose of the brief overview was to provide some idea of the large selection of optical crystals available for the optician to fashion into a variety of shapes and finishes. In most cases little or nothing is known about the properties of these materials, and the optician is usually left to his or her own experience and ingenuity to come up with a solution.

### References for Section 1.5 (Optical Crystals)

[139] Malitson, I. H., "A redetermination of some optical properties of calcium fluoride," *Applied Optics,* vol. 2, no. 11, November 1963.

[140] "Spectral curves of IR materials," *1981 Laser Focus Buyers Guide.*

[141] Hargreaves, W. A., "Magnesium fluoride: Update and summary of optical properties," *Laser Focus,* September 1982.

[142] "Crystals, infrared," *1983 The Optical Industry and Systems Purchasing Directory,* p. E-17.

[143] Truett, W. L., "IR optics revisited," *Photonics Spectra,* April 1984.

[144] Hunter, W. R., "Optics in the vacuum ultraviolet," *Electro-Optic Systems Design*, November 1973.

[145] Truett, W. L., "Optical materials—The role of impurities," *Lasers and Applications*, August 1984.

[146] Miles, P., "High transparency infrared materials—A technology update," *Optical Engineering*, vol. 15, no. 5, September/October 1976.

[147] Drexhage, M. G., "IR transmitting fluoride glasses," *Laser Focus*, October 1980.

[148] Newman, B. E., "Optical materials for high-power lasers: Recent achievements," *Laser Focus*, February 1984.

[149] "Ruby laser rods," *Union Carbide Crystal Products*, 1969.

[150] Glass, A. J., "The laser's impact on crystal technology," *Optical Spectra*, October 1972.

[151] Smakula, A., "Optical materials and their preparation," *Applied Optics*, vol. 3, no. 3, March 1984.

[152] Belt, R. F., "YAG: A versatile host," *Laser Focus*, April 1970.

[153] Staff report: "One-year delivery time for YAG rods is slowing shipments by laser makers," *Laser Focus*, January 1980.

[154] Staff report: "YAP output promising in rangefinder test," *Laser Focus/Electro-Optics*, September 1983.

[155] Staff report: "Emerald laser may find applications in machining," *Lasers and Applications*, December 1982.

[156] Buchert, J., "Emerald—A new gem laser material," *Laser Focus/Electro-Optics*, September 1983.

[157] Petricevik, V., "A new tuneable solid-state laser," *Photonics Spectra*, March 1988.

[158] Staff report: "Allied formally introduces alexandrite, pointing out wide range of applications," *Laser Focus*, April 1981.

[159] Detrio, J. A., "Trends in optical materials," *Technical Digest*, OSA Workshop on Optical Fabrication and Testing, December 1982.

[160] Jamieson, T. H., "Ultrawide waveband optics," *Optical Engineering*, vol. 23, no. 2, March/April 1984.

[161] Staff report: "Excimer specialists warn of straining ability to build alkali halide windows," *Laser Focus*, October 1976.

[162] Staff report: "KCl optics handles 20-kW industrial $CO_2$ laser output," *Laser Focus/Electro-Optics*, January 1985.

[163] Reichelt, W. H., "Metal optics in $CO_2$ laser fusion systems," *Optical Engineering*, vol. 16, no. 4, July/August 1977.

[164] Newman, B. E., "Damage resistance of $CO_2$ fusion laser optics," *Optical Engineering*, vol. 18, no. 6, November/December 1979.

[165] Smakula, A., "Optical materials and their preparation," *Applied Optics*, vol. 3, no. 3, March 1964.

[166] Miles, P. A., "Crystalline infrared transmitting materials," *OSA Optical Shop Notebook II*. Section XVI: *Infrared Optics*, OF&T, 1977.

[167] Saito T., "Diamond turning of optics," *Optical Engineering*, vol. 15, no. 5, September/October 1976.

[168] Ruderman, W., "Crystals: Non-centric for Lasers," *Optical Spectra*, February 1976.

[169] Tebo, A., "Manufacturers and suppliers of electro-optic modulators and Q-switches," *Laser Focus/Electro-Optics*, May 1988.

[170] Adhav, R. S., "Frequency doubling crystals—Unscrambling the acronyms," *Electro-Optical Systems Design*, December 1974.

[171] Adhav, R. S., "Materials for optical harmonic generation," *Laser Focus*, June 1983.

[172] Dennis, J. H., "Index of refraction of KDP," *Applied Optics*, vol. 2, no. 12, December 1963.

[173] "Spectral curves of IR materials," *1981 Laser Focus Buyers Guide.*

[174] Adhav, R. S., "Guide to efficient doubling," *Laser Focus*, May 1974.

[175] Staff report: "CLEO '83 Preview—Fusion," *Lasers and Applications*, April 1983.

[176] Cleveland Crystals, Inc., Cleveland, OH.

[177] Castelli, L., "Lithium niobate applications in optics and acousto-optics," *Laser Focus/Electro-Optics*, December 1985.

[178] Adhav, R. S., "BBO's nonlinear optical phase-matching properties," *Laser Focus/Electro-Optics*, September 1987.

[179] Tebo A., "Scientists develop useful optical materials," *Laser Focus/Electro-Optics*, August 1988.

[180] Tebo A., "KTP carves growing frequency-doubling niche," *Laser Focus/Electro-Optics*, September 1988.

[181] Belt, R. F., "KTP as a Harmonic Generator for Nd:YAG Lasers," *Laser Focus/Electro-Optics*, October 1985.

[182] Lin, J. T., "Choosing a non-linear crystal," *Lasers and Optronics*, November 1987.

## 1.6. MISCELLANEOUS OPTICAL MATERIALS

Most precision optical components are made from the materials that were described in the preceding sections. There are, however, a number of materials that have been successfully used for optical purposes that are usually not thought of in that context. Certain plastics, metals, and ceramics have found use as optical materials. Injection-molded clear plastic lenses have been produced in very large quantities for the opthalmic industry and for the low-end, high-volume consumer market. Reflective optics made from metals have long played a significant role in space exploration, surveillance, and high-energy lasers. New ceramic materials have also found use as a high-quality mirror material for hostile environments.

### 1.6.1. Optical Plastics

Plastic materials play a major role in high-volume, low-cost optics for non-critical applications. Although the optician will rarely be called upon to polish optical plastics, it is still a good idea to have some fundamental understanding of this branch of optics. Most optical plastics are injection molded into finished components, while larger parts are usually cast, then shaped, and optically polished. These methods are described in Chapter 2.

There are two basic types of optical plastics: thermoplastics and thermosets. Thermoplastics can be reheated and reshaped several times without appreciably affecting the quality of the material [183]. This property makes thermoplastics an ideal choice for injection molding. Thermosets, on the other hand, cannot be altered once set because during polymerization the polymers cross-link three-dimensionally, and this makes remolding impossible [183]. This feature makes thermosets the usual choice for casting.

Injection molding is potentially a very cost-effective way to produce commercial quality optics, as long as very high volumes of the same type of optic is required. The rule of thumb is that volumes must exceed 10,000 pieces of the same optic before the high cost of the necessary tooling and equipment can be amortized [184]. Injection-molded optics have a number of advantages besides cost over conventional glass optics of equivalent size and configuration. Among these are light weight, potentially good surface quality, better impact resistance, and the ability to accept complex shapes such as aspheric surfaces and integral mounting. On the down side, optical plastics are not stable under thermal conditions, and they are a lot less abrasion resistant than even the softest optical glass. An additional shortcoming is that all plastics shrink in volume as they cure. The typical 0.5% shrinkage for injection-molded optics prevents the attainment and maintenance of an accurate optical figure. It is therefore unlikely that there will be diffraction-limited plastic optics [184].

Shrinkage is even more of a problem for thermosets. Shrinkage factors of 13% to 14% have been reported [183, 185]. This much shrinkage makes optical figure control a practical impossibility, except for the most forgiving requirements. One such optically noncritically application is the casting of opthalmic lenses. Most eyeglass wearers use lenses produced this way. Casting of plastic optics is reserved for very large parts for which injection molding would not be practical or for low quantities of standard shapes and sizes for which injection molding would not be economically feasible. Massive optical components of up to several feet in diameter have been successfully cast. They had to be finished after casting and curing by using conventional machine shop and optical fabrication practices.

There are three primary thermoplastics that make up the bulk of injection-molded optics. These are acrylic polymers, polystyrenes, and polycarbonates. In addition to these there are several other plastic materials that were developed for special applications.

Acrylic polymers were the first plastic used for optical applications, and they have remained the most popular to this day. These materials are also known as acrylics or polymethyl methacrylates (PMMA), but they are even better known by their tradenames such as Plexiglass and Lucite. Acrylics are easily molded. They have a refractive index of 1.49 and excellent optical transmission in the visible region. Some special grades transmit well into the near UV part of the spectrum. Acrylics are, however, quite soft and not very scratch resistant [186].

Polystyrene is the lowest-cost optical plastic, and it is even easier to mold at high-production rates than acrylics. As a result it is used extensively for very high-volume and low-cost optics. Because of its relatively high refractive index of 1.59 and high dispersion, polystyrene lenses have been combined with acrylic lenses to form air-spaced achromatized doublets [186]. Although polystyrene is more scratch resistant than acrylics, it has relatively poor impact resistance. It can be easily shattered.

Polycarbonates distinguish themselves by a comparatively high heat resistance and excellent impact resistance. They are essentially unbreakable. But these useful properties make them difficult to mold. They are, however, easily scratched, and optical components molded from these materials may require a harder overcoat to protect the optical surfaces. Lexan is the tradename of a well-known polycarbonate [187].

Other optical plastics are TPX, SAN, and NAS. These are used for special applications. TPX (polymethylpentene) has fairly good near-IR transmission and good thermal stability and chemical resistance. It has, however, a tendency to degrade when exposed to UV radiation [184]. SAN (styrene acrylnitrile polymer) is characterized by a low coefficient of thermal expansion, but the surfaces are soft and vulnerable. NAS (methyl methacrylate styrene copolymer) has a refractive index of 1.56 and high transmittance. But like SAN, the material is soft and easily scratched. Another plastic that has found use in optics is polyphenylene oxide. It is an opaque plastic that is useful only for reflecting surfaces. Its primary characteristics are its good chemical resistance and its excellent strength, even at elevated temperatures [184].

The thermosets are derived from a number of clear epoxies or resins. One in particular is formulated specifically for optical applications. It is ADC (allyl diglycol carbonate), also known as CR-39, which is used extensively for opthalmic (eyeglass) lenses.

The refractive index of optical plastics ranges from about 1.45 to 1.60. Although the index homogeneity of optical plastics approaches that of standard optical glass, the change in index as function of temperature ($dn/dt$) is about 50 times greater than that of optical glass. This makes the optical performance of optical plastics much less stable in thermally fluctuating environments [183]. Optical plastics have only limited usefulness in the IR regions because they tend to exhibit very complex and deep absorption spectra [188].

Most plastics can be AR coated and receive reflective coatings. However, the adhesion of the coatings is not as good as it is on glass substrate. Hard antiabrasion coatings can also be applied [184].

There are currently efforts underway to extend the range and properties of optical plastics by combining organic polymers such as acrylics, polystyrene, and polycarbonates with inorganic network formers such as $SiO_2$, $TiO_2$, $ZrO_2$, $Al_2O_3$, and $B_2O_3$ [189]. These efforts are still in the R&D stage, but if successful, they could create a new series of optical plastics.

### 1.6.2. Metal Optics

Since metals and most ceramics are not transparent at any optical wavelength, their use for optical purposes are limited to reflective optics only. Aluminum and beryllium have been used as metal mirror substrate for some time. A new development is silicon carbide reinforced aluminum which improves on some of the less useful properties of aluminum without sacrificing those that make aluminum a preferred choice.Copper, molybdenum, and lately also silicon carbide make excellent mirrors for $CO_2$ and other high-energy lasers. Titanium and stainless steel are alternate mirror substrate materials, but their use has been limited. More recently new mirror materials such as germanium cordierite and other composite materials have been developed. Even liquid mercury has been proposed as a mirror material. These materials will be discussed in the last two sections.

Metals are used for reflective optics in optical systems that are exposed to high thermal loads. Therefore high thermal conductivity and high-heat capacity are essential properties. Thermal conductivity is a measure of how well a material dissipates heat and heat capacity defines the material's ability to accept and retain heat. Metals excel in both of these properties relative to other mirror materials. In addition the coefficient of thermal expansion should be as low as possible to limit deformations of the optical surface resulting from thermally induced stress. Metals do not perform as well as low-expansion glass ceramics.

Other required properties include low density or weight per unit volume and a high stiffness-to-weight ratio. These properties are particularly important for space applications where light weight is essential. A high stiffness-to-weight ratio permits extensive lightweighting of the mirror without sacrificing stiffness which is essential for maintaining the optical quality of the reflecting surface under mechanical loads. Metals have surface and bulk stress characteristics that differ considerably from the more traditional optical materials. Surface stresses result from the machining, grinding, and polishing operations. These can be locked in and cause deformation of the optical figure, especially in thermally fluctuating environments. Thermal cycling prior to finishing the mirrors can relieve much of the surface stress. Bulk stresses that remain in the material from the casting, forging, or hot-pressing operations can result in surface deformation by a long-term process of stress relief called *microcreep*. This condition is best avoided by choosing materials with high microcreep strength or by starting with stress-free materials [190, 191, 192].

The surface finish on conventionally polished metal mirrors is not nearly as good as that attainable on glass. This condition results from the fact that metals have a polycrystalline structure which does not polish as uniformly as amorphous materials such as glass. In addition impurities in the metals tend to aggregate in the grain boundaries between the polycrystals which further complicates the polishing process. But by applying a dense plating to the prepolished mirror, optical finishes equaling those on well-polished glass optics can be obtained on the plating [190].

Aluminum (Al) has been used as mirror material for many years. Since it is a fairly soft metal, it is difficult to obtain a good optical finish within economically acceptable polishing cycles. As a result aluminum is almost always overcoated with an electroless nickel (Kanigen) plating which accepts a better finish in much less time. Bare aluminum has been successfully diamond turned to a high degree of optical finish but larger mirrors are first Kanigen plated before diamond turning.

There are many aluminum alloys, and it is important to select the best ones for optical use. Up to now, the primary choice has been Al 6061-T6 which can be readily Kanigen plated and is also easily diamond turned [193]. Other wrought or forged aluminum alloys have also been used. Although aluminum is the lowest cost metal used for optical purposes, it has lower tensile strength and poorer long-term stability than other metal mirror substrate materials. This requires a stress relief treatment prior to Kanigen plating, diamond turning, or optical finishing. A treatment of thermal cycling and chemical etching has proved to be very effective [192].

Casting methods have now been perfected to produce uniformly finegrained aluminum alloys with greatly reduced impurity levels. The mirror blanks can be cast to near-net shapes which greatly reduces the cost of machining operations [194].

To overcome some of the stability limitations of aluminum while retaining its more useful properties, silicon carbide reinforced aluminum (SiC : Al) has been developed in recent years [195]. These alloys are also known as metal matrix composites (MMC). Aluminum alloys are reinforced with very fine SiC particles that increase the stiffness and strength but maintain the normal density of the alloy [196]. This greatly increases the stiffness-to-weight ratio which can make SiC : Al an economic alternative for lightweight optics. By changing the SiC concentration between 10% to 50% by volume, Young's modulus can be adjusted to suit the requirements, and the CTE can be closely matched to that of the electroless nickel plating [196, 197]. Unlike beryllium, SiC : Al is nontoxic and can be easily machined without special precautions [197]. The material is produced by a vacuum hot-pressing method.

Based on the impressive volume of papers written on beryllium, one must conclude that the majority of metal optics is produced from beryllium. This, however, is not the case. But there is no doubt that the development of beryllium (Be) as an optical mirror substrate has been very important in the development of metal optics. The main reason for the interest in beryllium lies in its low density, and thus low weight, and in its inherent stiffness which permits additional lightweighting. Be mirrors also perform well at very low temperatures such as those found in space. These considerations are of great importance for optical systems used on space missions.

Beryllium has a superior stiffness-to-weight ratio over other mirror materials. A Be mirror of equal stiffness has only 20% of the weight of a fused quartz mirror [198]. The superior microyield and microcreep strength of Be reduces deformation of the mirror under load. The *microyield strength* is a measure of the stress required to permanently deform the mirror, while the *microcreep strength* which is about half of the microyield strength is a measure of the long-term dimensional stability. Microyield strength tends to increase with increasing BeO content and finer grain sizes [199].

Beryllium has a relatively high coefficient of thermal expansion (CTE) at room temperature which is also anisotropic because of its hexagonal structure. But the CTE at the typically very low operating temperature is quite low. It has also high thermal conductivity which helps dissipate irregular heat loads [198]. Because of this thermal behavior, however, Be can exhibit thermal instability that can affect its optical performance [200].

Beryllium suffers from certain drawbacks that prevent its use on a broader scale. The most important of these is toxicity of Be dust, which is generated during the shaping of the mirrors. Long-term exposure to airborne Be dust can lead to beryllicosis. This is a chronic lung disease similar to the black lung disease of the coal miners or the white lung disease, or silicosis, of people exposed to silicate dust. But since it is also a potent carcinogen, Be can be machined and ground only under strict OSHA controlled working conditions, which include ample ventilation, air filtration of the work environment, and TVA limits on the exposure that a worker can be subjected to during a regular workday [201]. These requirements add substantially to the already appreciable cost of the material and the cost of machining lightweighted mirrors from a solid blank. In addition beryllium is not mined in the United States, and there is only one suitable source for this important material.

Most early Be mirrors were machined extensively from solid vacuum hot-pressed (VHP) billets to achieve the required level of lightweighting. But VHP Be typically contained material inhomogeneities, which created problems during the machining operations and caused instabilities in the finished mirror [202]. Some of these problems could be overcome by chemically stress relieving the machined mirror blank. In recent years there have been some advancements in powder metallurgy that have led to a successful process that produces Be mirror blanks to near-net shapes [202]. This process is called hot isostatic pressing (HIP) which is very similar to the hot-pressing methods used to produce certain IR materials. The HIP process can produce mirror blanks that are pressed to near-net shapes and use only 10%

of Be as compared to machining from the solid [200]. This approach reduces the impact of some of the less favorable aspects of Be, most notably minimum machining, and thus much reduces dust generation and cost. The toxicity problem and the constraints it imposes remains, however, because the starting material is a fine Be powder that must be handled with the utmost care.

Bare Be can be polished, but the success of the effort depends on the BeO concentration and on the porosity of the hot-pressed material [201]. The BeO concentration should be as low as possible, but a high microyield strength requires a relatively high BeO content. Obviously a trade-off is needed here. The density of the hot-pressed material should be as high as practical to limit its porosity. Even under the best conditions, obtaining a good polish and control of the optical figure at the same time is no easy task. Typically the reactive nature of Be makes it difficult to achieve a high optical finish. This is why Be optics are typically plated with a layer of a denser material that will accept a better finish.

The surface treatment is in most cases a low-stress electroless nickel plating. Such a plating not only prepares the surface for an optical polish, but it also protects the mirror substrate from the corrosive effects of humidity in the presence of chlorine ions [199]. In addition it provides a barrier that limits the toxicity problem during optical fabrication. For very high optical surface requirements, such as low surface scatter, which cannot be achieved on bare Be or the nickel-plated substrate, a thin but dense layer of amorphous Be is sputtered on prepolished surfaces. The best surfaces, measured at 10 Å rms surface roughness, have been produced on Be mirrors with this method [202, 203].

Electroless nickel plating has been applied to the Al and Be mirror substrate to protect it from environmental degradation and to prepare the surface so that it can accept a high polish. This plating, also known as Kanigen plating, is a hard, dense amorphous layer of nickel, between 25 and 250 $\mu$m thick, which can be easily ground and polished with standard optical fabrication methods. Kanigen plating on Al and Be has also been successfully diamond turned. A good optical polish can only be achieved if the phosphor content of the amorphous nickel plating lies between 7% and 10% [204]. Other platings have also been applied, and only recently it was reported that a beryllium layer deposited on a beryllium substrate resulted in even better optical finishes than those typically achieved on Kanigen plated surfaces [203].

Another important optical application of metals is for use as $CO_2$ laser mirrors. The primary materials used for this are copper (Cu), molybdenum (Mo), and silicon carbide (SiC). The latter will be discussed in the next section.

Copper has been used since the early 1970s for $CO_2$ laser mirrors. Initially copper mirrors were shaped using standard machine shop practice, after which they were ground and polished using standard optical fabrication methods adapted to metal polishing. The process was quite slow and unpredictable, and it required the expertise of highly skilled opticians to produce good surfaces. Improvements in the selection of very pure (99.99%) copper alloys made it a little bit easier to polish copper, although it was still a difficult task because copper is quite soft [205]. By the mid-1970s single-point diamond turning (SPDT) had been perfected to the point where polishing of copper mirrors using SPDT was very successful. Most copper mirrors for $CO_2$ lasers are now diamond turned and then postpolished to remove the characteristic grooves from the SPDT process and to improve on surface figure.

Oxygen-free high-conductivity copper (OFHC Cu) has been found to be the best for SPDT and subsequent postpolishing. Copper is a material that is well suited for use in high-energy lasers (HEL) such as $CO_2$ lasers. It has a high natural reflectivity and low absorption at 10.6 $\mu$m, excellent thermal conductivity, and, because of its cubic structure, no anisotropic CTE. It is not easily polished, however, and the mirror should be subjected to a stress relief heat treatment prior to SPDT. To simplify the polishing of copper mirrors prior to diamond turning, fine-grain copper is plated on the previously prepared Cu mirror substrate.

Molybdenum (Mo) has been used for HEL and particularly $CO_2$ laser mirrors since the early 1970s. Although Mo mirrors are much harder than Cu mirrors, it is still a slow process to get a good polish on them but easier to control the process than for copper. Diamond turning of Mo has not been successful. Mo has high thermal conductivity. It is also much harder than Cu. Reflectivity of bare polished Mo is high, but absorptivity is twice that of Cu. Therefore silver coatings are required for very high power levels [206].

Some alternate metals such as stainless steel, titanium, and brass, have been considered from time to time but have not found widespread applications mainly because they do not offer compelling advantages over other materials. Stainless steel (St.St.) is quite heavy, and its low tensile strength relative to its density makes it difficult to stabilize for optical purposes. Traditional, meaning slow and costly, polishing methods must be used on St.St. since diamond turning of this material has not been sufficiently successful. However, stainless steel, in particular 304 St.St., takes on a very high polish. Titanium (Ti) has good mechanical stability, but its anisotropic CTE makes figure control in thermally fluctuating environments difficult. Also diamond turning of Ti has not been successful. Brass has never developed into a useful mirror material.

One unusual application of metal mirror optics has been reported recently. An old idea was resurrected to produce a large metal mirror using a pool of liquid mercury. As the pool is rotated, the surface of the mercury assumes a parabolic shape. The $f$-number of such a mirror can be easily changed by changing the speed of rotation [207]. But as it is with everything that seems to be too good to be true, this approach suffers from serious limitations. The mirror cannot be pointed, which means that it can only look straight up. There is also considerable concern about the health hazard of the toxic mercury vapors [208]. The high cost of mercury and its considerable weight are additional serious drawbacks.

### 1.6.3. Ceramic Materials

Silicon carbide (SiC) is a fairly new $CO_2$ mirror material. Although it had been in use for many years as a ceramic material and a hard abrasive, large solid blanks with a surface layer of sufficient density for mirrors had to wait for the CVD process. As SiC does not melt at temperatures and pressures that can be economically achieved on earth, it cannot be grown from the melt like so many other optical materials [209]. It is a ceramic and as such, it needs to be fired to attain its ultimate hardness. Like many other ceramics, SiC can be readily machined into even complicated shapes before it is fired. This makes high-energy laser (HEL) mirrors with integral water cooling a possibility [210, 211]. SiC mirrors are now produced in appreciable sizes by the CVD process.

The inherent porosity of SiC, which is typical for ceramic materials, has made it difficult to obtain a sufficiently good optical polish for laser applications. One solution attempted to deposit a thin layer of polycrystal silicon on a previously shaped SiC blank. More recently a high-density layer of SiC was deposited onto a CVD SiC substrate.

SiC is a mirror material of much interest for high-energy laser (HEL) applications because it has excellent thermal conductivity (about 50% of Cu), a low CTE (nearly $\frac{1}{6}$ that of Cu), low-density (light weight), and high mechanical rigidity (maintains good optical figure under thermal and mechanical loads) [212]. In addition SiC is also chemically inert and can survive hostile chemical environments that other materials cannot survive.

SiC is most of all a very hard material. It is much harder than any other material used for $CO_2$ laser mirrors. This means that the optical surface is very tough and not easily damaged. But this also means that the grinding and polishing process is exceedingly slow, and as a result very costly. A thin and dense layer of chemically vapor deposited SiC on a SiC substrate has a sufficiently fine and pore-free structure so that 1 Å rms surface finishes have been reported [213]. The material removal process during grinding and polishing is strictly by mechanical abrasion which is a very slow process on such a hard material. It has been reported that removal rates are 29 times slower than on fused silica and 58 times slower than on BK 7 glass. The fine grind, however, was measured to be four times finer relative to fused silica and five times finer as compared to BK 7 [214].

A new mirror material has been developed in recent years that combines a relatively high thermal conductivity with a very low CTE. This material is germanium cordierite ceramic ($CG_4$) which is produced by hot isostatic pressing (HIP) of a powdery mixture of MgO, $Al_2O_3$, $SiO_2$, and $GeO_2$. Its thermal conductivity is twice that of ULE with a near-zero CTE that is characteristic of glass ceramics. The CTE can be adjusted by varying the proportions of $GeO_2$ and $SiO_2$. It also has superior flexural strength as compared to the traditional glassy mirror materials.

The HIP process can produce lightweighted blanks to near-net shape. This minimizes the time and cost to remove excess material. The polishability of this material still needs some improvement, since hot-pressed materials retain a residual porosity of typically $1$-$\mu$m pores. A high degree of polish is difficult to obtain [215, 216].

Since the mid-1980s several development efforts are under way to develop lightweight but high-stiffness mirror materials for space applications. The more important of those under consideration are graphite-fiber-reinforced epoxy, glass matrix composites, and solid foam ceramics. While some of these new materials are designed to serve only as structural members, others are designed to serve as mirror substrate. New methods of fabrication are required to work these novel materials once they emerge from the R&D labs. Mention is made of these materials to underscore the fact that the development of materials for optical applications is quite dynamic and continuous and that the successful optician and manufacturing engineer must be able to adapt to new and challenging requirements.

### References for Section 1.6 (Miscellaneous Optical Materials)

[183] Wolpert, H. D., "A close look at optical plastics," *Photonics Spectra,* February 1983.

[184] Byrd, R. E., "Molded-plastic optics are coming," *Electro-Optical Systems Design,* June 1972.

[185] Grendol, C. L., "Molded plastic optical elements come into their own," *Laser Focus/Electro-Optics,* October 1988.

[186] Bennett, R. C., "Design patterns in plastic optics," *Optical Spectra,* August 1978.

[187] Staff report: "Materials worth watching—Plastics stand up to be counted," *Optical Spectra,* July 1976.

[188] Truett, W. L., "Optical materials—The role of impurities," *Lasers and Applications,* August 1984.

[189] Schmidt, H., "Inorganic-organic polymers as materials for optical applications," *Technical Digest,* Series vol. 19, OSA Workshop on Optical Fabrication and Testing, OF&T 1987.

[190] Barnes, W. P., "Basic properties of metal optics," *Optical Engineering,* vol. 16, no. 4, July/August 1977.

[191] Detrio, J. A., "Trends in optical materials," *Technical Digest,* OSA Workshop on Optical Fabrication and Testing, Palo Alto CA, OF&T 1982.

[192] Pacquin, R. A., "Processing metal mirrors for dimensional stability," *Technical Digest,* Workshop on Optical Fabrication and Testing, Anaheim CA, OF&T 1981.

[193] Reichelt, W. H., "Metal optics in $CO_2$ laser fusion systems," *Optical Engineering,* vol. 16, no. 4, July/August 1977.

[194] Staff report: "Mirror castings in near-net shapes," *Photonics Spectra*, April 1988.

[195] Lipeless, J., "Design of ultralightweight mirrors," *Laser Focus/Electro-Optics*, March 1984.

[196] Mohn, W. R., "Recent applications of metal matrix composites in precision instruments and optical systems," *Optical Engineering*, vol. 27, no. 2, February 1988.

[197] Mohn, W. R., ". . . Made of aluminum alloys . . . ," *Photonics Spectra*, January 1986.

[198] Pacquin, R. A., "Hot isostatic pressed beryllium for large optics," *Optical Engineering*, vol. 25, no. 9, September 1986.

[199] Fullerton, R. C., "Using beryllium instead, Part II," *Optical Spectra*, August 1979.

[200] Levenstein, H., "Industry develops new IR technique," *Laser Focus/Electro-Optics*, July 1988.

[201] Fullerton, R. C., "Using beryllium instead, Part I," *Optical Spectra*, July 1979.

[202] Pacquin, R. A., "Advanced lightweight beryllium optics," *Optical Engineering*, vol. 23, no. 2, March/April 1984.

[203] Moreen, H. A., "Low scatter large beryllium mirror development," *Technical Digest*, Series vol. 19, OSA Workshop on Optical Fabrication and Testing, Rochester NY, OF&T 1987.

[204] Arnold, J. B., "Machineability studies of infrared window materials and metals," *Optical Engineering*, vol. 16, no. 4, July/August 1977.

[205] Reichelt, W. H., "Metal optics in $CO_2$ laser fusion systems," *Optical Engineering*, vol. 16, no. 4, July/August 1977.

[206] Hoffman, R. A., "Ion polishing of metal surfaces," *Optical Engineering*, vol. 16, no. 4, July/August 1977.

[207] Anderson, P. H., "Will future astronomers observe with liquid mirrors?" *Physics Today*, June 1987.

[208] Dickson, E. M., "Reflecting on liquid mercury mirrors," *Physics Today*, April 1988.

[209] Nieberding, W. C., "Researchers develop long-sought SiC crystal growth technique," *Industrial Research and Development*, September 1983.

[210] Sherman, G., "Optics for industrial $CO_2$ lasers," *Laser and Optronics*, September 1988.

[211] Sherman, G. H., "Latest optics for industrial $CO_2$ lasers," International Conference on Laser Material Processing—Science and Applications, Osaka, Japan, May 1987.

[212] Holmes, L., "Optics and Applied Technology Laboratory makes success of light work," *Laser Focus/Electro-Optics*, May 1984.

[213] Fuchs, B. A., "Polishing CVD silicon carbide," *Technical Digest*, OSA Workshop on Optical Fabrication and Testing, Seattle, WA, OF&T 1986.

[214] Fuchs, B. A., "Removal rates, polishing and sub-surface damage of chemical deposited silicon carbide," *Technical Digest*, Series vol. 19, OSA Workshop on Optical Fabrication and Testing, Rochester NY, OF&T 1987.

[215] Mehrotra, J., "New materials for precision optics," *Optical Engineering,* vol. 25, no. 4, April 1986.

[216] Mehrotra, J., "Germanium–cordierite ceramics for precision mirrors," *Technical Digest,* Series vol. 19, OSA Workshop on Optical Fabrication and Testing, Rochester NY, OF&T 1987.

# 2

# Material Production and Forms of Supply

It is often desirable for the optician to know how optical material is produced. Not only does this information satisfy the natural desire to know but, more important, it helps the optician to make better decisions about how to handle and process these materials. This knowledge can also explain the inherent material and size constraints that limit their use.

Of these processes, the making of glass has a long history. Some traditional methods of glassmaking are still being used today, but most optical glasses are made with modern day tank methods. High-quality optical glass is produced in large volume in a continuous tank method. Technical glass, much of which has found applications in optics, is either drawn sheet glass or cast float glass. The pressing of glass has been perfected to the point where large quantities of finish-pressed lenses are produced for consumer applications.

The greatly expanded role of infrared (IR) optics necessitates a close look how these materials are produced. The often significant cost of the raw materials is further increased by slow and energy-dependent growth processes that in many cases do not lend themselves to economic volume production. Vertical or horizontal crystal growth methods must often be used to produce these materials. Chemical vapor deposition during which a metal vapor reacts with a gas to form a solid material is used to produce several important IR materials. Some IR materials can also be cast but at the cost of material quality. That is why the casting method is reserved for large sizes that cannot be produced by other methods.

Fused silica, fused quartz, and a number of related low-expansion materials have found widespread use in optics. They are produced by a variety of methods including melting, casting, flame hydrolysis, and hydrothermal processes. Other optical crystals are grown by a variety of methods ranging from vertical crystal pulling to fusion casting. Water soluble crystals are grown from a saturated aqueous solution. Other crystalline materials are pressed into a solid from a powder under heat and pressure. Plastic optical components are produced in large quantities by either thermal compression molding or by injection molding. These and several other material production methods will be discussed in this chapter.

## 2.1.  OPTICAL GLASSMAKING

There are two basic ways to make glass and many variations on these two
methods. The first method is the more traditional method that has been in
use for a long time in its basic form. It is called the *single-batch process* or
the *day tank method*. It is used for small lot production. The second method
has evolved over the last 40 years into the current continuous tank method
for high volume manufacture of optical glass. The furnace used for either
method can be heated by oil or gas flame, resistivity heaters, or high-fre-
quency (inductive) heat sources.

### 2.1.1.  Day Tank Method

The single-batch process uses a relatively small melting pot made from clay,
fused silica, or platinum. Because such pots can be charged and melted in
one day, they are also called day tanks. The finished glass is typically cast
directly into a form from these tanks. The volume produced by the day tank
method can range from less than 50 kg (100 lb) of a special glass melted in a
platinum-lined vessel to more than 500 kg (1000 lb) melted in a traditional
clay pot. Figure 2.1 shows such a setup.

In preparation for melting glass with the day tank method, the chemically
pure ingredients are finely ground and thoroughly mixed. It is often necessary to reprocess the ground-up materials to remove traces of the grinder
mechanism. The mix is then melted in fireproof clay crucibles at 1300°C to
1500°C (2350°F to 2700°F). Fused silica or platinum crucibles are used for
glasses with constituents that are especially difficult to melt or that are
chemically aggressive.

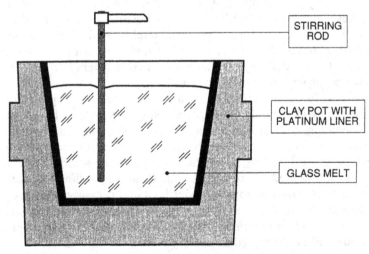

**Figure 2.1.**  Typical day tank.

Alkali serves as a flux medium during the melt. It makes it possible for the glass mix to melt before the melting temperature of the quartz sand has been reached. The addition of calcium makes the glass stable in the presence of water.  Alkali additives such as potash ($K_2CO_3$) and calcium carbonate ($CaCO_3$) disintegrate at the high melt temperatures, liberating carbon dioxide gas ($CO_2$) in the process. This causes the molten glass to develop bubbles which then must be driven off through a purification process. To facilitate the removal of the bubbles, certain substances are introduced to the bottom of the melt that tend to form large bubbles. These large bubbles rise rapidly and carry many of the smaller bubbles along to the top.

The molten glass is mechanically stirred to make it homogeneous. During the stirring, as the melt cools, it becomes more viscous. At that point, the glass is poured into rectangular iron forms and it will slowly cool at a preset rate. The form is placed into an oven in which the temperature is reduced according to an accurately controlled time plan. This critical step in the process is called *annealing*. The temperature region in which the glass passes from the liquid to the solid state is cooled at a particularly slow rate in order to greatly reduce any internal stresses.

### 2.1.2.  Continuous Tank Method

The continuous tank method has been in use since the mid-1950s for large volume production of optical glass. Figure 2.2 shows such an arrangement schematically [1]. In this method the premixed batch material is continu-

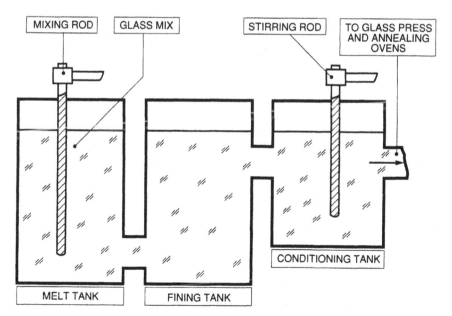

**Figure 2.2.**  Schematic of continuous tank method.

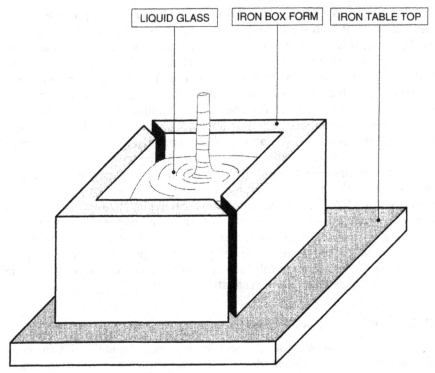

| LIQUID GLASS | IRON BOX FORM | IRON TABLE TOP |

**Figure 2.3.**   Casting of glass blocks.

ously fed into a tank. There the ground-up ingredients melt and are properly mixed as they slowly flow through the furnace. Bubbles are removed in the fining tank and striae are eliminated in the conditioning tank. When the process is complete, the molten glass can be fed into a casting form, but it is more typical to deliver it directly to a molding press.

In the case of cast blocks, the continuous tank is located at an upper level of the glass plant. The still molten but somewhat viscous glass flows from there as a stream the size of a finger into a form that rests on an iron table below the furnace (Fig. 2.3).The form consists of two iron angle pieces that merely interlock, creating a cubic form of approximately 300-mm (12-in.) sides. Because the arriving glass is already viscous, it immediately hardens upon contacting the iron form and does not run out despite the fact that the form is by no means tight. The resulting cubic glass blocks are fine-annealed and are then ground and polished on two opposite sides for inspection of internal flaws.

## 2.1.3.   Slab Glass and Block Glass

The glass plates from the day tank casting are approximately 1000 mm (40-in.) by 600 mm (24-in.) and about 250 mm (10-in.) thick. They are first broken

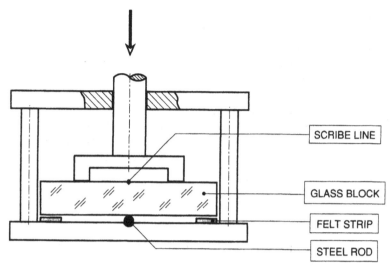

**Figure 2.4.** Glass press.

into manageable pieces. To that end, the slab is scored on one side at the location of the intended break. The glass plate is then broken into two pieces under a press (Fig. 2.4). The resulting break is clear and transparent. The outer surfaces of the pieces are ground and polished so that they are also transparent. The glass is then inspected for internal flaws through these transparent surfaces.

The cubic blocks from the continuous melt tank are too thick to be broken this way. They must be cut into smaller pieces with a large diamond saw. The method is similar to that used by stonemasons: Very large glass blocks are parted with a steel wire that is drawn through a coarse abrasive slurry.

### 2.1.4. Limitations of Either Method

Only a select number of glasses can be produced by the continuous melting process, and not all glasses can be directly pressed. Sensitive glasses must be melted in small batches in day tanks then cast and repressed later. The cast or pressed glass must be carefully annealed to ensure that the desired optical properties will not be lost and to reduce internal stress. The rule of thumb is that the annealing time, and therefore the cost, increases as the square of the glass thickness.

There is one significant difference between the day tank and the continuous tank methods. The molten glass in the day tank is continually stirred to properly blend all glass ingredients together. This blending action ensures glass of uniform composition throughout the entire melt in the tank. Therefore a certification of the melt based on a sample also certifies every piece of glass cast, molded, or pressed from that melt. Since the properties of glass can vary slightly but significantly enough from melt to melt, each batch receives a melt number that must be carefully maintained throughout the design and fabrication stages to ensure that the finished optic performs to the design specifications.

The continuous process, on the other hand, does not have an effective stirring or blending feature, despite the fact that the molten glass in the conditioning tank is continually stirred (Fig. 2.2). This means that the uniformity of the glass composition cannot be controlled as accurately as in day tanks. The uniformity of the glass composition will drift over the course of often long melt runs. This condition requires that samples of the glass be taken at appropriate intervals according to a strict schedule to be analyzed and classified for the optical properties of interest. The certification is then by date of manufacture.

### 2.1.5. Current Status of Optical Glass Manufacture

Most glass manufacturers use both methods, although the majority of glass types are still produced by the day tank method. For instance, Schott [2, 3] is the largest optical glassmaker. It supplies about 75% of all optical glass worldwide. Over 250 different types of glasses are produced by Schott, most of them by the day tank method. Hoya [4, 5] also produces a large number of different optical glasses. Hoya pioneered the development of the continuous glass-melting process and has a sophisticated continuous glass-melting facility for producing large volumes of BK 7 (Hoya BSC 7) optical glass. Corning [6] also uses the continuous tank process and has day tank capability through its wholly owned subsidiary in France [7].

The modern science of glassmaking must not only blend the correct ingredients to achieve specific optical properties but must also properly balance the mechanical, the thermal, and especially the chemical characteristics of the glass. It should then not be surprising that significant knowledge of the old art of glassmaking is still required today.

### 2.1.6. Inspection for Flaws in Optical Glass

After proper preparation, glass plates and glass blocks, must be inspected for internal quality. Typical glass flaws are entrapped stones, which are most often broken off pieces of the clay pot, large bubbles or zones of bubbles, and striae caused by local discontinuities in the refractive index. In addition optical glass is also inspected for strain in polarized light. Test pieces are taken from the glass melt and then optically worked to measure the precise refractive index and dispersion.

**Bubbles.** According to one standard [8] the bubble content of the glasses is divided into eight bubble classes of which actually only the first four can be used for optical components (Table 2.1). This specification states that a glass volume of 100 cm$^3$ may contain a certain number of bubbles. The sum of their cross-sectional areas is measured in square millimeters. Inclusions in

**Table 2.1.  Bubble Classes for Optical Glasses
per DIN 58 927**

| Bubble class | Sum of bubble crossection ($mm^2$) permissible in 100 $cm^3$ of glass |
|:---:|:---:|
| 0 | from 0.00 to 0.03 $mm^2$ |
| 1 | from 0.03 to 0.10 $mm^2$ |
| 2 | from 0.10 to 0.25 $mm^2$ |
| 3 | from 0.25 to 0.50 $mm^2$ |

the glass, whether small stones, knots, or crystals are evaluated as bubbles with equivalent cross section. Other standards exist that classify the bubble and inclusion content in optical glass in a very similar manner.

Typical bubble concentrations for individual glass types are indicated in the glass catalogs by bubble class. Because of greatly improved glass manufacturing processes, only glasses up to bubble class 2 are generally offered today. Bubbles are easily seen when they are viewed in backlighting through the prepolished sides of raw glass. Comparison tables are used to determine whether the allowable bubble content in the glass has been exceeded.

**Striae.** Local zones in the glass that have an abruptly changing refractive index compared to the surrounding glass are called *striae*. They are the equivalent to the peculiar swirls that one sees when sugar is dissolved in water, except that in glass they are fixed in place. Since striae can take on many different appearances, they are broadly differentiated as band striae, layered striae, string striae, or filament striae. This type of material flaw can be very detrimental in optical systems because it degrades the optical performance of lens systems. Striae appear in most common glasses in varying degrees, and minor striae can be accepted for many noncritical applications. To detect and locate striae in glass blocks and plates, a white screen is illuminated with an approximate point light source. When a glass block that contains striae is held into the diverging beam, the striae will show on the screen as dark shadows, as shown in Fig. 2.5.

**Strain.** Strain in glass creates an effect called *birefringence*. This means that the glass has different refractive indices in two different orientations. In other words, the speed of light differs depending on which direction the light travels through the glass. A measure of strain is then the path difference in nanometers (nm). This means that if the light travels 1 cm in one direction through the glass, it will travel during the exact same time interval a slightly greater or shorter distance in a direction orthogonal to the first. Figure 2.6 shows this relationship.

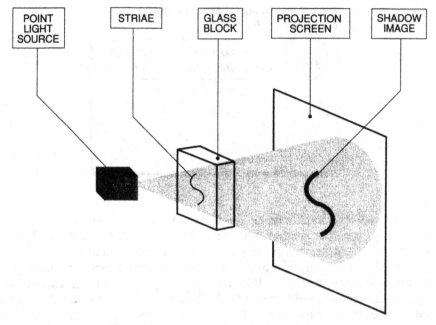

**Figure 2.5.**   Inspection for striae in raw glass.

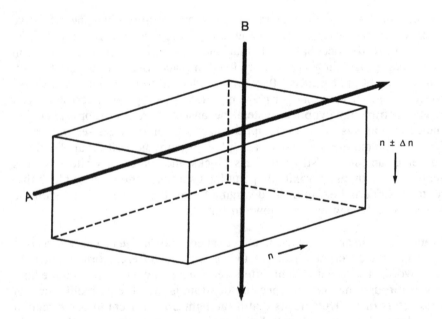

Since the index seen by ray B differs by Δ n,
ray B travels a longer or shorter
path length than ray A.

**Figure 2.6.**   Strain birefringence.

The glass blocks are tested for strain between two crossed polarizers. The strain reveals itself when viewed in transmission as light and dark spots or bands. Color transitions can be seen when a phase plate is introduced. This type of measurement is very sensitive. When the glass under test is touched, the warmth of the hand can readily simulate strain.

**Refractive Index.** The refractive index is measured accurately to the fifth decimal place on specially polished test pieces for each glass melt. A refractometer and the light of corresponding spectral lamps serves that purpose. These measurements also provide the values that are needed to calculate the Abbé number or dispersion value using Eq. (1.1).

### 2.1.7. Internal Quality Grades

Every glass manufacturer uses a grading scheme that is applicable only to its glasses. Some efforts in standardizing glass grades have been made. One such example [8] will be used here. According to this standard, optical glasses are divided into five quality grades and grading is based on the internal quality of the glasses. This standard is used by Schott [2].

The standard quality N has been tested in one direction for striae and strain. The refractive index $n_e$ must not vary by more than $\Delta = 0.0001$ within a melt. This quality grade is useful for photographic lenses and binoculars.

Quality grade NVH is a standard quality glass with improved homogeneity. The refractive index $n_e$ varies within the melt by less than $\Delta = 0.00002$, which makes this grade useful for high-quality photographic lenses, scientific instrumentation, and reproduction optics.

Quality grade NSK is a standard quality optical glass with special annealing characteristics. The strain in this type of glass is especially low or it has been brought deliberately to a specific value. This expensive glass is typically used only for astronomical refractive optics.

Grade NVS is a standard grade glass with critical striae selection. This glass is especially low in striae. It is used for high-quality objectives and for refractive astronomical optics.

The final optical glass quality grade is precision quality P. This select type of glass is tested in several directions for strain and striae. The refractive index varies within a piece by less than $\Delta = 0.000005$. This condition can be achieved only for a few glass types. Its application is for master and reference pieces.

### 2.1.8. Typical Raw Glass Configurations

According to the same specification [8], optical glass can be supplied in the following basic configurations:

*Bar glass* is cast into bars or is extruded. It is polished on two opposite sides. It can be supplied in quality grades N and NVH (Fig. 2.7).

**Figure 2.7.**   Bar glass. (Source: Schott Glass Technologies, Inc.)

*Block glass* is polished on two opposite sides. Two other opposing sides
are ground. This type is deliverable in quality grades N, NVH, and
NSK (Fig. 2.8).

*Slabglass* has six worked surfaces of which two opposite sides are polished. It is available in grades N, NVH, and NSK (Fig. 2.9).

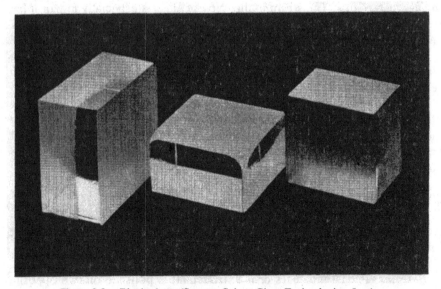

**Figure 2.8.**   Block glass. (Source: Schott Glass Technologies, Inc.)

**Figure 2.9.** Slab glass. (Source: Schott Glass Technologies, Inc.)

*Rod glass* are extruded with circular, rectangular, or triangular cross section. Discs are cut from round stock to serve as lens blanks. Triangular stock yields prism blanks. Rod glass can be delivered only in standard quality grade N (Fig. 2.10).

*Glass gobs* are approximately round blanks of equal weight that are pri-

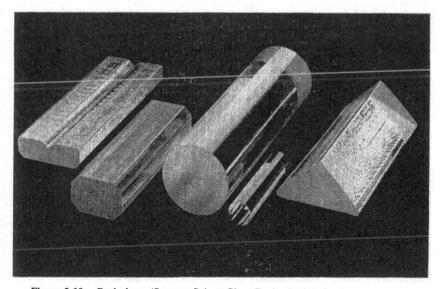

**Figure 2.10.** Rod glass. (Source: Schott Glass Technologies, Inc.)

**Figure 2.11.** Glass gobs. (Source: Schott Glass Technologies, Inc.)

marily reprocessed into lens or prism pressings. They are also available only in grade N (Fig. 2.11).

*Preshaped glass* is a rapidly emerging product as optical shops have realized the benefits of purchasing preshaped lens or prism blanks. These blanks are precision machined all around to customer specifica-

**Figure 2.12.** Preshaped glass. (Source: Schott Glass Technologies, Inc.)

tions. They are fabricated from previously carefully selected and properly certified glass. These prefabricated blanks can be delivered in any required quality grade (Fig. 2.12).

*Plateglass,* also called "semioptical" or "spectacle glass," has a refractive index of 1.523 and a dispersion value of 58. It is typically fabricated into less demanding parts such as eyeglass lenses or simple magnifiers. It is available from the glass manufacturers as pressings or in the form of glass plates. Plateglass is drawn and fire-polished in a continuous sheet by a method called the *Fourcault process.* A fireproof nozzle floats on the liquid glass in a tank. The shape and weight of the nozzle allow liquid glass to rise through the slotlike orifice. This overflowing and highly viscous glass is drawn through rollers, and it solidifies during this stage into a glass sheet with two fire-polished sides (Fig. 2.13). The glass is cut off above the rollers in the desired lengths. The plate thickness is controlled by the shape of the nozzle, the temperature of the liquid glass, and the draw speed.

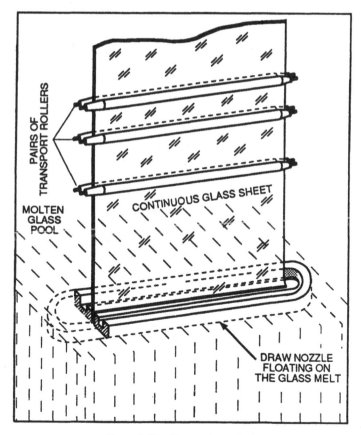

**Figure 2.13.** Fourcault method.

**Figure 2.14.**    Rolled or cast glass. (Source: Schott Glass Technologies, Inc.)

*Rolled or cast glass* is poured onto iron ways and rolled to the desired
thickness (Fig. 2.14). It is reheated later in an oven and slowly cooled
to reduce strain. It is not transparent on either side, for it typically
exhibits patterns that are impressions from the iron ways.

Plateglass as well as cast glass can have noticeable layered striae that do
not normally interfere when the viewing direction is perpendicular to the
plate. Consequently the viewing direction must be considered for parts made
from these glass types.

*Float glass* is plateglass that has solidified while floating on a tin melt. The
surfaces are almost as flat and as clean as if the glass had been optically
polished. It is often important to know which side was in contact with
the tin, since the opposite side tends to be smoother. This is usually
done by visual inspection.

### References for Section 2.1 (Optical Glass)

[1] Izumitani, T., *Optical Glass,* American Institute of Physics, 1986.
[2] Schott Glasswerke, Germany.
[3] Schott Glass Technologies, USA.
[4] Hoya Corporation, Japan.
[5] Hoya Optics, USA.
[6] Corning Glass Works, USA.

[7] Sovirel, France.

[8] DIN 58 927.

**Other Sources**

Lewis, William C., "Optical glass," Corning Glass Works, 1974 *Optical Shop Notebook I*. Section II: Optical Materials.

Scrivener, A. B., "Optical materials—Glass," Chance Pilkington, 1975 *Optical Shop Notebook I*. Section II: Optical Materials.

Bailey, Donald, "Future optical glass," Hoya Optics USA, Inc., 1976 *Optical Shop Notebook I*. Section II: Optical Materials.

## 2.2. GLASS PRESSINGS

Pressings are thermally formed raw glass blanks. They are produced either from reheated glass or from the viscous but still somewhat liquid glass of a melt. The main advantage of using glass pressings are greatly reduced material overages. This translates into faster shaping operations and reduced diamond tool wear.

The trend continues toward higher optical quality pressings made to increasingly exact dimensional tolerances while reducing cost at the same time. This requires that glass-pressing houses modernize their operations and make improvements in material analysis and control.

### 2.2.1. Pressings from Reheated Glass

Glass produced by the day tank or batch process is first cast into strips. The strips are cut, weighed, and placed into a mold inside a gas or electric furnace. The glass is then heated to the softening point and hot pressed either by manual forming and molding or by mechanical pressing into lens and prism blanks. The freshly molded or pressed blanks must first be preannealed and later fine-annealed to relieve all process-induced internal stresses and to precisely control the index of refraction.

Specifically, the precast glass strips or plates are cut into square pieces that correspond to the calculated volume of the required lens pressings. A carbide wheel glass cutter is rolled under pressure across the glass plate to score the surface. The parts are separated by applying light pressure along the scored scribe lines (Fig. 2.15).

It may occasionally be necessary to separate thicker pieces by tapping the surface opposite the scribe line with a small hammer or a mallet. Only the anvil of the hammer can be used because the claws tend to leave impact fractures in the glass that remain as flaws in the pressings. All sharp edges must be chamfered so that there will be no flaws on the surface of the pressings caused by fractured splinters or by an edge that is folded over. The

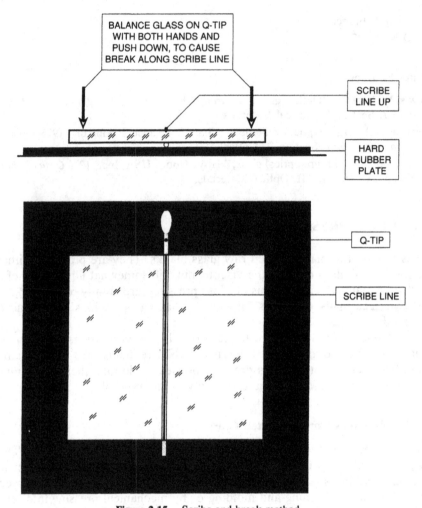

BALANCE GLASS ON Q-TIP
WITH BOTH HANDS AND
PUSH DOWN, TO CAUSE
BREAK ALONG SCRIBE LINE

SCRIBE
LINE UP

HARD
RUBBER
PLATE

Q-TIP

SCRIBE LINE

**Figure 2.15.**    Scribe and break method.

chamfers are often manually ground on a wet stone grinding wheel or on a diamond-plated beveling wheel when the quantities are low. For higher quantities it is more economical to tumble the pieces in an aqueous abrasive slurry to round the sharp edges automatically. Carefully premeasured gobs (Fig. 2.11) are used for the highest volume pressing jobs.

The glass pieces are heated in a high-temperature oven until they become pliable. They are then removed from the oven with two iron bars that are flattened at one end and placed into a preheated pressing mold. Larger square pieces are first formed between the flat ends of the two bars until they are nearly round. Care must be taken while this is done to prevent the glass from overlapping into folds, which would appear as flaws in the finished pressings.

The press is typically driven by compressed air, which is activated by a foot switch. It is equipped with an upper and a lower mold for pressings that do not exceed 30 g (1 oz). The lower mold slides up and down in a guide bushing that controls the diameter of the pressings. It receives the preheated piece of glass. During the pressing cycle, the upper mold descends into the bushing to form the top side of the pressing. The upper mold is retracted at the end of the cycle; the lower mold rises and lifts the finished pressing out of the bushing so that it can be removed or ejected. Figure 2.16 shows this process.

The bushing and the upper and the lower molds are heated just enough to keep the glass barely pliable. A thin layer of clay and fire brick powder is applied to the mold to prevent the glass from adhering to it. The same powder is also used in the oven. This release agent becomes part of a skin on the blanks that must first be removed during the early stages of shaping them into lenses or prisms. Depending on the size of the pressing, it may require a removal of up to 1 mm (0.040 in.) per side to ensure that all traces of this skin has been removed.

The completed pressings are placed into a flat tray filled with silicious powder, which is a poor thermal transmitter. This helps them to slowly cool

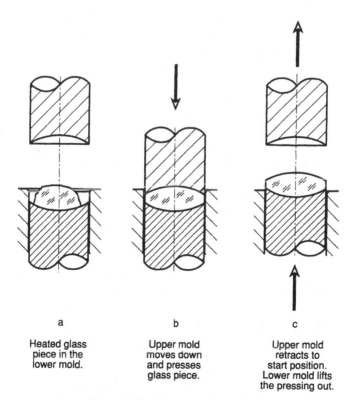

a

Heated glass
piece in the
lower mold.

b

Upper mold
moves down
and presses
glass piece.

c

Upper mold
retracts to
start position.
Lower mold lifts
the pressing out.

**Figure 2.16.** Lens-pressing method.

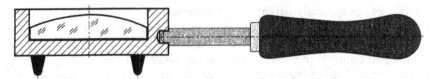

**Figure 2.17.**   Lower mold for large lens pressings.

without thermal fractures. The pressings are then heated up again in a carefully controlled oven to the transition temperature of the glass from which they are made. The oven temperature is slowly lowered according to a very carefully controlled schedule. As the pressings cool down, the oven temperature must be monitored to inhibit the formation of strain in the glass. This process is called annealing. The refractive index of the glass can be influenced to the third or fourth decimal place by controlling the rate of temperature change. It is easiest to bring the refractive indices to accurate values with electrically controlled thermal cooling.

Heavy pressings or molded prisms cannot be as easily ejected as the smaller pieces. They tend to deform or damage one another because of their greater heat content. To overcome these problems, the lower mold for larger parts is shaped like a pan (Fig. 2.17). The pressed part is removed from the press together with the lower mold, which is placed on a heated surface where it can slowly dissipate its heat. This way the pressing can solidify in the center without deformation. Because of the longer cooling time it is necessary to use several molds for the same type of part in rotation.

### 2.2.2.  Pressings Made from Glass Melt

Continuous process glass can be direct pressed as it comes out of the melt furnace. This is typically an automatic operation that is very cost-effective when the glass type and the shape of the pressings are suited for a high-volume process of 20,000 pressings or more of the same part.

The automatic sequence is shown in Fig. 2.18. The glass flows as a viscous stream through a heated aperture. A shearlike arrangement cuts off equal increments from the band. Each piece represents the correct amount of glass needed for a pressing. The soft glass drops into a lower mold that is mounted with several other molds on a conveyor system. The stepwise forward motion of the conveyor is synchronized with the cutting sequence. In a subsequent step as the process advances, the upper form comes down on the soft glass and forms it into a pressing. In this way one pressing is formed with each advancing step of the conveyor.

The glass pressing then passes through a cooling tunnel. The continuous belt of the conveyor system turns down at the end of this tunnel, and the pressing drops out of the mold into a box filled with silicate sand. It can be later fine-annealed if that becomes necessary. Even though this process

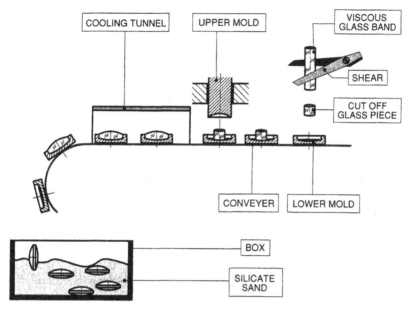

**Figure 2.18.** Automated lens pressing.

appears to be straightforward, there are technical difficulties that have to be overcome. A detailed discussion of the difficulties with the automatic lens-pressing approach cannot be done in this abbreviated description.

### 2.2.3. Fabrication Allowances for Pressings

Fabrication allowances are required for the grinding, polishing, and edging steps that must be performed to convert a lens pressing into a functional lens. These allowances must be considered when lens-pressing dimensions are specified. Figure 2.19 shows typical thickness and diameter overages. The diameter overage is a function of the lens shape. The center thickness overage is about 1 mm (0.040 in.) per side over the maximum thickness of the finished lens.

Smaller allowances are acceptable for smaller lenses; larger allowances may be required for larger lenses. The thickness overages for both sides can be approximated as follows:

$$A = 1.6 \text{ mm} + \frac{d}{100}, \qquad (2.1)$$

where:

$A$ = thickness overage in mm,
$d$ = diameter of pressing in mm.

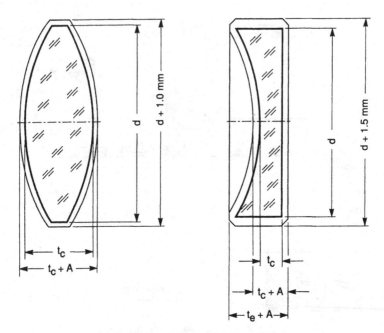

**Figure 2.19.** Overages for lens blanks.

Equation (2.1) is graphically represented in Fig. 2.20. As the figure shows, a 40-mm (1.6-in.) diameter pressing should have a 2-mm (0.080-in.) thickness overage to allow for grinding of both surfaces. The minimum overage can deviate from the values in the graph depending on the chosen fabrication method. For instance, only 0.6-mm (0.024-in.) thickness overage is needed for eyeglass lenses because they are worked as single surfaces.

The diameter overage required for edging after the lens has been centered is dependent on the shape of the lens. Lenses with convex surfaces require an overage of only 1.0 to 1.25 mm (0.040 to 0.050 in.). Lenses with plano or concave surfaces should have a diameter overage of 1.5 to 1.75 mm (0.060 to 0.070 in.). The inevitable small chips along the sharp edges of the lens necessitate working bevels, which may have to be cleared out. Additional overages are sometimes required for lenses with low refractive powers to ensure that the edge will clear out during the final edging operation. Pressings for such lenses often require an overage of 2.0 to 3.0 mm (0.080 to 0.120 in.). This is discussed in greater detail in Chapter 5.

When considering overages for lens pressings, the manufacturing tolerances must also be noted. One German standard [9] provides such standards. There is no comparable U.S. standard, but glass-pressing houses use their own specifications.

A general list of lens sizes is given in Table 2.2. For example, in this table a 75-mm (3.000-in.) diameter lens pressing will have a diameter tolerance of

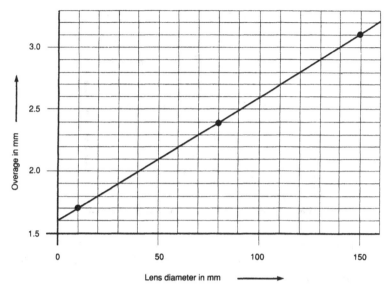

**Figure 2.20.** Lens thickness overage.

±0.3 mm (±0.012 in.). The tolerances of lens sizes must be considered during the design and manufacture of inprocess tooling and fixtures.

Table 2.3 compares the unnecessarily large center thickness tolerances of old methods with the much better tolerances that can be held with modern methods. The new series that has been proposed is half or even less than half of the old version.

### 2.2.4. Lens-Pressing Weights

Glass manufacturers issue special price lists for lens pressings. The prices are listed according to pressing weight and glass type. The cost of the molds

**Table 2.2. Diameter Tolerances for Lens Pressings per DIN 58 926**

| Diameter of pressing in mm | Permissible diameter tolerance in ± mm |
|---|---|
| 0 to 63 | 0.2 |
| 63 to 80 | 0.3 |
| 80 to 100 | 0.4 |
| 100 to 150 | 0.5 |
| 150 and over | 1.0 |

**Table 2.3. Center Thickness Tolerances for Lens Pressings**

| Diameter of pressing in mm | Permissible thickness tolerance in ± mm | |
|---|---|---|
| | old [9] | proposed [10] |
| 0 to 12 | 0.8 | 0.4 |
| 12 to 25 | 0.6 | 0.3 |
| 25 to 63 | 0.5 | 0.2 |
| 63 to 125 | 0.7 | 0.3 |
| 125 to 250 | 1.0 | 0.5 |

is usually quoted separately, but it can also be amortized over a specified quantity of identical lens pressings. The pressing weight can be calculated from the volume of the lens pressing and the specific weight of the optical glass used. The specific weight of a material is also called its *density*. It is usually expressed in terms of grams per cubic centimeter (g/cm$^3$). The density of each glass is listed in all glass catalogs (see Figs. 1.2, 1.3, and 1.4).

The basic equation for calculating the weight $W$ of an object (in g) involves the product of the object's volume $V$ (in cm$^3$) and its density $\rho$ (g/cm$^3$):

$$W = V\rho. \tag{2.2}$$

The volume of a lens pressing can be calculated on the basis of the following considerations: Each lens pressing is composed of two spherical segments and a cylindrical segment, as shown in Fig. 2.21. The volume of each segment can be calculated separately, and the volume of the pressing is then the the sum of the individual segment volumes. This relationship is mathematically expressed in Eq. (2.3). If the dimensions are entered in centimeter (cm), then the volume is calculated in cubic centimeters (cm$^3$) from which the weight in grams (g) can be calculated using Eq. (2.2).

$$V = \pi \left[ s_1 \left( R_1 - \frac{s_1}{3} \right) + s_2(R_2 - s_2) + \frac{d^2}{4} (t_e) \right] \tag{2.3}$$

where:

$V$ = volume of pressing in cm$^3$,
$d$ = diameter of pressing in cm,
$s_1, s_2$ = sag in cm, negative for concave sides,
$R_1, R_2$ = radius in cm, negative for concave sides,
$t_e$ = edge thickness of the pressing in cm.

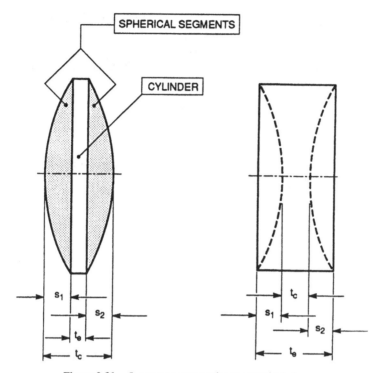

**Figure 2.21.** Lenses as composite geometrical shapes.

The sag heights ($s_1$, $s_2$) and the edge thickness ($t_e$) can be taken from a scale drawing or they can be calculated using Eqs. (2.4) and (2.5):

$$s_n = R_n - \sqrt{(R_n)^2 - \left(\frac{d}{2}\right)^2}$$  (2.4)

$$t_e = t_c - (s_1 + s_2)$$  (2.5)

where:

$t_c$ = center thickness of the pressing in cm (the concave sags are negative).

Equation (2.3) is too complicated for a quick determination of the lens pressing volume. It can be simplified by an approximation when the sag values are expressed as $s = d^2/8R$, which leads to Eq. (2.6). As before, concave radii are negative.

$$V = \frac{d^2\pi}{4}\left[\left(\frac{d^2}{R_1}\right) + \left(\frac{d^2}{R_2}\right) + 16t_e\right].$$  (2.6)

A further simplification of the approximation results when the surfaces of the pressings are not considered spherical but assumed to be parabolic [10]. Figure 2.22 compares these surfaces. When the curvature is relatively slight, the parabolic shape differs very little from the spherical shape. The volume of the parabolic segment is expressed as

$$V_p = \frac{\pi d^2 s}{8}. \tag{2.7}$$

The formula used to calculate the volume of a lens pressing with parabolic surfaces can be derived from Eq. (2.7):

$$V = \frac{\pi d^2}{8} (s_1 + s_2 + 2t_e) \tag{2.8}$$

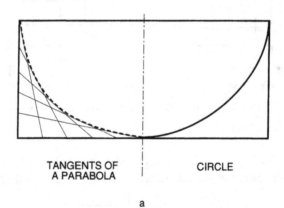

TANGENTS OF
A PARABOLA

CIRCLE

a

**Figure 2.22a.** Similarity of parabolic and spherical surface.

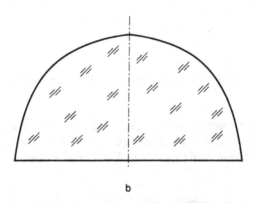

b

**Figure 2.22b.** Lens pressing with parabolic surface.

Since $(s_1 + s_2 + t_e) = t_c$, the center thickness of the pressing, Eq. (2.8) can be further simplified as

$$V = \frac{d^2}{8} (t_c + t_e).\tag{2.9}$$

In Eq. (2.9) all values are entered as positive, that is, without consideration of whether the pressing has concave or convex surfaces. To test the validity of this approximation, the volumes of two lenses are calculated using Eqs. (2.3), (2.6), and (2.9) in Figs. 2.23 and 2.24. The results appear in Table 2.4. Despite the strong curvatures of the lens pressings in this example, the results of the approximation remain close to the exact value. The error is less than 1% when using Eq. (2.9). Since the values used in the calculations, such as diameter, center thickness, radii, and edge thickness, can be taken directly from the design document, it is best to use Eq. (2.9) as a first-order approximation to determine the pressing weight. The necessary overages must first be added to these values (see Fig. 2.19).

Up to this point the discussion has been based on the use of the metric values of centimeter and gram. The use of the English system of measure is still quite widespread in the United States, however, and this can lead to an erroneous application of these equations. The conversion to metric cannot be avoided because the density of the glass is always expressed in terms of $g/cm^3$. It is possible to convert the density to English measure by remembering that 1 cubic inch ($in.^3$) equals about 16.39 $cm^3$ and that 1 pound (lb) equals 453.6 grams. This conversion yields a density in terms of $lb/in.^3$. It is nevertheless much simpler to convert linear English measure to linear metric measure and to use the density value right out of the catalog to calculate the weight in grams.

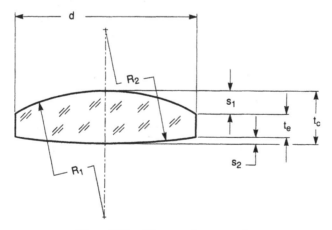

**Figure 2.23.** Biconvex lens example.

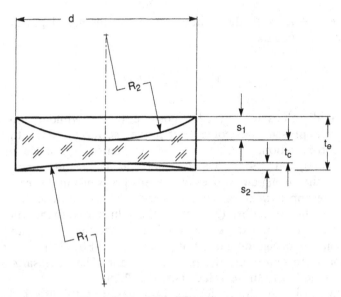

**Figure 2.24.** Biconcave lens example.

### 2.2.5. Clear Molding of Lens Surfaces

Curved, nonspherical surfaces are very difficult to grind and polish. Such surfaces are called *aspherics*. They offer the lens designer certain advantages, but they are very costly to produce with traditional methods. If the surface accuracy requirements are not too stringent, aspheric lenses can be economically produced in volume by a clear glass molding technique.

Clear surfaces can be created during the pressing process only when the mold surfaces themselves are highly polished and when the material is fed in as liquid glass. The typical raw material for clear molded lenses are fire-polished drawn round rods. The results of the process are particularly good when only one lens surface, for instance, the aspheric surface, needs to be

**Table 2.4. Comparative Results for Calculating the Lens Volume Using Different Approximations**

| Volume in $cm^3$ per Eq. | Positive lens (Fig. 2.23) | Negative lens (Fig. 2.24) |
|---|---|---|
| 2.3 | 11.07 | 8.34 |
| 2.6 | 11.79 | 8.57 |
| 2.9 | 11.00 | 8.40 |

clear molded. The other side can then be ground and polished using standard methods.

The glass rods are suspended in a gas-heated kiln with one end supported by a yoke. An opening in the kiln is designed in such a way that two glass rods can be inserted at the same time. They are then alternately removed for pressing one lens at a time. Handles which are about 0.5 m (20 in.) long are attached to the cool end of the rods. They serve as thermal protection for the operator and as support in the yoke. The operator determines by the color of the glowing rod and by the manner in which the rod end deforms under the heat if the required molding temperature has been reached.

The properly heated rod is then removed and quickly inspected for surface flaws at the softened tip. Such flaws would remain on the surface of clear pressed parts. It is often possible to position the rod in such a way that any residual surface flaws will end up on the surface that needs to be ground and polished later on. The heated rod end is then introduced into the mold in the most favorable attitude.

The polished mold is in the upper position. This facilitates the removal of the lens from the mold without the need to touch the clear molded side. The polished mold is made from a select grade of steel that does not oxidize at the high molding temperatures. The gas-heated press applies a force to the upper mold. The pressure is increased slowly so that entrapped air has a chance to escape. The press remains closed from one to three minutes, depending on the size of the part, so that it can solidify in the mold.

Before the press is opened, any protruding glass, also called *flashing,* can be cut off with a coarse shear because it is still soft during the early stages of the molding process. Once it has cooled down sufficiently, the freshly pressed lens is carefully removed and placed into the cooling oven for fine annealing. The typically plano second side of the lens is later ground and optically polished.

### 2.2.6. Tempering of Glass

Although the tempering of glass is not directly related to glass pressing, it is included in this section because it represents a small but potentially important aspect of the thermal treatment of glass. Silicate glasses can be tempered against chipping and surface damage, as well as for protection against thermal fracture. This treatment requires that the parallel surfaces of a part are intentionally under stress. The stressed condition must be well balanced between the two sides to avoid surface deformation and possible breakage. The stressed condition can be detected between crossed polarizers. The stress is introduced by one of two methods, namely, thermal tempering and chemical tempering.

For thermal tempering the glass part is heated up to the softening point of about 700°C (1300°F). It is then subjected to sudden cooling in an air flow at about 20°C (68°F). The shape of the glass surface can change slightly because of this thermal treatment; then accurate surface flatness cannot be maintained. The glass cannot be optically worked after tempering because it will fracture.

During the chemical tempering process, sodium ions on the glass surface are replaced by potassium ions. Since potassium ions are larger than sodium ions, a stress condition results in the surface, and it has a hardening or strengthening effect. The chemical ion exchange penetrates about 0.1 mm (0.004 in.) into the glass. For this process a potassium salt such as potassium nitrate ($KNO_3$) is melted at 420°C to 450°C (790°F to 840°F). The glass is suspended in this melt for 16 hours or more. The ion exchange takes place during this time. This type of tempering does not result in a noticeable change in surface shape.

A variation of this ion exchange process is used to produce a new type of optical material called *gradient index glass*. By subjecting preshaped glass blanks to carefully controlled ion exchange cycles, the index of refraction of the glass is modified so that it varies across the part in a plane normal to the axis (radial gradient) or in the direction of the axis (axial gradient). This index gradient can be utilized to make a lens with spherical surfaces that behaves as if it had aspheric surfaces.

### References for Section 2.2 (Pressings)

[9] DIN 58 926, June 1973.

[10] Harry Schade, 1984, unpublished manuscript translated from German.

### Other Sources

Baker, Terry, United Lens Co., "When to use and how to specify glass pressings," *Workshop on Optical Fabrication and Testing, Technical Digest*, 1982.

## 2.3.  INFRARED MATERIAL MANUFACTURING METHODS

Most infrared (IR) materials are crystalline in nature and a variety of crystal growth methods are used to produce them. Some of these materials can also be cast. The methods of production for germanium, zinc sulfide, zinc selenide, silicon, and gallium arsenide are discussed in this section. Other materials are amorphous like glass and are referred to as *infrared glass*. Casting methods similar to those described for glass are employed for IR glasses, but considerable differences exist between these two approaches.

### 2.3.1.  Germanium Manufacture

Germanium is a by-product that is extracted from zinc, copper, and lead ores. The extraction process [11] begins with the blending of the raw material which is a sludge left over from the smelting of the primary metals. The blend is chlorinated and purified to yield germanium chloride ($GeCl_4$). This intermediate product is then subjected to hydrolysis, filtration, and vacuum drying to create germanium dioxide ($GeO_2$). The impurity level at this stage is on the order of 1 ppm. The germanium dioxide is further purified in a hydrogen reduction step at about 650°C and then cast at 1100°C into an As-reduced Ge bar. The impurity level has been reduced by this treatment to about 0.01 ppm. The bar is then zone-refined to an impurity level of about 0.001 ppm (or1 ppb) to yield intrinsic Ge metal. The Ge metal is used to grow Ge ingots or to cast large Ge blanks. This process is schematically shown in Fig. 2.25, center.

Unlike the vertical zone refinement technique described for silicon in Section 2.3.4, the zone refining for germanium is done in a horizontal graphite boat. The boat passes through an induction heating coil which locally heats the germanium to a molten state. The molten zone of the ingot travels from end to end as the boat moves horizontally through the coil. The impurities in the raw Ge tend to congregate at the trailing end of the moving molten zone. They are thus literally stripped from the rod and end up in the tail section of the ingot which is cut off later (Fig. 2.25, lower left). This zone refinement must be repeated several times to get the impurity level reduced to 1 ppb [12].

The zone-refined germanium can then be Czochralski (CZ) grown or cast. The resistivity of zone refined Ge is <30 ohm-cm. The CZ growth method for germanium is nearly identical as the one described for Si in Section 2.3.3. The method yields either single (mono) or polycrystal Ge ingots. It requires precise control of the crystal growth parameters such as pulling speed, crystal rotation, and thermal gradients. The growth rate is about one inch per hour (Fig. 2.25, upper right). A several hundred pound charge can be grown into a 300-mm (12-in.) diameter monocrystal Ge boule or into a 375-mm (15-in.) diameter polycrystal boule [12].

Residual impurity levels (or the optical quality) are determined indirectly by a resistivity measurement. Typical resistivities are 50 ohm-cm for as-reduced Ge and 25 ohm-cm or less for zone-refined Ge, which would exhibit the lowest absorption.

Single crystal Ge is grown by the CZ method to about 200 mm (8 in.) diameter. Large Ge blanks can be cast to near net shape in sizes up to 650-mm (26-in.) diagonal dimension, but vertical casting can yield only polycrystal blocks of Ge [13]. This method is typically reserved for large IR window blanks for which the other growth methods would not work. The mold is a square or rectangular graphite crucible (Fig 2.25, upper right).

Because of the high cost of the raw material and the recoverability of waste cuts and even swarf, the primary manufacturers of Ge [14, 15, 16] offer preshaped blanks for windows and pregenerated blanks for lenses. These blanks are a very economical option for the shop because the pregenerated lens blanks can be supplied to near-net shape, which reduces the

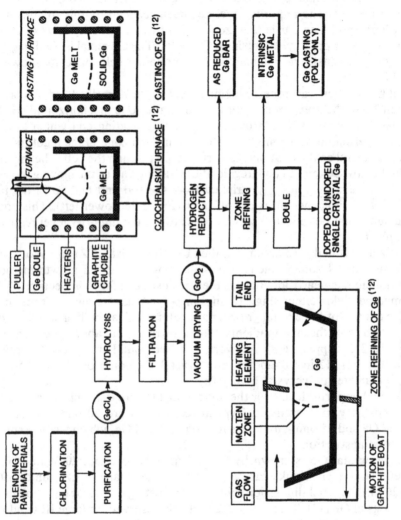

**Figure 2.25.** Schematic of germanium extraction [11].

cost of optical fabrication. The material manufacturers can afford to do this because the high cost of the material and the limited availability of Ge makes scrap and swarf reclamation economically feasible. As an example, the cost of germanium in cast or ingot form in 1982 ranged from \$300 to \$800/kg [17]. Circular plane parallel blanks cost approximately \$150/in.$^3$ [12] in 1987.

### 2.3.2. Chemical Vapor Deposition

The chemical vapor deposition (CVD) process is used to produce several important IR materials. The most important of these are ZnSe and ZnS. The CVD process for these two materials is described as follows [18]:

Chemical vapor deposition is a process that produces a solid from the chemical reaction of gases and vapors. This process was developed in the early 1970s in response to the need for increased volumes of ZnSe and ZnS and the improvement over the limited transparency of the hot-pressed Irtran materials in use at the time. Hydrogen selenide gas ($H_2Se$) or hydrogen sulfide gas ($H_2S$) react with the vapors from a molten pool of zinc to form ZnSe and ZnS, respectively. This processes can be described with the following notation [18]:

ZnS:   Zn vapor + $H_2S$ gas = ZnS (solid) + $H_2$ (gas).

ZnSe:   Zn vapor + $H_2Se$ gas = ZnSe (solid) + $H_2$ (gas).

The chemical vapor deposition process takes place in a chamber similar to an autoclave at temperatures of about 600°C and at pressures below 100 torr [19]. The solid deposits form on the inside walls of a graphite mandrel as shown in Fig. 2.26. Large sheets, which can be as large as 40 × 60 in., of either ZnSe or ZnS are produced this way. The typical thickness of the sheets is 1 in., but much thicker sheets have been produced for special requirements [20]. For instance, in 1986 one manufacturer [21] reported success in producing a ZnSe window blank that was 30 in. in diameter and 1.75 in. thick. Even larger thicknesses have been reported since then. The mandrils can be shaped with concave pockets to grow blanks for IR domes (Fig. 2.27). These mandrels have been referred to a "muffin tins."

Growth rates are on the order of 50 $\mu$m (0.002 in.) per hour. Therefore it takes three weeks of an uninterrupted and flawless growth cycle to produce a 25-mm (1-in.) thick substrate. Growth rates of 100 $\mu$m per hour have been achieved.

The transparency of CVD process zinc sulfide (ZnS) can be markedly improved by a postdeposition treatment at high temperature and pressure. This treatment removes zinc hydrate impurities and greatly reduces the typical microporosity of CVD ZnS [22]. The material becomes much more transparent in the visible region of the spectrum, and it can be described as being nearly water clear. This treatment, however, results in much larger

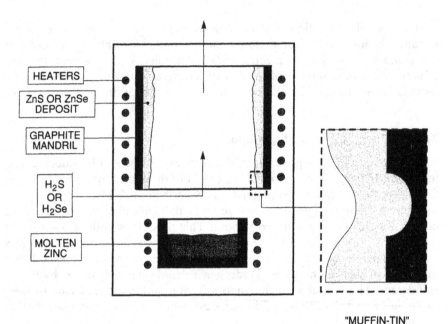

**Figure 2.26.** Chemical vapor deposition (CVD).     **Figure 2.27.** Graphite mandrel.

grain sizes than those typical for untreated CVD process ZnS, which reduce the flexural strength of the material by 50% [18]. One such material is available under the tradename "Cleartran" [21].

The original source for CVD ZnSe and ZnS [23] has since discontinued operation; another [24] decided not to commercialize their product after a brief entry into the market. This leaves only two viable suppliers for these materials [25, 26].

Although it is not easy to generalize prices of optical materials, it is useful to have some awareness of certain rule-of-thumb prices. With that in mind, it can be said that CVD process ZnSe costs about \$300/in$^3$, CVD process ZnS costs about \$200/in$^3$, and Cleartran ZnS costs about \$275/in$^3$. The actual prices are often higher depending on size, configuration, specification, and quantity. One should always get a firm quote from the material supplier before making a commitment to a customer.

A variant of the CVD technique is a sandwich structure composed of a thin layer of ZnS chemical vapor deposited on a previously polished ZnSe substrate [18]. The purpose of doing this is to combine the superior optical properties of ZnSe with the greater mechanical strength of ZnS. These sandwiched substrates are of interest as IR windows for high-performance aircraft. However, the unavoidable mismatch of the coefficients of thermal expansion (CTE) leads to internal stress.

### 2.3.3. CZ Method for Si

The Czochralski (CZ) crystal growth method is used to produce about 90% of all single crystal silicon grown for electronic and optical applications. A fused quartz crucible is charged with chunks of undoped, high-purity poly-crystal silicon. A precise quantity of highly doped monocrystal silicon is added to the charge. This provides the pulled crystal with desired resistivity. The doping elements are usually boron and phosphor [27]. The loaded crucible is placed in the bottom of a vertical CZ puller (see Fig. 2.28), and the chamber is closed. The growth environment, which is an argon atmosphere at several hundred torr, is established in the chamber.

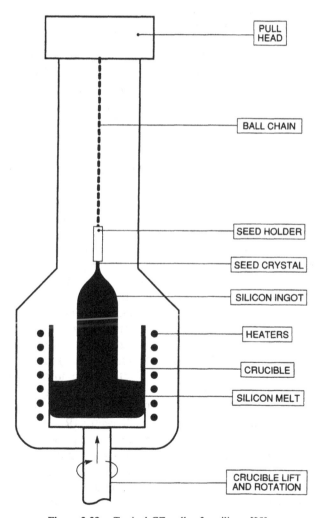

**Figure 2.28.** Typical CZ puller for silicon [28].

Resistivity heaters which surround the rotating crucible are energized, and the silicon charge will melt when the crucible temperature exceeds 1420°C. A small seed crystal, usually 3 to 4 mm (0.120 to 0.160 in.) in diameter, with the desired crystal orientation is dipped into the molten pool and slowly withdrawn as it is rotated in the opposite direction from the crucible rotation. Molten silicon solidifies as a single crystal and in this manner, a single crystal ingot of silicon is slowly pulled from the melt. The process ends when the molten pool in the crucible is exhausted [29]. The growth rate is about 25 to 50 $\mu$m (0.001 to 0.002 in.) per second. The diameter of the ingot is controlled by the rate of crystal pull and the temperature. The growth process is almost completely automated so that the various critical parameters can be properly controlled.

The CZ growth method is the most cost-effective way to grow silicon, but it tends to produce crystals with higher impurity levels compared to crystals grown with other methods. The impurities are unintentional by-products of the growth process. Since molten silicon is quite aggressive, it tends to vigorously attack most crucible materials, and this limits the choice to fused quartz for most applications. But even fused quartz will be attacked by the molten silicon, and consequently oxygen is liberated from the crucible to become the major impurity in silicon at a level of 10 to 20 ppm [30].

Carbon originating from the heat sources is introduced as an impurity as well. The original polycrystal charge also contains trace impurities of a variety of elements that become contaminants in the CZ grown silicon boule. Some reduction in the impurity level has been achieved when boules are pulled in a vacuum. CZ growth in a magnetic flux field has had similar results. Although metallic impurities in CZ grown silicon have been reduced to the ppb (part per billion) level, oxygen and carbon remain significant impurities. The oxygen impurity density is usually higher at the seed end, while the carbon impurity density tends to be higher at the opposite or "tang" end of the ingot. For the highest-quality oxygen-free silicon, silicon nitride crucibles have been successfully used.

The CZ method permits the growth of much larger ingots than is possible with other growth methods. The continued improvement of the CZ growth process has resulted in the successful pulling of 125-mm (5-in.) diameter, nearly dislocation-free silicon more than 900 mm (36 in.) long [27].

Generally silicon is grown for semiconductor and photovoltaic applications. However, one company has specialized in the optical application for silicon as $CO_2$ mirror substrate or for mid-IR optics. There are two companies specializing in growing and processing silicon for IR optics applications [31, 32]. One of these [31] has a reported capability to grow single-crystal Si ingots in a three-story crystal-pulling furnace up to 550 mm (22 in.) in diameter. The same firm also routinely pulls 375-mm (15-in.) diameter ingots by 250 mm (10 in.) long in a CZ puller. It takes two days to pull such a crystal. The 1990 cost of round lens or mirror blanks ranges from $10 to $40 per cubic inch depending on size and quantity.

### 2.3.4. Float Zone Refining of Silicon

One proven method to greatly reduce the impurity levels in silicon is the vertical zone refinement or float zone method. A high-purity polycrystal rod of silicon is slowly passed vertically through an RF heating coil inside a low-pressure chamber. During the initial stages of the process, a single-crystal seed crystal is added to the molten end of the ingot, which causes the polycrystal silicon to solidify as single-crystal material. The melt zone slowly moves along the ingot as it is lowered through the coil and the entire polycrystal rod is thus converted into monocrystal Si (Fig. 2.29).

Float-zoned silicon is almost entirely free of undesirable impurities since the melt zone tends to push impurities in the polycrystal toward the tail end of the rod [33]. Carbon and oxygen are the two most detrimental impurities

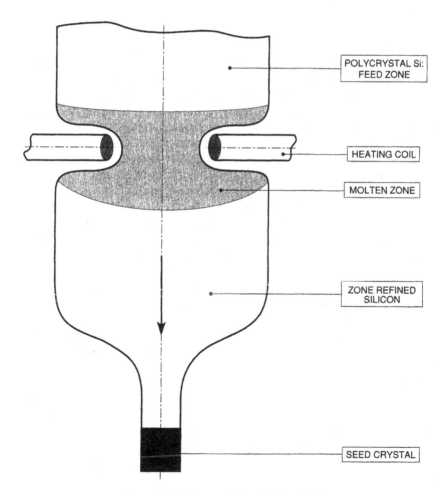

**Figure 2.29.** Zone refining of silicon [37].

in CZ grown silicon. In float-zoned silicon both impurities have been reduced to minimal levels. A reasonably complete reduction of the residual carbon is possible. The oxygen content of float-zoned silicon is less than 2 ppm, or better than ten times less than that of CZ grown material. The low oxygen level is possible because the molten material is not in contact with any crucible, and since the process is performed at near vacuum, any volatile impurities such as oxygen will readily outgas. The end product of this refinement process is a high-purity silicon as evidenced by the high-resistivity values attainable [34, 35]. This process is very costly, however, because of the high-power requirements. It is also limited in size to about 75-mm (3-in.) diameters, although the potential to go up to 150 mm (6 in.) in diameter exists. Because of these constraints only about 10% of all silicon produced is float zoned [36, 37]. There are plans to zone refine silicon in space as one of the Space Shuttle experiments.

### 2.3.5.  Casting of Silicon

The energy concerns of the 1970s encouraged the development of alternate energy sources, and solar energy emerged as one of the more promising schemes. Among the many different approaches explored, photovoltaic energy held the most promise for success if the cost of that energy could be reduced twentyfold to compete with conventional energy sources.

Photovoltaics use silicon as the substrate material, but accepted methods for producing silicon were much too expensive. Alternate low-cost methods had to be devised to grow polycrystal silicon in large sizes and in large volume. The methods explored at that time were casting, the heat exchanger method (HEM), and the ribbon method.

Casting of polycrystal silicon is possible, but the resulting material is not of sufficiently high quality for photovoltaic applications. The polycrystal structure begins forming at the surface of the mold and then degrades both in crystal size and impurity level as the silicon solidifies toward the center. This is shown in Fig. 2.30. The center is the last to solidify, and most of the impurities in the melt congregate there as they are swept ahead of the solidification line. Therefore very high purity starting materials are needed to successfully cast silicon. This requirement makes it a very costly method. Polycrystal silicon has been successfully cast into blocks up to 300 mm (12 in.) square [30].

Because of the cost and impurity problems, casting of silicon is now considered an obsolete approach. The HEM or heat exchanger method (described in Section 2.4.2 of this chapter) shows greater promise, but the abatement of the energy crisis has put the development of low-cost photovoltaics temporarily on hold. The silicon ribbon process that has been mentioned in the industry news produces material that is not useful for optical applications.

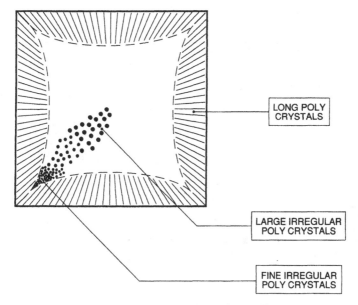

**Figure 2.30.** Polycrystal structure of cast silicon [38].

## 2.3.6. Horizontal Bridgeman

The Horizontal Bridgeman (HB) growth method is used to produce optical grade GaAs. The method is described based on discussions with the primary producer of optical grade GaAs [39].

A 600- to 900-mm (24- to 36-in.) long and 50- to 100-mm (2- to 4-in.) diameter high-purity fused quartz tubing, called an *ampoule,* which is closed at one end, is loaded at the closed end with chunks of elemental arsenic (As) of $6N$'s purity (i.e., 99.9999% pure or total impurities <100 ppm). The loading of this material is done in a laminar flow bench with toxic dust exhausts in the fill room by operators wearing cleanroom apparel and disposable respirators. A sand-blasted and acid-etched fused silica boat is then loaded with $6N$'s pure gallium (Ga) metal. A small seed crystal with ⟨111⟩ or ⟨110⟩ crystal orientation is fastened into a recess at one end of the boat. The loaded boat is then inserted into the ampoule with the seed crystal pointing toward the As. The ampoule is evacuated to $10^{-1}$ Torr vacuum pressure and then hermetically sealed with a fused quartz plug that is constrained by the narrowed neck at the fill end. This arrangement is shown in Fig. 2.31.

The typical charge is 2.5 kg (5.5 lb) total for single crystal GaAs and up to 6 kg (13.2 lb) for polycrystalline material. Ingots of up to 100 mm (4 in.) diameter are possible but 50 mm (2 in.) diameter are more typical.

The loaded and sealed ampoule is then placed into the HB growing furnace, which is designed to bring the reaction end to the required melt tem-

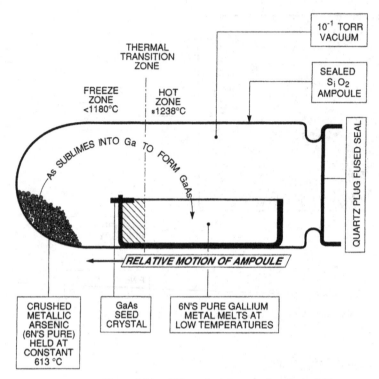

**Figure 2.31.**   HB method for growing GaAs [39].

perature of 1238°C and to keep the "freeze zone" at a sharp 60°C gradient
below the melt temperature. At these high temperatures much of the arsenic
vaporizes and sublimes into the molten gallium. As the boat is very slowly
passed through the thermal transition zone into the freeze zone, the material
solidifies as GaAs. This process is indicated in Fig. 2.31 by an arrow pointing
from right to left.

This growth process is not without its problems. The use of fused quartz
ampoules and boats leads to appreciable silicon contamination from the $SiO_2$
which requires intentional doping with chromium to neutralize the adverse
effects of silicon impurities in GaAs on its resistivity and absorption [35, 40].
For high resistivity, semi-insulating GaAs (which is the best choice for optical use), the Cr doping levels must exceed the Si impurity levels [41]. The
use of pyrolytic boron nitride (PBN) boats has been suggested for HB, but
that will not entirely solve the Si contamination problem since the ampoule
still has to be made from fused quartz [42].

The gallium and arsenic raw materials are expensive, as is the energy
required for this process. To this cost must be added relatively low growth
yields of 50% or less for single crystal and less than 75% for poly. One
reason for the low yield are occasional ampoule failures, called *blowouts,*

which require the evacuation of people from the growing room and then a complete air replacement in the room before anyone reenters it. Other factors contributing to low yield are single crystals going poly, excessive crystal dislocations, inclusions, and thermal fracturing during cool down.

More recently some success has been achieved with a vertical Bridgeman process that not only produces purer GaAs but also circular diameter ingots. This type of ingot is more appropriate for optical applications because most of the blanks are specified as round disks. It also reduces the significant waste when edging round stock from the typical trapezoidal cross section that results from the horizontal Bridgeman process.

### 2.3.7. Liquid Encapsulated Czochralski

Undoped semi-insulating single crystal gallium arsenide (GaAs) is grown by a variant of the standard CZ crystal growth method. The GaAs is formed by the reaction of molten gallium metal and arsenic vapors. To contain these volatile vapors of arsenic for the reaction process, a pool of molten boron oxide ($B_2O_3$) serves as an encapsulant, as shown in Fig. 2.32. This process, called *liquid encapsulated Czochralski* (LEC), yields high-resistivity GaAs in standard size CZ boules, which are at present limited to a diameter of 75 mm (3 in.). The resulting material is useful for a wide range of optical and microelectronic applications.

The growth process is nearly identical to the standard CZ method, but there are two significant modifications. The first is the encapsulation by molten boron oxide. The other is the use of a crucible made from pyrolytic boron nitride (PBN) instead of the fused quartz crucible that is used in the standard CZ or the HB growth methods. The high-melt temperature of nearly 1300°C leads to partial decomposition of the quartz crucible, liberating electrically active silicon, which can become a major impurity in GaAs that will lower its resistivity [42]. The use of PBN crucibles eliminates those problems; they do not contain silicon and do not decompose or break as readily.

LEC grown GaAs has a much lower Si impurity level than HB grown material. By adding small amounts of chromium dopant, single-crystal resistivities of $>10^8$ ohm-cm have been achieved [35]. The resultant GaAs ingot (boule) can be either arsenic rich or gallium rich. The benefits or detriments of either condition have not yet been determined for optical applications.

A hybrid process for growing GaAs boules that combine elements of the HB and LEC growth methods has been reported [40]. Elemental As and Ga are combined in premeasured proportions and placed in a PBN crucible. As the crucible is heated in a nitrogen atmosphere at low pressure, boron oxide is added. It readily forms a molten pool over the GaAs melt, thus trapping the As vapors because it is a low-melt glass. The arsenic vapors react with the molten gallium, forming GaAs in a fashion similar to that of the HB method. The crucible is then slowly raised vertically through a sharp temper-

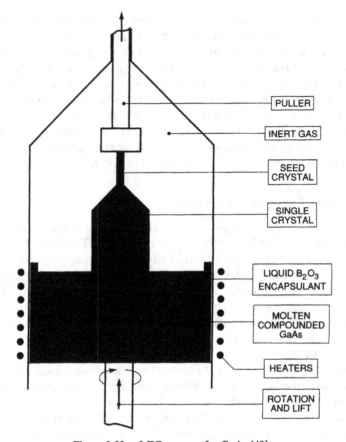

**Figure 2.32.**   LEC process for GaAs [40].

ature gradient (freeze zone) at which point the GaAs solidifies. This method
has yielded high resistivity ($5 \times 10^7$ to $10^8$ ohm-cm) GaAs without chromium
doping.

### References for Section 2.3 (IR Materials)

[11] Metzer, F. I., ed., *New Uses for Germanium,* Midwest Research Institute,
     1974.
[12] de Ruijter, I., "Germanium enters a new era," *Photonics Spectra,* July 1987.
[13] Cooper, W., "IR window advances: FLIRs and beyond," *Photonics Spectra,*
     October 1988.
[14] Eagle Picher Industries, Inc., Quapaw, OK.
[15] Vinten Penarroya, Inc., Fountain Valley, CA.
[16] Metallurgie Hoboken-Overpelt SA/NV, (MHO), Belgium.

[17] Hilton, A. R., "Production and fabrication of blanks from germanium and chalcogenide glasses," *Optical Shop Notebook*, vol. 2, OF&T 77, San Mateo, CA; Adams, J. H., "Infrared materials—Germanium," *Technical Digest*, OSA Workshop on Optical Fabrication and Testing, Palo Alto, CA, OF&T 1982.

[18] Taylor, R. L., "An infrared alternative: Vapor-deposited materials," *Laser Focus*, July 1981.

[19] Miles, P., "High transparency infrared materials—A technology update," *Optical Engineering*, vol. 15, no. 5, September/October 1976.

[20] Taylor, R. L., "Infrared materials—CVD materials," *Technical Digest*, OSA Workshop on Optical Fabrication and Testing, Palo Alto, CA, OF&T 1982.

[21] Morton Thiokol, Inc., Advanced Materials Div., Woburn, MA (formerly CVD, Inc.).

[22] Taylor, R. L., "CVD IR materials: New challenges," *Photonics Spectra*, July 1982.

[23] Raytheon Corp., Research Division, Lexington, MA.

[24] Metallurgie Hoboken-Overpelt SA/NV, (MHO), Belgium.

[25] Morton Thiokol, Inc., Advanced Materials Div., Woburn, MA (formerly CVD, Inc.).

[26] Two-Six, Inc (II-VI), Saxonburg, PA.

[27] Huff, H. R., "Chemical impurities and structural imperfections in semiconductor silicon—Part I," *Solid State Technology*, February 1983.

[28] Silicon crystal growning furnace, Ferrofluidics Corp, Solid State Technology, May 1982.

[29] Swaroop, R. B., "Advances in silicon technology for the semiconductor industry—Part I," *Solid State Technology*, June 1983.

[30] Adams, J. H., "Specifications for optic grade germanium and silicon blanks," *SPIE Proceedings*, vol. 406, April 1983.

[31] Silicon Castings, Inc., San Jose, CA.

[32] Lattice Corporation, Bozeman, MT.

[33] Melanio, D., "Spotlight on silicon," *Microelectronic Manufacturing and Testing*, August 1979.

[34] Wacker, Inc., product literature 1980.

[35] Liaw, H. M., "Trends in semiconductor material technologies for VLSI and VHSIC applications," *Solid State Technology*, July 1982.

[36] Swaroop, R. B., "Advances in silicon technology for the semiconductor industry—Part I," *Solid State Technology*, June 1983.

[37] Kramer, H. G., "Float-zoning of semiconductor silicon: A perspective," *Solid State Technology*, January 1983.

[38] Wacker, Inc., product literature on casting of silicon.

[39] Crystal Specialties, Inc., Colorado Springs, CO.

[40] AuCoin, T. R., "Liquid encapsulated compounding and Czochraiski growth of semi-insulating gallium arsenide," *Solid State Technology*, January 1979.

[41] Winston, H., "Semi-insulating gallium arsenide substrates for high frequency FET and IC fabrication," *Solid State Technology*, January 83.

[42] Finicle, R. L., "PBN material results in better GaAs crystals, IC devices advances," *Industrial Research and Development,* June 1983.

## 2.4.  GROWTH METHODS FOR OPTICAL CRYSTALS

There are many crystal materials that are transparent in the visible region of the spectrum. Some of them transmit well into the UV region, others are birefringent, while still others exhibit electro-optic effects. These properties make them useful for optical applications. The methods to produce these crystals vary over a wide range. For instance, synthetic crystal quartz is grown at high pressures and elevated temperatures by a hydrothermal process, synthetic sapphire is produced with a high-temperature heat exchanger method, alkaline earth fluorides are grown, cast or hot-pressed, while hygroscopic crystals are solution grown. These methods are described in this section.

### 2.4.1.  Hydrothermal Process

Synthetic or cultured quartz is grown by the hydrothermal process. Pieces of high-purity natural rock quartz are filled into the lower third of a high-pressure reactor vessel called an *autoclave*. A rack loaded with synthetic quartz seed plates is placed above the rock quartz. The vessel is then filled to about 80% of its total volume with an alkaline aqueous solution. At this point the autoclave is sealed, heated to 350°C, and pressurized to a high degree. These conditions cause the aqueous solution to expand to occupy the entire volume of the autoclave. The rock quartz, also called *nutrient quartz,* dissolves into the solution and precipitates from it onto the seed crystals in the rack. During a continuing process of dissolution and precipitation, the cultured quartz crystals grow around the seeds. This process continues until the supply of nutrient quartz is exhausted. The chamber is depressurized and brought down to ambient temperature, at which time the cultured crystals, also called *stones,* are removed form the rack.

The basic aqueous nutrient-rich solution at the high temperatures and pressures tend to attack the inner walls of the steel vessel. This results in the formation of silicate oxides [43]. Trace impurities of iron and other metals from the nutrient quartz form nucleation sites in the cultured crystal. This causes the formation of small but discernible silicate oxide inclusions which are quite typical for synthetic crystal quartz [44].

Depending on the size, shape, and crystal orientation of the seed bar, a variety of different cultured quartz crystals can be grown. By far most cultured quartz is grown for piezoelectric and other nonoptical applications. The type of crystals most suitable for optical applications are the so-called optical half sections. Other types of quartz crystals can also be used but at the expense of additional waste material that has to be removed.

The as grown cultured quartz bar has faceted ends and a rectangular cross section. The length of the bar is the direction of the $Y$-axis. The $Z$-axis (or optical axis) runs across its width, and the $X$-axis across the height, as shown in Fig. 2.33. The cross section of the bar reveals zones of different crystal orientation and thus different optical behavior. In the center of the cross section is the seed crystal, which runs the entire length of the stone in the $Y$-direction. This seed crystal is removed from the stone not only because it is expensive and reusable but also because the initial growth on the seed includes impurities and entrapped gasses. There is a $+X$ zone above and a $-X$ zone below the seed. On either side of the $+X$ zone there are transition zones called *S-zones*. The remainder is pure $Z$-material which is the only part of the crystal that is useful for optical applications. To produce the optical half section (Fig. 2.34), the undesirable orientations have been almost totally removed to leave only pure $Z$-material. The manufacturer will identify one of the sides of the bar and orient it to within 15 arc min to the $Z$-axis to serve as reference for subsequent cutting and shaping.

The standard seed crystal shown in Fig. 2.35 is 2 to 2.5 mm thick (0.080 to 0.100 in.) and from 150 to 190 mm long (6 to 7.5 in.). The height of the seed crystal determines the $X$-axis which can vary in dimension from 5.5 mm (0.220 in.) for the $Y$-bar to up to 15 mm (0.600 in.) for the $M$-bar. Since there is no growth in the $Y$-direction, the length of the as-grown bar remains the same as the length of the seed crystal.

## 2.4.2. Heat Exchanger Method

The heat exchanger method (HEM) of crystal growth was developed in the early 1970s [46] for growing large, up to 250-mm (10-in.) diameter, synthetic sapphire crystals. Silicon for solar cell manufacture and cobalt-doped magnesium fluoride for high-power laser applications have also been grown by the HEM method. There are now plans to grow GaAs by the HEM process for digital electronic devices and for optical communications.

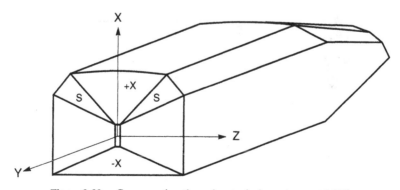

**Figure 2.33.** Cross section through a typical quartz crystal [45].

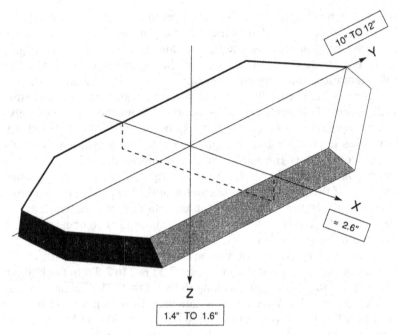

**Figure 2.34.** Optical half section [45].

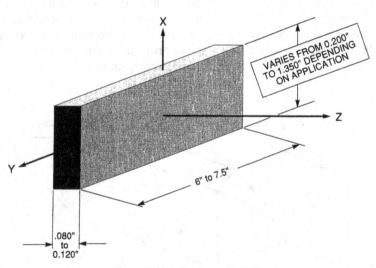

**Figure 2.35.** Seed crystal [45].

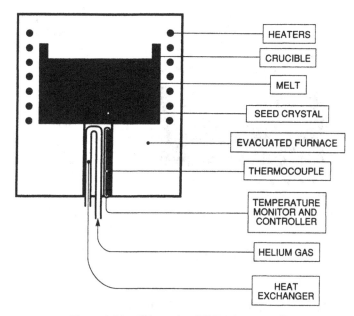

**Figure 2.36.** Schematic of HEM furnace [48].

As shown in Fig. 2.36 the crystal is grown from a seed crystal at the bottom of the HEM crucible. Thermal gradients are controlled by helium gas flowing through heat exchangers. This permits the control of the transition from the liquid to the solid phase without any motion, as is required for the Czochralski crystal growth method. The accurate control of thermal gradients during the HEM growth process allows the growth of large crystals that are essentially free from fractures or internal stress, since the process allows for annealing of thermal stresses immediately after solidification [47, 48].

### 2.4.3. Methods to Produce Alkaline Earth Fluorides

**Bridgeman-Stockbarger-Stober Method.** Synthetic calcium fluoride ($CaF_2$) has been grown by the Bridgeman-Stockbarger-Stober (BSS) method for some time [49]. Other alkaline earth fluorides such as high-quality single crystal $BaF_2$ and $SrF_2$ have also been synthetically grown by the BSS method. The size of the ingot is typically limited by the diameter and depth of the graphite crucible [50]. By growing the crystals in a reactive gas atmosphere, oxygen has been all but eliminated as a major contaminant in alkaline earth fluorides [51].

**Fusion Casting.** Fusion casting is a method to reduce or eliminate oxygen or hydroxyl ion impurities in alkaline earth fluorides such as $CaF_2$, $BaF_2$, and

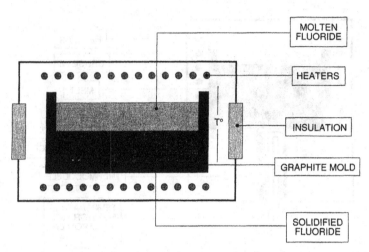

**Figure 2.37.**  Fusion casting.

$SrF_2$ [51]. Polycrystal ingots of up to 50 cm (20 in.) in diameter have been produced by this method [50], as shown in Fig. 2.37.

**Reactive Atmosphere Purification.** BSS grown alkaline earth fluorides are passed through a purification step in a reactive fluorocarbon atmosphere. This treatment, called reactive atmosphere purification (RAP), eliminates oxygen from alkaline earth fluorides; it increases their optical transmission and reduces the IR absorption [50, 53]. Prior to this development, low-absorption IR transparent materials were improved by a chemical purification process.

### 2.4.4.   Solution-Grown Crystals

Until the early 1970s KDP crystals were still size limited to about 100 × 100 × 300 mm (4 × 4 × 12 in.). To grow such a boule would take many weeks. Since deuterated KDP (made with heavy water) is even more difficult to grow, its size was then limited to 60 × 60 × 200 mm (2.4 × 2.4 × 8 in.) [54].

By the 1980s KDP crystal sizes had grown to 300 × 300 × 600 mm (12 × 12 × 24 in.). Since the optical quality of solution-grown crystals improves with slower growth rates, an optimum growth rate of 1 mm (0.040 in.) per day has emerged as an acceptable compromise between economics and quality. This means that to grow a 600-mm (24-in.) long crystal, it takes a growth period of nearly 20 months without a process failure or equipment breakdown [55].

Even larger boules were grown by the mid-1980s in support of Japan's fusion laser program. One U.S. grower [56] succeeded by growing boules weighing 250 kg (550 lb), which were 420 × 420 × 1000 mm (16 × 16 × 40 in.), to yield 390-mm (15.4-in.) diameter monolithic (made of one solid piece) KDP frequency doubling crystal.

The cost of KDP grown from supersaturated solutions is about $20/cm$^3$ or $330/in.$^3$ [57]. A polished and coated 15 × 15 × 25-mm KDP crystal with electrodes and mounted in a housing costs approximately $2000 [58].

The solution-growth method for KDP starts with large seed blocks of the crystal being placed into a nonreacting aqueous solution which is saturated with dissolved KDP. The temperature is then slowly lowered to create a supersaturated solution that forces the excess KDP to precipitate and crystallize on the seed block. Supersaturation and crystal growth can also be achieved by letting the water slowly evaporate [59].

At a growth rate of 1.5 mm (0.060 in.) per day, it would have taken about one year to grow a boule large enough for the 270 × 270-mm (10.5 × 10.5-in.) KDP crystals for the Nova laser [60] frequency converters. To speed matters up, the lab opted for a 3 × 3 array of smaller crystals that were faster to obtain at the desired level of quality. The material supplier improved the process and increased the growth rate by slowly agitating the growth solution with a turbine circulation arrangement. Increased growth rates of up to 25 mm (1.0 in.) per day have been reported. The growth process is very slow, and it is difficult to control and maintain over extended periods of time, but it is the only option for crystals that cannot be grown any other way.

**References for Section 2.4 (Optical Crystals)**

[43] Kinloch, D. R., "The culturing process: New life for quartz," *Optical Spectra,* December 1978.

[44] Kinloch, D. R., "Inclusions in cultured quartz crystal: New development and lower concentration," *Technical Digest,* OSA Workshop on Optical Fabrication and Testing, Palo Alto, CA, December 1982.

[45] Sawyer Research Products, Inc., Eastlake, OH.

[46] Crystal Systems, Inc., Salem, MA.

[47] Staff report: "New growth process yields large crystals," *Laser Focus,* March 1983.

[48] Schmid, F., "A new approach to high temperature crystal growth from the melt," *Solid State Technology,* September 1973.

[49] Malitson, I. H., "A redetermination of some optical properties of calcium fluoride," *Applied Optics,* vol. 2, no. 11, November 1963.

[50] Miles, P., "High transparency infrared materials—A technology update," *Optical Engineering,* vol. 15, no. 5, September/October 1976.

[51] Detrio, J. A., "Trends in optical materials," *Technical Digest,* OSA Workshop on Optical Fabrication and Testing, Palo Alto CA, December 1982.

[52] Truett, W. L., "Optical materials—The role of impurities," *Lasers and Applications,* August 1984.

[53] Detrio, J. A., "Action in the infrared," *Photonics Spectra,* September 1983.

[54] Adhav, R. S., "Frequency doubling crystals—Unscrambling the acronyms," *Electro-Optical Systems Design,* December 1974.

[55] Staff report, "Materials & chemicals—Optical spectra's annual report," *Optical Spectra,* July 1981.

[56] Cleveland Crystals, Inc., Cleveland, OH.

[57] Staff report, "Laser fusion crystals: Easing growing pains," *Photonics Spectra,* January 1986.

[58] Lin, J. T., "Choosing a nonlinear crystal," *Lasers and Optronics,* November 1987.

[59] Milstein, J. B., "Techniques for the growth of bulk optical crystals," *Laser Focus/Electro-Optics,* April 1988.

[60] Lawrence Livermore National Lab, Livermore, CA.

## 2.5.  SPECIAL METHODS FOR CRYSTALS, METALS, AND PLASTICS

A series of methods were developed over the years to compact crystalline or metallic powders into solids under heat and pressure. In some cases, low strength single crystal materials have been pressed into high strength polycrystalline materials. These methods are variably referred to as hot pressing, hot forging, press forging and hot isostatic pressing or HIP. Although there are some differences between these methods, they are sufficiently alike that for the purpose of this description, they can be treated together under one heading.

The method basically involves placing a powder of the raw material into a suitable mold and then compressing it under substantial pressure and heat to nearly 100% of its theoretical density. In the case of hot forging of certain crystals, the powder is replaced by a solid single crystal blank. The rest of the process is essentially the same.

### 2.5.1.  Hot Pressing of Crystals

The IRTRAN [61] materials are one group of IR crystals that have been successfully hot-pressed into a variety of shapes, including dome-shaped blanks. These materials were $MgF_2$, $CaF_2$, ZnS, ZnSe, MgO, and CdS. See Chapter 1, Section 1.4.7, for additional details on IRTRAN. Another crystal material that has been hot pressed is lithium fluoride (LiF). There is one problem with IR materials that are hot-pressed from powders. Because 100% of theoretical density cannot be achieved even under the most ideal conditions, there are residual microscopic voids evenly dispersed throughout the bulk. These voids cause scattering of shorter wavelengths. Even though some of these crystals transmit well into the visible region, the high scattering precludes their use at these wavelengths. Consequently most of these materials have now been replaced by more advanced materials, such as CVD ZnS and CVD ZnSe, and fusion cast alkaline earth fluorides, such as $CaF_2$ and $MgF_2$.

All hot-pressed optical materials are fine-grain polycrystals with an average grain size of about 20 $\mu$m. This fine-grain structure gives them greatly increased mechanical strength over that of single crystals of the same material. For instance, the yield strength of hot-pressed sodium chloride (NaCl) is six times greater than that of single-crystal NaCl. Hot pressing also tends to improve the resistance to microcreep and makes the crystal more water resistant as compared to equivalent single-crystal material. Some hot-pressed materials must be processed in a reactive gas atmosphere. The pressing of cadmium sulfide (CdS) in a hydrogen sulfur gas ($H_2S$) atmosphere is such an example.

Hot pressing of single-crystal materials has been most successful for alkaline halides. This process is also called *secondary hot forging*. High-purity single-crystal alkaline halide crystals are first grown either by the Stockbarger method in a reactive atmosphere or from a saturated solution (like most NaCl crystals). These single crystals are then pressed into polycrystalline material to about half of the initial blank thickness [62]. To improve the optical quality of the resulting material, the single-crystal material is first purified in a reactive gas atmosphere process (RAP) that eliminates oxygen impurities.

The single-crystal alkaline halide which, in the case of NaCl, have been grown up to 400 mm (16 in.) in diameter, is placed between two heated molds, and the ingot is heated until it becomes malleable. It is then pressed or extruded into the desired shape. These hot-pressed alkaline halides, particularly NaCl and KCl, are marketed under the tradename "Polytran" [63, 64]. Hot-pressed Polytran windows have been successfully produced in sizes up to 300 mm (12 in.) in diameter. The internal purity of the polycrystal material of the larger windows, however, is somewhat lower than in the smaller windows for which this parameter is easier to control. For some noncritical applications, some alkaline halides such as KBr and CsI can be cold pressed into usable shapes. These materials do not offer sufficient mechanical strength for most applications.

### 2.5.2. Hot Isostatic Pressing of Metals and Ceramics

Metals and ceramics can also be formed by hot isostatic pressing (HIP). Beryllium (Be) is now typically formed by the HIP process from a metal powder into intricate shapes. The advantage of the HIP process lies in the elimination of the extensive and very costly machining that is required to shape Be from the full into a lightweighted optical mirror blank. Beryllium is an expensive metal that is not easy to machine, but above all, it is a toxic material that can only be worked safely under strict OSHA-enforced environmental controls.

Although the HIP process helps overcome some of the problems associated with machining from the full, the toxicity of the beryllium powder
imposes even stricter controls on the handling of this material since Be
becomes particularly hazardous when inhaled as airborne particles. The HIP
process for Be is a two-step process. First, the Be powder is cold pressed at
50,000 psi to about 80% of theoretical density. The resulting raw blank is
then hot pressed at constant (or isostatic) pressure of 15,000 psi in a pressurized argon atmosphere at 600°C (1100°F). The final mirror blank is then at
near 100% theoretical density. Extensive lightweighting of mirrors for space
applications can be achieved with this method [65].

New ceramic materials such as germanium cordierite are HIP processed.
A blend of powdered oxides is first cold pressed, then sintered, and finally
hot pressed under isostatic pressure to very close of its theoretical density
[66].

### 2.5.3. Molding of Optical Plastics

There are basically three different ways in which transparent plastics can be
molded into optical components. The simplest, but also the most labor-
intensive method, is casting. Because of its high labor content, this method
is reserved only for large components for which the other methods would
not be practical, or for prototypes or short runs for which the high cost of
tooling cannot be justified. A glass mold is easily prepared using standard
optical shop practices. A slow curing liquid plastic material, such as allyl
ester resin, is poured into the mold which has been prepared with a suitable
release agent, and the resin is allowed to cure into a solid and optically
transparent plastic. The plastic materials used for casting suffer from significant shrinkage during the curing cycle and that fact makes the control of the
optical figure all but impossible [67]. This is why casting is used primarily for
lens blanks, which are then further machined and optically polished.

The second method is thermal compression molding. Starting with plastic
sheets, pellets, or even powders, optical components of limited quality are
formed between two heated dies in a vertical press [68]. The pressing and
curing cycle is typically 15 minutes per load. Fresnel lenses and other noncritical optics are produced this way.

Finally the most commonly used method is injection molding. Thermoplastic raw materials are used, and the process is fast and very low cost at
high volumes. Plastic powders are melted and injected at high pressure into a
temperature-controlled steel mold in a ram-type injection press. The pressure required for proper molding can exceed 10,000 psi. Precise control of all
molding parameters such as temperature, pressure, and time is essential for
a successful and repeatable molding process. The most expensive part of the
process, besides the molding machines, are the precision metal molds that
are required for producing the highly polished surfaces of plastic optics. The
high cost of making these molds requires high volume to amortize the front-

end investment. The molds, and particularly the critical inserts needed for each lens surface, are made from a carefully hardened and select grade of stainless steel that will accept a high degree of optical polish. CNC or SPDT machines are typically used to machine the critical mold surfaces (which can be plano, spherical, aspheric, or any optically useful shape) to a high degree of precision [69]. The molds can also be designed to provide for integrated mounting surfaces on the finished plastic optics.

### References for Section 2.5 (Crystals, Metals, and Plastics)

[61] Eastman Kodak Co., Rochester, NY.

[62] Miles, P., "High transparency infrared materials—A technology update," *Optical Engineering,* vol. 15, no. 5, September/October 1976.

[63] Staff report, "Chemical and materials—Quality is the watchword," *Photonics Spectra,* May 1984.

[64] Harshaw Crystal Products, Solon, OH.

[65] Pacquin, R. A., "Hot isostatic pressed beryllium for large optics," *Optical Engineering,* vol. 25, no. 9, September 1986.

[66] Mehrotra, J., "Germanium-cordierite ceramics for precision mirrors," *Technical Digest,* OSA Workshop on Optical Fabrication and Testing, Rochester, NY, October 1987.

[67] Benjamin, R., "Plastic optics," Bell & Howell product literature, 1979.

[68] Wolpert, H. D., "A close look at optical plastics—Part I," *Photonics Spectra,* February 1983.

[69] Grendol, C. L., "Molded plastic optical elements come into their own," *Laser Focus/Electro-Optics,* October 1988.

# 3

# Optical Shop Supplies

Optical shop supplies are essential elements for the manufacture of optics. It is a very broad term since it includes both major and minor items without which the manufacture of optics would not be possible. The following topics will be discussed in this chapter:

- Loose abrasives
- Polishing compounds
- Polishing pitch
- Polishing pads
- Blocking wax and pitch
- Cements and adhesives
- Coolants, protectants and solvents

## 3.1. LOOSE ABRASIVES

Optical materials are ground with hard mineral grains that have a Mohs hardness of seven or greater. Sufficient hardness is required to abrade a wide range of optical materials, but it is also desirable that the minerals resist cleaving, which means that the individual abrasive grains should not fracture easily. There are natural abrasives that are mined from ore and processed for use in optics. Some of them are no longer used because a number of very good synthetic abrasives have been developed for industrial use. In addition to abrasive hardness and resistance to grain breakdown, the shape and sizing of the individual abrasive grains determines the characteristic of the abrasive.

### 3.1.1. Natural Abrasives

The first abrasives used for optical manufacture were of natural origin. Emery, corundum, and garnet powders were the only available abrasives until the introduction of the first synthetic products. Natural diamond abrasives have also been in use for many years. Because of the increasing importance of diamond abrasives, especially as they relate to diamond tools, they are described separately in Section 3.1.3. of this chapter.

Because these abrasives were of natural origin, it was important to know the exact location from which the raw material was extracted from the earth for conversion into useful abrasives. Each source of a raw material had its own characteristics that imparted desirable or undesirable qualities to the abrasive. Even so the abrasives from the same source were not exactly uniform from batch to batch because the composition of the rock from which it was mined was not uniform throughout. There could be a great deal of variability which the optician had to recognize during the early stages of use and adjust the grinding method accordingly.

With the easy availability of synthetic abrasives of excellent quality and uniformity, the role of natural abrasives in the manufacture of optics has greatly diminished. They are mentioned here primarily for the sake of completeness and because all of them are still commercially available. The natural abrasives can offer some advantages over the synthetics under certain conditions, such as a milder cutting action with easier grain breakdown that can result in substantially finer finishes with some materials.

**Emery.** Emery is a natural mineral product that at one time was the primary abrasive used in optics manufacturing. There are two important sources. The oldest one is on the Mediterranean island of Naxos which yields Turkish emery. The other one is in the United States which yields an abrasive that is known as American emery.

Emeries are natural blends of aluminum oxide ($Al_2O_3$) and magnetite ($Fe_3O_4$). Their color ranges from light to dark grey. Turkish emery is darker than the American variety because of its higher aluminum oxide content. This difference makes it a more aggressive abrasive. However, both emeries have a milder abrasive action than the now more popular synthetic aluminum oxides because of softer minerals in their aggregate structure. The Mohs hardness of emeries lies between 8 and 9.

The grains of emery range in size from a coarse 54 mesh down to a medium fine 320 mesh. The grains are sharp and angular, and they tend to produce a relatively rough surface texture on most optical materials. These properties limits their current use to mostly nonoptical applications such as the lapping and buffing of metals. There may be occasional use for emery in the optical shop for special roughing operations.

There is a synthetic abrasive on the market today that is called *optical emery* [1]. The aluminum oxide content is similar to that of the natural emeries, but the synthetic version does not contain any magnetite. It is not a true emery. The light tan color of these products also distinguishes it from the natural emeries. Table 3.1 compares the composition of this abrasive with that of the natural emeries. This synthetic emery has been formulated and sized for double-sided lapping of semiconductor and glass materials.

**Table 3.1.  Chemical Composition and Physical Parameters for Natural Abrasives**

| Chemical compound | | US Emery | Turkish Emery | Corundum | Garnet | Optical Emery |
|---|---|---|---|---|---|---|
| Aluminum oxide | $Al_2O_3$ | 41.4% | 67.4% | 89.0% | 20.4% | 52.5% |
| Calcium oxide | CaO | | | | 3.0% | |
| Chromium oxide | $Cr_2O_3$ | | | <1.0% | | |
| Ferrous oxide | FeO | | | | 9.6% | |
| Ferric oxide | $Fe_2O_3$ | | | <1.0% | 12.6% | <1.0% |
| Magnetite | $Fe_3O_4$ | 25.9% | 24.5% | | | |
| Magnesium oxide | MgO | 3.6% | <1.0% | | 12.3% | |
| Manganese oxide | MnO | | | | <1.0% | |
| Silicon dioxide | $SiO_2$ | 25.3% | 3.5% | 6.6% | 41.2% | 17.6% |
| Titanium dioxide | $TiO_2$ | 2.8% | 3.5% | 3.0% | | 1.2% |
| Zirconium oxide | $ZrO_2$ | | | | | 28.5% |
| **Physical Parameters** | | | | | | |
| Specific gravity | $g/cm^3$ | 3.5 | 3.8 | 3.8 | 4.0 | 4.0 |
| Mohs hardness | | 8.0 | 9.0 | 9.0 | 8.5 | 9.0 |

**Corundum.** Corundum is naturally occurring aluminum oxide. Its composition is nearly 90% $Al_2O_3$ with small amounts of silicon dioxide ($SiO_2$), titanium oxide ($TiO_2$), and traces of other compounds. It has either a rhombohedral or hexagonal crystal structure. This means that the crystals tend to be barrel shaped and that they cleave at right angles to the long axis. Since the individual abrasive grains are fragments of the original crystals, this fracture mechanism gives the grains the desirable blocky shape. The predominant color of corundum is light tan. Corundum powders are available in the finer mesh sizes, which are classified by elutriation from 280 mesh (42 $\mu$m) down to 3600 equivalent mesh (4.5 $\mu$m).

Corundum has been used to fine-grind glass optics for well over one hundred years. It is still used today for high-volume ophthalmic manufacture. Its inherent hardness and toughness, and the fine finishes it can produce, makes corundum a good alternative to synthetic abrasives for fine-grinding silicon quartz and other hard optical materials.

**Garnet.** Garnet abrasives are mined in the Adirondacks in the state of New York. The raw material is a homogeneous iron aluminum silicate in which all oxides are combined to form a pink to deep red mineral. The general chemical formula that describes this mineral is $Fe_3Al_2 * 3(SiO_4)$. The primary constituents are silicon dioxide and aluminum oxide. Ferrous oxide (FeO) and ferric oxide ($Fe_2O_3$) are also important constituents. The presence of iron oxides in the mineral gives garnet its specific color. Table 3.1 gives a more detailed breakdown of the composition of garnet.

Garnet is a rather soft abrasive, with a Mohs hardness between 7 and 8. The abrasive grains are sharp, angular, and quite irregular. Garnet is one of the more friable abrasives. This tendency causes it to break down into smaller grains rather quickly at relatively low grinding pressures. Its softness and its breakdown characteristics make garnet a good choice for fine-grinding some of the softer optical materials. A skilled optician can take advantage of these characteristics to produce superior fine-ground surfaces. Garnet abrasive powders are available in most standard mesh sizes and can be classified down to less than one micron average particle size.

### 3.1.2. Synthetic Abrasives

Synthetic abrasives have been produced for over 50 years. The methods of manufacture have been perfected during this time to such an extent that almost all abrasives used in optical fabrication today are synthetic. They are typically of excellent uniformity and high purity. They are also more precisely graded and classified than the natural abrasives, and this results in predictable finishes each time.

The most widely used synthetic abrasives are aluminum oxide powders. They range in color from tan to pure white, depending on residual impurities. Several different quality grades will be discussed. Almost all free abrasive grinding done today in optical shops is done with synthetic aluminum oxide powders.

Two other important synthetic abrasives are silicon carbide (SiC) and boron carbide ($B_4C$). Both abrasives are very hard with the hardness of boron carbide approaching that of diamond. While silicon carbide may find occasional use in the rough grinding of hard optical materials, boron carbide is only used as an economical alternative for diamond. Both of these abrasives are briefly discussed for the sake of completeness.

**Synthetic Aluminum Oxide Abrasive.** Commercially produced aluminum oxide is the abrasive that is used in most optical shops today. Since sapphire is a pure form of aluminum oxide, it has also been called sapphire powder on occasion. The accepted term in use today is *alumina*.

The raw material is derived from bauxite, which is an argillaceous earth much like a fine clay. It is primarily composed of aluminum hydroxide. About 60% is aluminum oxide with iron oxide, titanium oxide, and calcium accounting for about 25%, while the remainder is made up of trace elements and bound water. The ore is first finely crushed and then calcined. Calcining is a purification process during which the crushed ore is heated to a temperature slightly below its melting point to drive off any volatile impurities such as organic residues and particularly entrapped water. The calcining step is followed by a milling operation during which the calcined bauxite is ground up into a coarse granular powder. Iron filings and carbon in the form of finely ground coke are blended with the bauxite powder. The blend is then placed into an electric arc furnace where it is melted into a solid mass at very high temperatures. The coke, which is derived from coal, is used as a reducing agent to liberate oxygen from the iron oxide; this converts it to metallic iron. The other undesirable constituents such as titanium, silicon, and calcium form an alloy with the iron and then settle as slag to the bottom of the furnace. Because of its distinctly darker coloration, the slag is later easily separated from the resulting furnace product which is a very hard material that is more than 95% aluminum oxide [2].

The raw aluminum oxide is ground into a coarse powder, with grain sizes typically between 0.5 and 2.5 mm. Depending on the refinement process used to improve the purity and hardness of the abrasive material, commercially available synthetic aluminum oxide is either a calcined alumina as produced or a fused alumina that has been subjected to additional heat treatment. The basic aluminum oxide is made up primarily of polycrystalline alpha alumina. The granular aluminum oxide powder is milled and finely ground up, and then it is graded and carefully sized by classification. The resulting product of this rather involved process is widely used in the optical shops. The basic process of synthetic abrasive manufacture is shown in Fig. 3.1. The grading and classification methods used are described in Section 3.1.4. of this chapter.

The color of aluminum oxide abrasive powders gives an indication of the purity of the material. Nearly 100% pure alumina is pure white. An off-white color indicates the presence of residual mineral impurities, and a tan abrasive powder contains appreciable percentages of these minerals. Table 3.2

**Table 3.2. Chemical Composition and Physical Parameters for Aluminas**

| Alumina | Regular | White | Precision |
|---|---|---|---|
| Crystal type | Alpha | Alpha | Alpha |
| Process | Calcined | Fused | Calcined |
| Color | Tan/grey | White | White |
| Composition(%) | | | |
| $Al_2O_3$ | 96.0 | 99.6 | 99.4 |
| $Na_2O$ | 0.0 | 0.2 | 0.4 |
| $TiO_2$ | 2.6 | 0.0 | 0.0 |
| $SiO_2$ | 1.1 | 0.1 | 0.1 |
| $Fe_2O_3$ | 0.3 | 0.1 | 0.1 |
| Knoop hardness | 2090 | 1980 | 2000 |
| Mohs hardness | 9 | 9 | 9 |
| Spec. gravity | 3.95 | 3.90 | 3.85 |

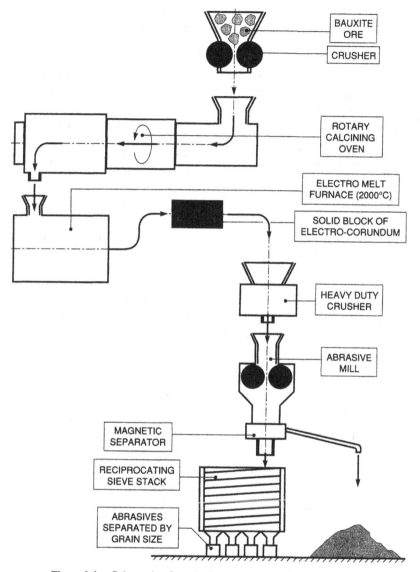

**Figure 3.1.** Schematic of synthetic abrasive powder manufacture.

shows the approximate composition of the three basic types of aluminum oxide powders.

There are several quality grades of synthetic aluminum oxide. Since there is no standardized way to describe these different aluminas, it is not always clear to the user which abrasive is of a specific grade. This ambiguity can easily lead to confusion. For instance, there are two basic crystal structures for synthetic alumina, namely, alpha and gamma. In addition there are fused, calcined, and hydrate aluminas. The technical information available from different suppliers appears to present these terms interchangeably. Although the differences in performance of the different aluminas might be small for some optical materials, they can cause some real problems for others. It is with these ambiguities in mind that the information in this section should be interpreted.

Regular fused aluminum oxide is the most economical synthetic grade. It is also known as optical emery, even though it shares only its color with its natural namesake. The color can range from brown to greyish tan (see Table 3.1). The crystal structure is primarily polycrystalline alpha alumina.

The alumina is crushed and sieve graded for mesh sizes to 280, or it is ball-milled and classified by elutriation to yield micron sizes from 45 $\mu$m on down (see Table 3.3). The abrasive grains are portions of the polycrystal material. They are blocky particles of high durability that present many sharp edges. The grading and elutriation processes are designed to eliminate all splintery particles that can cause scratching. The abrasive grains have an inherent capillarity which aids in proper wetting that increases the resistance of the grains to agglomerate in aqueous slurries. Most aluminum oxide powders are available with suspension treatment to ensure complete dispersal of the abrasive grains in the slurry. It also reduces settling and packing of the abrasive in recirculating slurry systems. The aluminas are nearly free of iron.

**Table 3.3.  Particle Size Designation for Alumina Abrasives**

| Average particle size in microns | Regular (grade) | White (mesh) | Precision (grade) |
|:---:|:---:|:---:|:---:|
| 40.0 |       | 280  | # 40 |
| 35.0 |       | 320  | # 35 |
| 30.0 |       | 400  | # 30 |
| 27.5 | # 275 |      | (28) |
| 25.0 | # 250 |      | # 25 |
| 22.5 | # 225 |      | (22) |
| 20.0 | # 200 | 500  | # 20 |
| 17.5 | # 175 |      | (18) |
| 15.0 | # 145 | 600  | # 15 |
| 12.5 | # 125 | 700  | # 12 |
| 9.0  | # 95  | 800  | # 9  |
| 5.0  | # 50  | 900  | # 5  |
| 3.0  | # 30  | 1000 | # 3  |
| 1.0  | # 10  | 1200 | # 1  |
| 0.8  |       | 1500 |      |
| 0.6  |       | 2000 |      |

This gives them very low magnetic properties, which is an important consideration when grinding semiconductor materials. Regular alumina is suitable for general industrial use and for noncritical optical applications.

White aluminum oxide abrasive is a high-purity version of the regular aluminum oxide powders. It is produced from very pure (>99%) fused alpha aluminum oxide stock. The abrasive powder is magnetically treated to remove ferrous contaminants. It is also chemically purified so that it can be used for applications where chemical contamination is a concern. Although the properties of white aluminum oxide are essentially identical to those of the regular aluminum oxide, the grains tend to be more friable because of their inherent porosity.

The individual grains are blocky with numerous sharp edges. The crushed raw stock is ball-milled to produce uniform grain sizes and close particle size distribution. The powder is then classified by elutriation into the standard micron sizes down to an average particle size as fine as 0.6 $\mu$m. Table 3.3 shows the typical grading schemes used for aluminas. The softer cutting action of white alumina makes it a good choice for grinding soft metals and soft nonmetallic optical materials.

Precision alumina is more commonly known by its trade name Microgrit [3]. It is also marketed by a number of suppliers of optical shop supplies under several different names. Precision alumina is made from calcined aluminum oxide. It is a pure white monocrystal alpha alumina that is more than 99% pure $Al_2O_3$. The most distinguishing characteristic of precision alumina is the irregularity hexagonal, platelet or disklike particle shape. The platelet-shaped abrasive grains have an aspect ratio of about one to five. This means that they are about five times as long as they are thick. During the grinding operation the platelets quite naturally orient themselves parallel to the surfaces of the grinding tool and workpiece. Figure 3.2 shows this behavior. This distributes the tool pressure over a larger area of abrasive particles than is the case with blocky grains. The platelet grains also resist breaking down and last much longer than other abrasive grains as a result. Unlike the stock removal action by rolling blocky abrasive grains, the platelet grains of precision alumina tend to remove material in more like a planing action. This greatly reduces the danger of scratching, and it typically produces finely ground surfaces with much reduced surface damage.

The unique particle shape of precision alumina required the development of special continuous flow elutriation methods for reliable and precise classification of the abrasive powders. The particle sizes range from 40 $\mu$m down to 1 $\mu$m. The grading is shown in Table 3.3.

The recommended slurry concentration is 25% abrasive by volume to water or other suitable vehicle. However, many users have found that much thinner slurries with concentrations as low as 10% work just as well. Thick slurries have proved to be quite ineffective. Like most other abrasives, precision aluminas are also available suspension treated for use in recirculat-

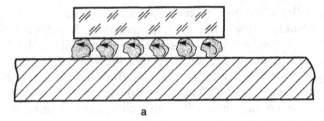

**Figure 3.2a.**    Standard grains: rolling or tumbling action.

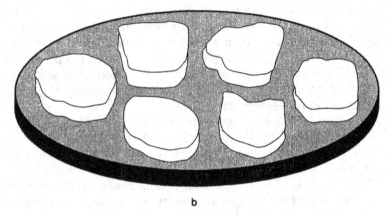

**Figure 3.2b.**    Platey calcined alpha alumnia in slurry during grinding: shaving or planing action.

ing systems. A couple of drops of glycerine or of a polyglycol solution can be added to aqueous slurries to act as extenders. A few percent of talcum powder added to oil-based slurries will do the same [4].

Precision aluminas are specifically formulated for precision optics to produce fine finishes on a wide variety of optical materials. Compared to other abrasives, precision aluminas are fast cutting on most optical materials, and they are also well suited for grinding medium hard metals and hard semiconductor materials. It is not by accident that these abrasives can be found in every optical shop.

**Silicon Carbide.** Silicon carbide is a very hard synthetic abrasive material that has found many uses in industry. It was once also known as carborundum. There is no naturally occurring form of this mineral.

The abrasive is produced at high temperatures from a blend of very pure quartz sand and powdered petroleum coke. The silicon, which is the main constituent in the quartz sand, combines with the carbon of the coke powder to form silicon carbide (SiC). Even though silicon carbide has a high melting point at 2200°C, it tends to slowly decompose through oxidation in air at temperatures above 1600°C.

There are two grades of silicon carbide. They differ in color, density, and hardness. One is black or dark grey, while the other is green. The black silicon carbide is tougher and slightly harder than the green silicon carbide. The reason for this difference is that the black variety has a much less porous structure which gives it a higher density than the green silicon carbide. However, since the physical parameters of both types are nearly identical, there is in all probability no real difference in their grinding efficiency under normal operating conditions. Some of the physical properties of silicon carbide and the composition of the two grades are listed in Table 3.4.

The raw material is crushed, ball-milled, and water classified by elutriation to yield blocky, sharp-edged, and closely sized grains of uniform shape. Although the crystal structure of alpha silicon carbide is a modified hexagonal form, the individual abrasive grains are conchoidal fragments of these crystals. The grain sizes of commercially available SiC abrasives range from a high of 6 mesh down to 280 mesh for the sieve graded sizes, and from 270 mesh (40 $\mu$m) down to an equivalent mesh of 2600 (<1 $\mu$m) for the elutriated grades. The most common sizes useful for optical grinding are listed in Table 3.4.

Silicon carbide is widely used in abrasive wheels for precision grinding of metals and other hard and brittle materials. Its high thermal conductivity and low thermal expansion characteristics permit the wheels to maintain their dimensional stability even under heavy working loads. Silicon carbide wheels are, however, rarely used in optics manufacture, since almost all bound abrasive tools are diamond wheels.

**Table 3.4.  Physical Properties, Composition, and Sizing of Silicon Carbide**

| Composition (%) | Black | Green | Mesh size | Micron size |
|---|---|---|---|---|
| SiC | 98.6 | 98.2 | 280 | 40 |
| SiO$_2$ | 0.5 | 1.0 | 320 | 35 |
| Si | 0.3 | 0.2 | 400 | 28 |
| C | 0.4 | 0.4 | 500 | 22 |
| Fe$_2$O$_3$ | 0.1 | 0.1 | 600 | 18 |
| Al | 0.1 | 0.1 | 800 | 12 |
| | | | 1000 | 8 |
| | | | 1200 | 6 |
| Hardness:  Knoop (kg/mm$^2$)  2480 | | | | |
| Mohs  9.3 | | | | |
| Specific gravity: (g/cm$^3$)  3.2 | | | | |

Silicon carbide loose abrasive powders are sometimes used when hard, brittle, and thermally sensitive optical materials have to be ground. It is the preferred abrasive for slurry saws in which crystal quartz and other similar materials are sliced. It is sometimes used as a free abrasive for the initial grinding step on silicon parts.

Its exceptional hardness makes SiC fast cutting, and it resists wear and grain breakdown. Silicon carbide grains will slowly break down during heavy duty grinding, but in doing so, they constantly present new cutting edges that keep the abrasive fast cutting. It is this tenacious resistance to grain breakdown that limits the use of SiC in optics to the initial roughing operation only.

**Boron Carbide.** Boron carbide is another exceptionally hard synthetic abrasive material. It is hard enough to serve as an economic alternative to diamond abrasives. Because of its costly method of production, boron carbide is many times more expensive than silicon carbide.

Boron carbide is produced in an electric arc furnace by fusing together ground-up boric acid glass and powdered petroleum coke. The product of this process is a very hard black substance, which is between 97% and 99% boron carbide ($B_4C$) with the remainder being trace elements such as free carbon.

This raw material is crushed, sieved, milled, and elutriated to yield uniformly blocky and closely sized abrasive grains. The specific gravity of this material is 2.51 $g/cm^3$, and the hardness of 9.5 on the Mohs scale. This is equivalent to a Knoop hardness that lies between 2800 to 3000 $kg/mm^2$, making boron carbide the hardest abrasive material next to diamond. The commercially available grain sizes range from 80 mesh on down to an equivalent mesh size of 1500.

The primary use of boron carbide is in ultrasonic machining of hard materials with a Mohs hardness of 8 or higher. Other applications of $B_4C$ is for those situations where SiC is not hard enough and where the cost of diamond abrasives powders cannot be justified.

### 3.1.3.  Diamond Powders and Compounds

Diamond abrasives are playing an ever-increasing role in the manufacture of optics and semiconductors. They are used in powder form as a diamond slurry or paste, or they are bound in a metal or plastic matrix for use in a great variety of diamond tools. The primary use of diamond has been up to now in diamond tools. Because of this importance, diamond tools are discussed separately in Chapter 4.

The use of diamond slurries has been limited so far to lapping and polishing of hard and difficult to work materials such as some metals, hard crystals like YAG, ruby, or sapphire, and ceramic substrate. But with the constantly increasing number of optical materials, it is quite likely that diamond slurries will play a greater role in the future. The majority of optical materials, however, are relatively soft and are best ground and polished with standard abrasives and polishing compounds. Not only are they much less expensive than diamond slurries, but they also produce better finishes on these softer materials.

Despite these limitations there is a sufficiently important place in optics for diamond abrasives to warrant a detailed description. This is especially true since the diamond powders also form the basis for the majority of diamond tools and an understanding of diamond powders will increase our understanding of diamond tools.

**Natural Diamonds.** Natural diamonds were formed at very high temperatures and under tremendous pressures deep within the mantle of the earth. They are found in various places on our planet, but by far the most important location is in South Africa. This country dominates the diamond trade through DeBeers, Ltd.

Diamond is pure elemental carbon in a specific crystalline form. The most flawless crystals are fashioned into precious gemstones by expert gem cutters. This accounts for only a small percentage of the total. By far the greatest percentage of all diamonds mined is used for a wide variety of industrial applications. Some of these industrial grade diamonds are specially processed and refined from what is called *diamond bort* for use in metallurgy, the semiconductor industry, and in optics manufacture. Bort is the collective term for small diamond grains in the micron range that are of a specific crystal structure and of high purity. Diamond bort is the raw material for diamond abrasives and for diamond tools. It is almost exclusively supplied by DeBeers, Ltd.

Good-quality natural diamond grains have a number of unique characteristics that are unmatched by any other substance known to humans. The most important of these is that diamond is the hardest known material. Its Mohs hardness is 10 (by definition), and the corresponding Knoop hardness exceeds 8000 kg/mm$^2$. The basic structure of diamond is elemental carbon that has crystallized into a cubic orientation. This produces octohedral (8-sided) and dodecahedral (12-sided) crystals. The blocky shape of these individual crystals in conjunction with the exceptional hardness gives diamond its unique abrasive characteristics. Not surprisingly diamond has also the highest abrasion resistance of all abrasives. This means that it does not readily wear, and it remains sharp and aggressive for a long time. Natural diamond has a brilliant clear color. It is also a much more efficient thermal conductor than copper. This makes diamond a most efficient heat sink for those special applications where the high cost of diamonds is outweighed by the benefits derived from its use.

The high thermal conductivity makes diamond also an excellent choice for machine tools where high loads from high stock removal rates would cause other abrasives to fail much sooner. Select grade single natural diamonds are used for diamond-turning operations because of their unmatched wear resistance. Despite all these superlatives diamond does have a drawback in that it tends to slowly decompose into graphite when it is subjected to temperatures that exceed 800°C for extended periods of time.

Previously unused diamond is called *virgin diamond*. High-quality natural virgin diamond powders are produced from diamond bort that has been crushed under carefully controlled conditions, then ball-milled, and finally sieved and classified to be used as loose abrasive or to be used for diamond tools. The sieving and classification removes the fines, which are very small and hence inefficient fragments, and eliminates any irregular shapes that can lead to scratching.

**Synthetic Diamonds.** Synthetic diamonds have been produced for more than 30 years by several different processes that were developed in response to the uncertain supply and the high cost of prime-quality natural diamond bort. The cost difference between natural and synthetic diamonds is now only 10% to 15%. But that is enough to economically produce sufficient quantities of synthetic diamond, so many diamond tools are made exclusively with synthetic diamond powders. Various blends of synthetic and natural diamonds, both virgin and reclaimed, are also often used for making tools. It is not clear from the user's standpoint what the advantages are of doing this.

The majority of diamond tools that use the synthetic variety or the blends are resinoid bonded tools. Metal bonded tools are made whereby the diamond grains are captured in a metal matrix by a high-temperature sintering process. While natural diamonds stand up well to this process, the synthetic diamond varieties remain stable only at temperatures below 600°C. Therefore most metal bonded diamond tools, especially those with hard steel and alloy bonds use natural diamonds.

Synthetic diamond has been commercially produced since the late 1950s. General Electric and Du Pont developed two different methods to produce diamonds through high-pressure and high-temperature synthesis. Synthetic diamonds are sometimes also called "man-made" or manufactured diamonds. These methods have been perfected over the years to the point where they produce industrial diamonds of the highest quality that are superior for some applications to natural diamonds. The two methods produce two different types of diamond. The GE process yields very strong single crystals with blocky shapes that are quite similar to select grade natural diamond. The only differences are that the synthetic diamond is not quite as tough and that its color is a dark green rather than the clear white of natural diamonds. The Du Pont process yields black, irregularly shaped, and randomly oriented polycrystalline diamonds. Their multifaceted shape produces many sharp cutting edges. The inherent friability of these crystals gives them a self-sharpening effect that is a very desirable attribute when working with hard materials such as sapphire and ceramics.

A wealth of information can be gleaned from the rather extensive technical information that is made available by the suppliers of diamond abrasives and tools. The following paragraphs are a selection of excerpts adapted from the trade literature.

Synthetically produced micron-sized diamond powders are made up of strong monocrystals that are typically in blocky octohedral or dodecahedral shapes. These GE process crystals have multiedged cutting surfaces that resist breakdown under high working stresses and elevated temperatures. This grade of diamond exhibits all the desirable qualities of select grade natural diamond. The regular shapes gives these diamonds excellent metal bond retention properties [5].

The standard synthetic grade will consist of irregularly shaped, partially blocky grains that range from moderately to highly friable under normal working stress. The friability causes the diamonds to fracture, thus exposing constantly new and sharp cutting edges. These grades are most often used for resin bonded tools, and they are a good choice where a softer cutting action is desirable [6].

Diamond powders (bort) are available in many types, shapes, and sizes. They are categorized by being either natural or synthetic, processed or unprocessed, tough or friable, and block or splintery. More than 60 distinctly different sizes and size ranges have been developed for use as a free abrasive or as bound abrasive in diamond tools. Tough and blocky diamonds are preferred where high pressures are generated during grinding operations. Hard and brittle materials such as fused quartz, sapphire, and fired ceramics are most effectively ground with more friable, self-sharpening diamond. For such applications synthetic diamonds perform better than natural diamonds. Natural diamonds are the best choice for metal bonds since synthetic diamonds do not perform nearly as well [7].

Synthetic diamonds are available in a variety of sizes and shapes for use in saw blades and other diamond tools. Diamond powders are available in three basic quality grades:

- Natural virgin diamond
- Synthetic virgin diamond
- Reclaimed diamond

Virgin diamond has never been used before. Reclaimed diamond is an inferior grade that is not useful for grinding and polishing of optics or other critical surfaces. Since it is composed of natural or synthetic, and often a mix of both, its color ranges from light to dark green. The most efficient free abrasive powder will have uniformly sized, blocky grains with sharp cutting edges. This way the majority of diamonds will be engaged as working particles during grinding at any given time [8].

**Table 3.5.   Relative Cost of
Diamond Powder**

| Diamond Grade | Relative Cost |
|---|---|
| Natural premium grade | 1.00 |
| Natural standard grade | 0.90 |
| Synthetic (Du Pont) | 0.81 |
| Synthetic (GE) | 0.87 |
| Reclaimed diamond | 0.70 |

Diamond powders are measured in carats. One carat equals 0.2 grams. Therefore 1 gram equals 5 carats, and there are 140 carats per ounce. The cost of diamond powder is dependent on its weight, type, and grade. Since prices for diamonds fluctuate like any other commodity, it is very difficult to generalize on the cost of diamond powder. One can, however, look at the relative cost of the various grades of diamond powders, where the cost of premium grade natural diamonds is unity. This is shown in Table 3.5.

Cubic boron nitride (CBN) is also produced by methods similar to synthetic diamond manufacture. CBN is not quite as hard as diamond, but with a Mohs hardness greater than 9 it comes pretty close. It is used as a very aggressive free abrasive or bound in vitrified grinding wheels to work steel alloys. CBN is rarely, if ever, used in optics manufacture.

**Abrasive Premix Slurries.** Premixed diamond slurries and pastes are most often the preferred way to use diamond powders. A discussion of these leads to the concept of diamond concentration which must be clearly understood. The great variety of diamond sizes and the manner in which they are sized are also of great importance.

Although diamond abrasives are commercially available as dry powder, almost all diamond abrasives are sold as premixed slurries or pastes. This is primarily due to the diamond powder's difficult wetting characteristics, which can cause problems of agglomeration if it is mixed improperly. Agglomeration is a problem for all sizes, but it is particularly unacceptable for the finer sizes of one micron or less that are used for polishing. Any agglomerating particles would cause serious scratches on polished surfaces.

Premixed slurries in which the diamond grains are suspended and uniformly dispersed are usually based on a vehicle of high-purity deionized water to which a synthetic mineral oil has been added to reduce surface tension and eliminate the mutual attraction between diamond particles. Other vehicles or carriers are water-soluble oils (emulsions), as well as high- or low-viscosity additives that are soluble in oil. However, diamond powders resist uniform dispersal in oil and because of this, there is a trend toward the exclusive use of water-soluble carriers [9]. These liquid premix suspensions are typically bottled in 400-g bottles that contain from 10 to 100 carats of diamonds, depending on size and concentration. The premix slurries are typically applied to the lap with eyedroppers.

Diamond compounds or pastes are the most prevalent form of supply for diamond abrasives. The diamond particles are mixed into a thick paste which is soluble in either water or oil. To prevent agglomeration, the paste must surround each particle and suspend it. There are three basic types of such paste carriers. Water-soluble pastes provide some cleaning action in addition to good lubricity. They are also used for applications where oil-based slurries cannot be used. Oil-soluble pastes offer the optimum lubricity which makes them the preferred choice for the polishing of metals. The third carrier type is a combination of both which takes advantage of the benefits that each of them has to contribute. These diamond compounds and pastes are color coded to clearly identify a particular diamond particle size. However, there is no industrywide color-coding standard that is used by every supplier. It is best to get clear information on the color-coding scheme used by the chosen supplier.

The compounds or pastes are packaged either in 5-g or 18-g jars, or in similar-sized dispensing guns. Packaging the compound in disposable cartridges for metal applicator guns or directly in disposable plastic dispensers makes the application of the diamond to the lap much more convenient and much less wasteful than any other method of application. This type of packaging protects the compound from external contamination and ensures that its quality does not degrade during use. The vehicles used on the lap in conjunction with these compounds or pastes must continue to prevent individual grains from agglomerating with others. Only this way will it be possible to obtain predictable results. It is best to rely on the manufacturer's recommendations on how to use these products most advantageously.

Another form in which diamond abrasives can be dispensed to the lap is from pressurized aerosol spray cans. These use diamond powders from submicron sizes up to 45 $\mu$m that are dispersed in a synthetic low-viscosity water- or oil-soluble vehicle. These sprays are typically used in labs where metallurgical samples are prepared.

**Diamond Concentration.** Diamond concentration is an important parameter that can greatly affect the results obtained. It is also one of the prime factors that determines the cost of the diamond abrasive. Since there is no accepted standard that clearly specifies the abrasive content in slurries or compounds, we find that the concentration can vary considerably from supplier to supplier for the same size and grade of diamond. Because of the lack of such a standard, even the manner in which the concentration is designated varies greatly as well. Low or light, medium or standard, strong or heavy, and even superior are the terms typically used to describe the concentration of diamond abrasives. A better and probably more objective way would be to define the concentration as carats of diamond per gram of compound or slurry. Because of these ambiguities it is prudent to stay with a reliable supplier and to resist the temptation to switch to an apparently cheaper new source. If a switch is necessary, it is important to make sure that the product of the new supplier is indeed equivalent to that of the old one.

The cost of diamonds tends to fluctuate as supply and demand changes. It is possible to discuss cost only on a relative basis. It is generally true that natural virgin diamond costs between 10% and 40% more than synthetic virgin diamond. However, much depends on sizes and quality grades. It is probably safe to say that on average natural virgin diamond costs 20% more than an equivalent quality grade of synthetic virgin diamond. See Table 3.5 for details. Upon close inspection, it can be further stated that finer sizes and lower concentrations cost less than larger sizes and higher concentrations. This relationship is shown in Fig. 3.3.

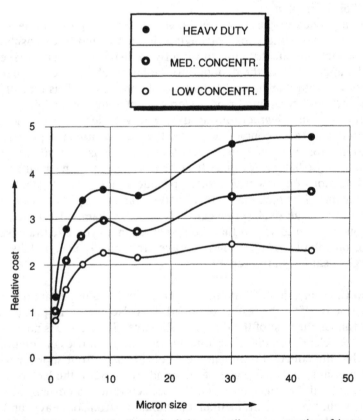

**Figure 3.3.** Cost of synthetic diamond relative to medium concentration of 1-$\mu$m size.

**Grading and Sizing.** The grading and sizing of diamond abrasives are two additional parameters of importance. Grading describes the type and quality of the diamond, whereas sizing describes the manner by which the powders are separated into specific ranges of particle sizes. Grading often assigns a letter code to the diamond type. The choice of that code is determined by the manufacturer or supplier because there is no standard for that. As a result it is nearly impossible for the user to determine if two diamond abrasives from two different suppliers are equivalent. Typical grades are listed in Table 3.6.

Synthetic diamonds are also electroplated with either nickel or copper in preparation for use in metal bonded or electroplated diamond tools. These will be discussed in greater detail in Chapter 4. Metal-coated diamond is not used as a free abrasive.

One abrasive product manufacturer [10] has come up with a simple scheme that makes it possible to have the customer specify the best diamond abrasive formulation for a specific application. The specification scheme uses two types of synthetic diamond, two different diamond concentrations, and two types of vehicles:

- Diamond types: polycrystal (D) and single crystal (A).
- Concentration: medium (M) and strong (S).
- Vehicle: water soluble (W) and oil soluble (O).

Together with diamond sizes ranging from 0.5 to 10 $\mu$m, a large number of different diamond slurry formulations are possible.

Sizing is a very important parameter for all abrasive products. It is particularly important for micron-sized abrasives that are often used for polishing critical surfaces. Careful sizing not only results in a powder with a specific average particle size but, more important, with an ideal narrow particle size distribution. Much more is said about that in the next section. Sizing is especially critical for very fine diamond powders below 1 $\mu$m where a few

**Table 3.6. Typical Diamond Grade Designations**

| Diamond type | Characteristic | Grade designation |
|---|---|---|
| Natural | medium tough | D or A |
| Natural | blocky, tough | BD |
| Natural | blocky, processed | ED |
| Synthetic | medium friable | ID or F |
| Synthetic | very friable | MD or E |
| Synthetic | nickel coated | ND, C or CD |
| Synthetic | copper coated | CDC or D |

slightly larger particles in the abrasive can create unacceptable scratching on polished surfaces. This is not so much of a problem for other abrasives, since the individual particles will break down very quickly long before they can do too much damage. Diamonds, on the other hand, are very tough, and they strongly resist particle breakdown. The grains are also much harder than any other material, so it takes only one oversize particle in a poorly sized abrasive to create major problems on the workpiece.

**Table 3.7. Diamond Powder Sizes**

| Avg. size (µm) | Range (µm) | Equiv. mesh | Color code | Diamond concentration | | |
|---|---|---|---|---|---|---|
| | | | | Light | Medium | Heavy |
| 1/10 | 0.0 - 0.2 | 240 000 | grey or white | • | • | • |
| 1/8 | 0.0 - 0.3 | 200 000 | white | • | • | • |
| 1/4 | 0.0 - 0.5 | 100 000 | grey or white | • | • | • |
| 1/2 | 0.0 - 1.0 | 60 000 | grey or white | • | • | • |
| 3/4 | 0.5 - 1.5 | 20 000 | white | • | • | • |
| 1.0 | 0.0 - 2.0 | 14 000 | blue or ivory | • | • | • |
| 1.5 | 1.0 - 2.0 | 13 000 | white | • | • | • |
| 1.8 | 0.5 - 2.5 | 12 000 | pink | • | • | • |
| 2.0 | 1 - 3 | 11 000 | clear or white | • | • | • |
| 2.5 | 2 - 3 | 9 000 | pink | • | • | • |
| 3.0 | 2 - 4 | 8 000 | yellow | • | • | • |
| 3.5 | 2 - 5 | 7 000 | yellow green | • | • | • |
| 4.0 | 3 - 5 | 6 000 | clear | • | • | • |
| 4.5 | 3 - 6 | 4 500 | yellow green | • | • | • |
| 5.0 | 3 - 8 | 4 000 | orange | • | • | • |
| 6.0 | 4 - 8 | 3 000 | yellow/orange | • | • | • |
| 7.0 | 5 -10 | 2 700 | green | • | • | • |
| 7.5 | 5 - 10 | 2 500 | green/orange | • | • | • |
| 8.0 | 5 - 12 | 2 300 | green | • | • | • |
| 9.0 | 8 - 12 | 1 800 | green | • | • | • |
| 10 | 8 - 16 | 1 700 | green | • | • | • |
| 12 | 9 - 15 | 1 500 | clear | • | • | • |
| 13 | 10 - 16 | 1 500 | green | • | • | • |
| 14 | 12 - 17 | 1 400 | blue | • | • | • |
| 15 | 10 - 20 | 1 200 | light blue | • | • | • |
| 16 | 12 - 22 | 1 100 | light blue | • | • | • |
| 17 | 15 - 20 | 1 100 | dark blue | • | • | • |
| 18 | 12 - 25 | 1 000 | blue | • | • | • |
| 20 | 15 - 25 | 900 | dark blue | • | • | • |
| 21 | 18 - 24 | 875 | dark blue | • | • | • |
| 22 | 20 - 25 | 850 | red | • | • | • |
| 25 | 20 - 30 | 800 | red | • | • | • |
| 26 | 22 - 30 | 700 | red | • | • | • |
| 30 | 22 - 36 | 600 | red | • | • | • |
| 35 | 30 - 40 | 500 | red or brown | • | • | • |
| 40 | 30 - 50 | 450 | brown | • | • | • |
| 45 | 36 - 54 | 330 | brown | • | • | • |

The particle size that identifies a particular diamond abrasive is almost without exception the average particle size in microns. It is an important parameter, but as pointed out in the preceding paragraph, it is equally important to know the particle size distribution. Table 3.7 is a composite list of commercially available diamond abrasives from a number of sources. It shows the nominal size and the associated typical particle size range. It also lists the equivalent mesh sizes. The entry in the color code column is something of a consensus compromise since numerous departures from the listed colors exist. The intent of the color coding is to clearly identify the diamond abrasive compound to avoid mixing sizes inadvertently. Since there is no generally applicable color code standard, manufacturers and suppliers have developed their own codes. A problem of misidentification can develop when the user switches to another supplier who uses a different color-coding scheme. Finally the availability of the diamond abrasives in various concentrations is indicated in Table 3.7 as well. With the considerable variability in defining different diamond concentrations, this list also represents a compromise.

Table 3.8 provides a more detailed listing of the diamond sizes and their typical particle size distribution, since these conditions are of interest in optics manufacture. The narrowest range represents the best-sized diamond abrasives, as advertised by the leading suppliers. The broadest range column lists the distribution of the not so carefully sized abrasives that are offered by

**Table 3.8.   Size Ranges for Diamond Abrasives**

| Nominal size | Narrowest range Min. - Max. | Range (µm) | Broadest range Min. - Max. | Range (µm) |
|---|---|---|---|---|
| 1 | 0 - 2 | 2 | 0 - 3 | 3 |
| 2 | 1 - 3 | 2 | 0.5 - 3 | 2.5 |
| 3 | 2 - 4 | 2 | 1 - 5 | 4 |
| 4 | 3 - 5 | 2 | 2 - 6 | 4 |
| 5 | 4 - 6 | 2 | 3 - 7 | 4 |
| 6 | 5 - 7 | 2 | 4 - 8 | 4 |
| 7 | 6 - 9 | 3 | 5 - 10 | 5 |
| 8 | 6 - 10 | 4 | 5 - 11 | 6 |
| 9 | 8 - 12 | 4 | 6 - 12 | 6 |
| 10 | 8 - 12 | 4 | 8 - 16 | 8 |
| 12 | 9 - 15 | 6 | 8 - 16 | 8 |
| 15 | 12 - 20 | 8 | 10 - 20 | 10 |
| 20 | 18 - 24 | 8 | 15 - 25 | 10 |
| 25 | 22 - 30 | 8 | 20 - 30 | 10 |
| 30 | 25 - 35 | 10 | 22 - 36 | 14 |
| 35 | 30 - 40 | 10 | 25 - 45 | 20 |
| 45 | 36 - 54 | 18 | 30 - 60 | 30 |
| 60 | 50 - 70 | 20 | 50 - 80 | 30 |

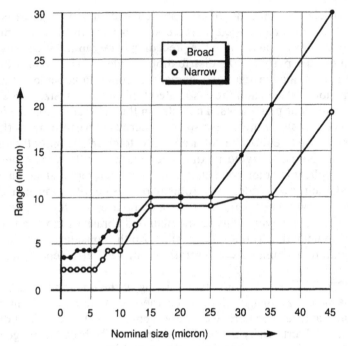

**Figure 3.4.** Diamond sizing.

the same supplier. As shown by the particle size range, significant differences exist that can have a marked effect on the quality of the surface finish. These typical diamond size ranges are graphically shown in Fig. 3.4.

Sizing is done either by a very slow sedimentation or elutriation process that can take as much as a month to complete for some of the submicron sizes, or by using electronically controlled centrifugal grading machines that are much faster but require several cycles until the desired size distribution has been achieved. Prior to sizing, the unprocessed diamond powders are mechanically shaped and sieved to remove splintery and flaky fragments, so only the desirable blocky grains remain. Repeated centrifugal sifting or elutriations will remove the fines, which are the very small particles that do not contribute to the abrasive action.

The results of the sizing processes are monitored regularly by centrifugal particle size analyzers or by Coulter counters. These specially designed particle analyzers measure the percent distribution of particles by size, count the number of particles per gram, or determine the number of particles per unit volume of liquid suspension.

### 3.1.4. Sizing, Grading, and Classifying

The grading and sizing of diamond abrasive powders were discussed in the preceding section. While there are many parallels between diamond and abrasive grinding powders, there are important differences in the sizing, grading, and classification of nondiamond abrasive powders to warrant a separate discussion.

The sizing methods and the way abrasive sizes are designated are the most difficult topics to write about when discussing abrasives. Sizing and classification methods can only be discussed in general terms, since the specifics are proprietary secrets that are closely guarded by the abrasive manufacturers. Not only are the sizing methods a mystery to the user, but the many different ways by which the various abrasive sizes are designated are a constant source of confusion. This chaotic situation is the result of the absence of a generally agreed-on functional standard. A theoretical standard does exist, as is shown in Fig. 3.5 which also shows some representative commercially available abrasive sizes. The theoretical standard is difficult to enforce because of the secrecy surrounding sizing and classification methods and because the shape of the particles strongly affects the result of sizing.

The abrasive size designation does not follow a uniform standard in the United States. However, some countries have such uniform standards. For instance, the German industry standard for free abrasives is DIN 58 751. It specifies not only the nominal sizes but also the permissible range of sizes from the smallest to the largest particle. The range is expressed in terms of a percentage of the nominal particle size. It also differentiates between sieve

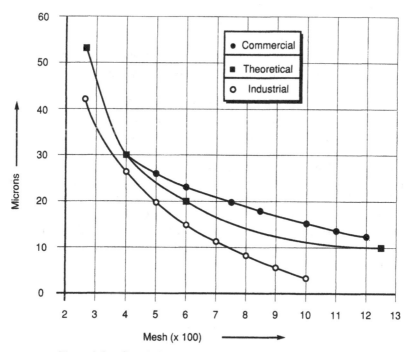

**Figure 3.5.**  Correlation between mesh sizes and micron sizes.

**Table 3.9.  Abrasive Grain Sizes Used in Optics per DIN 58 751**

| Mesh size (nominal) | Average size (µm) | Range in % | Elutriated grain size | Average size (µm) | Range in % |
|---|---|---|---|---|---|
| OS 230 | 230 | 20.0 | OS 53 | 53.0 | 3.0 |
| OS 163 | 163 | 14.0 | OS 45 | 45.5 | 2.0 |
| OS 115 | 115 | 13.0 | OS 37 | 37.5 | 1.5 |
| OS 75 | 75 | 10.0 | OS 29 | 29.2 | 1.5 |
| | | | OS 23 | 22.8 | 1.5 |
| | | | OS 17 | 17.3 | 1.0 |
| | | | OS 13 | 12.8 | 1.0 |
| | | | OS 9 | 9.3 | 1.0 |
| | | | OS 7 | 6.5 | 1.0 |
| | | | OS 5 | 4.5 | 0.8 |
| | | | OS 3 | 3.0 | 0.5 |

sizing (mesh) and sizing by elutriation (microns). Table 3.9 summarizes the standard abrasive sizes.

This variety of abrasive grain sizes is used in the optical shops for the following processes:

· Grain sizes from 45 to 75 µm are used for roughing.
· Grain sizes from 17 to 37 µm are used for pre-fine-grinding.
· Grain sizes from 3 to 13 µm are used for fine-grinding.

In addition to the classification of grain sizes per DIN 58 751, abrasive grain designations used in the older systems and those used in the United States are listed for comparison in Table 3.10. The particle size correlation is only approximate in some cases. The correlation between micron size and mesh size also differs from other systems (see Fig. 3.5).

Unlike the confusing specification requirements for ordering just the right type of abrasive in the United States, DIN 58 751 provides a scheme to unambiguously specify the needed optical abrasive. In accordance with this scheme, the order can be expressed with standardized identification letters. For instance, a nominally 13-µm synthetic aluminum oxide for optical use is specified as

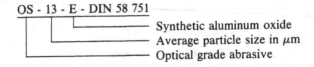

OS - 13 - E - DIN 58 751
————————— Synthetic aluminum oxide
————————— Average particle size in µm
————————— Optical grade abrasive

**Table 3.10.   Approximate Correlation of Different Abrasive Sizes**

| DIN 56 751 abrasive | Settling time (min) | Mesh sizes | Barton Garnet | AO Emery | KC Optical |
|---|---|---|---|---|---|
| OS 163 |     | 100   |      |          |     |
| OS 115 |     | 120   |      | M301     |     |
| OS 75  |     | 280   |      |          |     |
| OS 63  |     | 320   | W 04 |          |     |
| OS 45  |     | 360   | W 03 |          |     |
| OS 37  |     | 400   | W 02 |          |     |
| OS 29  | 15  | 600   | W 1  |          | 275 |
| OS 23  | 40  | 900   | W 3  | M302     | 225 |
| OS 17  |     | 1 100 | W 4  | M302 1/2 | 175 |
| OS 13  | 60  | 1 600 | W 6  |          | 125 |
| OS 9   | 90  | 2 800 | W 9  | M303     | 95  |
| OS 7   | 120 | 2 000 | W 7  | M303 1/2 | 50  |
| OS 5   |     | 3 000 |      |          |     |
| OS 3   |     | 3 200 | W 10 |          |     |

The stock removal efficiency and the surface finish of the ground surface is not determined by the type of abrasive and by designated mesh or micron size alone. The width of the particle size distribution, the uniformity and shape of the abrasive grains, and the abrasive's friability and hardness also contribute to the quality of finish. Since it is difficult to deduce from these known properties the efficiency of an abrasive for a specific optical material, optical shops in Japan and Germany routinely perform functional tests to qualify an abrasive for its intended application. Such tests are rarely, if ever, performed in the American shops. Such tests are considered sufficiently important in Germany that DIN 58 751 contains a section that describes the method to be used. The test to determine the grinding efficiency of an abrasive measures material removal from a standard glass as a function of time.

$$S = \frac{m}{t}, \tag{3.1}$$

where

$m$ = amount of material removed in grams,
$t$ = grinding time in seconds,
$S$ = grinding efficiency in grams per second.

When using this test, one can objectively state that the grinding efficiency of garnet is the least of all abrasives used in optics and that of aluminum oxide is the highest. That is the reason why garnet is especially valued by some opticians for fine-grinding of critical optics; the grinding pits have less depth than other abrasives for equivalent grain sizes. It is also the reason why aluminum oxide is the most efficient abrasive used for a wide variety of optical materials. Silicon carbide, boron carbide, and especially diamond would have much higher removal efficiencies than aluminum oxide, but they are rarely used in optics fabrication.

Grading and classification are two very important operations performed during the manufacture of abrasive products. Mesh sizes are graded while micron sizes are typically classified. Grading is done with successive sieves down to 280 mesh. Particle sizes smaller than that are classified by water separation, which is also called *elutriation*. Successive elutriation steps are required for closely sized abrasives down to one micron or less. These methods are described in greater detail in this chapter. But first, the relationship between nominal grain size and particle size distribution for both mesh-sized abrasives and for abrasives classified by elutriation must be examined.

Figure 3.6 shows typical particle size distributions for both mesh and micron sized abrasives. Both curves have the same nominal correlation line that relates theoretical mesh size to the average particle size in microns. Figure 3.6a shows the broad sizing of abrasives designated by standard mesh sizes, while Fig. 3.6b shows the micron-sized abrasives which are narrower sized. The sizing is determined by the difference between the upper and lower limits of the particle size distribution. The distribution of both curves

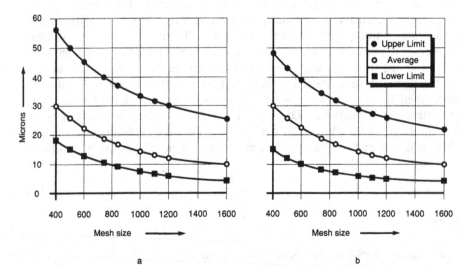

**Figure 3.6a.** Abrasive A (mesh)—broad sizing.   **Figure 3.6b.** Abrasive B (micron)—narrow sizing.

is strongly skewed toward the larger particle sizes, which means that there are appreciable percentages of particles in the abrasive that are larger than the nominal size. In addition to a slightly reduced lower limit, the micron sizes have also a substantially reduced upper limit. This says that micron-sized abrasives are better sized than the mesh-sized abrasives for the same nominal size. The curves also show that the micron-sized abrasives have a larger percentage of finer particles that do not contribute to the grind.

To examine this problem further, let us take a close look at the particle size distribution for three commonly used fine-grinding and pre-fining abrasives. Figures 3.7a and 3.7b compare the typical particle size distributions for 9 micron, 12 micron, and 20 micron abrasives. Figure 3.7a shows the upper and lower limits for a composite of ten commercially available micron-sized abrasives. Figure 3.7b shows the distribution of the closest sized abrasives in the sample of ten. The nominal particle sizes are common to both figures. The skewed distribution that favors the larger particles is clearly shown. In other words, the nominal or average line lies much closer to the lower limits than to the upper limits. The narrower-sized abrasive represented in Fig. 3.7b has fewer fine particles and much fewer larger particles than the composite distribution in Fig. 3.7a. The reason for this difference is that it is easier to remove more of the larger particles through repeated elutriation steps, while it is more difficult to extract the finer particles at the same rate. The information in Fig. 3.7a and 3.7b can also be represented by a bell-shaped distribution curve. Two such curves are shown in Fig. 3.8 for a standard 12 micron abrasive and for a narrowly sized 12 micron abrasive. The skewed distribution and especially the difference in the width of the two

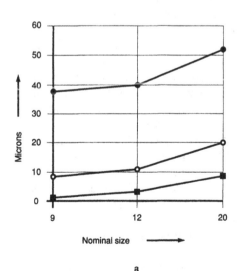

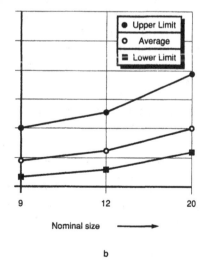

**Figure 3.7a.** Broadly sized fine-grinding abrasive.

**Figure 3.7b.** Narrowly sized fine-grinding abrasive.

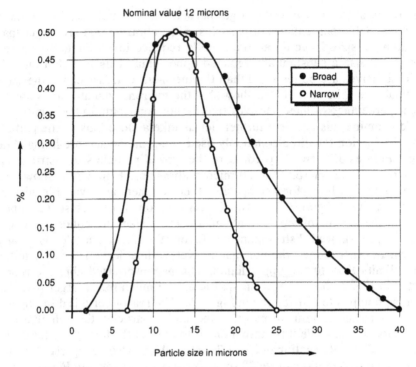

**Figure 3.8.** Narrow versus broad size distribution for 12-$\mu$m abrasive.

curves is quite evident. The narrowly sized abrasive is distinctly better than the standard abrasive, although both are designated as 12 micron abrasives.

In the literature of almost all abrasive manufacturers, the sizing of abrasives is represented by an S-shaped cumulative percent curve, as shown in Fig. 3.9, and not as a skewed bell-shaped distribution, as shown in Fig. 3.8. The bell curve has two orthogonal linear axes which clearly bring out the skew in the distribution. The apex of this curve is at the 50% point at which the mean particle size is 12 microns in this example. However, the industry prefers to represent the quality of their abrasives by the cumulative percent curve which is laid out on a semilog scale that gives the appearance of a much more symmetrical distribution than there is in reality. The point of inflexion at which the S-curve reverses direction is the 50% point which occurs in this example at the mean size of 12 microns. This means that half of all particles are larger than 12 microns, while the rest are smaller. In addition to appearance, however, this representation of the particle size distribution is fully compatible with the commonly used particle-counting methods.

Particle sizing is done by sedimentation based on Stoke's law, which says that the larger and therefore heavier particles will settle out faster than the smaller and thus lighter particles of the same material. Stoke's law, how-

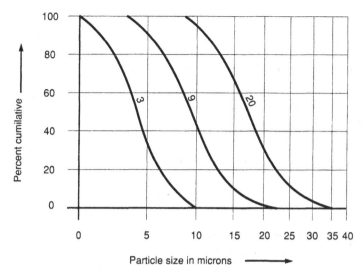

**Figure 3.9.** Particle size distribution curves.

ever, depends on the assumption that the particles are spherical in shape. This condition is usually not fulfilled for abrasive grains or most other powders. Some of the blocky abrasive grains approach this ideal shape, while the platey shapes are far removed from it. The difference in the grain shape often leads to different results when the abrasives are measured with the same instrument, even though they may have basically the same particle size distribution. This effect is shown in Fig. 3.5 which correlates micron sizes with equivalent mesh sizes. There is a theoretical curve that provides that correlation. In reality, however, this ideal is rarely met, as shown by the other two curves that approximate it but do not meet it. The commercial grade abrasives are actually coarser than the ideal, while the industrial grade abrasives are considerably finer for identical mesh sizes.

The separation of grain sizes is done by mechanical sieving, by elutriation, or by centrifugal sizing. The crushed and milled abrasive material is separated into different grain sizes through a succession of reciprocating sieves. The coarsest sieve is on the top and the finest on the bottom of the sieve stack. The finest sieve has a mesh size of about 50 microns. Only the finest abrasive particles pass through this sieve, and they must be further classified through the process of elutriation or centrifugal separation.

The elutriation and the centrifugal process depends on Stoke's law. Assuming a spherical shape of the grains, one can calculate the settling time at a specific settling depth $h$ as follows:

$$t = \frac{18\eta h}{g(d^2)(\rho_1 - \rho_2)}, \tag{3.2}$$

where

$t$ = settling time in seconds,
$\eta$ = viscosity of the liquid in g/cm-sec or poise,
$h$ = settling depth in cm,
$g$ = gravitational constant = 981 cm/sec$^2$,
$d$ = grain size in cm,
$\rho_1$ = density of the abrasive material in g/cm$^3$,
$\rho_2$ = liquid density in g/cm$^3$ (for water $\rho_2 = 1$)

**Example**

How long will it take for an abrasive grain of 0.01 cm diameter to sink 100 cm in water at 20°C? The values are

· water at 20°C has a viscosity of $\eta = 0.01$ poise,
· settling depth $h$ is 100 cm,
· density of water is $\rho_2 = 1$ g/cm$^3$,
· density of the abrasive is $\rho_1 = 3.9$ g/cm$^3$,
· grain-size $d = 0.01$ cm,

$$t = \frac{18(0.01)100}{981(.0001)(3.9 - 1)},$$

$$t = 63.3 \text{ sec.}$$

Two different elutriation processes are known. The first is a pure settling process in static water; the other is a forced settling process that is aided by flowing water. In the first case the abrasive is subjected to different settling times in containers that are arranged in a row in steplike fashion (Fig. 3.10). The unsorted abrasive is introduced into the top container and stirred up in water. This elutriant is allowed to stand for a prescribed time interval, after which time the water with the remaining suspended finer grains is poured into the next lower container. It is then poured through a fine sieve to strain out gross contaminants. The second container is assigned a longer settling time so that the next finer grit can settle out. Meanwhile more water and abrasives are added to the upper container, and the process continues until the lower drums are also filled.

This way every successive container has a longer settling time than the preceding one. In addition to those abrasive grains that correspond to the settling time, there is an amount of finer grains in each of the containers that property belong in one of the lower containers. This is unavoidable because during the time that a large grain sinks from the surface to the bottom, a finer

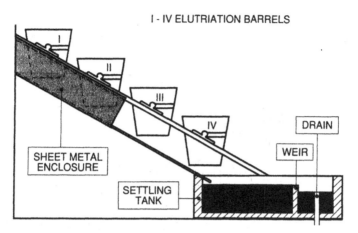

**Figure 3.10.**   Elutriation system for abrasive grain sizing.

grain that was suspended at the middle of the container will reach the bottom at the same time. To overcome this problem, only water is added to the top container and assigned to each container according to the settling time. The elutriation process is repeated, whereby the contents of every container are thoroughly stirred each time. In this manner the proportion of incorrectly sized fine grains is approximately halved each time and becomes insignificant after the fifth elutriation cycle. This process is well suited for small quantities, and it can be easily set up in any shop. It is also utilized to reclaim usable grains from previously used slurries. The used abrasive, however, must be treated with a suitable chemical agent prior to elutriation to remove all organic contaminants.

The second sizing process works continuously, and it serves for the preparation of larger quantities. The abrasive is mixed with water and flows directly from the mill into the first funnel-shaped container that has an overflow on top (Fig. 3.11). Grains that settle at a slower rate, than the velocity of the upward directed flow permits, reach the overflow and run together with the overflowing water into the next lower conical container. The larger grains settle out faster than the upstream flow and collect in the lower part of the cone, where they can be tapped off during the continuous process.

Besides the two wet processes there is also a dry process called *wind sifting* which uses centrifugal force to separate the coarser grains from the finer ones. The different grain sizes are separated with this method in a high-velocity rotating airflow. Good results have been achieved in recent times with this method. Fine abrasives can be classified in production volume with a centrifugal air classifier.

A manufacturer of such a device [11] describes the process as follows: The fine particles enter an annular high-velocity zone in which larger and thus heavier particles are separated outward by centrifugal forces, while the

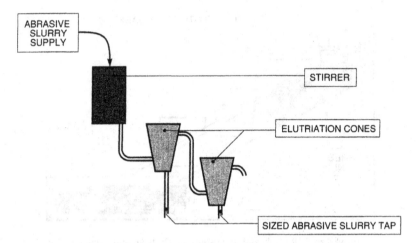

**Figure 3.11.** System for elutriation sizing using vortex separation.

finer particles are separated inward by drag forces. This method can rapidly separate a batch of fine abrasive powder into two separate batches of somewhat overlapping, but otherwise distinctly different, particle size distributions. Similar to the common water elutriation methods, the density of the particles is a major factor that determines the lower limit of sizing. Higher-density particles can be separated out to lower-micron sizes than lower-density powders. Closely sized batches in the range from 1 to 20 microns can be separated out at production rates of several thousand pounds per hour.

The particle size distribution for classified abrasives must be measured and verified before the product can be sold. Special instruments are needed for that purpose. Sedimentometers are often used with standardized procedures for the analysis of fine particles. It is, in its simplest form, a graduated vertical glass cylinder that is filled with methanol into which the abrasive sample has been stirred to form a light slurry. As the abrasive sample settles out, the increasing height of the sediment is timed as it reaches certain index marks. The sedimentation height as function of elapsed time provides the data points for the accumulated percent distribution curves. However, this method is inherently inaccurate. Other instruments used are centrifugal separator, photodensitometer, and the Coulter counter. Centrifugal methods still have a fairly high uncertainty range. The photodensitometer and the Coulter counter have proved to yield the most accurate and reliable data. Therefore any precision classified abrasives should be analyzed by one of these two methods.

### References for Section 3.1 (Loose Abrasives)

[1] Fujimi Corporation, USA.

[2] Adapted from Universal Photonics catalog 1976, USA.

[3]  Micro Abrasives Corp., USA.

[4]  Ibid.

[5]  Kay Industrial Diamond Corp., USA.

[6]  Ibid.

[7]  Rogers & Clarke, USA.

[8]  Ernst Winter & Sohn, GmbH, Germany.

[9]  Kay Industrial Diamond Corp., USA.

[10]  Fujimi Abrasives Co., Ltd., Japan.

[11]  Donaldson Corp., USA.

## 3.2.  POLISHING COMPOUNDS

Polishing compounds are used in conjunction with wax, pitch, or synthetic pad polishers on a broad range of optical materials to convert a fine-ground surface to a highly reflecting or highly transmitting surface. Since an accurate surface shape and a high surface quality is also typically demanded, the material composition and grain sizing of polishing compounds must be very uniform.

Polishing compounds are made up of a variety of synthetically refined oxides of rare earths and other elements. One of the oldest polishing compounds is the bright red to dark red iron oxide which is more commonly known as rouge. It is suspected that the ancient Egyptian artisans were already aware of the polishing characteristics of these naturally occurring compounds, although they used them to polish gemstones and metals. Rouge is hardly used anymore, although it was in widespread use only 50 years ago. Zirconium oxide was used as a lower-cost alternative to rouge, mainly in the opthalmic industry. But the use of zirconium oxide began to increase in the optical industry when the supply of rouge dried up during World War II. However, zirconium oxide was not as aggressive as rouge, and the search for an effective replacement for rouge culminated in the development of cerium oxide and other rare earth oxide polishing compounds.

These rare earth compounds are prepared either as individual oxides, such as cerium oxide of varying degrees of purity, or as blends of a number of oxides. The most popular of these blends was the rust-colored Barnesite [12] that became the standard of the precision optics industry until its production was abruptly discontinued in the early 1970s. Efforts to develop an effective substitute have been only partially successful. The green chromium oxide has proved to be most suitable for polishing certain metals, while the usually cream-colored tin oxide is sometimes used for polishing optical crystals.

The preferred polishing compounds for today's precision optical industry are high-grade cerium oxide and premix formulations containing zirconium oxide, alumina, or precipitated colloidal silica. The use of chemically active premixed slurry concentrates, which were originally developed for polishing semiconductor wafers, is on the increase as more applications are found for the multitude of requirements in precision optics.

### 3.2.1. Polishing Alumina

Very fine aluminum oxide powders with an average particle size of two microns or less are called *polishing alumina*. These aluminas are very similar to the aluminas used for grinding except that they are purer, finer, and much more carefully sized. The alumina powders are mixed with water to form a slurry that is used for precision polishing of a great variety of optical materials. The powders can also be mixed with light oils or other compatible liquid vehicles for special applications.

There are two basic types of alumina: alpha and gamma. Their basic difference lies in the crystal structure of the raw material. Alumina powders are made from high-purity calcined aluminum oxide. Calcining is a carefully controlled thermal treatment that drives off entrapped water. The degree of calcination relates directly to the hardness of the resulting material. Mild calcination yields a relatively soft alumina, while a high degree of calcination will produce a fairly hard alumina.

The alpha alumina has a hexagonal monocrystalline structure that fragments into platey shapes. These tiny platelets have a length-to-thickness-aspect ratio between 3 : 1 and 6 : 1. The gamma alumina has a cubic crystal structure, and it is available only in submicron particles sizes. It is somewhat softer and slightly less dense than alpha alumina which is probably due to a suspected porous structure. The superfine (0.05 $\mu$m average) particles of gamma alumina tend to agglomerate and aggregate.

In the fundamental alumina production process [13], aluminum pellets are subjected to a spark discharge in water. This treatment produces aluminum hydroxide which leads to the formation of very fine aluminum particles and to their hydroelectrolysis. The resulting hydrate alumina is quite soft, and it must be subjected to a certain degree of calcination to give it the desired hardness. This basic process has been in use to produce alumina powders since the 1950s. It results in an alumina powder with highly uniform grains of essentially 100% pure aluminum oxide with total trace impurities less than 100 ppm. Table 3.11 summarizes the composition and other typical properties of alumina powders.

The finer the alumina powders, the more is the tendency to agglomerate. Agglomerates are formed when a large number of individual particles are attracted to one another by their individual charges to compact into larger clusters that act as if they were large individual particles. The obvious result of agglomeration is the formation of numerous sleeks on polished surfaces, since the slurry acts as if it were much coarser until the agglomerates break

**Table 3.11. Typical Properties of Alumina Powders**

| Parameter | Alpha | Gamma |
|---|---|---|
| Purity (% $Al_2O_3$) | > 99.6 | > 99.9 |
| Hardness (Mohs) | 9.0 | 8.0 |
| Specific gravity (g/cm$^3$) | 3.9 | 3.6 |
| Crystal structure | hexagonal | cubic |
| Particle shape | platey | rounded |
| Composition (%) | | |
| $Al_2O_3$ | 99.6 | 99.9 |
| $SiO_2$ | 0.1 | trace |
| NaO | 0.2 | trace |
| $Fe_2O_3$ | 0.1 | |
| $TiO_2$ | trace | |
| Typical sizes (µm) | | |
| | 1.0 | 0.05 |
| | 0.5 | |
| | 0.3 | |
| | 0.1 | |

down. An even more undesirable result of agglomeration is that it leads to the particularly troublesome comet tail scratches that take a long time to polish out. This tendency to agglomerate makes it necessary to ensure complete wetting of all particles so that they remain uniformly dispersed in the water or oil vehicle and are not attracted to each other. Ultrasonic agitation of the slurry can greatly facilitate wetting and dispersion of all particles.

The alumina powders can be also specially treated by the manufacturers to prevent, or at least greatly reduce, the tendency to agglomerate. Treated alumina powders have a mean residual agglomerate size between three and ten times the nominal particle size. Such powders are typically marketed as "agglomerate free," which is obviously misleading because plenty of agglomerates remain. However, compared to untreated aluminas which can have agglomerate sizes that range from six to over two hundred times the nominal size, the advantages of treated aluminas over the untreated product should be obvious.

The harder alpha alumina will cut fast while introducing only minimal surface damage to the workpiece. This is due to their unique platelet shape which causes a higher concentration of contact points between the lap and the workpiece than is true for more spherically shaped particles. The polishing load is more uniformly distributed, so the workpiece is subjected to much lower polishing stresses. Also the platelets provide more of a planing action that shaves off surface material without leaving deep fractures or furrows in the surface that are so typical of other abrasive polishing compounds.

The particle sizes for polishing aluminas range from 1 micron down to 0.05 micron. The grading and sizing is verified by electronic Coulter counter and by microscope comparison methods that include the use of a scanning electron microscope (SEM). The recommended polishing slurry concentration is 20° Baumé which is equivalent to a specific weight of 1.16 g/cm³. The pH of a freshly mixed alumina slurry with a solids concentration of 10% is about 8.5.

Aluminas are useful for polishing optical materials that have a Mohs hardness of 8 or less. Table 3.12 lists typical polishing applications for the once popular Linde [14] alumina powders which have now been replaced by a number of new but very similar products. Aluminas are particularly effective for polishing a wide variety of optical infrared (IR) materials, including silicon, germanium, gallium arsenide, zinc selenide, zinc sulfide, and other II-V and II-VI materials. Although aluminas polish these materials fairly fast, other polishing compounds such as zirconium oxide and colloidal silica produce better finishes for critical applications. These will be discussed in Sections 3.2.3 and 3.2.4, respectively.

### 3.2.2. Cerium Oxide

Blends of rare earth oxides and specifically cerium oxide ($CeO_2$) are the primary polishing compounds used today to polish a great variety of optical materials. Other polishing compounds used are zirconium oxide ($ZrO_2$), alumina, and colloidal silica ($SiO_2$). To a much lesser degree, tin oxide (SnO), iron oxide ($Fe_2O_3$), and chromium oxide ($Cr_2O_3$) are sometimes used for special applications. Blends of cerium oxide and zirconium oxide have also yielded good polishing results. Cerium oxide and the rare earth oxides

**Table 3.12. Typical Applications for Linde Alumina Powders**

| Designation (Linde) | Alumina Type | Avg. size (μm) | Polishing Application |
|---|---|---|---|
| A | alpha | 0.3 | for standard alumina polishing |
| B | gamma | 0.05 | for fine polishing on soft materials |
| C | alpha | 1.0 | for aggressive polishing of hard materials |

are discussed in this section. The other polishing compounds are covered in Section 3.2.3.

Prior to World War II red rouge was the standard polishing compound for most glass optics. Red rouge is also known by its chemical name iron oxide. When the supply of red rouge from Germany stopped during the war, American manufacturers responded by searching for substitutes. The use of zirconium oxide, which had found some prior acceptance in the opthalmic industry, was greatly expanded. But zirconium oxide was not a very good substitute for the much more aggressive iron oxide. The search continued until the discovery of the excellent polishing qualities of cerium oxide. An effective substitute for red rouge had been found in cerium oxide and some of the other rare earth oxides. After this discovery red rouge not only lost its dominance as optical polishing compound, but it was almost totally replaced by the much cleaner cerium oxide compounds.

The designation of a family of elements as "rare earths" is actually a misnomer, since rare earths are really quite abundant. They constitute about one-sixth of all naturally occurring elements. Of the rare earths, cerium oxide is by far the most abundant. Natural cerium oxide and other rare earth oxides such as lanthanum oxide ($La_2O_3$) are found in ores called *monazite* and *bastnasite*. Initially cerium oxide was extracted only from monazite. But cerium oxide was merely a by-product of the extraction of thorium so it was not always available because the supply was dictated by the demand for thorium. Since then large deposits of bastnasite in the United States and elsewhere in the world have been found to yield high-grade cerium oxide in sufficiently large quantities to alleviate any supply constraints. Most cerium oxide compounds and rare earth blends are now extracted and refined from bastnasite ore, although a small percentage still continues to be extracted from monazite.

The extraction and refinement process is quite involved. A fairly broad description of the process must suffice here. The monazite or bastnasite ores are finely crushed and subjected to a process called *cracking* that involves high temperatures and pressures. The resulting product is filtered and chemically refined to remove unwanted mineral substances such as silicates and iron compounds which have proved to be detrimental to the polishing effectiveness of the finished compound. The intermediate product is then thoroughly dried during a calcining step and finally sized and carefully classified into a cerium oxide polishing powder.

Most rare earth polishing compounds contain about 50% cerium oxide. Only very few of the cerium oxide polishing compounds approach 100% pure $CeO_2$. Almost all of them contain varying percentages of some of the other rare earth oxides and other oxides, namely, silicon dioxide ($SiO_2$), sodium oxide (NaO), barium oxide (BaO), and strontium oxide (SrO), among others which can be present in appreciable quantities. These added ingredients determine the color of the compound, which can range from dark rust brown, light brown, tan, pink, cream, to nearly pure white. Although it is difficult to generalize, it can be said that the lighter the compound, the higher is the percentage of cerium oxide. Nearly pure cerium oxide is pure white. Also the higher the cerium oxide content, the higher is the cost and, presumably, the better are the polishing characteristics. However, like all generalizations, this one too has its glaring exceptions. One of the most effective and most popular polishing compounds was Barnesite which was made by Lindsay Chemical Company and discontinued in the early 1970s for economic reasons. This technically very successful product had a cerium oxide content of only about 50%, and its color was a light rust brown, owing to its other oxide constituents. Despite several attempts by other producers to capitalize on its past acceptance, Barnesite has not been successfully reproduced.

The above example illustrates the difficulty that a user of these products faces when deciding on the best product for a specific application. The selection must be often made on the basis of conflicting claims for a confusing variety of products. Not only are there several producers of polishing compounds with each producing a variety of grades, but there many more supply houses which often blend their own compounds for frequently no other reason than to offer a "unique" product. The lack of enforced standards for grading the quality of polishing compounds becomes quite obvious upon closer inspection. In support of this argument, the products and the associated claims from ten sources, both producers and suppliers, were closely scrutinized. Between them, they produced or marketed 48 different cerium oxide polishing powders. Although this study was extensive, it was by no means complete or exhaustive. The actual situation is probably even more confusing. For these 48 products there were descriptive terms for 24 different quality grades, ranging from "superior quality" to "most economical." These various quality grades can be roughly divided into three groups:

1. *Superior or precision quality.* These compounds are primarily formulated for polishing of precision optics on pitch or pad polishers used with brush feed or recirculating slurry systems. They typically have high cerium oxide content and range in color from pink to pure white. They are carefully sized with average particle sizes ranging from 0.5 to 1 $\mu$m. Centrifugal sizing limits the largest particles to 5 $\mu$m.

2. *Premium or opthalmic quality.* These compounds are specifically formulated for high-speed polishing of opthalmic lenses or for medium-quality high-volume optics. They are especially effective in recirculating slurry feed systems. Their color ranges from rust to pink, with a corresponding cerium oxide concentration ranging from moderate to high. Careful sizing methods limits average particle sizes to be between 1 to 3 $\mu$m.

3. *Commercial or economy grade.* These compounds are best for high-speed pad polishing of low-end, high-volume optics. They are also used for soft pad polishing of commercial grade mirrors. The cerium

oxide content ranges from low to moderate, which is reflected in the colors that range from dark rust brown to tan. The particle sizing is pretty broad ranging from 1 to 10 $\mu$m with the average somewhere between 3 and 5 $\mu$m.

Although it is very difficult to generalize on prices, it is safe to state that the opthalmic grades of cerium oxide are about 2.5 times and the precision grades are more than 5 times the cost of the average economy grades. While there is only a small difference in the polishing efficiency between cerium oxide and zirconium oxide compounds when they are used on pitch laps, cerium oxide has proved to be far superior over any other compound when the polishing is done on synthetic pad polishers. The pH of freshly mixed cerium oxide slurries ranges from 6.8 to 8.2, depending on the specific compound used. A higher pH may actually be detrimental to some of the more sensitive glasses or optical materials and the choice of compound must take this into account.

For polishing machines that automatically recirculate polishing slurry through pipes and direct it to the polishing tools, the compound is mixed with ample water. To prevent the slurry from settling out in the pipes which would clog them up, it is best to use a fine-milled compound, if at all possible, and choose a high slurry flow rate. On occasion certain materials are added to the polishing slurry to keep the compound well suspended in water and thus prevent it from settling out in the supply lines. Such polishing compounds have often the added designation "AC," which means anticaking. Also foam preventative substances are often added as well.

Most compounds used today are available suspension treated so that they can be used in recirculating slurry systems. Some compounds have natural suspension characteristics and do not need additional treatment. Other slurry additives such as antifoaming agents, rust inhibitors, and pH inhibitors are also available from most suppliers. The use of these additives must be carefully monitored and controlled; they do not directly contribute to the polishing efficiency and could have a detrimental effect on the polished surfaces under unfavorable conditions.

The mixing of the slurry must be done under controlled conditions to achieve maximum efficiency and the most economical use of the compound. The minimum recommended starting concentration for a cerium oxide slurry lies between 5% and 15% concentration. This means 50 to 150 gs of compound per liter or 8 to 24 oz/gal of water. Another way to measure slurry concentration is to use a Baumé scale hydrometer. This commonly used method is described in Section 3.2.6. For a low starting concentration, the Baumé reading should be between 7° and 12°. This range must be periodically monitored and small amounts of premixed slurry concentrate must be added from time to time to keep the slurry constant throughout the polishing cycle. For high stock removal pad polishing, the cerium oxide slurry should be used at temperatures between 42°C and 56°C (75°F and 100°F).

### 3.2.3. Other Polishing Compounds

In addition to the alumina and cerium oxide polishing compounds, there are several other oxides used for polishing optical surfaces. Zirconium oxide, long used in the opthalmic industry, has found fairly broad acceptance in many precision optics shops. Chromium oxide is often preferred for polishing metals, whereas tin oxide is sometimes used for polishing optical crystals. Iron oxide, once the primary polishing compound, is rarely used today except for some specific application. Additionally there are many unspecified polishing compounds available that have been specifically formulated for the polishing of optical plastics.

Zirconium oxide ($ZrO_2$) compounds are very fine powders ranging in color from pure white to a creamy white. The initial use was in the opthalmic industry more than 50 years ago, and zirconia is still widely used there. In the meantime specially prepared zirconia powders that are often blended with alumina or ceria have been successfully used in the precision optics industry as well.

The low zirconia content economy grades have average particle sizes in the range of 5 $\mu$m with the largest particles exceeding 8 $\mu$m. These grades are most often used in recirculating slurry systems for high-speed pitch or pad polishing of low-end opthalmic or glass optics. The better grades with fused zirconia content of about 90% are usually much better sized, with average particle sizes of less than 1 $\mu$m and with no particles larger than 5 $\mu$m. These compounds are particularly effective for polishing good-quality opthalmic lenses or standard-quality glass optics on high-speed equipment using hard impregnated polyurethane polishers and recirculating slurry feed.

The best grades of zirconium oxide are wet-milled, uniformly sized, agglomerate-free versions of the standard grade, or they are based on chemically precipitated zirconia. These fine powders with a 0.3-$\mu$m average particle size are specifically prepared for precision polishing on pitch or synthetic pad of germanium, silicon, ceramics, plastics, and metals. The precipitated variety is 99.5% pure zirconium oxide with small fractions of a percent of silicon dioxide, aluminum oxide, and iron oxide.

The pH of aqueous zirconia slurries can range from mildly acidic at pH 5.5 to moderately alkaline at pH 9.5. The correct pH level must be matched to the chemical sensitivity of the optical materials that must be polished. All zirconium oxide slurries can be used on pitch or pad polishers, and the slurry application can be manual or by automatic slurry feed. The suspension properties for recirculating slurries must be checked, and suspension agents may have to be added to avoid settling of the compound. The most commonly recommended starting concentration is around 10° Bé which corresponds to a 10% by weight solid concentration. The concentration can range from a low of 5° Baumé at 4% zirconia to a high of 20° Baumé at more than 20% $ZrO_2$.

Chromium oxide ($Cr_2O_3$) is a bright green oxide powder. It is a very effective compound for polishing some metal surfaces. Oil-based slurries made from select grades of fine-milled powders are preferred for polishing of high-carbon steel and stainless steel. Aqueous chromium oxide slurries have also been successfully used for polishing very hard crystals such as ruby and sapphire.

Tin oxide (SnO) is a relatively soft polishing compound that is nearly 100% pure. It is occasionally used to polish optical crystals when other more commonly used compounds have failed to get the desired surface quality. Tin oxide powders that have been prepared for precision optical polishing are characterized by good particle dispersal qualities in aqueous slurries.

Iron oxide ($Fe_2O_3$) powders, often called *polishing rouge,* are synthetically produced from iron sulfate under carefully controlled oxidizing conditions. There are several different grades that differ not only in their color but also in their polishing characteristics. Red rouge is better than 99.5% iron oxide. Black rouge is a very dark red magnetic iron oxide. These polishing powders are very fine with average particle sizes in the 1 micron range. Their high purity makes them useful for precision polishing of glass optics. Iron oxide powders do stain everything they get in contact with and that includes the opticians. For that reason alone these compounds are used now only rarely and then only in very small quantities.

Plastic polishing formulations have been specifically prepared for the polishing of opthalmic plastic optics. There are a large number of such special formulations on the market, both as dry powders or as aqueous premix slurries. The ingredients of these compounds are generally not identified. They are intended to polish hard resin eyeglass lenses and contact lenses made from methylmethacrylate and polycarbonate. Even though the manufacturers, and especially the suppliers, are quite secretive about the basic ingredients in these formulations, there can be little doubt that they are the same compounds that have been discussed in this chapter. Some are based on alumina, ceria, silica, or zirconia, and others may be blends of a number of them. It seems that a bit of mystery coupled with an attractive package and a catchy name can justify a sometimes substantially higher price. These practices, however, do nothing to impress the technical user.

Another way of looking at polishing compounds is from the perspective of applications that have proved to be useful:

1. *Polishing compounds for plastics and crystals.* For polishing acrylics such as Plexiglass and Polystyrol, polishing compounds with a tin base have proved to be effective. Plastic optics can also be polished with very fine cerium oxide, which is especially effective when soft synthetic polishing pads are used. Most of the optical crystals can be polished with ceria compounds as well. Chromium oxide is often preferred for polishing calcite and calcium-containing crystals. Sapphire and other hard crystals can be polished quite efficiently with a select grade diamond powder that has a grain size of less than 1 micron.

2. *Polishing compounds for metals.* Metals are polished with chromium oxide or with a highly purified form of argillaceous earth. Cloth polish-

ers are typically used with these compounds. Steel can be polished free of sleeks only when it is hardened. These polishing compounds are almost never commercially available of a sufficiently high level of quality that is required for optical polishing. Therefore they must be first specially prepared in the optical shop prior to actual use. Small quantities of the compound must be rubbed between two fine-ground glass disks to break down any large particles and agglomerates.

3. *Polishing compounds for hard materials.* Hard metals and hard crystals such as sapphire are best polished with fine diamond powder mixed with water or olive oil. Diamond paste can also be used successfully for especially hard steel. By contrast, optical glass and other soft optical materials cannot be polished sleek free with the diamond mix. The diamond can be obtained in grain sizes D 0.25 $\mu$m to D 1.0 $\mu$m. It is also available embedded in water-soluble paste that comes in tubes or in syringes that allow for controlled dosages. Pitch polishers can be used quite effectively with submicron diamond compounds. This approach can produce surfaces that must meet testplate fit requirements. Also the expensive diamond remains on the surface of the polisher, whereas with the cloth and pad polishers much of the diamond migrates to the underlying layers where they become effective only when the tool is used for a long time.

### 3.2.4. Colloidal Silica

Coloidal silica powders are made from precipitated amorphous silicon dioxide ($SiO_2$) of the highest purity. These submicron powders are mixed with chemically active aqueous solutions to form stable dispersions that are specifically formulated for the chemo-mechanical polishing of semiconductor substrates. This type of polish produces nearly flawless surfaces not only on silicon but, with some modifications of the chemistry, also on germanium, gallium arsenide and a number of other optical materials.

In chemo-mechanical polishing of silicon, the chemically active solution dissolves a few atomic layers which turn into silicates. This chemically altered layer is then adsorbed by the colloidal silica particles and is carried away by the constantly flowing slurry. Since the polishing is typically done on synthetic polishing pads called *poromerics* and the individual silica particles are very fine, there is almost negligible abrasive action and thus no scratching of the surface. Depending on the chemical characteristics, as represented by the pH value, the polishing action on silicon wafers can result in high stock removal rates exceeding several mils per hour.

One of the characteristics of premixed dispersions is that individual particles remain in suspension almost indefinitely. If some settling should occur, the particles are easily redispersed. The spherically shaped submicron particles do not agglomerate because they are specially treated to neutralize any residual charge [15]. This is accomplished by stabilizing the partially ionized, and thus negatively charged, silica particles by a positively charged sodium ion ($Na^+$). The compound is in equilibrium, and attraction between particles that normally leads to agglomeration is eliminated. The stabilization can also be accomplished by an ammonium ion, which is done in those cases where sodium could become an unacceptable contaminant.

Any agglomerates that do form, either due to insufficient stabilization or to destabilizing external influences, tend to be rather soft and break up easily without causing much surface damage, even though they can be several microns in diameter. Some of the first chemically active colloidal silica dispersions have shown a tendency to form hard, glasslike, white agglomerates that can cause considerable scratching. The currently manufactured products, however, do not exhibit this problem for the most part. The small average particle size of less than 0.1 micron and the relatively high particle concentrations in the premix of up to 50% by volume gives the slurry exceptionally high surface area for optimum adsorptivity of waste products from the polishing process.

Colloidal silicas are stable in anionic or nonionic wetting agents or emulsifiers and in water-miscible solvents such as alcohols, glycol, and other polar solvents. They are, however, incompatible with nonpolar solvents or solvents that are not soluble in water. The pH level also greatly influences the long-term stability of colloidal silica dispersions. The stability is greatest around pH 3 (acidic) and pH 10 (alkaline). The dispersions become unstable and tend to jell below pH 2.5, between pH 5 and pH 7, and above pH 11. It is therefore very important to constantly monitor and control the pH level of the slurry [16].

When colloidal silica dispersions are subjected to freezing temperatures (whether in transit, storage, or use), irreversible precipitation of the silica particles takes place, and the slurry becomes useless. In the opposite case where slurry temperatures locally approach the boiling point of water, rapid gellation often occurs which also renders the slurry ineffective. In a well-controlled polishing environment neither of these two extremes will be approached. The newer colloidal silica products will not precipitate at the freezing point or dry prematurely in air to form gritty agglomerates.

Colloidal silica is most often marketed as a specially formulated premix called a *permanent dispersion*. Since these are typically concentrates, they must be diluted prior to use. The most often recommended dilution ratio is one part premix to ten parts of deionized (DI) water. Colloidal silica is also available in powder form. This gives the user the opportunity to create a variety of different slurries to better match the particular polishing requirement. One manufacturer [17] provided the following recommendation on how to prepare an active polishing slurry for polishing silicon wafers: The colloidal silica powder is gently stirred into deionized water at a ratio of 5% to 6% by weight. This does not sound like much, but the powder being so light creates a rather concentrated slurry at that ratio. A sodium hydroxide solution made up of half an ounce of NaOH to one gallon of DI water is slowly added to the mix until the pH meter reads somewhere between pH 10 and pH 11. Potassium hydroxide (KOH) or sodium metasilicate can also be used. The slurry pH must be constantly monitored during polishing, and pH corrections must be periodically made to maintain the pH level between 10 and 11.

Alkaline colloidal silica premixed concentrates are used primarily for the polishing of semiconductor substrate. One part of the concentrate must be diluted with ten parts of DI water to create the optimum slurry, which has been proven to produce essentially flawless surfaces on single crystal silicon wafers and other semiconductor materials. They have also been used successfully to polish single-crystal silicon parts for IR applications. The best results are achieved when a two-step polishing process is used. A three-step process is preferred by some manufacturers. The first step uses a slurry with an elevated pH of 11 and higher than usual solid concentration of 10% at fairly high pressures of 70 to 100 kg/cm$^2$ (about 5 to 7 psi). This prepolishing is done at optimum machine speeds on a moderately hard poromeric pad. These conditions create frictional heat that raises the temperature of the slurry. To accelerate this process, the slurry temperature can be purposely raised with suitably designed immersion heaters to between 100°F and 120°F. Under these polishing conditions rapid stock removal of about 2 mils per hour is quite typical. This rapid polish, however, is achieved at the expense of haze formation and orange peel effects. It requires a second polishing cycle at reduced speeds, pressure, pH, and solid concentration on a softer poromeric pad to remove these undesirable surface effects. Typical operating parameters for the second polish are polishing pressure between 40 and 70 kg/cm$^2$ (or 3 to 5 psi), a pH of about 10.5, and a solid concentration of no more than 5%. This finishing cycle produces the desired superior surface quality on silicon.

For optimum effects the slurry must be supplied to the pad at rates between 2 to 4 liters per hour (about 0.5 to 1 g/hr). Although recirculation of chemically active slurries is not always recommended, the slurry for the final polish should never be recirculated because of its unavoidable contamination, which can lead to surface damage. Figure 3.12 shows a typical single flow slurry supply system [18].

Other semiconductor materials that have been successfully polished with colloidal silicon dispersions are single-crystal germanium and gallium arsenide. The chemistry of the polishing concentrate must be changed, however, to suit the intended application. For instance, the alkalis used for silicon polishing such as sodium hydroxide (NaOH) must be replaced by sodium hypochlorite and potassium hydroxide for polishing germanium and by sodium hypochlorite alone for polishing gallium arsenide. These two materials are also used for IR optics, but they are almost always polycrystalline. Active slurries can be used under certain conditions to polish them if they are sufficiently diluted. If they are used at the strength recommended for single-crystal polishing, preferential etching of the polycrystal structure

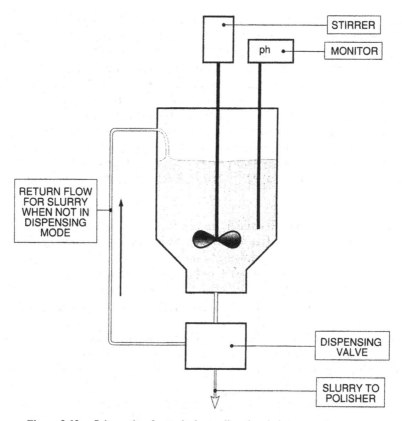

**Figure 3.12.** Schematic of a typical one-directional slurry supply system.

would result in an unacceptable surface. Neutral colloidal polishing solutions produce good surfaces on polycrystal germanium and gallium arsenide, although the polish is quite slow. Gallium gadolinium garnet (GGG), sapphire ($Al_2O_3$), lithium niobate ($LiNbO_3$), and surfaces of bare aluminum and electroless nickel plated aluminum have all been successfully polished with neutral colloidal silica solutions.

Chemically active premix concentrates must be stored in proper containers [19]. They can be safely stored in stainless steel containers or in steel containers that have been treated with rust inhibitors. Plastic containers also are a good choice, but they must be reinforced with fiberglass to minimize breakage, rupture, and subsequent spillage. Aluminum containers must not be used because the hydroxides attack the aluminum. This would reduce the strength of the container and contaminate the slurry to make it unfit for use.

When dry colloidal silica powders are used to formulate customized polishing slurries, great care must be taken to avoid inhalation of the fine particulates of silica. These easily airborne particles can lead upon prolonged exposure to pneumoconiosis, a lung disease similar to but not as deadly as the dreaded black lung disease of the coal miners. Therefore safety goggles and approved respirators must be worn when these powders are used even only on occasion. For more frequent use in larger volumes, proper ventilation must also be provided, and dust collectors may have to be installed as well to ensure that the airborne particle concentrations for long-term exposure do not exceed permissible limits. In addition alkaline premix concentrates can cause eye burns and skin irritation, which is medically known as contact dermatitis. In conclusion, proper care dictated by common sense must be used when working with these and all similar materials.

Referring to the slurry types listed in Table 3.13, the following uses are typical:

A     Standard colloidal silica premix for polishing silicon wafers; ready to use after a 10 to 1 dilution in deionized water.

B1    Higher solids concentration than the standard colloidal silica premix; more abrasive and adsorbing action to allow for higher stock removal. The low sodium oxide concentration is essential to prevent sodium contamination of polished silicon wafers.

B2    Lower pH and higher viscosity but requires the addition of suitable alkali solutions for customizing pH values.

C1    A neutral pH premix for polishing alkali sensitive materials such as aluminum, GGG, and sapphire.

C2    Premix formulated for polishing rare earth garnets and other such materials where particle size and not the pH of the slurry is the more important parameter. The polishing action is predominantly mechanical (i.e., abrasive).

C3, C4   Colloidal silica premix slurries formulated for polishing silicon and designed for manufacturers that prefer to adjust the pH

**Table 3.13. Properties of Commercial Colloidal Silica Premix Slurries [20]**

| Type | $SiO_2$ (%) | pH | Avg. size (mµ) | Spec. wt. (g/cm$^3$) | $Na_2O$ (%) | Viscosity (cp) |
|------|------|------|------|------|------|------|
| A | 30 | 10.5 | 15 | 1.2 | 0.4 | 5 |
| B1 | 40 | 10.4 | 20 | 1.3 | 0.1 | 10 |
| B2 | 40 | 9.2 | 20 | 1.3 | 0.1 | 30 |
| C1 | 50 | 7.5 | . | 1.4 | 0.1 | 15 |
| C2 | 50 | 8.5 | 60 | 1.4 | 0.3 | 10 |
| C3 | 50 | 9.0 | 20 | 1.4 | 0.4 | 40 |
| C4 | 50 | 9.5 | 50 | 1.4 | - | 35 |

value to suit their needs. The high silica concentration makes them a more economical alternative to the standard silicon premix. Like all of the concentrates, they must be diluted with DI water to a ratio of at least 10 to 1. These formulations are also available with a low-sodium content.

### 3.2.5. Liquid Premix Concentrates

Premixed concentrated slurries were initially developed with colloidal silica for polishing silicon wafers. They were also used as alumina-based suspensions for polishing plastic contact lenses. While some of these premixed slurry concentrates are chemically neutral, the majority of them are chemically active. This means that they are either alkaline with a pH greater than 7 or acidic with a pH less than 7. The alkaline slurries have a pH value between 10 and 13, whereas the mildly acidic solutions have a pH value between 5 and 6.

The semiconductor industry and the ophthalmic industry have been the leading users of premixed concentrates. Precision optics manufacturing has been very slow at first to consider these new products. However, over the past decade some of the concentrated polishing slurries have been tested with good results. The chemically neutral slurries have found many uses, and some of the chemically active slurries are mainly used for polishing certain IR materials. This initial reluctance has now been overcome to the point where there are formulations on the market that have been specifically designed for use in precision optics. The trend points toward an expanded use of these products in the near future, especially as a result of anticipated changes in the manner by which precision optics will be produced.

Liquid premix slurries are also called *permanent suspensions* because the individual particles are uniformly dispersed in the liquid. This prevents the formation of agglomerates which can cause sleeks and scratches on polished surfaces. The suspension properties are provided by the addition of water-soluble and environmentally benign substances that ensure the uniform wetting of all particles without creating unacceptable hazards. With very few exceptions, all of the previously described polishing compounds have formed the basis for premixed liquid concentrates. Some formulations are made up of blends of two or more standard polishing compounds. While the neutral slurries are usually mixed with deionized water, the alkaline solutions have additives such as sodium hydroxide, ammonium hydroxide, or sodium hypochlorite. Mildly acidic premix solutions, using acetic or similar acids, have also been prepared for use in optical shops to polish optical glasses that have a tendency to stain.

There are several advantages that concentrated premix slurries have over the traditional polishing slurries. The particles are uniformly dispersed to prevent agglomeration and permanently suspended to prevent settling. The chemical nature of some of these premixes can substantially speed up the polishing process by dissolving or softening a few atomic or molecular layers of the substrate surface, which then the compound can easily remove and the flowing slurry can carry away. Since not all optical materials can be chemically polished—such as most optical glasses, polycrystalline materials, many plastics, and most metals—neutral premix slurries using DI water as the vehicle are commercially available. A further advantage lies in the way these products are prepared. Unlike the mixing of traditionally polishing slurries, these products are formulated and bottled under controlled conditions. This ensures their uniformity and purity.

The premixed slurry are concentrates that can be used uncut for some applications, but for most uses they must be first properly diluted. Typical dilution ratios are two to one or three to one. Even higher ratios may be required at times. However, too much dilution will cause the dispersal and suspension properties to fail, and it can weaken the chemical action sufficiently to render the slurry ineffective.

The majority of commercially available premix concentrates for use in precision optics are based on zirconium oxide. Products that have found acceptance in optical shops are Zox PG [21], Lustrox [22], and Glanzox [23]. These permanently suspended dispersions are available with solid concentrations ranging from 20% to 50%. The average particle sizes range from 0.2 $\mu$m for wet-milled zirconium oxide to 1.5 $\mu$m for the chemically precipitated $ZrO_2$. The wet-milled compound is recommended for the final polishing of Si, Ge, and GaAs, and the precipitated compound is used for precision polishing of high-index optical glass, optical plastics, optical crystals, metals, and a variety of IR materials. The chemistry of these premix solutions ranges from neutral to alkaline up to pH 13. The chemically active agents of the alkaline solutions are sodium and ammonium hydroxide for high-rate polishing of silicon and sodium hypochlorite for precision polishing of single-crystal germanium and gallium arsenide.

Near-neutral concentrated premix solutions that range from pH 6 to pH 8 can be used full strength for polishing plastic contact lenses, as well as a variety of other optical plastics. The dilution ratio is 2 to 1 for the polishing of most optical glasses, 3 to 1 for many optical crystals, and 4 to 1 for most common IR materials. Properly diluted neutral solutions are also useful for polishing metals used in optics.

While the majority of premix concentrates use zirconium oxide as the abrasive medium, other oxides such as aluminas, silicas, and rare earth oxides can also be used. Alumina-based neutral premixed slurries [24] with particle sizes ranging from 0.05 to 3 $\mu$m are recommended for polishing optical plastics such as methylmethacrilate (better known as Plexiglass or Lucite), metallographic samples, and a variety of optical and laser crystals. Another product [25] is based on rare earth oxides, mostly cerium oxide. This tan liquid premix is specifically formulated for brush-on pitch polishing of optical glass components. It can also be used for pad polishing in a recirculating slurry feed mode. Since it is also a concentrate, it must be diluted with water at a ratio of 8 to 1. This dilution ratio corresponds to a slurry concentration of 10° Bé.

A number of acidic premix slurries [26] with pH 4 are based on alumina, ceria, silica, zirconia, or a combination of these. They are intended for use on most IR materials and other optical crystals. They work also quite well on most silicate optical glasses and on fused silica. These slurries should not be used on sensitive glasses and on crystals that may be etched by the mildly acidic nature of the slurry.

Whether the slurry is acidic or alkaline, commonsense precautions must be exercised when working with chemically active solutions just as one would with any other potentially harmful substance. Particularly the chlorine-containing formulations must be handled with caution because under extreme adverse conditions highly toxic chlorine gas can be liberated. It is always advisable to read, completely understand, and then follow the manufacturers precautions on use and disposal. It is also highly recommended to request a Material Data Safety Sheet (MSDS) when ordering chemically active products. This is your right under the law.

Chemically active and permanently suspended slurry premixes should be used as prescribed by the manufacturer. Unfortunately, many opticians are compulsive experimenters, and even the best among them have great difficulty using the same process twice. Since they often lack the necessary background in chemistry and physics, these opticians tend to be more like the alchemists of old by using the approach: "Let me add a few drops of this stuff and see what happens." Sometimes positive results are achieved this way, despite the fact that there is only a very slim chance for that to happen. These successes are then used to justify the process, and they encourage further aimless experimentation. But unless the optician has kept accurate records or clearly remembers what was done, it is highly unlikely that the results can be duplicated next time or any other time. This type of haphazard experimentation must be discouraged, even for normal polishing practice. It should be strictly avoided, however, with the use of chemically active slurry premixes. The uncontrolled addition of incompatible substances can easily upset the critical chemical balance of the dispersion. Such an imbalance can lead to the formation of gels and the irreversible precipitation of the particles. Although less likely, it is not out of the question that chemical reactions with unpredictable results can occur that could lead to unsafe operating conditions or can have detrimental effects later in the process. The best approach is to use these products as intended by the manufacturers and to leave the experimentation to those who are qualified to do it correctly.

Although this paragraph belongs rightfully in Section 3.1 on abrasives, it is included here because the same premixed suspension technology has been applied to standard abrasives as well [27]. Fine-grinding aluminas from 3 to 15 $\mu$m, as well as fine silicon carbide, have been successfully dispersed in permanent suspensions. There are even claims that sizes up to 320 grit (45 µm) have been suspended successfully indefinitely. Diamond abrasives are also available in permanent suspensions as an economic alternative to the more traditional diamond pastes.

### 3.2.6.  Monitoring the Slurry Condition

There is usually no need to monitor the condition of the slurry during normal polishing. This is especially true when the slurry is manually applied from a squirt bottle, since it is used only once. When polishing slurries are recirculated, however, it is essential that the condition of the slurry be monitored regularly to ensure that it remains in the range of optimum efficiency. It is even more important to keep chemically active slurries that are used in a recirculating mode under close watch, for relatively small changes can upset their chemical balance.

Two slurry parameters must be measured to determine the need for corrective addition to the slurry. One is the slurry concentration, and the other is the pH value. The slurry concentration is measured with a hydrometer, whereas the pH value is determined either with litmus paper or, more accurately, with a suitable pH meter.

Maintaining the slurry concentration within an optimum range is important because it directly affects the polishing efficiency of the slurry. If the slurry becomes too weak because of settling or carry out, then the removal efficiency will be reduced. This leads to greatly increased polishing times. If the slurry becomes too thick because of rapid evaporation of the water, the polishing effect may cease altogether as the workpieces merely slide on the thick slurry. When the hydrometer measurement shows that a critical level has been reached, an appropriate amount of makeup slurry or water is added to bring the slurry back into the optimum range.

Monitoring the pH value is important because the suspension properties of the slurry depend on maintaining the proper chemical balance. A change in the pH value can cause the particles to precipitate out of the slurry and leave it weak and ineffective. Some materials, particularly stain-sensitive optical glasses, can be adversely affected by slurries that have been allowed to become either too acidic or too alkaline. In a recirculating slurry system for high-speed polishing of standard glass optics, appreciable amounts of the glass are removed and get into the slurry. Since almost all glasses have alkaline compositions, the introduction of that polished off glass makes the slurry increasingly alkaline, and this eventually leads to a marked decrease in polishing efficiency of the slurry and a corresponding degradation of the surface quality obtained. Other optical materials, such as germanium or silicon, require elevated pH levels for optimum stock removal during the polishing, which is necessary to limit the length of polishing cycles. Controlling the pH values of a slurry is also done with specially prepared makeup slurries that must be added in response to a change in the slurry pH as determined by the pH-monitoring steps.

The best way to measure the solid concentration in a slurry is to measure its specific gravity. The simplest way to do that is with a hydrometer. A hydrometer set consists of a standard measuring cylinder and a bulbous glass float that is weighted at one end to make it float upright when it is lowered into a liquid that fills the cylinder. A paper strip that is graduated on a Baumé degree scale is sealed in the long and slender glass neck of the float. This arrangement is shown in Fig. 3.13. A sample of the slurry is poured into the cylinder, and the float is immersed. When the float comes to rest, it will have displaced an amount of the slurry that weighs exactly as much as its own weight. The corresponding Baumé value can then be directly read off the scale at the liquid surface. This value is a measure of the liquid density, and indirectly it is a measure of the solid concentration in the liquid. The hydrometer sinks deeper into less dense liquids and not as deep in denser liquids. Aqueous slurries are always denser than water since solids are added to it to create the slurry. The conversion from Baumé values, which are in Baumé degrees, to specific weight for liquids heavier than water is $144.3/(144.3 + n)$ where $n$ is the reading in Baumé degrees [28]. Figure 3.14 represents this relationship. The conversion factor for liquids lighter than water is $144.3/(144.3 - n)$.

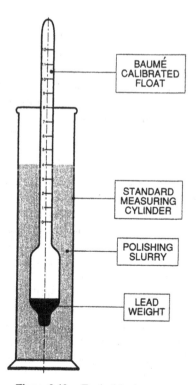

**Figure 3.13.** Typical hydrometer.

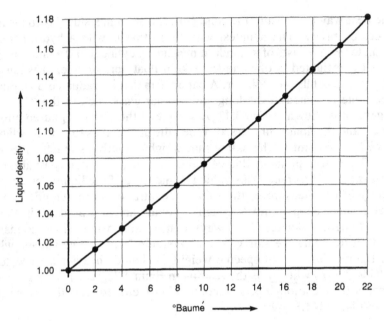

**Figure 3.14.** Slurry concentration.

Graduated Baumé floats for liquids heavier than water are usually sold in two ranges. One is calibrated from 0° Bé to 12° Bé, and the other overlaps with a measurement range from 9° Bé to 21 Bé. These are available from optical shop supply houses or from any laboratory supply company.

The acidity or alkalinity of the slurry is determined by measuring its pH value. What is really measured is the concentration of hydrogen ions in the liquid. The pH scale ranges from 0 to 14, with pH 7 being neutral (water). It is a scale that increases in decades to the alkaline side and decreases in decades to the acidic side. The crossover point lies at the neutral pH of 7. This means that the solution becomes increasingly more acidic as the pH readings decline toward 0, or becomes increasingly more alkaline as they increase toward the maximum of pH 14. Every step in the pH value represents a tenfold increase (or decrease) in its chemical potential. For instance, a pH value of 5 is ten times more acidic as a pH 6 reading, a pH 8 reading is 10 times less alkaline as a pH value of 9, and so on.

The measurement of the pH values can range from a simple test that uses pH-sensitive test paper strips to sophisticated multichannel electro-chemical pH meters with digital readout that can cost several hundred dollars. The test paper changes color when immersed into a chemically active aqueous bath. The color is then matched to reference colors that are graduated in pH numbers. Two different paper strips are used. One is calibrated in even pH numbers, the other in odd numbers. This approach gives the measurement a bracketing effect that can, in theory, determine the pH value to an accuracy of 0.5 pH number. While this kind of accuracy is probably quite sufficient for most slurries, the fact that the solids in the slurry frequently obscure the colors makes this approach much less reliable. The use of a suitable intermittent immersion-type pH meter is most likely the best bet. It need not be an expensive model, but it should read reliably to an accuracy of 0.1 pH number. There is a wide variety to choose from. Laboratory supply companies are a good source.

**References for Section 3.2 (Polishing Compounds)**

[12] Lindsay Chemical Company, USA.
[13] Dr. Iwatami, Fujimi Abrasives Co., Ltd., Japan.
[14] Union Carbide Corp., Linde Division, USA.
[15] "Nalcoag colloidal silicas," Nalco Chemical Company, USA.
[16] Ibid.
[17] "QUSO microfine silica," Philadelphia Quartz Company, USA.
[18] Ibid.
[19] "Polishing semiconductor wafers with Nalcoag colloidal silica," Nalco Chemical Company, USA.
[20] Philadelphia Quartz Co. and Nalco Chemical Co., USA.
[21] Ferro Corporation, Transelco Division, USA.
[22] Tizon Chemical Corporation, USA.
[23] Fujimi Corporation, USA.
[24] "Baikalox," Baikowski International Corp., USA.
[25] "Zox CE-89," Ferro Corporation, Transelco Division, USA.
[26] "Ultra-Sol," Solution Technology, Inc., USA.
[27] Solution Technology, Inc, USA.
[28] Brockhaus, F. A., *Brockhaus der Naturwissenschaften und der Technik*, Wiesbaden, Germany, Fourth Edition, 1958.

## 3.3. WAX AND PITCH

Waxes and pitches play a very important role in optics. Blends of natural or synthetic waxes and natural pitch products are used for holding lenses and other optical components to tools when they are processed through grinding and polishing operations. Natural pitch is used to make polishing tools for precision polishing of a wide variety of optical materials. Some wax blends can also be used for making polishers that are used for special applications. These materials will be discussed in this section.

### 3.3.1. Blocking Waxes and Pitches

Blocking is a term used in optics that describes the process by which individual optical parts are held in place during manufacture. Much more will be said about blocking in Chapter 5. Unlike polishing pitches, which are relatively soft and must flow, blocking waxes, blocking pitches, and the various blends of these must have very low creep. This is a very important property, since blocking materials must hold parts firmly in place under constantly changing forces and elevated operating temperatures. They must be rather stiff and have a fairly high melting temperature. Since the most important objective in optical fabrication is the attainment of a highly polished surface while at the same time achieving a high degree of precision, it is also essential that the blocking wax or pitch be nondistorting. High strength seems to preclude a low distorting hold, but that is exactly what a good blocking wax or blocking pitch must be able to do. In addition none of these qualities would mean a great deal if the blocking material could not be removed after the optics is finished. Therefore blocking waxes and pitches must also be readily demountable and be readily removed from the optics with simple and safe methods that can be performed on the shop floor. In response to these constraints, many different blends and special formulations have been developed for this purpose.

Blocking waxes are used where ease of application and good holding power are the primary concerns. Most waxes are quite stiff to resist flow and creep. Since they tend to shrink as they solidify from the molten state, blocking wax can set up sufficiently large stresses to deform thin and delicate parts. A study [29] has shown that some blocking waxes can distort relatively thick parts with a 1 : 8 aspect ratio between 2 to 4 fringes depending on the flatness of the back surface. As a result the use of blocking waxes is limited to blocking noncritical thin parts or for precision blocking of very thick parts.

Blocking waxes are usually blends of either beeswax, paraffin, rosin (colophonium), and to a lesser degree carnauba wax, microcrystalline wax, ceresine wax, ozokerite, bayberry wax, and montan wax. Most blocking waxes, whether ready-made or mixed in the shop, are a blend of beeswax and rosin with the other waxes used in small proportions as additives to give the formulation specific qualities. Table 3.14 lists these waxes and their important attributes which were gleaned from the product literature of a major wax supplier [30].

Beeswax is a natural product that varies in quality and properties not only by source, but also from batch to batch from the same source. Like most natural products, the quality of beeswax is very dependent on weather conditions and other regional effects. Neither the cost nor the availability can be predicted with some degree of certainty.

A synthetic variety of beeswax based on montan wax was developed to overcome these problems. This variety is now used almost exclusively for most technical applications. These products are quite pure and of uniform quality from batch to batch. There are basically two types of synthetic beeswax that differ in color. The white beeswax is merely a refined version of the yellow commercial grade beeswax. The refined grade also has a lower melting point than the yellow grade. The commercial grade beeswax that comes in raw bricks or slabs contains a small amount of residual impurities

Table 3.14.  Waxes Used in Optics and Their Properties

| Wax type | Origin | Melt point range °C | °F | Color |
|----------|--------|:----:|:----:|-------|
| Beeswax | insect | 61 - 65 | 142 - 149 | white/yellow |
| Rosin | pine | 76 - 80 | 168 - 176 | amber/yellow |
| Paraffin | petroleum | 52 - 74 | 125 -165 | white |
| Carnauba | palm | 82 - 86 | 180 - 187 | light tan |
| Microwax | petroleum | 60 - 93 | 140 - 200 | white |
| Ceresine | petroleum | 53 - 79 | 128 - 175 | white |
| Ozokerite | petroleum | 63 - 91 | 145 - 195 | white/yellow |
| Bayberry | berry | 46 - 49 | 116 - 120 | tan |
| Montan | coal | 85 - 88 | 185 - 190 | white/yellow |

that gives it its characteristic yellow color. These impurities are removed by melting the wax and straining it through a series of filters. The result is the refined white beeswax that is one of the main ingredients in most blocking waxes.

Rosin, also called *colophonium*, is not a wax but rather the processed sap from pine trees. It is described here because its primary use is as main ingredient for a common blocking wax for which it is blended with beeswax or other waxes. The most commonly used blocking wax is made up of one part beeswax or ozokerite to four parts rosin. Other mixtures with a higher rosin content are also used for stacking and other operations. After the melt has been properly blended through careful stirring, it is strained through a filter and cast into manageable bars or into waxed paper cups.

Rosin or colophonium is a natural resin that is quite hard and appears to be like glass where it fractures. It is a by-product of the distillation of turpentine from pine tree sap from which the volatile constituents have been extracted. Rosin comes in several different grades depending on the percentage of residual volatiles. The least distilled and softer varieties are called green pitch that is used with polishing pitch. The hard and brittle rosin used for blocking wax is often triple distilled. Yellow pitch lies between these extremes in both hardness and level of distillation. Rosin has a fairly high melting point (see Table 3.14), but that is strongly dependent on its hardness. The harder versions tend to be brittle. Optical grade rosin is nearly 100% pure. To attain this level of purity, the raw rosin must be refined repeatedly. The raw rosin has a dark amber to dark orange color. When it gets refined repeatedly, its color lightens to a slightly transparent yellow as impurities are removed to a level of 0.1%. It is available most often in broken chunks.

There are a number of rosins on the market. For instance, WW wood rosin, FF wood rosin, WW gum rosin, terpene resin, and ester gum are all terms used to describe rosin products. While some of these names may describe the same material, there can be significant differences between them. Of these listed, WW gum rosin is typically used for optics purposes after it has been triple distilled and refined through filtration. The regular WW gum rosin has a melting point of 71°C (160°F), while the triple distilled optical grade has a melting point of 82°C (180°F) [31].

Paraffin is a mixture of petroleum-based hydrocarbons which are by-products of crude oil distillation. The refined paraffin is translucent white in color, odorless, and tasteless. Some blends of these hydrocarbons have a very low melting point at about 50°C (120°F), but other blends will melt only after temperatures exceeding 75°C (167°F) have been reached. Whether one uses paraffin only as an ingredient or one uses it by itself, it is always important to know which type of paraffin it is.

Paraffin is one of the few waxes that can be used directly without additional blending with other waxes. It has a well-defined melting point rather than the broad melt temperature region that is typical of other waxes. This, and the additional fact that paraffin shrinks noticeably when it solidifies, can set up unacceptable stress deformation on thin and critical parts. The bond is also not very strong. However, the nearly waterlike viscosity of melted paraffin makes it possible to press it out into a very thin bond. This is why it is often used for holding exact parallelism on thick windows and critical angles on prisms.

Carnauba wax is a hard and brittle natural wax. This high melting point (see Table 3.14), yellow to light tan wax is extracted from South American palm tree leaves. Despite repeated attempts no suitable synthetic substitute for this wax has been found. Carnauba wax is not used directly in the optical shop but rather as an ingredient of commercially available wax formulations.

Microcrystalline wax is a petroleum-based wax that is available in two basic forms. One is a low melting point wax that remains plastic at room temperature, and the other is a hard and brittle wax with a much higher melting point (see Table 3.14). These synthetic waxes are refined and specially formulated for holding semiconductor wafers or similar thin and large area substrate on the block. They can be pressed out to a uniformly thin layer that makes accurate control of thickness and parallelism possible. The high-temperature version is intended for parts that are polished by high-speed and high-temperature methods. The low-temperature wax is designed for use on more critical parts.

Ceresine and ozokerites are both mineral waxes that are by-products of petroleum distillation. Ceresine is refined from ozokerites. Most ceresine wax is now produced from additional refinement and blending of paraffin and other miscible constituents. Like carnauba wax these waxes are used only as additives for special blocking wax or blocking pitch formulations.

Shellac is a natural product that is refined from certain trees in India and Indonesia. It is available in shiny dark brown, partly transparent flakes. It is a very hard natural resin that imparts special qualities to waxes and pitches. Sealing wax, which has found use in the optical shop as centering wax, contains a fair percentage of shellac. If shellac is mixed with certain blocking pitches, it yields a very hard but resilient pitch that is preferred for manual centering and edging of finished lenses. Shellac also makes an excellent protectant for finished optics. Enough of the flakes are dissolved in a small amount of alcohol until a dark brown, relatively thick syrup is created. This syrup is brushed on the polished surfaces and allowed to completely dry. The resulting protective layer is as hard as some of the better photoresists in use today.

Blocking wax is a collective term that covers a variety of different formulations for often widely differing applications. The preceding description of the various waxes is not entirely complete since other ingredients are also used in wax formulations, but the list covers the most important ingredients to make a wide variety of blocking waxes.

Standard blocking waxes are blends of rosin and beeswax. Common rosin to beeswax ratios are from 4:1 to 8:1, with some special formulations having even higher ratios. The higher the rosin content, the harder and more brittle is the wax and the higher is the melt temperature. These type of waxes are used for many routine blocking tasks and for stacking optical blanks in preparation for milling, dicing, and edging. These and many other operations will be discussed in Chapter 5.

There are numerous variants to these basic formulations that have been developed for specific tasks. For instance, there are low melting point waxes for holding thin, easily deformable substrate such as semiconductor wafers without introducing blocking stresses. Low-melt waxes with good holding power have been developed for blocking thermally sensitive crystals. The melting point of some of these is low enough that the wax can be removed in hot water. There are transfer waxes that have a lower melting point than the standard blocking waxes. These are suitable for transfer blocking of flat parts. Some wax formulations have been specially made for stacking of flat glass pieces in preparation for a variety of shaping tasks. Blanchard wax is a special formulation of rosin and beeswax with some special additives to give it proper holding power for milling operations. This wax comes in three hardness grades: soft (154°F), medium (160°F), and hard (165°F). Finally, there are a number of wax-based, high-temperature, high-strength, thermoplastic mounting cements that are used to hold optical crystals and other materials during heavy duty operations such as sawing, coring, and milling. Since these operations are done with diamond wheels, it is important that the holding wax will not load up the wheels. These thermoplastic cements fulfill this requirement, thus keeping the diamond wheels free cutting.

In more recent times some opticians have had good success in using a modified formulation of synthetic cements that cure when exposed to ultraviolet (UV) light. These UV cements are used for stress-free blocking of critical optical components during the various grinding and polishing operations. Unlike the standard UV cements, which are discussed in Section 3.5, these specially modified cements do not set up quite as hard. They form a firm but temporary bond that can be released when the last operation has been performed. The cured cement must readily dissolve in solvents that can be safely used in the shop.

In addition to the preceding list, there are resinous cements that are specifically made for blocking lenses into spot blocks. These are made from rosin, green pitch, and a small amount of shellac. The molten mixture must be strained through a sieve or a cloth filter after which it is poured into wax paper tubes to solidify into conveniently sized wax sticks. Semiround sticks are easily formed in a corrugated sheet metal mold.

Another application for special wax and pitch formulations is for optically centering lenses. Lenses of low refractive powers or lenses that are too sensitive to be centered by other means are cemented to centering spindles with such a cement. They are typically made from roughly equal parts of pine tar pitch or green pitch and shellac. The commercially available stick sealing wax is often preferred for smaller cementing tasks such as mounting lenses on wooden dowels. It is composed of a mix of rosin, shellac, and fine filler materials that make it stronger than the standard centering wax.

Other blocking waxes and cements are used in many variations that differ from shop to shop. Some of these are used for reasons of tradition and preference, others are formulated for special applications. The raw waxes and many of the special blends are commercially available either in cakes or in conveniently sized bars or sticks. The relative cost of the raw waxes varies quite a bit. Paraffin is typically the lowest priced wax material. Refined beeswax costs about six times as much. Low-temperature wax that has a high paraffin content is three times as expensive as paraffin, while centering wax with a high shellac content costs at least twice what the low-temperature wax costs. Specially formulated high-temperature waxes, often called *crystal cement,* can cost 20 times as much as paraffin.

Blocking pitch is used for pitch button blocking of lenses and similar parts that cannot be wax blocked because of their shape or their tendency to deform. It is made from pine tar pitch with significant percentages of hard waxes like shellac and inert filler materials such as powdered chalk blended into the molten pitch. Blocking pitches are much harder than polishing pitches with melting points usually above 70°C (158°F). The pitch hardness can be further increased by increasing the amount of fillers. This also raises the pitch viscosity, making it more resistant to flow.

Blocking pitch must have excellent holding strength, while retaining sufficient resiliency to reduce stress deformation of sensitive parts to a minimum. It must also be uniform from batch to batch. It should have a coefficient of expansion that differs sufficiently from both the metal tool and the optical glass so that cold deblocking, also called *chilling,* can be done easily. It must strongly resist creep and fatigue caused by the often significant and constantly changing shear forces acting on the parts during polishing. The lenses must not move or slide even under severe polishing conditions; otherwise, controlling the optical figure and obtaining the required surface quality may be all but impossible.

The color of blocking pitch is typically black, but green-colored blocking pitch is also available. The change in coloration is achieved by the color of the inert filler. The filler material must be fine enough so that it will not damage sensitive polished surfaces during blocking and deblocking. One supplier of optical shop supplies [32] offers three blocking pitches (either black or green) that are formulated for use at different ambient temperatures. The softer grade is called *winter grade,* while the harder grade is called *summer grade.* The difference in hardness compensates for the differences in temperatures experienced in those optical shops that are not equipped with an efficient air-conditioning system. For the modern air-conditioned shop, there is a year-round grade that is a compromise between the other two grades. Another supplier [33] offers a blocking pitch that is harder than most others. Although it is designed to block lenses for high-speed polishing operations, its exceptional strength makes it an excellent choice for blocking small lenses with steep curves to lens blocks. These and all other blocking pitches are commercially available in bars, cakes or bulk cans. One exception is a blocking foil [34] that comes in rolls. It is a coarse loose-weave fabric that has been impregnated with a soft blocking pitch. If a suitably large piece of this blocking foil is placed on a heated metal blocking tool, a uniform very thin pitch layer of 0.02 mm (0.0008 in.) can be transferred to the tool when the fabric is slowly removed. This material comes in two hardnesses. The soft pitch should be used for thin or for large lenses, while the hard version is best for thicker and for small lenses. The uniform pitch thickness lessens the tendency for thin lenses to spring or distort. It also allows for much better center thickness control.

Although they don't seem to fit precisely under the heading of this section, there are two types of blocking materials that must be mentioned here. One is plaster and the other are low melt metallic alloys. Plaster of Paris is chemically $CaSO_4:2H_2O$. It is a natural mineral that is extracted from mines. The naturally entrapped water is driven off by heating the raw mineral to about 130°C (266°F). The resulting plaster is then marketed as a white powder. When water is added to this powder, it is reaccepted into the chemical composition, which then returns into a solid mass. Plaster is used to hold prisms or other otherwise difficult to hold parts in special plaster blocks. How to make and use plaster blocks is described in greater detail in Chapter 5. Alabaster plaster, dental plaster, model plaster, and stucco plaster are some of the varieties preferred in optical shops.

Low-melt metallic alloys have been successfully used to firmly hold ophthalmic lenses during polishing. These alloys, which are composed of low-melt metals such as indium and gallium, are melted and literally cast around the lenses. The melt temperature is very low, so there is little danger to the glass lenses. This material has not found widespread use in precision optical shops, but it may be an alternative for special jobs where the usual blocking methods are insufficient for the task.

### 3.3.2. Polishing Pitch and Wax

The most important use of pitch and wax in the optical shop is to make polishers. A uniform layer of wood pitch, sometimes plain wax, and also blends of various pitches and waxes is applied to a metal or glass tool to serve as polishing surface. The surface of the polishing tool can be flat (plano) or curved (convex or concave). The majority of polishers is made from specially refined wood pitch that can be used as supplied. Some shops prefer to blend several pitch varieties, often with the addition of beeswax or other waxes to achieve the desired consistency. Wax polishers are used rarely and then only for plano work.

Polishing wax must be made only from the highest-quality wax. It is used in most cases only for polishing plano surfaces. The reason is that it is very difficult to press out a still-warm wax polisher into a spherical shape because wax, unlike pitch, is not a viscoelastic but a plastic substance. This means that it does not flow and will not readily conform to accept the shape of a mold unless it is poured in a molten state. To make a plano wax polisher, beeswax and rosin (colophonium) are melted and blended together in varying proportions. The beeswax to rosin ratio in these blends can range from 4 : 1 down to 1 : 20. Much depends on the specific application and the experiences of the shop. A wax polisher must be broken in first with a dummy block. Once such a plano wax polisher has successfully taken on a flat surface, it will make an excellent polishing tool that can produce outstanding surfaces both in terms of flatness and surface quality for a long time. The making and use of a wax polisher are described in Chapter 5.

Polishing pitches are either of natural origin or they are synthetically produced. There are several basic types:

| | |
|---|---|
| Wood pitch | Deciduous (beechwood) |
| | Coniferous (pine or fir) |
| Rosin based | Green pitch |
| | Yellow pitch |
| Petroleum-based pitches | |
| Asphalt tar pitch (coal based) | |

Of these, the black or dark brown wood pitches are most commonly used for polishing optics. Green and yellow pitch polishers have been made, but they have proved to be too hard for most applications mainly because of their insufficient flow characteristics. Some opticians prefer to use formulations of petroleum-based pitches. Asphalt tar pitches are not useful as polishers; they do not flow readily, although they are very soft. The rosin-based pitches (green and yellow pitch) and the asphalt tar pitch, however, play very important roles as additives to the other pitches for modifying their hardness and flow characteristics.

The most common blended polishing pitches are melts of pine tar pitch and yellow or green pitch. Depending on the desired hardness, one part of pine tar pitch is mixed with three parts of yellow pitch or six parts of green pitch. On occasion, a soft pine tar pitch is combined with a harder beechwood pitch. The softening point for most polishing pitches lies near 50°C (122°F). For the preparation of polishing pitch, the basic ingredients are weighed according to the desired amount and required mix ratio. The harder ingredient, typically the colophonium pitch, is melted first. The softer pitch can then be added to the melt and stirred in without additional heat. When all the pitch is melted and thoroughly blended together, it is strained through a sieve or a suitable filter. When mixing pitch to create a pitch with modified characteristics, it must be kept in mind that the lower viscosity pitch (usually the softer pitch) will have a much greater effect on the mixture than the harder pitch. This means that only small amounts of the softer pitch needs to be added to modify the harder pitch.

Since some of the pitch constituents are quite volatile at relatively low temperatures, the pitch must not be unnecessarily heated, and it must not remain in the molten state too long. These volatile constituents tend to evaporate when the pitch is heated which adversely affects the pitch viscosity. In shop terms, the pitch will be harder or more brittle and it will not flow as readily. These conditions may make the pitch unfit for the intended application. Therefore the temperature should be set just high enough to melt the pitch. Higher temperatures may melt the pitch faster but only at the expense of viscosity. Most pitches have melting points between 50°C and 100°C (120°F to 212°F). The best way is to heat the pitch slowly by leaving the set temperature between 10°C to 20°C (18°F to 36°F) above the melting point of the pitch. This will reduce the danger of driving off valuable volatiles but still leave the pitch pourable.

Just prior to pouring the pitch into molds or to casting a lap, the pitch mixture should be stirred thoroughly to blend all ingredients of the pitch. The stirring action must be very gentle to prevent the introduction of air bubbles. A freshly poured lap must be permitted to stand for a while to give the entrapped air bubbles an opportunity to rise to the surface. The bubbles can then be easily eliminated by a quick flash with the open flame of a bunsen burner or a similar open flame. The polisher can then be pressed out. If the pitch is not used right away, it can be poured into wax cups which separate easily from the hardened pitch. Most shops prepare a supply of pitch blends of varying hardness. Some suppliers offer ready-made polishing pitch formulations of different hardness.

There are numerous commercially available pitches and pitch formulations on the market, but only a few are widely accepted in the precision optics industry. Although we want to keep the use of tradenames to a minimum in this book, it is not possible to avoid using them altogether. Some products are so generally used in the industry that their tradename can be considered a generic designation. This is true for two lines of polishing pitch products. One of them are the Gugolz [35] wood pitches from Switzerland, the other are the petroleum based Cycad [36] pitches from the United States.

Wood tar pitches are extracted from resinous pinewood tar and from the tar of deciduous trees. They are extracted during dry distillation of pinewood and beechwood, respectively. The two pitch types differ in both odor and appearance. Deciduous wood tar pitches are deep black and usually hard and brittle. Resinous pinewood tar pitches are softer, often sticky, and are dark brown in color. Beechwood smells strongly of creosote. Pinewood pitch has a similar but milder smell and at the same time is resinous. Pinewood dissolves in alcohol much faster than beechwood pitch. Wood tar pitches differ from all other pitches by their characteristic odor, and they tend to exude acidic reactive fumes when heated.

In the past the first choice of most opticians was black Swedish pitch. It was the traditional standard of a quality polishing pitch. Other pitches, such as the 450 pitch from Universal [37] in the United States, have also enjoyed widespread popularity. These wood tar pitches have now been all but replaced by the series of Gugolz pitches. These pitches are carefully graded by melt temperature and hardness into five distinct grades. Table 3.15 lists these and other pitches with some of their more important characteristics. The main reason for broad acceptance in the optical shops are the carefully controlled quality and their excellent polishing characteristics, which are kept uniform from batch to batch over long periods of time. That and the fact that these pitches are supplied in convenient 1 kg (2.2 lb) easy-to-use cardboard containers and that they can be used directly as supplied without straining and additional mixing has made these pitches so popular in the optical shops.

Petroleum-based pitches are also used in optical shops. The series of Cycad pitches is preferred by many opticians in the United States. These pitches are blended with proprietary additives to give them their specific polishing characteristics. The blending also affects their color that can range from a dark amber to a clear pale yellow. Table 3.15 lists some of these pitches. There are three basic pitch formulations, each of which is available in four different hardness grades, namely, soft, medium, medium-hard, and hard. The first of these is designed for general precision polishing of a wide range of different optical materials. The second pitch type exhibits a small change in hardness as function of temperature. It is specifically formulated for polishing IR materials and metals . It can also be used on glass optics. The third type is characterized by a large change of hardness with changing temperatures. This permits the control of pitch hardness and thus flow characteristics by varying the temperature of the slurry. This type of pitch is recommended for use on low-expansion materials.

Rosin-based pitches are primarily green pitch, yellow pitch, or blends of these and some other pitch additives. Table 3.15 lists some of them. Al-

**Table 3.15.** **Basic Properties of Commonly Used Polishing Pitches**

| Source or supplier | Pitch properties Designation | Hardness | Melting point or range °C | °F | Viscosity[40] (poise) |
|---|---|---|---|---|---|
| Gugolz | # 55 | very soft | 52 - 55 | 126 - 131 | $1 \times 10^8$ |
|  | # 64 | soft | 68 - 72 | 154 - 162 | $3 \times 10^8$ |
|  | # 73 | medium | 77 - 80 | 171 - 176 | $2 \times 10^9$ |
|  | # 82 | hard | 79 - 82 | 174 - 180 | $5 \times 10^{10}$ |
|  | # 91 | very hard | 84 - 87 | 183 - 189 | $1 \times 10^{11}$ |
| Cycad | Optical | soft | - - - - - - | - - - - - - | $2 \times 10^9$ |
|  | Cleargold | medium | - - - - - - | - - - - - - | $3 \times 10^9$ |
|  | Blackgold | med. hard | - - - - - - | - - - - - - | - - - - - - |
|  |  | hard | - - - - - - | - - - - - - | $9 \times 10^{10}$ |
| Tivoli | Yellow #37 | soft | 37 | 99 | - - - - - - |
|  | Yellow #45 | medium | 45 | 113 | - - - - - - |
|  | Yellow #50 | hard | 50 | 122 | - - - - - - |
| Universal | Burgundy | soft | 57 | 135 | $8 \times 10^7$ |
|  | Burgundy | medium | 60 | 140 | $2 \times 10^8$ |
|  | Burgundy | hard | 63 | 145 | $2 \times 10^9$ |
|  | # 450 | med. hard | 71 -74 | 160 - 165 | $4 \times 10^9$ |
|  | Green pitch | soft | 63 -65 | 145 - 150 | $2 \times 10^{12}$ |
|  | High speed | very hard | 110 | 230 | - - - - - - |
| Zobel | - - - - - - | soft | - - - - - - | - - - - - - | - - - - - - |
|  | - - - - - - | medium | - - - - - - | - - - - - - | $2 \times 10^9$ |
|  | - - - - - - | hard | - - - - - - | - - - - - - | $1 \times 10^{11}$ |

though they are rated quite low in melting point, these pitches are relatively hard and do not flow readily. Only the softest pitch grades can be used as supplied. The harder grades cannot be pressed out to make a well-fitting polisher. To make a polisher with these pitches, the harder grades must be mixed with the softer ones. These pitches are also reluctant to take on a charge which is an essential property that a good polisher for precision work should have. One way to make a hard polisher is to cast it and then rub the desired surface into shape with a rough ground break-in tool. More is said in Chapter 5 on how to make a pitch polisher.

The flow characteristics of rosin-based pitches are such that even the softest grades appear stiff, which is another way of saying that the pitch resists flow and thus the polisher will retain its shape quite well for a long time. This fact causes great difficulty when making and breaking in a polisher, but it is a very beneficial feature once a good surface of the required shape has been established on the polisher. These type of polishers are very effective for medium to low-end optics which are produced in high volume by semiskilled labor.

In addition to the basic pitch types, optical shop supply houses offer a variety of polishing pitch products under many different names. Some fall into one of the three basic pitch types already discussed; others are blends or mixes of these and other ingredients such as waxes. For instance, there is burgundy pitch [38] that comes in different hardness grades. The semihard 450 black pitch that has already been mentioned can be softened by adding a few drops of linseed oil or by mixing about 5% of beeswax by weight to the pitch. There are also very high pitch formulations with high melting points that are used for high-speed polishing. Even a blend of organic and mineral pitches is offered as a substitute for the once favored Swedish pitch. Some suppliers, such as Zobel [39], will formulate high-quality custom blends to suit individual requirements.

All in all, the optician has a lot to choose from, and with careful selection based on keen observation, a suitable polishing pitch can always be found for even the most difficult requirements. This process of elimination takes patience and attention to detail. A series of aimless tests will most often lead nowhere. A carefully planned and faithfully executed program of mixing and blending will often lead to a customized pitch formulation that will be better for the intended application than anything that is commercially available. If all fails, the optician has always the opportunity to discuss any specific problem with the real experts, the people that make and supply the pitch.

The cost range of polishing pitches can exceed 1 : 20. This means that the most expensive pitch can cost 20 times more than the least expensive pitch. Of course volume greatly affects the cost of pitch, and there is a very steep premium to be paid when one selects the pitch packaged in small convenient sizes. If a large quantity of pitch is needed, it is much more cost-effective to buy it in bulk rather than prepackaged. The cost savings more than offset the added cost and inconvenience of extracting the pitch from the bulk containers. Premium pitches that are specifically formulated for precision work are two to three times more expensive than the best standard polishing pitches. The higher cost is presumably due to the more careful processing and the stricter quality control that promises to yield much better uniformity than the standard grades provide.

### 3.3.3. Theoretical Considerations

While the melting point of polishing pitches is often used as the measure of its hardness, it is its viscosity that is the single most important property. Izumitani [41] writes: "Pitch is a visco-elastic material that readily conforms to the shape of the workpiece through viscous flow. The polishing process is possible only when a relatively soft polishing compound is used in conjunction with a visco-elastic polisher." This says that the pitch viscosity determines how easily a polisher will flow or how well it will conform to the shape of the workpiece and maintain intimate contact with it. Proper flow characteristics of the polisher permit the correction and control of the shape of the workpiece surface, while its ability to contact the workpiece surface intimately ensures a clean and sleek-free polish. The viscosity of the pitch also determines how well the polisher will take on a surface charge of the polishing compound. The process of charging is essential for efficient polishing. This means that numerous individual polishing compound particles can embed themselves in the surface of the polisher where they can remain active for some time.

The next few paragraphs are a summary of some of the points described in a remarkable paper on polishing pitch [40]. Anyone who wants to know more about this topic should study this unique paper. Although the extensive use of mathematics and physics may be intimidating, it is well worth struggling through this paper to gain a much better understanding of what this seemingly mysterious substance is.

According to the paper, the pitch viscosities range over five orders of magnitude from $10^7$ to $10^{12}$ poise at or near room temperature. It does not necessarily follow that the low-viscosity pitch is the softest and that the high-viscosity pitch is the hardest, nor is it true that the melting point is related to the pitch viscosity. For instance, in Table 3.15, the lowest- and the highest-viscosity pitches are both quite soft. The low-viscosity pitch is a soft burgundy pitch with a melting point of 57°C (135°F), while the highest-viscosity pitch is a rosin-based green pitch with a melting point between 63°C and 66°C (145°F and 150°F); that is, it is only slightly higher than for the burgundy pitch. The viscosity of most pitches varies greatly with a change in temperature. This behavior is shown in Fig. 3.15 which has been adapted from the referenced paper. It shows a viscosity change for wood pitches by a factor of two for every 1.6°C to 2°C change in temperature and for every 3°C temperature change for petroleum-based pitches.

Although viscosity is the predominant property of pitches, it is the pitch elasticity that most affects surface quality. The elasticity of pitch is influenced by polishing pressure and pitch thickness. The pitch flow characteristics that determine how well the pitch lap will accept a shape and maintain it and how well it will remain in contact with the work are its viscosity and elasticity, respectively. These pitch lap characteristics are termed *compliance* in the paper.

For instance, a very thin lap, especially if not grooved, would have very low compliance even with the softest pitch. The polisher would be considered hard or stiff. If the pitch thickness is made too thick and the grooves cut into the pitch too wide and too deep, such a polisher would have very high compliance. It would be considered a soft lap, even if it was made with the hardest pitch. Neither of these extremes produces a lap that the optician would want to use, since the control of figure would be next to impossible. Therefore the best lap is one that lies halfways between these extremes. It

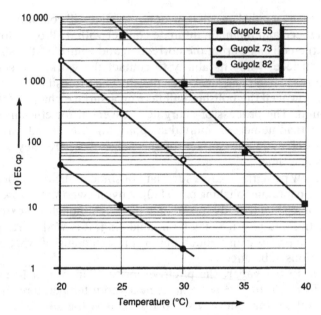

**Figure 3.15.** Pitch viscosity versus temperature [40].

would be a lap with moderate compliance and properly spaced and configured grooves of sufficient depth to allow for the controlled flow of the pitch surface.

In theory the grooves in the pitch must remain constant in volume, while the shape of the grooves will change as a result of a shearing force acting on the surface of the pitch lap. But the predominant flow characteristics of pitch give rise to another shear force that is a function of the pressure acting on the pitch lap normal to the surface. Not only does the shear force parallel to the lap surface deform the grooves, but the pressure-dependent shear force causes the pitch to flow from the center of the facets into the grooves. If this process is permitted to continue for some time, the grooves will be filled in and the lap thickness will be reduced by a corresponding amount. This explains why in the center of a heavily loaded lap, the original grooves are not only twisted but also completely filled in.

Although there are diverging views and preferences on the grooving patterns on pitch laps, some generalizations can be made. Square cuts are preferred for plano laps. The square facets can additionally be scored diagonally to reduce conditions that can lead to an astigmatic surface shape. Spherical laps usually receive circular grooves, which are concentrated either near the center or the edge to drive the lens block surface more convex or more concave, respectively. A central hole is essential to prevent pitch buildup at the center that would generate a concave zone (hole) in the lens block surface. If the hole is cut too large and the overarm stroke of the machine is too short, the opposite effect can result. Many opticians prefer a spiral cut most likely because it is the easiest one to cut in a volume production environment. They can be used on both plano and spherical polishers. It requires some practice to make spiral cuts correctly to affect the desired change in the shape of the polisher. Sometimes other, more complex, groove patterns are used. Grooves can be cut in an equilateral pattern on plano laps. Hexagonal or circular patterns require the use of molds or netting to achieve the desired results. These groove patterns do not lend themselves to induce changes in the shape of the lap surface. They are best for plano laps that do not change shape once the surface has been established.

By carefully selecting pitch lap thickness, groove pattern, groove depth, groove angle, and facet size, the optician can affect a change in polishing lap compliance by a factor of eight, assuming that all other parameters such as pitch viscosity and lap pressure can be held constant over the range of dimensions that are commonly used in the shop. By varying viscosity and pressure as well, the lap compliance can be modified over a broad range. Therefore the optician has a large number of options and it is not always an easy matter to choose the best option for a given condition.

### 3.3.4. Testing of Pitches and Resins

The natural products are processed by a number of chemical companies which generally offer them with data on the softening or flow points. These points provide information on the consistency of the pitches. The unit of measurement is centigrade (°C).

The softening point defined by Krämer-Sarnow is that temperature in centigrade at which a pitch plug which has been allowed to solidify in a glass tube is pushed out of the tube by the weight of a mercury column. The glass tube, open at both ends, 100 mm long with a 6-mm internal diameter, is sealed at one end with a 5-mm-high plug of the pitch. After the pitch has hardened, 5 grams of mercury are poured on top of the pitch plug. When heated in a water bath, the pitch plug and the mercury flow out of the tube when the softening temperature has been reached.

The flow point and the drop point are determined with a standardized apparatus [42]. A plug of pitch is melted into a tube and is kept flush with the tube end. After the pitch hardens, it is heated again in a water bath. When the flow point is reached, a hemispherical pitch deformation will protrude from the tube end. If the water is further heated, the deformed pitch will develop into a drop, which falls away from the tube when the drop point has been reached. The measured temperatures in centigrade denote the flow and drop points.

The ductility is a measure of the stretchability of the pitch. The pitch is cast into a bar in a standard mold. When at room temperature, the bar is stretched at 25°C at a speed of 5 cm/min until it fails. The distance stretched in cm until failure is the measure of ductility.

The penetration hardness of pitches is measured by a standardized method [43]. The pitch is melted into a test cylinder that has a diameter of 25 mm and is about 20 mm high. It is warmed to 25°C in a water bath, while a needle weighted down with 100 grams penetrates into the pitch. The point of the needle has a cone angle of 9° 10′ and it is flattened at the point to a diameter of 0.15 mm. The penetration, which is measured in 0.1-mm increments, is the depth of the needle reached in 5 seconds. This method is very similar to the Twyman Penetrometer [44]. It measures the penetration depth into the pitch sample of a 14° conical point that has a 0.5-mm truncated end. The measurement is made over a duration of 5 minutes with a 1 kg load on the point.

A method that is similar to the penetration measurement is frequently used in optical shops (Fig. 3.16). In general this measurement does not yield quantitative information. Only reference samples of proven polishing and blocking pitches are compared to new formulations.

A more sophisticated method to determine the viscosity of pitch is shown in Fig. 3.17 [40]. It was used to determine the viscosity values in Table 3.15 and in Fig. 3.15. The liquid pitch is poured to fill the void between the outer bushing and the inner shaft. After the pitch has solidified, it is immersed into a temperature-controlled water bath, as shown in Fig. 3.17. When the set temperature has been reached, a weight is attached to the torsion wheel. This causes the pitch to slowly yield as the shaft wants to turn. The angular displacement of the shaft as a function of time can be translated mathematically into pitch viscosity at the test temperature.

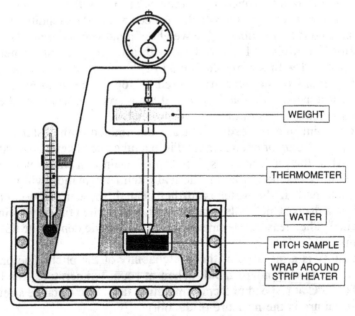

**Figure 3.16.**  Pitch penetration tester.

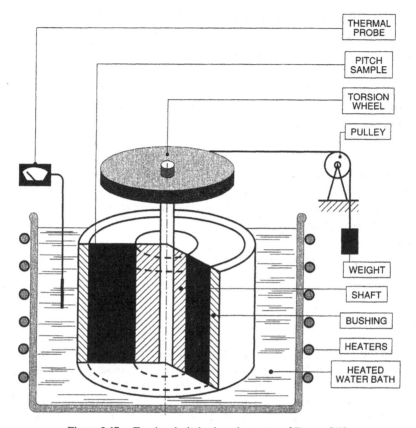

THERMAL PROBE

PITCH SAMPLE

TORSION WHEEL

PULLEY

WEIGHT

SHAFT

BUSHING

HEATERS

HEATED WATER BATH

**Figure 3.17.**   Torsional pitch viscosity tester of Brown [40].

Since optics is only a small user of pitch as compared to other industries, the customary trade information on pitch properties has not been developed for optics. Therefore additional tests are necessary. A sure way to test its usefulness is to subject a pitch sample to the intended application. In some countries, such as in Germany, there are official standards (DIN) to describe and specify the properties and behavior of optical shop materials like pitch. There are no such standards in the United States.

One test that is based on experience consists of essentially drawing a heated pitch sample into pitch strings. A pitch that is useful for optics applications should permit the drawing of long strings. The torn strings must not pull back in rubberlike fashion. Pitches with equal measurement values could therefore be absolutely useful for different applications.

### References for Section 3.3 (Waxes and Pitches)

[29] Karow, Hank H., "The importance of blocking," *Workshop on Optical Fabrication and Testing, Technical Digest,* Seattle, WA, 1986.

[30] International Wax Refining Co., USA.

[31] Ernst Zobel Co., USA.

[32] Universal Photonics, Inc., USA.

[33] Cycad Products, USA.

[34] Loh Optical Machinery, USA.

[35] Gugolz, Switzerland, marketed in US by Meller Optics, Inc.

[36] Cycad Products, USA.

[37] Universal Photonics, Inc., USA.

[38] Ibid.

[39] Ernst Zobel Co., USA.

[40] Brown, Norman J., "Optical Polishing Pitch", LLNL.

[41] Izumitani, T. S., *Optical Glass*, American Institute of Physics, 1986.

[42] DIN 51 801, Sheet 2.

[43] DIN 51 579.

[44] Twyman, F., *Prism and Lens Making*, Hilger & Watts, Ltd., London, Second Edition, 1952.

## 3.4.  POLISHING PADS

Although most precision polishing is still done on pitch polishers, the majority of commercial polishing is done on natural or synthetic polishing pads. Natural pads are made of felt, woven wool cloth, cotton, or silk. Some synthetic pads are made of synthetic fabrics such as nylon or rayon. But most synthetic polishing pads are made from materials that have been specifically developed for polishing. Pellon [45], polyurethane foam, polyester fibers in polyurethane, poromerics, and Polytron [46] are the primary pad materials that are in use today.

### 3.4.1.  Felt and Cloth Pads

Felt and cloth polishers have been used for polishing commercial glass products and metallurgical samples for a long time. These are elastic polishers that differ from the viscoelastic pitch polishers in that they do not flow to conform to the shape of the workpiece surface. The workpiece surface is forced to some degree to conform to the shape of the felt or cloth polisher. This makes the control of optical figure almost impossible, although a high degree of polish can be achieved in a relatively short time. It is for these reasons that felt and cloth polishers are rarely used in the precision optical shop. They are described here primarily for the sake of completion and because they represent an economic alternative to more traditional polishers. They can be very effective where rapid polish rather than high precision is desired. For instance, felt polishers are very useful when edges and bevels must be polished. Cloth polishers are preferred for diamond polishing of metals.

Felt is a nonwoven material that is composed of tightly interlocking natural wool fibers. It is produced without the use of fillers or binders under the combined effects of pressure, moisture, and heat which are aided by chemical reactions. Felt most often consists of a blend of virgin wool, reprocessed wool, or reclaimed wool. Virgin wool is sheep or lamb wool that has never been used in felt before. Reprocessed wool is composed of wool fiber that has been recovered from previously unused felt products. Reclaimed wool is made up of low-grade wool fiber that is derived from used felt products.

The typically off-white polishing felts are hard-rolled wool felts that are available in large sheets of varying thicknesses. They range in thickness from 1 mm (0.040 in.) to 5 mm (0.197 in.), and they come in large rolls up to 1.5 m (60 in.) wide and up to over 90 m (100 yd) long. Felt sheets are typically 0.9 m square (36 × 36 in.) and range in thickness from 3 mm ($\frac{1}{8}$ in.) up to 75 mm (3 in.). The most popular felt thicknesses are 3 mm ($\frac{1}{8}$ in.), 6 mm ($\frac{1}{4}$ in.), and 12 mm ($\frac{1}{2}$ in.). Note that the metric values are approximate.

The quality grades of felt are descriptive by origin of the wool. The best grade is S1 (fine white Spanish wool), the standard grade is S2 (white Spanish wool), and the economy grade is S3 (fine white Mexican wool). The wool fiber content ranges from 100% for grade S1 to 95% for grade S3.

Felt density is an even more important parameter for felt polishers. By one method of grading, there are four densities [47]:

Soft        Used for light polishing of glass at low pressure; tends to generate rounded corners and edges.
Medium      Used for general industrial polishing; commonly used for commercial quality flat glass polishing.
Hard        Primarily used to polish large area flat glass surfaces.
Rock hard   Used for high-pressure, high-efficiency polishing where sharp edges and corners are important.

A more definitive way to characterize felt hardness is according to a standard [48] method that defines it in terms of quality group and density (g/cm$^3$); see Table 3.16. Felts of group H (medium hard) are best suited for polishing.

**Table 3.16. Felt Density Designation**

| Felt designation | Density (g/cm$^3$) |
| --- | --- |
| H 1 | 0.52 |
| H 2 | 0.56 |
| H 3 | 0.60 |
| H 4 | 0.64 |
| H 5 | 0.68 |

**Table 3.17. Swelling of Felt Polishers**

| Hardness | % Swelling |
|----------|------------|
| Soft | 25 to 35 |
| Medium | 30 to 50 |
| Hard | 40 to 60 |
| Rock hard | 50 to 70 |

Felts are highly absorbent and as such tend to swell appreciably as they absorb water from the slurry (see Table 3.17). The absorbency and swelling is a function of hardness or density. The denser felts absorb more water and swell much more than the less dense felts.

The tendency to swell so appreciably, which causes any polisher to change its shape, as well as the inability of felt to conform to the shape of the workpiece surface precludes the use of felt for any precision polishing. However, by impregnating felt with polishing pitch, some of the benefits of felt polishers such as aggressive polishing are combined with the optical figure control capability of pitch polishers. Pitch-impregnated felt polishers have been successfully used for low-cost, high-volume polishing of opthalmic lenses using high-speed machines with recirculating slurry systems.

Felt with randomly oriented interlocking fibers is often used in polishing wheels. This three-dimensional random orientation gives felt wheels uniform strength and density that leads to optimum polishing efficiency at the wheel periphery. The recommended peripheral speeds for felt wheels is a function of wheel diameter and face width. The manufacturer's recommendations should be carefully followed.

Besides felt which is a nonwoven or nondirectional polishing material, a variety of woven materials are used as polishers and are often preferred for special applications. Wool, cotton, silk, rayon, and nylon fabrics are sometimes used for polishing. Finely woven silk cloth, only 0.003 in. thick, is often preferred as polishing surface for polishing metals or metallurgical specimen with diamond compounds that are miscible in water. Conveniently sized pads with a moisture barrier and pressure-sensitive adhesive backing are commercially available from optical shop supply houses.

Blue 100% wool cloth, also called *Navy cloth*, has long been used for polishing of better-grade commercial optics. They are particularly popular in the opthalmic or eyeglass lens industry. Like most woven materials, Navy cloth has a smooth and a wooly side. It is the smooth side that is used for polishing. There are three grades that are differentiated based on thickness and weight (density):

| | | |
|---|---|---|
| 16-oz cloth | 1.0-mm (0.040 in.) thick | Light grade |
| 22-oz cloth | 1.5-mm (0.060 in.) thick | Medium grade |
| 30-oz cloth | 2.0-mm (0.080 in.) thick | Heavy grade |

The 30-ounce cloth is most widely used for opthalmic and commercial sphere polishing. The 22-ounce cloth is popular with opthalmic houses for both cylinder and sphere polishing. The light grade 16-ounce Navy cloth is used for the preparation of metallurgical specimen. Navy cloth is also available with pitch impregnation for semiprecision polishing of low-end optics. This material can be an economic alternative to the more expensive synthetic polishing pads, especially for noncritical polishing.

Two other woven wool products must be mentioned here since they have enjoyed some popularity for opthalmic, commercial lens and plastic optics polishing applications. One is Blue Streak wool cloth and the other is green billiard cloth. Both are woven from 100% domestic virgin wool. They are about 1 mm (0.040 in.) thick and come in rolls about 1.42 m (56 in.) wide. Precut pads are available with or without pressure-sensitive adhesive backing. Blue Streak can be pitch impregnated and is also available with a polishing compound filled plastic impregnation. These pads are specially formulated for high-speed commercial polishing. The plastic-impregnated pads attempt to combine the advantages of plastic and wool polishing. The billiard cloth is usually not impregnated with either pitch or plastic.

Napped rayon cloth that has a velvetlike polishing surface is used for polishing metals and other difficult to polish materials with either oil- or water-based diamond compounds. The pads are 1.2 mm (0.048 in.) thick, and they are available with a plastic barrier layer and pressure sensitive adhesive. A rayon on cotton fabric is preferred for polishing of opthalmic and commercial plastic lenses. It makes also an effective polisher for light polishing of soft optical materials. In addition to the polishing cloths mentioned, there are felt like woven wool cloths 2.5 to 5.7 mm (0.095 to 0.225 in.) thick for rapid stock removal in commercial volume production and for heavy-duty edge and bevel polishing of glass products.

### 3.4.2. Synthetic Pads

Pellon is one of the first synthetic polishing pads. They have been in use for nearly 40 years. The newer, more efficient synthetic polishing pads have all but replaced this product. Since it is in many ways a forerunner of today's pads, it is useful to take a closer look at this material.

Pellon has been made by a patented process [45]. It is a densely compacted material that is composed of many randomly oriented synthetic fibers that are held together by a plastic binder. The nondirectional arrangement of the fibers creates a microporous surface that traps polishing compound particles and holds them near the surface where they can be most active. The fibers and binder will absorb and hold slurry liquids, such as water, oil, or kerosene, quite readily to keep the pad surface well lubricated for efficient polishing.

Pellon pads were originally developed for high-speed polishing of opthalmic lenses. It has also been particularly useful for polishing plastic lenses. Over the years it has become a very effective pad material for a wide variety of optical crystals, although it has now lost its leading role in this field to the large number of modern high-quality synthetic pad products that have been specifically formulated for this purpose.

Pellon pads are available precut into circular pads, die cut into special shapes, or in sheets. They come in three basic pad thicknesses: light—0.4 mm (0.016 in.), medium—0.6 mm (0.022 in.), and heavy—0.9 mm (0.034 in.). They come with or without adhesive backing that is designated (W) for water- or alcohol-based slurries or (K) for kerosene- or oil-based slurries. A universal adhesive backing is also available that can be used with either type of slurry. This adhesive is either pressure sensitive or it is a dry layer that must first be activated by either methyl ethyl ketone (MEK) or by heat. The pads can also be perforated which is often preferred when working with diamond compounds.

Polyester fiber reinforced polyurethane pads are a modern version of the Pellon pads. They are produced and marketed under a variety of different trade names to serve as polishers for the semiconductor, optical, and other related industries. Surfin [49], Durlon [50], Politex [51], Newcor [52], Poro-Pol [53] are but a few of these trade names. The construction of these pads is essentially the same for all of them with polyester fibers or similar synthetic fibers compressed and in a polyurethane resin matrix.

These synthetic pad materials are characterized by their relatively high hardness, resulting in rather rigid and fairly abrasion-resistant polishing surfaces that will hold their shape for a long time once the shape has been established. This holds true even under severe working conditions. The shape of the polisher may have to be broken in with special pad dressing methods that are described in one of the following paragraphs. The non-woven structure of the pad surfaces creates numerous pores and voids that capture and retain individual slurry particles for effective polishing. The pore structure can range from open to dense, depending on pad hardness. This relationship is shown in Table 3.18. Some of the recommended uses for the pads in Table 3.18 are:

- Low-density pads are best for high-volume production where rapid stock removal is essential. They are especially useful for prepolishing of silicon wafers and for final polishing of soft glasses.
- Moderate-density pads are good for prepolishing of semiconductor wafers and for polishing of hard glass at moderate pressures and temperatures.
- High-density pads are best for high-rate stock removal production polishing of hard materials at high pressure (up to 20 psi) and high operating temperatures (up to 50°C).

Maintaining the flatness on a plano polisher or the curvature of a spherical polisher requires a hard pad (Shore A hardness of 70) that must also have excellent pad thickness uniformity. These hard pads may have to be grooved

**Table 3.18.   Pad Parameters for Fiber Reinforced
Polyurethane Pads**

| Parameter | Pad density | | |
|---|---|---|---|
|  | Low | Moderate | High |
| Hardness | soft | medium | hard |
| Pore structure | open | dense | very dense |
| Rigidity | resilient | rigid | compacted |
| Permeability | high | moderate | low |
| Liquid (CFM) | 12 | 5 | 1 |
| Construction | more fiber less binder | fiber nearly equals binder | less fiber more binder |

to ensure good slurry flow for optimum polishing and to prevent the glazing over of the pad surface. A copious and uninterrupted slurry flow must be supplied during the polishing cycle. Perforation of the pads often help to properly distribute the slurry.

Polyurethane is a thermoplastic material that is viscoelastic like pitch but has a much higher viscosity. However, it cannot be pressed out like pitch can. It therefore is not easy to make a good plano polisher or an accurately fitting curved surface. Diamond dressing tools are often used to establish the correct surface shape. However, once the surface is established, it will hold its shape for a long time.

Polishing pad dressing tools are essentially diamond pellet laps that have been specifically prepared to keep hard polishing pads flat and to keep the pad surfaces free of compacted slurry that often leads to glazing of the pad surface. Spherical pellet laps can also be used to maintain the correct shape of spherical polishers and to keep the surfaces properly dressed. The design and making of diamond pellet laps are described in Chapter 4. These pad dressing tools are periodically used to grind pad polishers at moderate pressures and slurry flow rates for 15 to 30 seconds. This treatment will quickly reestablish the optimum polishing surface.

This process can be further refined for spherical pad polishers according to one account [54]. The condition of the cast iron tool that holds the pad also contributes to the accuracy of the polishing surface. By carefully matching the radius of the tool to the thickness of the pad, a very accurate polishing surface can be established. This was accomplished by lapping the cast iron tool surface with a diamond pellet tool until the desired radius of curvature had been established. After the pad was applied, the pad surface was also lapped in with a corresponding pellet tool to create the precise radius of curvature. The same correction process was also performed on the fine-grinding tools so that a matched set of tools would be available. Once these rather extensive preparatory steps were done, the results were "fantastic."

The generation of the correct surface figure was very predictable as long as the match of the curvatures of the grinders and the polishers was maintained by periodic reconditioning with the pellet tools.

Most synthetic polishing pads are temperature resistant far beyond the normal operating temperatures generated during polishing. They are also chemically resistant between pH 4 and pH 13. They will be attacked, however, by strong acids (<pH 4) and by strong bases (>pH 13).

Polyurethane-based polishing pads perform best with water slurries of $CeO_2$ and $ZrO_2$ to which some detergent has been added to improve wetting. Many of them work equally well with the oil- or organic-solvent-based slurries that are often used for polishing with diamond compounds. They also work quite well with alcohols or ethylene glycol, which are used as the slurry vehicle for polishing water-soluble crystals.

The raw stock of most pad materials is available in rolls to about 1.42 m (56 in.) in width. Typical pad thicknesses range from 0.75 mm (0.030 in.) to 1.5 mm (0.060 in.), although much thicker pads can be cut on request. The relatively high fluid absorbency of these materials causes swelling of the pad between 20% and 40% of the original pad thickness.

Most commercially available polishing pads come in a variety of colors. These colors are used at the manufacturers or suppliers discretion. They do not represent a color code. Almost all polishing pads can be purchased with pressure-sensitive (PS) backing that can be used with either water or oil slurries. The pads are also available without backing and can be glued to polishing tools with a contact cement for maximum life. Strong bonds can also be formed with beeswax, rosin, or pitch. Methyl ethyl ketone (MEK) activated dry adhesive backing is really more of a convenience than an advantage, since the bond will not be as strong as any of the others mentioned. Such polishing tools should be used only for light pressure applications.

Porous polyurethane pads have found widespread use for many polishing applications in optics and in related fields. They are made entirely of polyurethane plastic that has solidified into a rigid, microcellular foam structure. Some of these pad materials have been impregnated with polishing compounds such as cerium oxide and zirconium oxide. This treatment is thought to increase the polishing efficiency of the pads. The slurry supply must be continuous, and the slurry must be liberally applied directly to the pad surface to avoid dry spots that can lead to a poorly polished surface and can also damage the pad. The slurry concentration should be kept fairly low, somewhere between 3° to 8° Bé, to reduce the tendency of the pad to glaze over.

Microcellular polyurethane foam pads are graded in several densities, ranging from soft to extra hard. Another important property is pad hardness which is measured with a Shore durometer on scales A or D. The values for these parameters for some commonly used pads are listed in Table 3.19. Polyurethane foam pads typically come in 23 × 55 in. sheets (about 584

**Table 3.19.  Density and Hardness for Polyurethane Pads [55]**

| Pad | Filler | Density Descript. | (lb/cu.ft) | Hardness Shore (A) | Shore (D) |
|-----|--------|-------------------|------------|--------------------|-----------|
| LP - 13 | $CeO_2$ | soft | 21 - 25 | 53 - 65 | 13 - 19 |
| LP - 46 | $ZrO_2$ | medium | 20 - 30 | 67 - 82 | 20 - 30 |
| LP - 66 | $CeO_2$ | medium | 22 - 32 | 67 - 82 | 20 - 30 |
| LP - 77 | $CeO_2$ | med./hard | 22 - 32 | 77 - 88 | 22 - 37 |
| GR - 15 | $CeO_2$ | med./hard | 27 - 32 | 70 - 80 | 22 - 30 |
| GR - 25 | none | med./hard | 25 - 30 | 75 - 85 | 25 - 33 |
| LP - 57 | none | hard | 30 - 40 | 81 - 91 | 30 - 40 |
| GR - 35 | $ZrO_2$ | hard | 34 - 42 | 85 - 93 | 33 - 43 |
| LP - 26 | $ZrO_2$ | hard | 34 - 42 | 85 - 93 | 33 - 43 |
| LP - 87 | none | extra hard | 46 - 53 | 94 - 97 | 45 - 50 |
| TLP - 88 | $CeO_2$ | extra hard | 65 - 75 | NA | 60 - 67 |

mm × 1400 mm) or in precut discs in standard or custom diameters. They range in thickness from a low of 0.5 mm (0.020 in.) to 12.7 mm (0.500 in.). The material is also available in 64-mm (2.5-in.) thick slabs that can be machined directly into spherical shapes.

While pads thicker than 2.5 mm (0.100 in.) can be mounted to the tool with pitch or wax, the thinner pads are best mounted with rubber cement to avoid bleeding of liquid pitch or wax through the pores of the pad to the polishing surface. There are many good rubber cements that are useful for this application. It is best to have the pad manufacturer or supplier suggest the best pad glue for the intended use.

The recommended use of rubber cement involves coating both the pad and the tool with a thin layer and letting it dry. A second coat is then applied to both. When the cement is no longer tacky, the pad is mounted on the tool and pressed out for half an hour. It is important to make sure that the pad is correctly oriented to the tool; it cannot be shifted once contact has been made. The thinner polyurethane pads can be preshaped between a matching set of heated sphere tools prior to cementing them to spherical polishing tools.

The polishing efficiency of polyurethane pads as measured by the rate of removal is shown in Fig. 3.18. The information has been adapted from a description of a commercial product [56]. It is quite obvious that the data for the graph were intended to convey the impression that brand A was superior to brand B. The use of these data in this book does not mean that such claims are established fact. The data were used only to illustrate the typical removal efficiencies for two types of commercially available pads. While the removal rate of a new brand A pad is very high initially, it rapidly decreases in

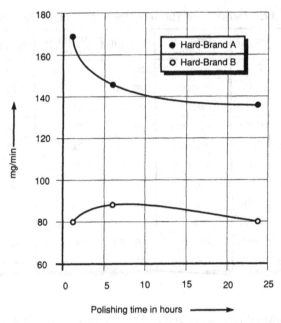

**Figure 3.18.**   Removal rates for two types of polyurethane pads [56].

removal efficiency and levels off at a slowly declining rate after 6 hours of continuous polishing. The brand B pad first increases in removal efficiency as the pad charges up and steadily declines after about 6 hours as the pad begins to glaze over. The pad surface should be reconditioned when its removal efficiency has declined to a predetermined level.

Wafer-mounting pads are variants of the polyurethane foam pads. They are not directly useful for polishing but intended for waxless mounting of silicon wafers. These pads are characterized by a smooth, rubberlike surface that holds the wafers by friction and suction. The pads are constructed from a microporous polyurethane material that has a pronounced density gradient from one side to the other. The rough side is glued to the tool, while the smooth side holds the wafers by friction. The numerous micropores on the pad surface act as miniature suction cups that hold the wafers in place just as if they were waxed down. Wafers mounted on these pads are processed with the same methods as the waxed wafers. They have proved to be quite suitable for high-pressure polishing.

This type of waxless blocking is useful only for very high aspect ratio circular pads such as silicon wafers. Shear moments created during polishing of thicker parts can slide them right off the friction pads. Mylar templates with circular cutouts can be permanently bonded to the friction pads to prevent lateral sliding and shifting of the parts. The cavity depth ranges in 0.025 mm (0.001 in.) increments from 0.15 mm (0.006 in.) to 0.5 mm (0.020

in.). The pad thickness is typically 1.5 mm (0.060 in.), although other thicknesses are also available. Circular pads up to 1 m (39.4 in.) in diameter can be precut. Otherwise, this material is also available in sheets up to 1.37 m (54 in.) wide.

Poromerics are a family of synthetic materials that was originally developed under the trade name Corfam [57] for use as packing for hydraulic systems. The primary use, however, is in the shoe industry where poromerics serve as artificial leather. In time the semiconductor industry discovered that poromerics make a very effective polishing surface for polishing silicon wafers. This led to the development of a variant of the original Corfam product as a thin polishing pad that has found widespread use in semiconductor wafer polishing and for polishing a wide variety of optical materials.

Poromeric pads have the look and feel of fine suede leather. This is why they are considered coriaceous (or leatherlike) in nature. These pads are made from a porous microcellular polyurethane foam that is bonded to a backing material. One such backing is a urethane polymer that is reinforced by polyester fibers. Other polymers are also used as backing to serve as barrier layer either to prevent the slurry from attacking the adhesive or to prevent the adhesive from penetrating to the polishing surface.

The structure of a typical poromeric pad can be described as follows: A porous but relatively dense sublayer penetrates 0.15 mm to 0.30 mm (0.006 to 0.012 in.) into the backing layer to form a tenacious bond with it that will not delaminate even under the heaviest workloads. This sublayer accounts for about half of the total pad thickness, which lies typically between 1.0 mm to 1.5 mm (0.040 to 0.060 in.). The porosity of the sublayer gradually increases away from the bond region. A distinctly directional, vertically oriented, highly porous outer layer is then coated onto the sublayer. This napped outer layer which accounts for about one-third of the total pad thickness serves as the polishing surface. The pore density, pore depth, and wall thickness between adjacent pores greatly determines the polishing qualities of the pad. Figure 3.19 [58] shows the pad structure schematically.

The uniquely elongated pore structure of poromeric pads makes them highly permeable which means that they readily permit the passage of fluids. This hydrophilic nature allows them to hold twice their weight in water. According to Du Pont [59] there are about 1 million micropores per square inch. Another source [60] reports 50,000 micropores per square centimeter.

Poromeric pads are strong enough that the higher-density pads can be used very effectively for high-speed, high-pressure polishing. While the standard pads and the less dense pads should be used at polishing pressures below 5 psi, the higher-density pads function best at pressures above 5 psi and at temperatures well above 100F. Highly densified poromerics do not have the unique porosity gradients of the standard material. They are more in the nature of very dense felt with a very homogeneous porosity throughout. These pads are made of randomly oriented polyester fibers that are

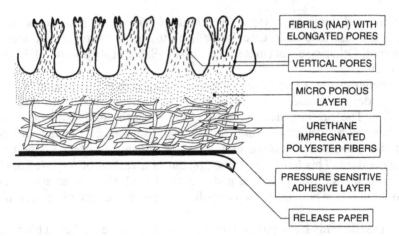

**Figure 3.19.**  Cross section through a poromeric pad [58].

uniformly dispersed in a porous polyurethane matrix. Highly densified poromeric pads up to 4 mm thick (about 0.150 in.) have been successfully used as
an effective substitute for polishing felts. Under the same operating conditions they exhibited much longer life than did the felt pads.

Poromeric pads are very useful for mechanical polishing of a great variety
of optical and semiconductor materials. They are particularly well suited,
however, for chemo-mechanical polishing of silicon, gallium arsenide, and a
host of other materials. Since these pads are chemically resistant over a
range from pH 4 to pH 13, they are not affected by mild acids and strong
alkaline solutions.

Poromeric polishing pads are available as precut discs up to 122 cm (48
in.) in diameter and in sheets up to 138 cm (54 in.) in width. Most common
size pads come with pressure-sensitive adhesive backing. Pads with an impervious barrier backing can be bonded to the tool with suitable spray-on
adhesives. The color of most commonly used pads is a flat black, but they
are also offered in many other colors. Since there is no industry standard
that governs color coding of pads, the various colors are merely used by the
pad manufacturers to highlight the unique properties of their products.

The typically flat black or rust brown standard poromeric polishing pads
have moderate porosity and are specifically designed for chemo-mechanical
polishing of silicon wafers and other semiconductor and optical materials.
Some pads are intended for use on specific materials such as gallium arsenide and gallium phosphide for which the pore structure and density must be
closely controlled for optimum pad performance. Other poromeric pads are
made to act softer than the standard variety. They are designed for use on
soft and difficult to polish crystals and metals. A much harder version of the
standard pad is designed for chemo-mechanical prepolishing of silicon wafers. The pores are smaller and shorter, and the pore walls are denser. This

makes the pad less flexible and tougher, allowing it to be used at higher than normal speeds, polishing pressures, and temperatures.

Hard and standard poromeric pads are usually bonded to flat (plano) polishing tools. The thinner and softer, and thus more resilient, pads can be made to conform to irregularly shaped and spherical surfaces. Although poromeric pads make efficient polishers, it is possible to control flatness only to a commercial quality level. This is true for all polishing pads. To combine the polishing efficiency of poromerics with the figure control of pitch polishers, some pad manufacturers vacuum impregnate standard poromeric pads with fine optical polishing pitch.

### 3.4.3. Plastic Polishing Surfaces

Polishing foils are very popular in Europe. Desmopan foil [61] is a new synthetic polishing pad material that operates on a unique principle that separates it from all other polishing pads. Desmopan foil has been marketed in the United States under the name Polytron [62]. It has been in use at Carl Zeiss in Germany for a number of years, and almost all precision polishing is done on this material. Excellent surface quality and $\frac{1}{10}$ wave flatness are commonly achieved on flat work. Results on lenses are quite similar. The average reduction in polishing time over the more traditional pitch polishing is better than 50%.

The quality of polish and polishing efficiency are only two advantages of this material. Another significant advantage is the exceptional life of the Desmopan foil. Plano polishers can last up to one year of constant use, while spherical polishers can survive daily use for several months. This means that tools need to be made only occasionally, which more than makes up for the costly and time-consuming task in making a Desmopan polishing tool. The making and breaking-in of Polytron polishers is not an inexpensive matter.

The following is a thumbnail sketch of the tool-making process which has been condensed from an application note for Polytron [63]. The pad must be carefully bonded to the previously prepared polishing tool. After the bonding is done properly, the polisher surface must be trued in (i.e., ground in) with a diamond pellet lap of the exact but opposite radius before the polisher can be used. It is this difficulty and the associated cost in making a Polytron polisher that has limited its use to high-volume production runs. The specific methods used to prepare and break in a Polytron polisher will be discussed in greater detail in Chapter 5.

Polytron (or Desmopan) is a translucent plastic sheet of very uniform thickness of about 0.5 mm (0.020 in.). It is available in standard disks of 125 mm (5 in.) diameter. The polishing side of this plastic sheet is covered with numerous parallel and uniformly spaced fine grooves. The opposite side, which is bonded to the polishing tool, is either smooth or circularly grooved. It is used on standard optical polishing machines with standard optical tools. The machine settings are very similar to those used for standard high-speed production runs. The preferred polishing slurry is based on high-purity cerium oxide powders. One such compound that has provided excellent results is Opaline [64], a nearly 100% pure cerium oxide. The slurry must be constantly supplied to the polisher through an effective recirculating system. Polishing cycles as short as 10 to 15 minutes per side on standard optical glass have been reported. The actual time depends on block size, part size, and material.

Thermoplastic synthetic materials have been used for high-volume polishing of lenses for some time. A fibrous filler material is intermixed with the thermoplastic. This provides numerous small voids that retain the polishing compound particles. In this sense these polishers are closely related to the polyester-filled polyurethane pads discussed in Section 3.4.2. However, the making of the polisher differs greatly from the use of the ready-made pad. It follows closely the making of a pitch polisher. The still-hot material is shaped into a spherical shell, which is glued to the matching spherical tool when cooled to room temperature. A similar synthetic material without filler is especially recommended for fine polishing. It must be impressed with a fine pattern so that it can hold the polishing compound. This approach is hardly in use today because there are so many excellent products available that are much easier to use.

## References for Section 3.4 (Polishing Pads)

[45] Pellon Corporation, Lowell, MA.

[46] Carl Zeiss, Oberkochen, Germany.

[47] James H. Rhodes & Company, Chicago, IL.

[48] DIN 61 200.

[49] Fujimi Corporation, Elmhurst, IL. (manufactured in Japan).

[50] J. I. Morris Company, Southbridge, MA.

[51] Geos Corporation, Stamford, CT.

[52] George Newman & Co., Santa Clara, CA.

[53] J. I. Morris Company, Southbridge, MA.

[54] Personal correspondence from Gerhard Wolf, Santa Rosa, CA, 1990.

[55] James H. Rhodes & Company, Chicago, IL, product catalog 1985.

[56] Ferro Corporation, Transelco Division, PennYan, NY.

[57] E.I. du Pont de Nemours & Company, Wilmington, DE.

[58] Geos Corporation, Stamford, CT, product note 2 "Polishing Media," 1982.

[59] Du Pont report "Corfam", No. A 56767-1968.

[60] Geos Corporation, Technical Report TR-21A.

[61] Carl Zeiss, Oberkochen, Germany.

[62] Optical Manufacturers International, Downers Grove, IL.

[63] Ibid. (applications note on Polytran).

[64] Rhodia, Inc., Monmouth Junction, NJ (Rhone-Poulenc, France).

## 3.5. CEMENTS AND ADHESIVES

The optician must be familiar with a wide variety of cements and adhesives. Cements are typically used to permanently join two optically transparent surfaces. Therefore the cement must be very pure, and its refractive index must match those of the glasses that are to be cemented. Achromatic doublet lenses and a number of different prism assemblies such as cube beam splitters are cemented. Adhesives are often used to bond optical components into mounts, but for the majority of applications, a number of different adhesives are used to assist the optician in certain shop tasks such as bonding polishing pads to tools or to provide for temporary bonds for some critical blocking methods.

### 3.5.1. Optical Cements

Optical cements can be of natural origin, but synthetic cements are in most common use today. Natural resin refined from the sap of evergreens was originally used as optical cement. Synthetic substitutes were developed to overcome availability problems and the problems associated with the fluctuating quality of this natural product. The first of these were thermoset optical cements that required the cemented parts to be heated to initiate polymerization. In time these cements were replaced with catalyst-activated synthetic resin cements that require the mixing of a liquid catalyst to the base resin just prior to the cementing operation. Although these cements represented an improvement over the thermosets, there were sufficient drawbacks to encourage further development. The current favorites are the photopolymer cements that cure rapidly when exposed to ultraviolet radiation. Although the UV cements have all but replaced the other cements for general optical shop use, there are still many applications where the thermosets and the catalyst-activated resins can provide advantages.

Canada balsam was the original optical cement. It is refined from the sap of a Canadian evergreen through chemical purification and processing. This product is still available in stick form, which comes in different hardnesses determined by the extent to which the solvent has been driven off. This cement has to be melted directly onto a preheated lens. It requires a great deal of skill to achieve a defect-free bond. Even though it is hardly used anymore, Canada balsam has an advantage over most other cements in that it can be easily decemented. This is a useful attribute when only a temporary bond is required. With a refractive index of 1.53 [65], it is well matched to most commonly used optical glasses.

Cellulose caprate is a synthetic substitute for Canada balsam that is used today for some special applications. It is a highly purified reaction product of cellulose and capric acid. It was developed in England where it is marketed as Watson Optical Cement [66]. The refractive index of the fully cured cement is 1.48, which limits its use to cementing lower-index glasses only.

Disturbing interference bands will be seen in the bond when high-index glasses are cemented with this low-index cement.

This product is a pale yellow thermoplastic cement that behaves like Canada balsam in that it melts at about 130°C (266°F) and resolidifies without change in composition as the temperature of the lenses return to ambient. This behavior permits the repositioning of misaligned lenses or prisms by simply reheating them to the melt temperature of the cement, correcting the error and allowing them to return to ambient undisturbed. The process of decementing is much more difficult for the other optical cements, which have to be totally separated, recleaned to remove the polymerized cements, and recemented.

If optical components cemented with cellulose caprate must be separated, it is easily done by heating them to above the softening point at 100°C (212°F) but no higher than the melting point. Residual cement is then readily removed with ketones or in chlorinated solvents. Cellulose caprate is insoluble in most alcohols, and acetone will merely soften it.

Compared with the other synthetic optical cements, cellulose caprate is surprisingly expensive, and that fact alone will eliminate it as a choice for any volume production. However, it is a good choice for particularly difficult cementing jobs.

Thermoset optical cements were developed by Eastman Kodak [67]. These are relatively high-viscosity thermosetting and partially prepolymerized methacrylates. The polymerization is completed by a heat-curing cycle. The prepolymerized state of these cements requires that it be refrigerated when not in use. It must be purchased as needed since the guaranteed shelf life, even if refrigerated, is only one month.

Lenses and prisms must be held in alignment by suitably designed fixtures to prevent parts from shifting during the relatively long curing cycle. The cement can be precured to a highly viscous consistency at 50°C (122°F) for about 15 minutes. A final alignment correction is possible during this state. Excess cement can be wiped off the edges with an acetone-dampened tissue or wipe. During the initial thermal curing cycle, the parts are held at 70°C (158°F) for 1 to 3 hours, after which the parts can be removed from the alignment fixtures. The removal must be done very carefully since sudden mechanical shock can delaminate the precured bond. Finally the bond curing is completed in a thermally controlled oven set at 70°C (158°F), and the parts are held there for 16 hours. The initial cycle can be accelerated by raising the oven temperature to 85°C (185°F) and holding the parts at that temperature between 1 or 2 hours. The same is true for the final cure, which can be done in less than 8 hours at the same temperature. However, this rapid curing can set up unacceptable bonding stresses in sensitive parts due to the significant shrinkage of the bond volume as the cement cures. For thermosets the shrinkage can approach 20%.

Decementing of fully cured thermoset bonds requires long immersion of the parts in hot xylene. A faster, but much riskier, way is to shock the lenses apart in near-boiling (200°C or near 400°F) castor oil, peanut oil, or any other suitable oil. There is the potential for considerable danger with these methods. Adequate ventilation must be provided, and commonsense safety precautions must prevail whenever heated solvents or oils are used. Open flames must be extinguished, since hot oil can become highly flammable. Hot lenses and prisms must only be handled with tongs. A face shield, heavy apron, and, at very least, safety gloves should be worn when performing this dangerous task.

Although thermoset optical cements represented an improvement over Canada balsam, it it easy to see that with the long thermal curing cycles and the difficulties in decementing, an easier to use cement needed to be developed. Catalyst-activated resins were the answer to some of the difficulties encountered with thermosets. These are commonly known as Lensbond [68], although other manufacturers market similar products under different trade names.

Lensbonds are thermoset two-part optical cements that are composed of a polyester resin base and a methyl ethyl ketone peroxide catalyst. The catalyst to resin ratio is about 1 : 50. Other ratios can also be used as shown in Fig. 3.20. The resin base is mixed with styrene to form a clear, viscous liquid. More than one-third of its volume is composed of volatile solvents, much of which must be driven off during a thermal curing cycle or during an

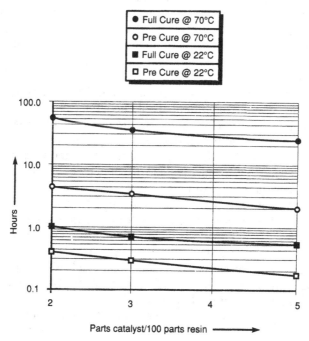

**Figure 3.20.** Effect of mix ratio and temperature on curing time for M-62 [69].

extended curing period at room temperature. This loss of volume leads to shrinkage, which can be almost 10% for this type of cement. Polymerization of the resin base is initiated and sustained by the catalyst. The process can be accelerated by specific thermal cycles, as shown in Fig. 3.20.

Room temperature curing is recommended for thin lenses that could be easily distorted by the rapid shrinkage of the bond volume when the curing is accelerated by a thermal cycle. For most other lenses and all prisms for which stress deformation is not a primary concern, a properly chosen thermal cycle can speed up the curing time. An increase in the catalyst ratio can speed up the process still more. These effects are shown in Fig. 3.20 for Lensbond M-62 [69]. The short precure times have made this cement a popular choice. Lensbond C-59 conforms to military specifications and is often the required cement on government contracts. It requires curing times that are two to six times longer than the corresponding times for M-62. Another version of the cement is Lensbond F-65, which requires much shorter curing cycles than any of the other Lensbond cements. It is formulated for prototype and small production runs or for those shops where suitable ovens are unavailable. Thermal curing is not recommended for this fast-setting cement. It may also be an unsuitable choice for thin lenses that can be easily distorted. Other manufacturers offer similar products.

While the curing times for these cements are shorter than for the thermosets, alignment fixtures are still required to prevent shifting of the aligned lens or prism pairs during the initial curing cycle. But unlike the thermosets, which require a thermal precure cycle of several hours, the catalyst-activated resins require only a fraction of an hour of precuring before the parts can be safely removed from the fixtures. This means that fewer and simpler fixtures are needed.

Lensbond and all other two-part cements should be kept away from open flames, since the volatile constituents could be ignited. Lensbond also emits a pungent, though not offensive, odor. It should always be used with adequate ventilation along with commonsense safety precautions. The resin component does not need to be refrigerated like the prepolymerized thermosets. But because it is photosensitive and affected by elevated temperatures, it is best to store it in a dark and cool place.

UV-curing cements are the optical cements of choice today. They are optically clear single-component liquid photopolymers that set up in seconds and fully cure in a few minutes when exposed to ultraviolet light. The cured cement forms a strong, resilient bond between two optically transparent surfaces. There is little or no loss of volume that can lead to stress deformation, since there are no volatile solvents in the cement. Any residual stresses in the bond are readily dissipated by the inherent resiliency of the fully cured cement. These cements offer additional advantages for optical assembly because there is no need to mix anything that can generate bubbles, which have to be removed before the cement can be used.

However, the primary advantages of the UV-curing cements lie in the ease with which they can be used and in the speed at which they polymerize. The nearly instantaneous precure greatly reduces the need for expensive fixturing because only one or two fixtures will be needed for each job rather than the much larger number of fixtures required for the longer curing cycles of the other cements. Cemented lenses or prisms aligned in a fixture need only be exposed very briefly to UV light to initiate polymerization to a sufficient degree that the parts can be immediately removed and handled. This will free up one fixture for a next assembly. This way only two fixtures will suffice for each job. Anyone who has done some optical cementing will understand why the UV-curing cements have all but replaced all other cementing methods.

The curing time of the cement is a function of bond thickness, the UV transparency of the optical material through which the light must pass to get to the bond, and the intensity of the UV source which is dependent on the energy of the source and its distance to the bond line. The maximum UV absorption for Norland cements (NOA 60 and NOA 61) lies in the range from 354 to 378 nm [70]. A number of UV-curing Lensbond cements (UV-69, UV-71, and UV-74) are also available with similar characteristics. The UV cements are quite transparent between 0.4 $\mu$m and 2.5 $\mu$m. The refractive index is 1.55. There are also other UV-curing cements on the market that are used for stress-free bonding of optical components into lens and prism mounts.

The curing time can be controlled by changing the UV intensity or the distance of the source to the bond line, or both. After the initial precure, a verification of proper alignment, and the removal of excess cement with acetone, the parts are final cured in 5 to 10 minutes with a 100-watt mercury lamp at a distance of 8 inches. Long wave fluorescent lamps can also be used at a distance of a few inches, but the curing time is about 1 hour. The preferred method is a 10- to 20-second precure exposure of individual assemblies with a short black light fluorescent lamp at the assembly station, followed by a 60-minute full cure with a 15-watt black light (366 nm) long fluorescent lamp which can illuminate a larger number of parts. Complete curing may take several weeks. During this aging period, the process of polymer crosslinking is completed, and a strong chemical bond will have been established between the optical surfaces and the cement.

Excess precured cement can be easily removed with a tissue lightly dampened with acetone. Precured parts should not be immersed in or flushed with acetone because the incompletely cured bond can be attacked by the solvent. However, precured parts that have bonding flaws can be separated by immersing them in methylene chloride for several hours. The actual debonding time depends on the degree of curing and the surface area of the bond. Small parts that are only lightly precured debond much quicker than larger parts that have been precured in a more intense light. The solvent resistance of fully cured bonds is very good.

### 3.5.2. Assembly Cements

Adhesives are used for many different shop tasks, and the optician can draw from a wide variety of commercially available adhesive products. Since in most cases there are no clearly defined tasks for which a specific adhesive has proved superior over any other, the optician is often forced to make choices based on insufficient data. The only reasonable approach is to have an understanding of the basic types of adhesives and their properties and to use this understanding to make choices for the required tasks. Some limited experimentation may refine the choices, but once an adhesive has been found that works and that is cost effective and readily available, it is best to stay with that product. Networking, which is a contemporary term for contacting other colleagues in the industry for the sharing of information, is another way to limit the choices. Calling an applications engineer of an adhesive manufacturer is always a good idea.

The purpose of this section is to examine the basic characteristics of the commonly used adhesive types and to provide some examples of how and where to use them in the optical shop. The adhesive types that are discussed are rubber cements, cyanoacrylates, anaerobics, silicone rubber, epoxies, casting compounds, and UV-curing mounting cements.

Rubber cements are most often used in the shops to adhere polishing pads to cast iron or aluminum tools. Many polishing pads are available with an adhesive layer bonded to the backing surface. The adhesives are either pressure sensitive, or they have to be activated by heat or a solvent. Some polishing pads that do not come with an adhesive backing can be fastened to the tool with wax or pitch. However, there are a number of pad materials that are best glued to the tool because the wax or pitch does not bond well with the pad or because the melted wax or pitch locally penetrates the porous pads to make them useless for quality polishing. For those situations and for all pads larger than 12 inches in diameter, a brush-on or a spray-on adhesive is preferred.

Pad adhesives must provide a strong bond with the pad to preclude separation and slippage under heavy polishing loads. The bond must retain some resiliency to cushion the effects of constantly changing forces during polishing. The adhesive must be easy to apply and readily removable once fully set. It must also form a uniform adhesive layer; lumps in the layer will pattern through the pad to cause polishing problems. The brush-on rubber adhesives are the best choice for heavy duty applications, while the more convenient spray-on adhesives are the preferred choice for less demanding applications.

Since these type of adhesives cure only when solvents can evaporate, the pads must be porous to permit the escape of the solvent. A polishing pad with a sealed backing layer will not form a good bond because most of the solvent cannot evaporate. For these, other bonding methods must be used. The majority of polishing pads are sufficiently porous to permit the use of rubber cement adhesives. Spray-on adhesives can be applied much more uniformly than is possible with the often viscous brush-on variety. If the brush-on adhesive is diluted with an appropriate solvent and several thin layers are applied, a very uniform adhesive layer will also be created. However, the dilution reduces the strength of the adhesive. This is also true for the spray-on adhesives.

The pad adhesives must be fast drying to facilitate the making of pad polishers but must not cure so fast to prevent the correct positioning of the pad on the tool. Care must also be taken to match the adhesive to the pad material. Some adhesives are specifically formulated for cloth or felt pads, while others are recommended for use on polyurethane or similar synthetic fiber products. One proven brush-on adhesive is 3M 1357 Scotch Grip contact cement, and 3M #77 is a very popular spray-on glue [71]. Other manufacturers produce similar adhesives, and sometimes consumer products will work just as well as the more expensive technical products.

Cyanoacrylate cements form a family of very fast setting and very high bond strength single-component adhesives. There are two basic types. One is ethyl resin based and the other is methyl resin based. The latter type has a much higher bond strength when fully cured.

Cyanoacrylates cure by an ionic process that is activated in the presence of mildly alkaline moisture. They permanently bond any close fitting, smooth surfaces of most solid substances. The surfaces that are to be bonded must be clean and must be in close proximity to one another. Rapid curing begins when the cyanoacrylate is trapped in a narrow gap between the two surfaces. The precure cycle lasts only a few seconds. The two bonded parts cannot be moved relative to one another after that. It will take another 8 hours to achieve a full cure of the bond.

Little or no shrinkage occurs during curing of this type of adhesive because it is composed of nearly 100% fully reactive constituents. This could make cyanoacrylates useful for stress-free and high-strength bonding of optical components. There are, however, a number of serious problems with using them for this purpose. First, the refractive index of cyanoacrylates is less than 1.5, which limits their use to low-index glasses. If they are used with higher-index glasses, distracting interference bands will be seen on the cemented surface. Second, the rapid curing will make it nearly impossible to press out entrapped air bubbles and to bring the two parts into proper alignment. Finally, debonding even of briefly precured parts is a very difficult task, requiring strong chemical solutions that can damage the polished optics. Therefore cementing lenses and prisms with cyanoacrylates is a possibility that has a low probability of success.

Some cyanoacrylate formulations have been developed specifically for temporary bonds. They can be debonded in a heated alkaline solution that is ultrasonically agitated. Debonding takes about 5 minutes, but much depends on the size and material of the parts and the temperature of the solution. The high temperature and the high alkalinity of the solution needed to break down the bond makes this method an unlikely choice for use on polished visible and IR optics.

This adhesive material is useful for rapid bonding of fixtures and in-process jigs. High-viscosity cyanoacrylates must be used to bond surfaces that are separated up to 0.25 mm (0.010 in.). For most work for which the gap between the surfaces is about 0.13 mm (0.005 in.), a lower-viscosity adhesive is used. A very low viscosity product may be necessary when the gap must be limited to 0.025 mm (0.001 in.). This would allow for accurate angular alignment of fixtures and jigs. The use of cyanoacrylates has been suggested for blocking lenses into spot tools for generating, grinding, and polishing of lenses. Since the bonds can only be broken in hot alkaline solutions, by mechanical shock, or by heating the blocks above 150°C (300°F), it is clear that neither method shows much promise in optics.

Cyanoacrylates are soluble in methy ethyl ketone (MEK), acetone, nitromethane, or dimethyl formamide. When these adhesives are used regularly and in volume, it is necessary to use safety glasses and polyethylene gloves. Rubber gloves or cotton gloves will not be useful since they will be readily bonded together. It is also necessary to provide for adequate ventilation. Any spilled adhesive must be flooded with water to accelerate curing. Spills must not be wiped since that will lead to exothermic polymerization and the liberation if irritating vapors [72].

Single-component anaerobic cements are a variant of cyanoacrylate adhesives. They cure very quickly when deprived of air while pressed out to a very thin bond between two nonporous surfaces. There is little or no shrinkage because there is no solvent or heat involved in the curing process. Eastman 910 adhesives [73] are a family of adhesives that represent this type. Extremely strong bonds result when the bond thickness is 0.025 mm (0.001 in.) or less. The curing time ranges from a few seconds to a few minutes, depending on the nature of the bonded surfaces. This adhesive can be useful in the optics shop for bonding metals, plastics, or rubber to fixtures and various fabrication aids. Glass to glass bonds are possible but difficult because the surfaces must be very clean. Anaerobic adhesives, however, are not useful for optical bonding of lenses and prisms.

Since the curing is so rapid, it is essential that the parts are properly positioned before they are bonded together. No movement is possible once the curing begins. The adhesive readily bonds skin to skin. The best response is to immediately flush the affected area with plenty of water. Never attempt to force separation since painful injury can result. Dimethylformamide (DMF) and nitromethane are solvents for cured adhesive.

Silicone rubber compounds and elastomers, more commonly known as *RTV*, can be a useful aid in the optical shop. These products and the related silicone elastomers have been used to make pitch button molds, pitch lap molds, as well as molds to hold lenses and other optical parts for specific operations. They have also been fashioned into block separators and into linings for optical part storage trays. RTV can also be used for low-stress blocking of critical parts. It must be pressed out to a thin layer for this purpose because its inherent elasticity may be more than is desirable for low-stress blocking. Besides, deblocking and cleaning of RTV-blocked parts is no easy matter. RTV has also found use as staking and sealing compound for shock-proof mounting of lenses and prisms.

RTV silicon rubber adhesive cures at room temperature when exposed to moisture in the air. It forms a skin after about 30 minutes, becomes tack free after a few hours, and is fully cured after 24 hours. The single component RTV, usually white in color, remains quite resilient after curing with a Shore A durometer hardness of 28 has been reported for one type [74] and 37 for another [75]. RTV gives off a characteristic acetic acid odor as it cures. Although it is not considered corrosive, a polished sample of glass or other optical material should be tested first before exposing finished optics made of the same material to RTV silicone rubber adhesives. Since RTV needs atmospheric moisture to cure, the bond thickness must not exceed 6.35 mm (0.25 in.). Otherwise, the moisture would be unable to penetrate the cured outer region to complete the curing on the inside. Bond surfaces must be cleaned with methyl ethyl ketone (MEK) to ensure a good bond. Excess RTV can be cleaned off with MEK. However, alcohols should not be used; fluorinated and chlorinated solvents will merely cause it to swell up.

Silicone elastomers are two-part silicone rubber compounds that do not require moisture for curing. This permits uniform curing throughout the material even in very thick layers. Shrinkage is minimal and not a factor in most cases. The silicone rubber base component is mixed with a catalyst at a ratio of 10 : 1. The curing is slow at room temperature, but curing time can be reduced at slightly elevated temperatures. The elastomers set up much harder than RTV. Shore A durometer hardness readings range from 32 to 65 points, depending on manufacturer and type of product. Besides the translucent variety, elastomers come in the colors white, red, tan, or black. Most of these are typically used for making molds, while others have been developed for encapsulating electronic circuitry. Some of these have also found use as protectant for optical parts.

Epoxies must be mentioned here because they are used on occasion in the optical shops. The permanent nature of the bond, the often significant shrinkage factors, and the extraordinary hardness of cured epoxies makes them unsuitable for most optical work.

There is one application for which epoxy is essential. Single- and two-component epoxies are used to permanently bond diamond pellets to metal tools to make pellet laps used for high-speed and high-volume lens grinding. The single-component epoxy is preferred for plano laps, while the two-component variety has proved to be a better choice for convex or concave laps.

The single-component epoxy is typically a thermoset that will slowly cure at room temperature. The curing cycle can be greatly accelerated by subjecting the bonded parts to a thermal cycle. Rapid curing, however, can set up significant stresses in the bond, which can cause shifting of the diamond pellets. This condition will require an often significant effort to correct the pellet tool to ensure that all pellets engage equally. A short precure cycle at a slightly raised temperature of 50°C (122°F) will increase the viscosity of the epoxy to the point where the individual pellets can still be positioned without shifting afterward. A subsequent thermal cure at 65°C (149°F) will then encourage final curing without setting up undue bonding stresses.

The curing of the two-part epoxy is initiated when a hardener or catalyst is mixed with the epoxy resin. The catalyst to resin ratio is very important and can range from 1 : 1 to 10 : 1. Equally important is the homogeneity of the mix. The catalyst must be completely mixed with the resin. Some manufacturers market premeasured plastic packs in which the carefully premeasured catalyst and resin is separated by a clamp. When the separator is removed, the two components combine. Vigorous kneeding of the pack results in a homogeneous mix, which is a precondition for a strong bond. One or both of the components may be color coded to assist the user in deciding when a homogeneous mix has been achieved, for the mix must have a uniform color.

The higher-viscosity epoxy mixes must be used within a few minutes, while low-viscosity formulations can have a pot life of several days. On average, full cure occurs in about 24 hours. This can be reduced to 8 hours by a thermal cycle at 95°C (203°F). Some of the two-part epoxies are highly exothermic, which means that they give off significant amounts of heat as they polymerize. This is especially the case when the polymerizing epoxy is contained in a sealed dispenser. Care is recommended when bonding thermally sensitive materials.

Significant reduction in epoxy volume occurs as polymerization progresses. This typical shrinkage characteristic makes epoxies a poor choice for bonding optical components either to each other or into mounts. It would set up unacceptably high bond stresses in the parts, which would lead to surface distortion, possibly stress birefringence, and even fracturing of the components.

It is always advisable to get detailed instructions for use of the epoxy from the manufacturer and then judiciously follow them. This is true for all supplies mentioned in this chapter. If there is any doubt about safety, health and environmental impact, a Material Safety Data Sheet (MSDS) should be requested from the manufacturer or supplier who has a legal obligation to comply with such a request.

## References for Section 3.5 (Cements and Adhesives)

[65] Schade, H., *Arbeitsverfahren der Feinoptik*. VDI-Verlag, Düsseldorf, Germany, 1955.

[66] Product literature from the Optical Division of The M.E.L. Equipment Company, Ltd., Barnet, Hertfordshire, England.

[67] Eastman Kodak Co., Rochester, NY (Kodak publication U-1).

[68] Summers Laboratory, Inc., Fort Washington, PA.

[69] Ibid.

[70] Norland Products, Inc., New Brunswick, NJ.

[71] 3M Company, Adhesives Division, St. Paul, MN.

[72] Loctite Corporation, Newington, CT (adapted from TDS 65A).

[73] Eastman Chemical Products, Inc., Kingsport, TN (publication R-192).

[74] Dow Corning Corporation, Midland, MI (Silastic 739 RTV, 1983).

[75] General Electric Co., Waterford, NY (RTV Silicones, S-41).

## 3.6. COOLANTS, SOLVENTS, AND OTHER SHOP SUPPLIES

In addition to the shop supplies discussed in the preceding sections, there are a number of other shop supplies that play an important role in the optical shop. Coolants are essential for the most efficient diamond tool operations, while solvents and detergents are needed to remove all soils from the optics that result from the various shop operations. Other chemical agents such as bases and acids are also briefly discussed.

### 3.6.1 Coolants

Coolants are required for those machining operations on optical materials that generate sufficiently high temperatures at the point where the tool contacts the workpiece to cause damage to both. Since almost all machining in optics is done with diamond tools, these coolants are specifically formulated to work most efficiently with such tools. Diamond tools are discussed in detail in Chapter 4.

The ideal coolant must have good heat dissipation to carry heat away from the cutting zone. It must also have good wetting capabilities so that it can reach the point where the heat is generated. Furthermore it must have good lubricating properties to reduce wear on the diamond tools and to keep them cutting free. Besides cooling the contact between the tool and the workpiece, the coolant must fulfill another important function by carrying away the abraded material. This material is commonly called *swarf*. The coolant must have a sufficiently low viscosity to allow the swarf to quickly settle out. If the abrasive swarf is left suspended in the recirculating coolant, it can erode the diamond bond and will lead to rapid wear of the diamond tool.

The coolant must be either not flammable or have very low flammability because a temporary interruption in the coolant flow can lead to sparks that will easily ignite the vapors of a flammable coolant. The ideal coolant should not cause health problems to the operator such as dermatitis or respiratory irritation. There should also be no offensive odor associated with its use. There are a number of oil-based and water-based coolants that meet most or even all of these requirements.

Oil coolants are low-viscosity oils that combine good heat dissipation with effective wetting of the contact area and excellent lubricating qualities. The use of these oils results in very efficient stock removal, while at the same time producing surface finishes that are clearly superior to those typically achieved with water-based coolants. The stock removal efficiency with oil-cooled diamond tools lies between two and three times that of the same tools cooled with a water-based coolant. Oil coolants also extend the life of expensive diamond tools significantly because of the superior lubricity of oils. They act as a rust preventative as well, so optical machines look like new even after many years of daily service. This is why many optical manufacturers, particularly in Europe, prefer the use of oil coolants.

Oil coolants facilitate the settling of swarf although not quite as efficiently as water-based coolants can because of their higher viscosity. It may require a larger volume of coolant in circulation and a larger settling tank to give the swarf a better chance to settle out before the recirculation cycle is complete. The combination of oil and settled swarf compacted on the bottom of the settling tank is a very messy affair and most operators dislike cleaning it out. The disposal of the oil sludge must follow safety rules and environmental laws, and this can make it a difficult and costly task.

Most oils are flammable, and although they may not ignite by themselves even under the most adverse conditions during accepted use, their vapors can readily ignite, especially in a high-temperature environment. It is one of the drawbacks of using oil coolants that oil vapors are generated by the typically high rotational speeds of diamond tools. These are not only hazardous from a flammability and health standpoint, but they are very difficult to contain. The oil vapors will coat every surface near the machine and that includes the floor. If several machines are operated with oil coolants in the same room, it may be necessary to install an expensive ventilation, filtration, and oil recovery system to avoid being fined for releasing oil vapors into the air. The oil film also attracts and holds airborne dust particles and fibers, and unless the machines are wiped down regularly, they will become quite unsightly in a very short time.

Despite these negative aspects there are many compelling advantages for oil coolants. Schade [76] recommends the use of Esso "Sommentor 28", while an American source [77] recommends Pella Oil made by Shell. Both oils meet the requirements for a good oil coolant.

At first glance water would seem to be an ideal coolant medium, since it has very good heat dissipation properties, wets most surfaces well, is neither toxic nor flammable, and its low viscosity aids in the rapid settling of swarf. However, water lacks one very important attribute that makes it a very poor choice: It has insufficient lubricating qualities. This means that the diamonds in the tools will wear and become dull very quickly. The cutting efficiency of the tool will decrease rapidly and soon degrade to the point where either the tool or the workpiece is damaged.

Water-soluble or miscible oils are added to the water to overcome this lubricity deficiency. Since these oils tend to foam in the recirculating systems used to pump the coolant to the work, antifoam agents must often be added as well. The wettability of the coolant can be improved by the addition of small amounts of detergents. Finally, rust inhibitors are also added to reduce rust formation on tools and exposed machine parts. The resulting mix of these various agents is called an *emulsion*. It should be fully miscible in water; this is just another way of saying that it must be water soluble. There are a number of such emulsions on the market under a variety of trade names. Some are formulated for general use, while others are specifically formulated for certain operations such as diamond lapping, pellet grinding, slicing, and generating. These operations are described in detail in Chapter 5.

Most coolants used in the U.S. optical industry are water based. One emulsion used quite often is Quaker 101 [78]; another is HMMO [79]. These emulsions are mixed with water at ratios of about 1 : 30. The best mix ratio depends on the intended use. For instance, the recommended mix ratios for HMMO to water are

| Heavy duty | Drilling and sawing | 1 to 20 |
| Medium duty | Slicing and dicing | 1 to 40 |
| Light duty | Precision grinding | 1 to 50 |

Water-based coolants solve some of the problems associated with the use of oil coolants. They do, however, have their own set of disadvantages. One of the positive differences is that the swarf can be separated more easily from water-based coolants and that it does not pose a disposal problem unless of course the swarf itself is toxic. Water-based coolants can be used for most diamond-machining operations but only at the expense of removal efficiency, increased diamond tool wear, and rougher surfaces on the workpiece. The feed rates on machines using water-based coolants must be reduced to about one-third of those possible with oil coolants. The reduced lubricity of water-based coolants, as compared to that of oil coolants, increases the wear of diamond tools and shortens their effective life. Because the diamond is not cutting as free, the resulting surface quality is also not as good as can be readily obtained with oil coolants.

Oil coolants are still necessary for centering machines because the sharp lens edges tend to chip easily with emulsion coolants. The oil also protects the delicate centering machines much better than a water coolant can. There are pros and cons to either approach. The best solution is the one that gives the most reliable and cost-effective results.

For very precise diamond work, such as fine generating, fine milling, and precision pellet lapping, one supplier [80] offers coolant filter paper. The coolant is passed through this paper as it recirculates back to the tool and workpiece to remove any residual suspended swarf particles. This filter is only useful with emulsion coolants. Since the filter tends to remove some of the emulsion, an occasional addition of makeup emulsion may be required.

Some of the additives that are found in emulsions are also available individually for use in recirculating grinding or polishing slurry systems. These are suspension agents that prevent the premature settling out of individual particles, foam suppressants that counteract the tendency of excessive foam generation in most recirculating systems, and wetting agents that ensure that there is no slurry film breakup that can lead to nonuniform distribution of the slurry and uneven grinding and polishing.

One special oil must be mentioned here because it is used in optical shops that use multiwafer slicing machines. It is marketed as PC Oil [81], and it serves as oil vehicle to hold and suspend abrasive grains as they flow through the reciprocating blades. The oil is the industry standard for this type of process. More is said about abrasive multiwafer slicing and slurry saws in Chapter 5.

### 3.6.2. Miscellaneous Shop Supplies

Polished surfaces on optical parts must be protected during optical manufacture to prevent stains and surface damage. Shellac dissolved in alcohol has already been mentioned as an effective and time-honored protectant. In response to this need, several optical shop suppliers offer a variety of ready-made products. Some of these are strippable plastic coatings that can be removed with adhesive tape. Some are synthetic coatings that are removable in hot water. Still others must be soaked in cleaning solvents to remove the coating. One of these is sold under the tradename Protectocote [82]. It is a fast drying brush-on protective coating that goes on fairly thick and remains resilient after it dries. This provides for a very good protective layer. Most of these protective coatings can be either brushed on or sprayed on. Household spray-on lacquers are also used in many optical shops, but the convenience of use is offset by their insufficient protective qualities. Water encroachment under the coating is a particular problem with this type of protectant.

Photoresist forms a very hard protective layer when allowed to dry properly. It can be brushed on or sprayed on. The spraying must be done in a properly designed spray hood because the volatile propellants can be toxic. If they are also flammable, the spray hood may have to be equipped with explosion-proof switches for the exhaust fan. These precautions apply to the use of any spray-on product that uses potentially hazardous propellants. Photoresist can be dissolved in acetone, but it can resist removal very tenaciously if it is baked on.

Chemical solvents are used in optical shops to dissolve pitch, wax, rosin, oils, and other organic substances that are used during the manufacture of

**Table 3.20. Solvents Used in Optical Shops [83]**

| Solvent type | BP (°C) | BP (°F) | TLV | Flam. |
|---|---|---|---|---|
| *Alcohols* | | | | |
| Ethanol | 78 | 172 | | Yes |
| Anhydrous ethanol | 78 | 172 | | Yes |
| Methanol | | | | |
| Isopropanol | | | | |
| *Aliphatics* | | | | |
| Heptane | | | | |
| Kerosene | | | | |
| Mineral spirits | | | | |
| *Aromatics* | | | | |
| Tolulene | | | | |
| Xylene | | | | |
| *Ketones* | | | | |
| Acetone | 56 | 133 | | |
| Methyl ethyl ketone | | | | |
| *Chlorinated hydrocarbons* | | | | |
| Trichloroethylene | 86 | 188 | 100 | No |
| Trichloroethane | 74 | 165 | 350 | No |
| Perchloroethylene | 121 | 250 | 100 | No |
| Methylene chloride | 40 | 104 | 200 | No |
| *Fluorinated hydrocarbons* | | | | |
| Trichlorotrifluoroethane | 47 | 118 | 1 000 | No |
| FC-113 blends and azeotropes | | | | |

optical components. This requirement is fulfilled by alcohols, ketones, aromatics, aliphatics, and especially the liquid fluorinated and chlorinated hydrocarbons. Table 3.20 lists the solvents that are used in optics.

The most commonly used solvents are alcohol and acetone. These solvents are used for most of the light cleaning required during the manufacture of optics. The final cleaning of finished optics is also done frequently with an alcohol or acetone wipe. Some opticians prefer to use a mix of acetone and alcohol, but the actual benefit derived from this practice is not entirely clear. Methyl ethyl ketone (MEK) is also used in some shops instead of alcohol and acetone. The strong odor and the toxicity of its fumes requires good ventilation that makes MEK a less favorable choice.

These solvents are usually not effective enough for dissolving heavy soils such as pitch and wax. Stronger solvents must be used for those conditions. Chlorinated or fluorinated hydrocarbon solvents are needed to do the heavy work. The soiled parts are either soaked and later washed in a static room temperature bath or they are cleaned in a specialized piece of equipment in which the solvent is vaporized. Such a unit is called a *vapor degreaser*. It is discussed in greater detail in Chapter 5. Contemporary environmental laws make it illegal to use these solvents in an unsafe manner. Health and pollution concerns have already put severe restrictions on the use of some chlorinated solvents. For instance, the once very prevalent trichloroethylene is hardly in use at all today because compliance with all applicable legal requirements makes its use prohibitively expensive. This is especially true since there are alternative solvents such as trichloroethane and methylene chloride that are almost as effective and much less of a problem. Serious environmental concerns about the ozone depletion problem could lead to a total ban on the use of chlorofluorocarbons (CFC) before the end of this century.

For some especially stubborn and troublesome cleaning problems other solvents may have to be used. Tolulene, xylene, heptane, and kerosene are sometimes needed. These solvents must also be used within the guidelines of safety, health, and environmental laws. It is strongly recommended to obtain a Material Safety Data Sheet (MSDS) for each solvent or other chemical used in the shops, and have these readily available for quick reference. Manufacturers and suppliers are required by law to furnish a MSDS when it is requested by a customer.

There are several different kinds of alcohols. Each has its own cleaning characteristics. Understanding the basic differences will help to make the best choices for a specific cleaning approach.

Alcohol made from grain or similar natural materials is called grain alcohol, ethyl alcohol, or ethanol for short. It is the stuff that we find in vodka, whiskey, and gin. Ethanol for technical use is usually denatured to make it unfit for human consumption. Ethanol is a good solvent for many different organic soils. When the solvent must absolutely leave no residue after cleaning, a very pure ethanol must be used. Such highly purified alcohols are also called absolute ethanol or anhydrous (water-free) ethanol. Sometimes these premium grades are also referred to as reagent grade or electronic grade alcohols. Although there are slight differences between these select grades, it is sufficient to know that one of these will do a good job when a surface must be very clean and free of stain.

Alcohols made from wood or similar organic matter are known as wood alcohol, methyl alcohol, or methanol. Since methanol is toxic and has a very offensive taste, it is used to denature ethanol. It is the most polar of all alcohols which makes methanol quite effective in removing insoluble particulates from lens surfaces. However, this high polarity tends to promote oxidation of white metals such as aluminum.

Isopropanol or isopropyl alcohol is very useful for the final cleaning of optics that have been cleaned in an aqueous bath. Isopropanol binds, absorbs, and displaces water. This property makes it an important additive to other solvents that are used in water displacement systems.

Acetone and methyl ethyl ketone are very effective cleaning agents for light soils. Their high rate of evaporation that does not attract and condense moisture permits the spot-free cleaning of optically polished surfaces. Acetone has a strong odor but is relatively harmless in a properly vented work area. MEK, however, should be used in an approved environment, since its vapors can be harmful to the health of operators. Alcohols and acetone can be mixed with other solvents that have evaporation rates that are similar to their own. Blends and azeotropes of alcohols and acetone with FC-113 (Freon) are common examples.

Chlorinated hydrocarbon solvents are best suited for dissolving pitch and wax. Trichloroethane solvents can be used in suitably designed vapor degreasers. The use of the once-prevalent trichloroethylene is now severely restricted. Perchloroethane is rarely used in optics because of restrictions and because its high boiling point makes the cost of vaporizing it quite prohibitive.

Most vapor degreasers that are used in optical shops today are charged with the much lower boiling point chlorofluorocarbon FC-113 solvents. Although pure FC 113, which is known better as Freon TF [84] and Genesolv D [85], can be used in vapor degreasers, the more effective solvents are blends or azeotropes of FC-113 and other solvents such as alcohol, acetone, and methylene chloride. A good substitute for the chlorinated hydrocarbon solvents is an azeotrope of FC-113 and methylene chloride which is marketed under the trade names Freon TMC and Genesolv DMC.

A variety of cleaning agents are designed for use in multistage aqueous cleaning systems. Technical detergents are often highly alkaline which limits their use to the cleaning of chemically insensitive optical materials. Only specially formulated neutral detergent solutions should be used for the aqueous cleaning of most optical glass and most infrared (IR) components.

Acids are used against stubborn stains on glasses. For instance, a few drops of acetic acid added to the slurry during the final stages of polishing will remove most polishing stains. Lens edges that are stained with compacted polishing compound can be cleaned with nitric acid to which a small amount of hydrogen peroxide has been added.

Failed coatings can often be chemically removed with concentrated sulphuric acid which is mixed with a small percentage of crystallized boric acid. Glass can be etched with hydrofluoric acid (HF) or with its salts. Hydrofluoric acid applied in liquid form results in clear etching, while the vapors of highly concentrated HF creates diffuse etching on glass. Solutions of fluoric ammonia ($NH_4$) $HF_2$ serves as etching ink to which acids such as HCl or salts like potassium sulfate ($K_2SO_4$) have been added. The texture of the resulting etch can be influenced by varying the concentration of the solution. Greatly diluted hydrofluoric acid also dissolves dried-on polishing compound. This method should be used only as last resort since even highly diluted HF can downgrade polished surfaces.

Concern for safety must always govern the use of these or any other chemicals. Solvents must be used only in properly designed equipment which must be operated in an approved manner. Adequate ventilation must be provided where solvent vapors may become a health hazard.

**References for Section 3.6 (Coolants, etc.)**

[76] Harry Schade, "Arbeitsverfahren der Feinoptik," unpublished manuscript in German, 1984.

[77] Universal Photonics, Inc., Hicksville, NY (1978).

[78] Process Research Products, Inc., Trenton, NJ.

[79] Rust-Lick, Inc., Anaheim, CA (1987).

[80] Loh Optical Machinery, Inc., Morton Grove, IL (1982).

[81] Process Research Products, Inc., Trenton, NJ (1985).

[82] Universal Photonics, Inc., Hicksville, NY (1976).

[83] Karow, Hank H., *Cleaning of Optics*, short course notes, Spring Conference on Applied Optics, Rochester, NY, 1982.

[84] E. I. du Pont de Nemours & Co., Freon Products Div., Wilmington, DE.

[85] Allied Chemical Co., Morristown, NJ.

# 4

# Tools and Fixtures

Precision optics was originally a branch of the mechanics trade and it became a separate craft later. The tools and fixtures that were required for the specific tasks in optical fabrication were continuously perfected. The traditional tools are used during the blocking, grinding, and polishing operations that are essential for producing optical surfaces. As machinery became more sophisticated, increasingly precise and more versatile special fixtures were developed. During the past 50 years diamond tooling has evolved to the point where the majority of optics produced in the world depends at least in part on the use of specially designed diamond tools. New fabrication aids have been developed in recent years to complement novel fabrication processes. In this chapter will be described tools and fixtures that require a basic or more detailed explanation. Other tools will be described in Chapter 5 as part of the description of machines and processes. This is done for clarity and to avoid repetition.

## 4.1. SPHERICAL AND PLANO TOOLS

Almost all optical surfaces are either flat (plano) or curved (spherical). Tools are needed to hold the optical parts in order to grind and polish the surfaces. Therefore blocking tools, grinding tools, and polishing tools are needed for each surface.

Blocking tools are used to hold lenses or other optical components during the grinding and polishing stages. Smaller blocking tools are usually made from cast iron. Aluminum is a better choice for larger blocking tools because cast iron tools would become unacceptably heavy, especially when the tool has to run on top. However, aluminum castings are more expensive than cast iron blanks. This makes them an economical choice only when sufficient quantities of optical components have to be produced. Another reason for choosing aluminum would be the reduction in weight that can provide better control of the surface figure during polishing.

Grinding tools are either spherical or plano tools made out of cast iron. They are used for the various stages of free abrasive grinding which prepares the lens surfaces for the polishing stage. These tools are typically precast into a rough shape and machined to the desired shape on a lathe or a milling machine. Some of the precast tool blanks are standardized [1], but most tool blanks used are made up as needed. Tool blanks for smaller grinding tools under 65 mm (2.6 in.) are cut from cast rod. The cast iron must be free of voids and inclusions, and it must be of uniform hardness. Occasionally grinding tools are made from brass or some other material. These are used for specific applications only.

Polishing tools support the polishing surface which is most often made of polishing pitch. Wax, synthetic polishing pads, and polishing foils can also form the polishing surface. These tools, which are spherical for lenses and plano for windows, and mirrors are usually made from cast iron when the tool runs on the bottom spindle of the polishing machine and preferrably from aluminum when they run on top.

### 4.1.1. Tool Threads

The tool blanks must be first provided with a thread so that the tools can be mounted on the lathe for further machining. Small tools receive male threads; larger tools get female threads (Figs. 4.1a and 4.1b). Small concave polishing tools can be made without threads since they typically run on top. There is no accepted standard for tool threads in the United States, although threads such as 1-8 and 1/2-13 are often found. In some countries, such as Germany, standardization of optics tooling has long been an accepted fact [2]. Standard AO tapers instead of threads can still be found on antiquated equipment in many U.S. shops. Special mounting techniques such as spanner chucks, vacuum chucks, and magnetic chucks are sometimes used in high volume production shops because they can reduce the time to mount and demount lens blocks.

The threads cut on optical tools should be equipped with centering bosses for male threads (Figs. 4.2a and 4.2b) and centering bores for female threads (Figs. 4.2c and 4.2d). Both types of threads should have a true running bearing surface that seats against the machine spindle flange. This ensures a true running tool, which is one of the prerequisites for precision grinding and polishing. A steel flange ought to be provided for aluminum tools to serve this purpose. The flange should be made from a corrosion resistant steel since the tools are constantly exposed to water. The flanges could be cast into the aluminum blank, but they must be prevented from turning. This can be accomplished by milling flats on opposite sides of the flange or by turning the two cylinder surfaces eccentric to each other, as shown in Fig. 4.3.

### 4.1.2. Making Spherical Tools

The spherical surfaces on optical tools are precast and cut to the desired radius on a sphere lathe or on an ordinary shop lathe with a sphere cutting attachment. Surfaces with a large radius of curvature are best generated in a

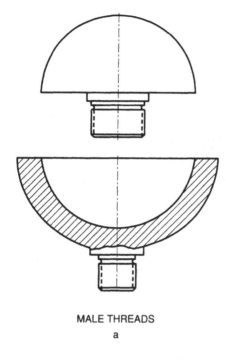

MALE THREADS

a

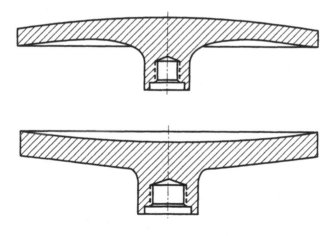

FEMALE THREADS

b

**Figure 4.1.** (*a*) Optical fabrication tool threads. (*b*) Optical fabrication tool threads.

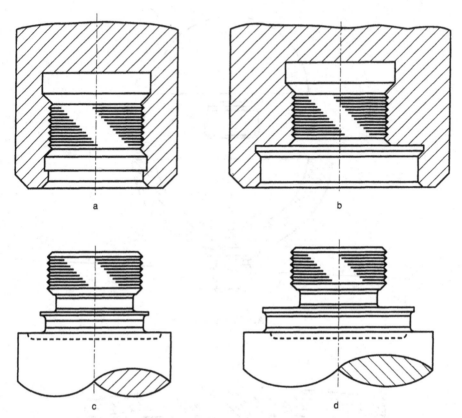

**Figure 4.2.** (*a*) Tool threads with small bearing flange. (*b*) Tool threads with large bearing flange. (*c*) Tool threads with small centering cylinder and bearing flange. (*d*) Tool threads with large centering cylinder and bearing flange.

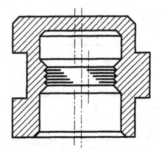

**Figure 4.3.** Example of an eccentric threaded steel flange.

fly-cutting mode on a vertical mill. Plano tools can be shaped on a facing lathe or on a mill.

The sphere lathe is especially designed to generate spherical surfaces which are so common in optical fabrication. The threaded tool blank is accepted by a threaded chuck, and the lathe is set up according to the accepted operating practice. The turning process is then automatic. The desired curvature is quite accurately achieved so that only minor corrections are required. Such a sphere lathe is not found in every optical shop.

Ordinary lathes that are equipped with a sphere-cutting attachment can also be used. Some of these attachments are commercially available from any machine shop supply house. An in-house modification of a lathe works often just as well. The sphere attachment is fastened to a mounting plate that is attached to the lathe bed. A horizontally mounted direct drive motor turns a large worm gear that is mounted to the plate. A linearly moveable stage that supports the cutting tool holder is mounted to the center shaft of the gear. A concave radius is cut on the rotating tool when the point of the cutting tool lies in front of the worm gear axis (Fig. 4.4a). Conversely, a convex radius results when it lies behind the gear axis (Fig. 4.4b). The radius

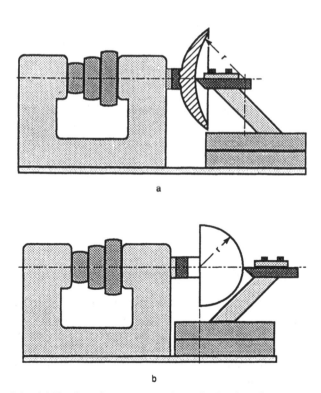

**Figure 4.4.** (a) Turning of a concave surface. (b) Turning of a convex surface.

of curvature can be set and corrected quite accurately with the tool bit stage. When turning concave surfaces, the starting radius must be smaller than the finished radius. It must be larger for convex surfaces. The final radius can be accurately approached by advancing the stage in one or several small increments. The gear shaft must be automatically driven because irregular feed rates will leave unacceptable grooves on the turned surface. It is also imperative that the point of the tool bit intersects the workpiece center exactly. A spherical surface can be generated only when this condition has been met. Size constraints on the lathe typically limit the radius of curvature to 300 mm (12 in.). Other methods must be used to generate longer radii.

Spherical surfaces can also be generated on a vertical mill with either a ring cutter or a single-point fly cutter. The head spindle of the mill is tilted at an angle $\alpha$ while the tool blank is screwed to a rotary table under the mill head (Figs. 4.5$a$ and 4.5$b$). The rotary table should be driven at a constant speed so that the resulting surface is uniform. The tool radius can be calculated from the diameter of the cutting tool and the head angle $\alpha$. Typically the head tilt angle $\alpha$ must be calculated for the selected cutter diameter $d$ to yield the desired tool radius $R$ by using Eq. (4.1):

$$\sin \alpha = \frac{d}{2R}. \tag{4.1}$$

**Example**

A radius tool with a radius of $R = 1719$ mm (67.677 in.) must be milled with a ring cutter of $d = 135$ mm (5.315 in.) What is the required head tilt angle $\alpha$?

$$\sin \alpha = \frac{135}{2(1719)} = 0.039267, \quad \alpha = 2.25°,$$

or

$$\alpha = 2°15'.$$

The same angle results when the inch values are used.

The divisions on the head tilt scale are usually too coarse to accurately set the correct head angle setting. Several cuts must be made after slight adjustments of the head tilt angle to generate the correct radius on the tool.

In modern optical shops, spherical tool surfaces can also be generated on computer numerically controlled (CNC) lathes. The cutting point does not desribe a continuous circular arc on these machines but is moved in very small incremental steps in the $x$ and $y$ coordinates. A circular cutting tool (Fig. 4.6) cuts evenly in every position.

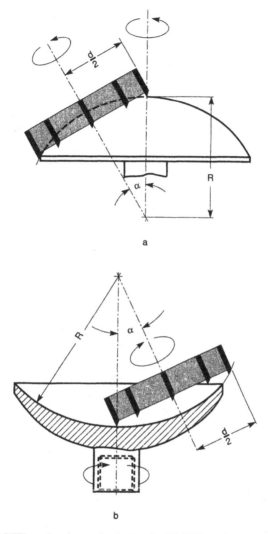

**Figure 4.5.** (*a*) Milling of convex spherical tools. (*b*) Milling of concave spherical tools.

### 4.1.3. Measuring the Tool Radius

The radius of turned or milled spherical tool surfaces are typically measured with radius templates or, more commonly, with hand-held spherometers. The radius templates are made from approximately 2 mm (0.080-in.) thick brass or steel sheet stock. They can be either turned or milled. The reference edge is beveled to a width of about 0.3 mm (0.012 in.). When the template can be turned as a complete circle, it can be made almost as accurately as a test plate. A template pair consisting of a convex and a matching concave template is required for each specific radius (Fig. 4.7).

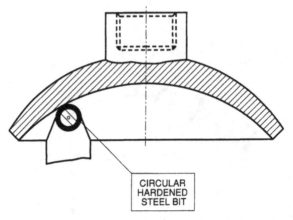

**Figure 4.6.**  Cutting tool set up for CNC generation of spherical surfaces.

The radius templates are used as comparison references that show the tool radius to be either too short or too long as compared to the template radius. By placing the concave template directly over the center of the convex tool, light can be seen through a fine gap caused by the difference in the tool radius relative the template radius. If the gap is toward the edge of the tool, the tool radius is too short; if it is in the center, then the tool radius is too long. This gap cannot be easily seen on concave tools. The convex template is also positioned directly over the tool center, but the radius mismatch is detected by slightly tipping the template to determine if it rests on

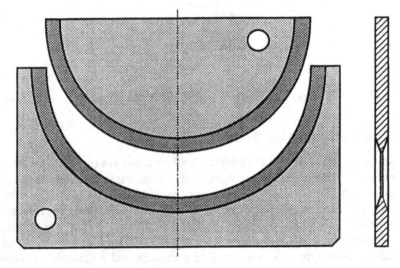

**Figure 4.7.**  Radius gauges.

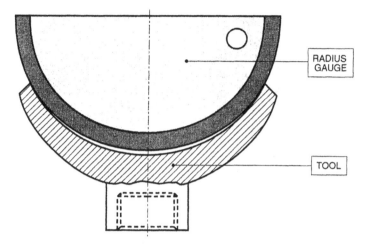

**Figure 4.8.** Testing with radius gauge.

the center or on the edge of the tool. An experienced optician can feel the difference in curvature between tool and template to an accuracy of less than a few tens of microns (Fig. 4.8).

Although there are standard precision straight edges available for checking the flatness of plano tool surfaces, highly accurate straight edges can be made in the shops from strips of glass. Glass strips are stacked together into two separate stacks and the stacks are ground flat on all four sides. These stacks are then fineground with one stack against the other. After separating the stacks one has a number of good plano straight edges.

Most modern shops do not use radius templates but rather rely on hand-held spherometers, which must be calibrated against a reference glass of the correct curvature. A hand-held spherometer is made up of a precision bell-shaped base and a high-resolution depth indicator (Fig. 4.9). The curvature of spherical surfaces can be compared to a calibrated reference surface with this device. A test plate can serve as reference, but this is not a recommended approach because the polished surface of the test plate can be easily damaged by the ring base or by the point of the indicator. A much better approach is to use a specially made fineground reference plate of the correct radius. It need not be polished, although a light shine may be required to test the curvature against a test plate.

When the curvature of the tool surface deviates from that of the reference, the position of the depth gauge pointer represents the deviation in terms of the measured sagitta. The *sagitta,* or *sag,* is a geometric term that defines the height of a section of a circle. Using this indirect method of measurement, we can deduce the accuracy of the radius of curvature from the deviation of the measured sag. Allowable or desired deviations can be set with the tolerance limit indicators on the depth gauge.

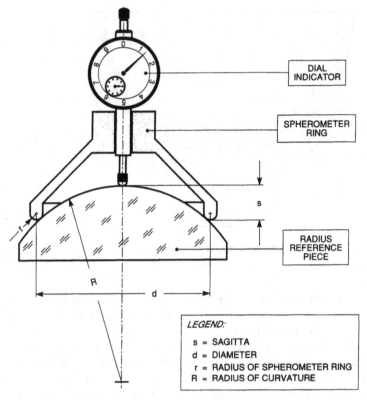

**Figure 4.9.**  Calibrating a hand spherometer.

For the measurements to have sufficient accuracy, the chosen ring base diameter must be as large as possible. A favorable condition exists when the ring diameter is nearly as large as the tool radius under test. It can be shown mathematically that the radius difference for this case is approximately six to seven times larger than the sag difference read off the indicator. The sag difference can be calculated per Eq. (4.2):

$$\Delta s = \left[ R_1 - \sqrt{R_1^2 - \left(\frac{d}{2}\right)^2} \right] - \left[ R_2 - \sqrt{R_2^2 - \left(\frac{d}{2}\right)^2} \right], \qquad (4.2)$$

where

$R_1$ = radius of reference,
$R_2$ = radius of tool,
$d/2$ = half the ring diameter,
$\Delta s$ = sag difference,
$\Delta R$ = radius difference = $|R_1 - R_2|$.

The dimensions must be either in millimeters or in inches. Mixing dimensions will lead to meaningless results.

**Example**

With a spherometer ring of $d = 100$ mm (3.937 in.) and a concave glass reference with $R_1 = 100$ mm (3.937 in.), the gauge of the hand-held spherometer is calibrated to read zero. How much deviation $\Delta s$ will the gauge show when the handspherometer is placed into a concave radius tool with $R_2 = 101$ mm (3.976 in.)? Equation (4.2) permits the calculation of the sag deviation that is read off the gauge:

$$\Delta s = \left[ 100 - \sqrt{\left\{ 100^2 - \left(\frac{100}{2}\right)^2\right\}} \right] - \left[ 101 - \sqrt{\left\{ 101^2 - \left(\frac{100}{2}\right)^2\right\}} \right],$$

$$\Delta s = 0.153 \text{ mm } (0.006 \text{ in.}).$$

The gauge will show 0.153 mm less sag than the reference when a concave radius tool with $R_2 = 101$ mm is measured.

When the radii of both surfaces are within 0.5% to each other, much simpler approximations can be used:

$$\Delta s \approx \Delta R \left(\frac{d^2}{8R^2 - d^2}\right), \tag{4.3a}$$

$$\Delta R \approx \Delta s \left(\frac{8R^2 - d^2}{d^2}\right). \tag{4.3b}$$

Hand-held spherometer sets of high precision are commercially available. Despite this many shops prefer to make their own spherometers to suit their particular needs. The in-house made rings are most often provided with a rounded edge (Fig. 4.9). The purchased ring sets are typically precision ground to a nearly sharp edge (Fig. 4.10). Tripod spherometers are specifically made to make measurements on large diameter surfaces with large radii. One such type is shown in Fig. 4.11. They are particularly useful for measuring the flatness of plano tool and workpiece surfaces.

### 4.1.4. Lens Blocks and Blocking Tools

It is desirable for economic reasons that several lens surfaces are ground and polished at the same time. The lenses or optical components are fastened for that purpose to blocking tools by a process called *blocking*. There are several different blocking approaches in use today, such as wax blocking, pitch button blocking, and blocking into spot blocks. Each one of these has advantages over the others for specific applications.

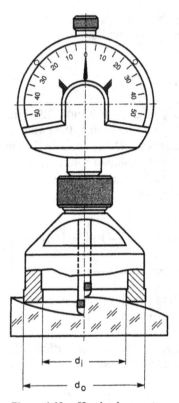

**Figure 4.10.**   Hand spherometer.

Wax blocking is generally used for holding noncritical plano parts to common blocks, tools, or fixtures for a variety of operations that include milling, coring, slicing, dicing, shaping, rounding, grinding, and polishing. The blocking tools are most commonly round plano tools made from a low expansion material such as Pyrex. Fixtures include L-jigs, flip-jigs and jigs for special shapes as are commonly found on prisms.

For pitch button blocking, the lenses are fastened onto or into spherical blocking tools with pitch buttons that are made from a shellac-based blocking pitch (Fig. 4.12). This is the preferred blocking method for low- to moderate-volume production of lenses. It usually does not pay to invest in expensive lens blocks with recessed lens seats when the volume of production is low or if there is no assurance of follow-on orders for the same parts. The economic limit lies today between a few hundred to several thousand lenses of the same type before expensive tooling makes economic sense. The exact break-even point depends on size, configuration, material, and required quality. Because of new developments in milling, generating, and blocking, the use of pitch button blocking has experienced an economic revival even for production volumes. Pitch button blocking is often the only way to hold

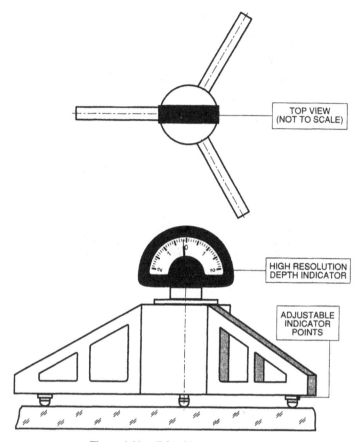

TOP VIEW
(NOT TO SCALE)

HIGH RESOLUTION
DEPTH INDICATOR

ADJUSTABLE
INDICATOR
POINTS

**Figure 4.11.** Tripod base spherometer.

sensitive or easily deformed lenses during manufacture. Another important application for this type of blocking is for reworking of defective lens surfaces.

Because it is generally much easier to obtain the tools and supplies needed for this blocking method, even high-volume orders are often started with the pitch button blocking approach until the special lens blocks, typically called *spot tools,* are made available. For this reason the term "provisional lens block" is sometimes used for pitch button blocks. The supplies for this method are then later utilized for reworking lenses rejected for dimensions, figure, or surface quality.

Once the pitch buttons have been attached to the lenses, the surfaces that must be worked are cleaned with a solvent. They are then set in a predetermined pattern into a concave lay-in tool, or onto a convex setup tool, and wrung down firmly with the help of a small amount of water which acts as a temporary adhesive. To achieve a good temporary bond this way, the radius

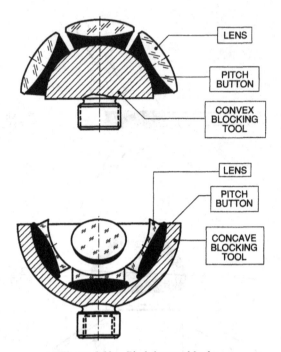

**Figure 4.12.**   Pitch button blocks.

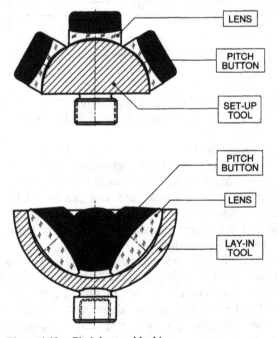

**Figure 4.13.**   Pitch button blocking.

of a convex lens surface must be larger and that of a concave lens surface smaller than the tool radius by a small amount (Figs. 4.13 and 4.14). This ensures that the lenses contact the tool only around their edge. A small void is formed between the lens surface and the surface of the tool which creates sufficient suction to hold the lenses in place during the blocking operation. A separation of one to 2 mm (0.040 to 0.080 in.) must be left between the individual lenses. This process step is known as setting the lenses either into the lay-in tool or onto the setup tool, depending on whether the block is concave or convex.

The actual blocking follows this important step. For lenses with concave surfaces the convex setup tool with the lenses attached is carefully lowered into the preheated concave blocking tool (Fig. 4.14a). For convex lens blocks the preheated convex blocking tool is allowed to sink into the pitch buttons of the lenses attached to the lay-in tool (Fig. 4.14b). The entire assembly can be quenched in a room-temperature water bath when the

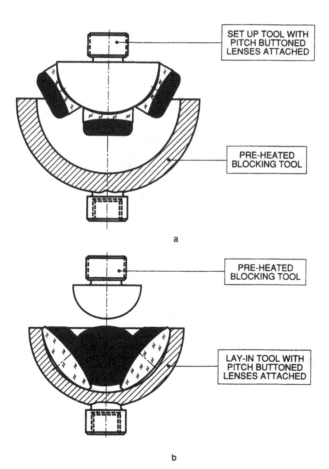

**Figure 4.14.** (a) Concave lens block. (b) Convex lens block.

blocking pitch has melted down to a predetermined depth. Sensitive blocks may become distorted during this quenching operation, and it is strongly recommended for such blocks that the heating of the blocking tools is carefully controlled in an accurately controlled oven so that their heat is consumed by the blocking process when the blocking tool has sunk to the desired depth into the pitch buttons. This is probably the best approach for most critical lens blocking.

The pitch buttons should be melted down only far enough so that the lenses in the center of the block will sit about 1 mm (0.040 in.) higher than those on the edge (Fig. 4.15c). The concentric form may be considered an acceptable lower limit (Fig. 4.15d). When the lenses seat deeper in the middle than on the edge because the radius of the blocking tool is too short, the edge row will not polish out because the blocking pitch will yield to the pressure exerted by the polisher (Fig. 4.15b). When the blocking tool radius is too long as shown in Fig. 4.15a, the lenses in the center will sit too high, so it will be difficult to control the figure in the central region of the block. These constraints also apply to concave blocks when the appropriate difference in curvature is taken into consideration.

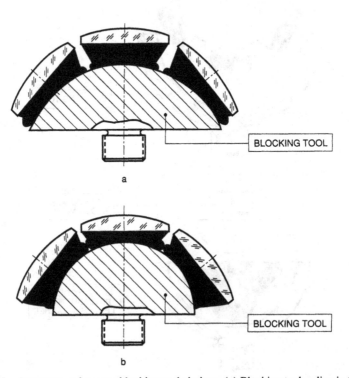

**Figure 4.15.** Improper and proper blocking tool choices. (a) Blocking tool radius is too long. (b) Blocking tool radius is too short.

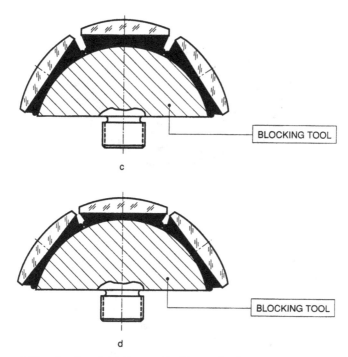

**Figure 4.15.** *Continued.* (*c*) Correct blocking tool. (*d*) Acceptable blocking tool.

The finegrinding tool may be used as a temporary lay-in or setup tool for rework of lenses that must be reground. When the surfaces only need to be repolished, the lenses can be set up in the polishing tool. Such blocks, however, must be repressed before the blocking pitch has hardened. These lens blocks tend to deform, since the previously preheated blocking pitch will shrink as it cools. Convex blocks tend to become more convex, while concave blocks become more concave. These blocks will retain the shape of the lay-in tool or setup tool quite accurately if they are pressed out again.

The required radius of the blocking tool can be calculated from the lens radius, the lens thickness, and the thickness of the pitch layer after blocking by using Eq. (4.4):

$$R_T = R_L - (t_e \text{ or } t_c) - t_p \tag{4.4}$$

For concave blocks (Fig. 4.16*a*), the concave radius $R_T$ of the blocking tool equals the lens radius $R_L$ minus the lens thickness ($t_e$ for negative lenses or $t_c$ for positive lenses) minus the pitch thickness $t_p$ which can range from 2 to 5 mm (0.080 to 0.200 in.) depending on the size of the lenses. Concave radii have negative values. Negative lenses are either biconcave, plano-concave, or negative meniscus lenses, while positive lenses are biconvex, plano-convex, or positive meniscus lenses.

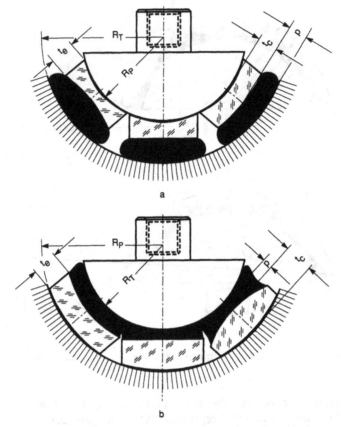

**Figure 4.16.** (*a*) Concave lens block. (*b*) Convex lens block.

## Example

Calculate the blocking tool radius for a concave lens block with a 35-mm radius. The edge thickness of the biconcave lens is 8.5 mm. The desired pitch thickness is 2.5 mm.

| Lens radius | −35.0 mm |
|---|---|
| Edge thickness | − 8.5 mm |
| Pitch thickness | − 2.5 mm |
| Blocking tool radius | −46.0 mm |

The required blocking tool radius is 46 mm concave.

For convex blocks (Fig. 4.16*b*), the convex radius $R_T$ of the blocking tool equals the lens radius $R_L$ minus the lens thickness ($t_e$ or $t_c$) minus the pitch thickness $t_p$, which can range from 2 to 5 mm (0.080 to 0.200 in.) depending on lens size.

**Example**

Calculate the blocking tool radius for a convex lens block with a radius of 48 mm. The center thickness of the plano-convex lens is 6.5 mm. The desired pitch thickness is 3.5 mm.

| | |
|---|---|
| Lens radius | +48.0 mm |
| Center thickness | − 6.5 mm |
| Pitch thickness | − 3.5 mm |
| Blocking tool radius | +38.0 mm |

### 4.1.5.   Pitch Buttons

Pitch buttons must be prepared in different shapes depending on the shape and curvature of the lens surface. Cone-shaped pitch buttons must be formed for short convex radii, while flat pitch buttons are preferred for plano and concave surfaces and for convex radii that are not strongly curved. This is shown in Fig. 4.17.

Concave lenses with thin centers are easily deformed by the blocking pitch. This deformation can be avoided for large lenses when the pitch button is recessed in the center (Fig. 4.18a) or when a small disk of a resilient

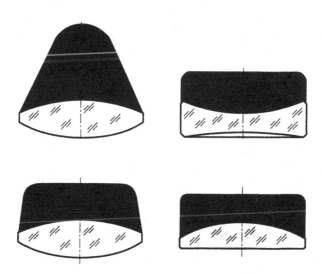

**Figure 4.17.**   Pitch button configurations.

**Figure 4.18.** (*a*) Low-stress blocking, using recessed pitch button. (*b*) Low-stress blocking, using absorbent paper.

material such as blotting paper is placed in the center under the pitch button (Fig. 4.18*b*).

Pitch buttons can be attached and shaped by hand, or they can be cast. The lenses must be prewarmed on a hot plate or in an oven for either method prior to attaching the pitch buttons. While the lenses warm up, the blocking pitch is melted in an electrically heated pan. For manual pitch button forming, the pitch should be heated only to the point where it becomes kneadable. For casting, the pitch must be in a readily flowing near-liquid state. When forming pitch buttons by hand, one withdraws from the pan a sufficiently large blob of blocking pitch for one lens. It is then pressed on the prewarmed lens, and the pitch button must be quickly formed by hand before the pitch hardens. This step requires some practice because the hot pitch and the heated lenses are difficult to handle. An experienced blocker can put pitch buttons on 90 lenses of 25 mm (nearly 1.0 in.) diameter in 1 hour with this method.

The buttons will be more uniform when they are cast in a metal mold rather than being formed by hand. The mold is made of either steel or brass, which is turned on a lathe and polished on the conical inside so that the warm pitch cannot stick to it. The metal molds can also be coated with a spray-on mold release to prevent the pitch from adhering them. Several molds are needed in rotation so that the buttons can cool down in the interim (Fig. 4.19).

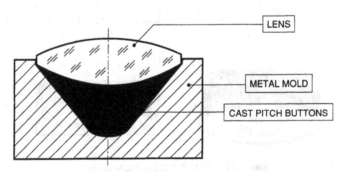

**Figure 4.19.** Molding of pitch buttons.

In volume production, pitch buttons can be attached to lenses most economically with a pitch button fixture called a *blocker*. A manually activated pump supplies the pitch to the mold, which is then exchanged for a cooled mold after each step. The prewarmed lens lies on the mold, and the pitch adheres to it to form a pitch button. The blocking pitch will adhere especially well when the lens surface has been spray painted with a lacquer. The lacquer must be allowed to dry completely before the pitch button is applied. This can be ensured by spraying the lenses before they are warmed.

### 4.1.6. Deblocking

Deblocking is a process step that follows grinding and polishing. The lenses must be removed from the blocking tool, and the pitch button must be removed from the lenses. The simplest method to remove lenses from the tool is to firmly rap the tool with a small mallet or hammer. This, however, is not always the best approach, since it tends to damage the tool and can cause the lenses to clash together causing unnecessary chipping. In any event fragile lenses or lenses that are too closely spaced must be hot deblocked, which involves heating the tool and pulling the lenses off one by one.

The blocking pitch can be separated from the lens with a sharp rap with a light wooden hammer. Obviously one must avoid hitting the lenses. To minimize the danger of damaging the lenses, a wedge of plastic or of hard wood is sometimes preferred. The wedge is set against the pitch button, which is then separated from the lens by an indirect blow from a wooden mallet. These manual deblocking methods are too time-consuming and tend to endanger the expensive lenses. For these reasons most modern optical shops use cold deblocking techniques.

Cold deblocking is initiated by placing the lens blocks in freezers or refrigerators which are set at approximately −15°C (5°F). Care must be taken to let the lens blocks cool to room temperature before they are put into the freezer. Thermal fracturing can occur when warm lenses are suddenly exposed to a decidedly colder environment. The process of failure is very similar to the thermal fracturing that often occurs when lenses that are at room temperature come in contact with a much hotter object.

After a time in the freezer or refrigerator, the blocking pitch separates cleanly from both the lenses and the blocking tool because it contracts more than either the lens material or the metal of the tool. The lenses can be simply lifted off. They must be submerged in either a solvent or water at once, since condensation forms quickly when the cold lenses are exposed to room-temperature air. Condensation can lead to severe stains on sensitive glasses.

Refrigerators can be loaded quite easily with lens blocks. The blocks can also be easily removed. However, cold air constantly escapes from the refrigerator because of the necessary frequent opening of the doors to load and remove lens blocks. Room-temperature air is introduced each time. This causes a layer of ice to form quite rapidly on the blocks. Top loading freezers, on the other hand, are more difficult to load effectively, but they tend to retain the cold air better since only small amounts of warm air are allowed to flow in each time the lid is opened. Because of this, freezers are preferred for short turnaround and reduced frost cycles for the blocks.

### 4.1.7. Blocking Small Lenses

Small lenses must be blocked with methods that differ from those described so far. The main reason for this is that pitch button blocking is too time-consuming for such lenses. Small lenses are those that have a diameter of less than 12.7 mm (0.500 in.). There are many special cases that require unique fabrication approaches for small lenses. In general small lenses are made from slab glass or from rod glass. From this raw material the blanks are shaped into small round disks, which are frequently finished to an accuracy of 0.01 mm (0.0004 in.) in thickness as well as in diameter. For nonglass materials the preferred starting shape is also a similarly shaped circular disk of suitable precision. The shape and the dimensional accuracy of the blanks determine the final dimensional accuracy of the finished lenses. Therefore accurate preparation of the blanks is essential. Figures 4.20a and 4.20b show a convex and a concave small lens block.

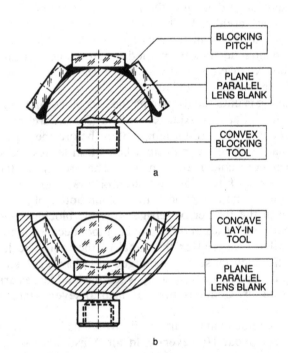

**Figure 4.20.** (a) Lens block with plane-parallel round blanks. (b) Lay-in tool for plane-parallel round lens blanks.

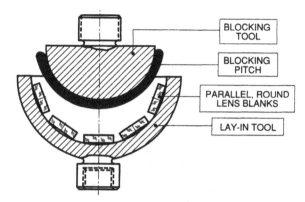

**Figure 4.21.**  Blocking tool covered with blocking pitch is pressed into lens blanks in lay-in tool.

For the fabrication of small lens blocks with convex surfaces, the round blanks are set into the lay-in tool with paraffin. The blocking tool is covered with a layer of blocking pitch that is made with hard sealing wax. The shaping of the pitch is done in a manner very similar to the making of a polisher. The pitch is heated in the open flame of a Bunsen burner or a bottled gas torch, and the blocking tool is then pressed into the lay-in tool which holds the lens blanks (Fig. 4.21). The lens blanks only adhere to the blocking pitch, and they easily separate from the lay-in tool. Excess blocking pitch must be removed by wiping the block with a solvent-soaked rag. The block must then be scrubbed in a strong detergent solution to counteract the effect of the solvent on the pitch. If this is not done, the pitch will remain sticky for some time.

A variant of this blocking technique involves pouring the viscous blocking pitch into the lay-in tool to which the lens blanks are fastened. The preheated blocking tool is then allowed to sink into this pool of viscous pitch to form the lens block. The inconvenience of cleaning off excess pitch and environmental concerns over the use of solvents have prompted some innovative opticians to come up with interesting solutions. One particularly innovative approach is to pour a uniform layer of salt between the lens blanks stuck to the lay-in tool to prevent the blocking pitch from penetrating to the surface of the lay-in tool. After removing the properly cooled lens block from the lay-in tool, one simply washes away the salt, and the messy clean-up step is not required.

The radius of the lay-in tool is calculated from the right triangle that is formed by the lens radius $R_L$, half the lens diameter $d/2$, and the radius $R_T$ of the lay-in tool. This relationship is shown in Fig. 4.22.

$$R_T = \sqrt{(R_L)^2 + \left(\frac{d}{2}\right)^2}. \tag{4.5}$$

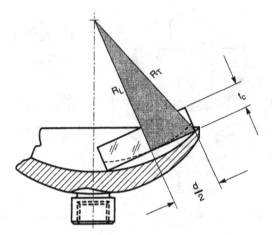

**Figure 4.22.** Calculating the radius $R_T$ of the lay-in tool for a small lens convex block.

## Example

Calculate the concave radius of a lay-in tool needed to block 4-mm (0.158-in.) diameter lenses that must receive a 10-mm (0.394-in.) convex radius.

$$R_T = \sqrt{(10)^2 + \left(\frac{4}{2}\right)^2} = 10.2 \text{ mm.}$$

The blocking tool radius can be calculated by using Eq. (4.4). The pitch thickness in this case is about the same as the center thickness of the finished lenses.

A somewhat different blocking method is used for blocking small lens blocks with concave surfaces. The round lens blanks are blocked directly into the concave blocking tool with hard blocking pitch such as sealing wax. Figure 4.23 shows this arrangement. The lens blanks are pressed carefully into the blocking pitch so that their edge rests directly on the surface of the blocking tool.

The radius of curvature of the concave blocking tool can also be calculated from a right triangle, as shown in Fig. 4.24. This triangle is formed by the lens radius $R_L$ plus the center thickness $t_c$, half the lens diameter $d/2$, and the blocking tool radius $R_T$. The center thickness is defined here as the thickness of the finished lens to the plano backside. For this geometry the following relationship can be established:

$$R_T = \sqrt{(R_L + t_c)^2 + \left(\frac{d}{2}\right)^2}. \tag{4.6}$$

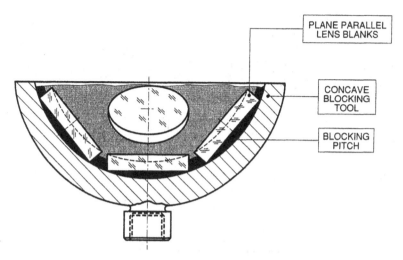

**Figure 4.23.** Plane-parallel round lens blanks blocked into concave blocking tool to receive concave radius.

If the lens radius is large relative to the lens diameter, the following approximation will often prove to be sufficient:

$$R_T = R_L + t_e, \tag{4.7}$$

where $t_e$ is the edge thickness of the finished lens.

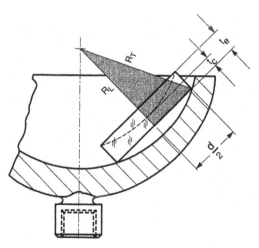

**Figure 4.24.** Calculating the radius $R$ of the blocking tool for small concave lenses.

**Example**

Calculate the radius $R_T$ of a concave blocking tool using Eqs. (4.6) and (4.7). The concave lens radius $r$ is 12.7 mm (0.500 in.), the lens center thickness $t$ is 1.0 mm (0.040 in.), the lens diameter $d$ is 7.0 mm (0.276 in.), and the corresponding edge thickness e is 0.5 mm (0.020 in.).

$$R_T = \sqrt{(12.7 + 1.0)^2 + \left(\frac{7.0}{2}\right)^2} = 14.1 \text{ mm.}$$

With the simpler approximation we calculate

$$R_T = 12.7 + 1.5 = 14.2 \text{ mm.}$$

**References for Section 4.1 (Spherical Tools)**

[1] Precast tool blanks per DIN 58 730 and DIN 58 734.
[2] Optics fab tooling threads per DIN 58 725 and DIN 59 726. These threads are matched to other standards for machine spindle threads.

## 4.2.  SPOT BLOCKS

Special blocking tools with either screw-on or recessed lens seats into which the lenses are blocked with a rosin-based blocking wax are known as *spot tools*. They are expensive precision tools that must be used and handled accordingly. Because of their high design and manufacturing cost, spot tools become an economic alternative only when a large number of the same lens must be fabricated over an extended period of time. Most jobs do not fall into that category, but when they do, spot tools can simplify the manufacturing process and greatly reduce manufacturing cost. The main reason for the limited use of spot tools in most optical shops is that they represent a significant up-front investment that requires several months of preparation for the design, manufacture, qualification, and correction of the tools before they can be used in production. Two separate spot tools are needed for each lens, even if they have identical radii on both sides. To gain the optimum economic benefit, a sufficiently large number of identical spot tools must be made to keep several runs of blocks in rotation. It has become easier today with the use of personal computers (PCs) and the availability of CNC five-axis mills to design and make spot tools right in house. But even under the best conditions, the real cost of such a tool often exceeds $1000.

### 4.2.1.  Spot Tool Threads

The accuracy of the tool design and the machining precision for the spot tools determines the shape and dimension of the lenses that are produced in these tools. This applies especially to the control of the lens center thickness. It is very important that the curvature of the block and the depth of the lens seats are made as accurate as possible. As a general rule a tolerance of ±0.1 mm (0.004 in.) is adequate for the lens seat depth for the first lens side. For the second side the lens seat depth tolerance must often be held to 0.01 mm (0.0004 in.). Unlike pitch button blocks which permit the reseating of lenses and other corrective steps, spot tools imprint their inherent precision (or inaccuracies) directly on the lenses made in them. There is little or no correction possible.

In addition to an accurate design and precise manufacture of the spot tools, the lens manufacturing process must be carefully planned in every detail and the plan must be faithfully executed. This is another reason why spot blocks make sense only for high-volume production; this kind of effort is often impossible for most jobs in todays optical shops. Despite these limitations spot tools have and will continue to play a very important role in the manufacture of lenses.

Spot tools that are lighter than 1 kg (about 2.2 lb) are usually made from cast iron or from a corrosion resistant steel. Cast rods or standard steel rod are the preferred raw material for spot tools of up to 60 mm (2.4 in.). Blanks for larger blocks must be cast to order. Cast aluminum is the preferred choice for large spot tools because of its ease of machining and its reduced weight. However, not only must the higher cost of cast aluminum be considered, but the mounting thread must be made out of steel and permanently fastened to the tool (Fig. 4.25). Some proposals have been drafted to standardize casting models for spot block blanks [3].

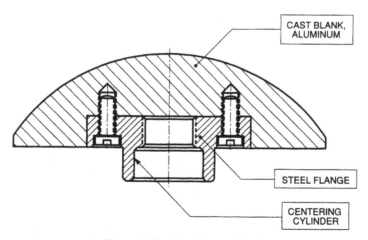

**Figure 4.25.** Blank for spot tool.

The first step in the manufacture of spot tools is the turning of the mounting thread with the centering boss or bore and the bearing surface. This is a very important step, since all other dimensions of the spot tool are dependent on the reference surfaces established during this operation. Male threads are preferred for blocks up to 80 mm (3.15 in.) in diameter, while female threads are needed for any blocks much larger than that. These typically standardized threads [4] are shown in Figs. 4.26a and 4.26b. However, these standards are not yet fully accepted, even in Germany where optics standards are highly developed because some of the older firms had already accumulated a sizable tool collection prior to the standardization. Any new spot tools produced should follow the standardization guidelines.

The mounting threads on spot tools are cut or chased on the loose side, which is verified through the use of suitable thread gauges to ensure that the centering boss or bore and the bearing shoulders can become fully effective. This is important because all other measurements for the spot block are referenced to the bearing shoulder. The true-running centering boss or bore prevents the blocks from running out of round, which would cause the lenses on one side of the block to be thicker than lenses on the opposite side. This effect is particularly noticeable on high-speed production lines.

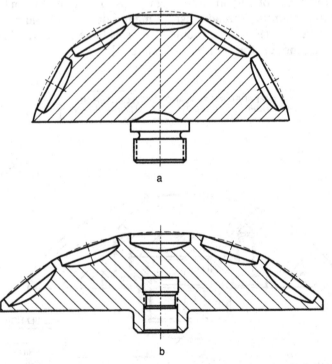

a

b

**Figure 4.26.** (a) Spot tools up to 100 mm in diameter have male threads. (b) Large spot tools over 100 mm in diameter have female threads.

A threaded steel insert must be pressed into the stem of aluminum spot blocks because the bore or boss and the bearing surface tends to wear or get damaged easily, and the tool would quickly loose its precision. Hard anodizing is often thought to be a sufficient treatment to prevent wear, but that only retards the wear process for a while. Since aluminum spot blocks are typically large, a large centering bore must be provided for proper guidance and stability. Often a specially designed insert or adapter can be attached to the spot tool so that it can be mounted on a sphere lathe or on a mill during the tool manufacture (Fig. 4.27). The same adapter can then later be used to mount the fully loaded spot block on a curve generator.

Magnetic chucks have also been used to mount small spot blocks on grinding or polishing spindles. This type of mounting is not useful for holding the tools on curve generators, so mechanical means to hold them must be provided. However, the loading and unloading of spot blocks can be greatly facilitated with the use of magnetic chucks. One such system [5] permits automation of load and unload cycles.

Vacuum chucks have also been used for quick mounting of lens blocks. This approach, however, is only useful for large blocks since the strength of

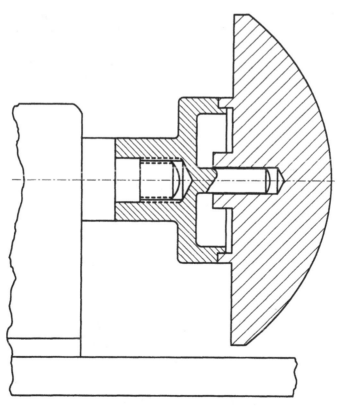

**Figure 4.27.**   Arrangement for mounting large spot blocks on radius lathe or mill.

a vacuum hold-down for a given pressure is directly proportional to the surface area of the vacuum chuck. Since too much vacuum pressure can distort the tool, the surface area of the vacuum chuck must be sufficiently large in order to keep the pressure necessary for safe hold-down low. Another potentially effective hold-down method uses compressed air rather than vacuum pressure. Only little interest has been shown so far for this approach.

### 4.2.2.   Lens Seat Design

The many variations in lens shapes require correspondingly designed lens seats. Only few difficulties can be expected when designing seats for fabricating the first lens surface. Unsuitably designed or improperly made lens seats for the second lens side can deform the lenses, cause wedging and center thickness errors, and contribute to stains on the already finished first lens surface.

A simple recess is generally sufficient for the fabrication of the first side. The recess is shaped in such a way that the lens rests on the edge and that a 0.1- to 0.2-mm (0.004- to 0.008-in.) void remains in the center for the blocking pitch (Fig. 4.28a). The lens seat recess may also have the matching curvature to the backside of the lens with a relief around the edge and the center. This way the lens is uniformly supported by an annular ring near the periphery (Fig. 4.28b). In either case the outermost edge of the recess should have a 0.5-mm (0.020-in.) deep relief channel, since lens pressings occasionally have flashing from the molding process around the edge. The relief

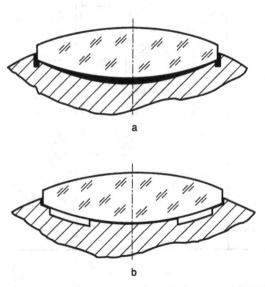

a

b

**Figure 4.28.**  (a) Lens seat in spot block with recessed center and edge support. (b) Lens seat in spot block with uniform relief and ring support.

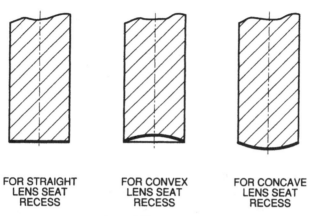

FOR STRAIGHT
LENS SEAT
RECESS

FOR CONVEX
LENS SEAT
RECESS

FOR CONCAVE
LENS SEAT
RECESS

**Figure 4.29.** End mills.

channel also serves as a reservoir for excess blocking pitch. Depending on the blank diameter, the diameter of the recess must be 0.2 to 0.3 mm (0.008 to 0.012 in.) larger than the lens blank.

The recesses are milled with end mills (Fig. 4.29) on a vertical mill or a five-axis CNC mill. The downfeed must be measured accurately to ensure that all recesses have the same depth. The recess depth is measured at the edge because that is where the lens will seat. The best approach to measure the depth of lens seats is to insert a dummy lens blank made of metal or glass into the recess and make a comparative measurement with a spherometer-type depth gauge (Fig. 4.30).

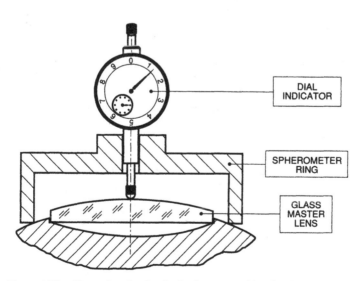

DIAL
INDICATOR

SPHEROMETER
RING

GLASS
MASTER
LENS

**Figure 4.30.** Measuring the depth of a lens seat with a hand spherometer.

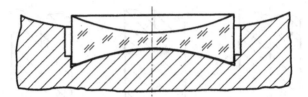

**Figure 4.31.** Lens seat with edge relief.

The maximum depth of the lens seat is determined by the shape of the lens. For instance, concave lenses with very large edge thickness should seat only 1 mm (0.040 in.) deep in the recess to avoid blocking stress deformation. Either the recess is held that shallow, or it is relieved around the edge to a depth that ensures that the correct seat is created (Fig. 4.31). A recess in which the lens rests only on its center should be used only after suitable protective lacquers have been found that can protect the already finished first side (Fig. 4.32). With this design the lens center thickness can be controlled to an accuracy of ±0.05 mm (0.002 in.). Even sensitive lenses will not deform with this kind of support. Since the danger of surface damage is great, the lenses can only be pressed down during the blocking stage. They must never be rotated or slid in the seat.

Screw-on lens seats can be made on a lathe without the need for a mill. The turned lens seats can be made with more accurate reference surfaces than those milled (Figs. 4.33a and 4.33b). It is customary for this design to turn the spot block body from cast iron and to make the screw-on lens seats from aluminum round stock. The seats must be fastened to the body with strong screws that must be secured well because considerable forces can act on the lens seats when the block is processed. As an additional feature, locking pins are used to prevent rotation of the seats in case the screws should come loose.

Lenses with nearly sharp edges can be effectively fabricated with screw-on lens seats. The recessed edge is replaced by a removable split ring (Fig. 4.34). After the blocked lenses have cooled down, the split rings are removed, and the block is processed in the usual fashion.

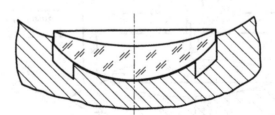

**Figure 4.32.** Lens seat with center support.

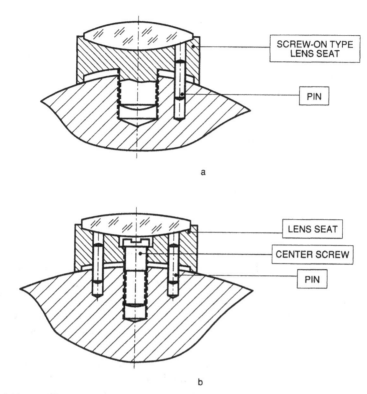

a

b

**Figure 4.33.** (a) Screw-on lens seat with lock pin to prevent turning. (b) Screw-on lens seat with lock pins to prevent turning.

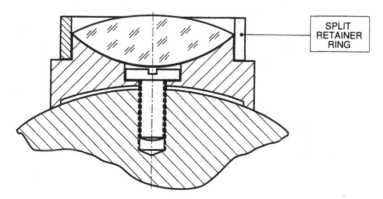

**Figure 4.34.** Screw-on type of lens seat with removable ring for sharp-edged lenses.

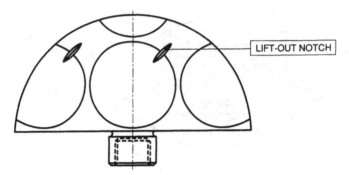

**Figure 4.35.** Lift-out notches that help in removal of lenses from lens seats of spot blocks.

Since spot blocks are precision tools, the milled or turned lens seats must be carefully deburred. This is especially important for second side spot blocks so that the already polished first side is not damaged. Spot blocks or lens seats made of aluminum must be black anodized. This will offer some protection against corrosion, and it will also reduce the light reflected off the blocked surface, which makes the inspection of polished second side surfaces easier.

Lens seats in which the lenses rest on the edge must be provided with a lift-out notch. When hot deblocking a finished block, one inserts a sharpened stick of wood into the notch to lift out the lenses. The notch should be cut to the right of the seat and slightly above center, as shown in Fig. 4.35. The notch must not be placed above or below the lens because otherwise the liquid pitch will run out during the blocking stage.

### 4.2.3.  Pin Seats

Another important feature of spot blocks, as well as standard spherical tools that are designed to run on top, is the incorporation of pin seats to engage poker pins that transfer the oscillating overarm motion to the tool on grinding and polishing machines. For tools with threaded stems, the pin seat can be milled directly into the tool (Fig. 4.36). Such seats are usually not very durable. Screw-in type runners or pin seats (Fig. 4.37) are much better because they can be easily replaced when they are worn out. A special runner has proved to be useful for large blocks (Fig. 4.38). Pins seats and poker pins have been standardized in Germany [6]. These standards recommend four ball diameters: 4 mm, 8 mm, 12 mm, and 18 mm. Two depths are provided for the pin seat. The deeper seat is used for blocks that are run on swing beam machines for which the pin always remains perpendicular to the pin seat (Fig 4.39).

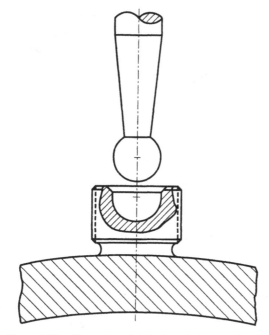

**Figure 4.36.** Pin seat sunk into threaded stem of tool.

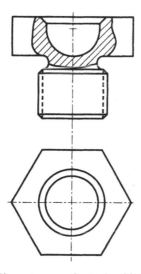

**Figure 4.37.** Pin seat runner for tools with female threads.

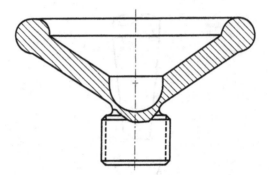

**Figure 4.38.** Handwheel-type runner for large tools.

### 4.2.4. Spot Block Design Considerations

There are two points of view regarding the choice of the first lens surface that is to be ground and polished on a spot block. One view prefers that the more strongly curved surface of a lens, whether convex or concave, should be polished first. This view is justified by reasoning that the required center thickness accuracy for the second lens surface is achieved for the larger blocks on many lenses at the same time. This is especially true when the second side is plano. This position is opposed by the argument that the lens can become a reject after polishing the first side. When the more strongly curved side is worked first, the value of the reject part will be higher than for the reverse case. Besides it is possible with the currently used machines and

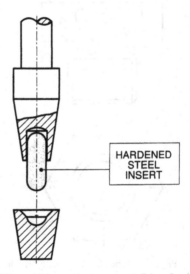

**Figure 4.39.** Pin and pin seat arrangement for swing beam machine.

tooling to hold the center thickness even on strongly curved lens surfaces. Either approach will yield good results provided that the tooling used is exactly as designed and that the processes are faithfully executed per plan.

The contemporary approach is to fabricate the longer radius first. This is usually not a problem with pitch button blocks where one has a free choice. When the lenses have a concave side, that side is usually done first. With the finished side facing down, the semifinished lenses will not get dirty in storage, or damaged in transport from station to station.

Before the spot blocks for a lens can be designed, the dimensions of the raw lens blanks must be determined. Raw lens blanks are used in the form of lens pressings for glass optics, pregenerated lens blanks for IR optics, or round blanks for any material. The diameter of these blanks is almost always larger than the finished lens, and a centering and edging step is required after both lens surfaces have been polished to bring the lens within its designed diameter tolerance. The center thickness is also larger than the finished thickness tolerance to allow for sufficient stock removal from each side during the grinding and polishing operations. These blank dimensions must be considered when designing a spot tool.

The spot block, which is simply called a *block* in the following text, is designed on the drawing board. The starting point for the design is the finished lens surface from which the center thickness of the lens is then subtracted and the back side of the lens is drawn. For first side blocks the overage allowance for the second side must be added to the lens center thickness. The recess of the lens seat is drawn by referencing to the back side of the lens. The depth of the lens seat recess corresponds to the chosen base radius of the block and the edge thickness of the lens (Fig. 4.40). These parameters depend on one another according to Eqs. (4.8):

For convex blocks,

$$t_s = t_e + R_T - R_L, \qquad (4.8a)$$

and for concave blocks,

$$t_s = t_e - R_T + R_L, \qquad (4.8b)$$

where:

$t_s$ = depth of the recess as measured near the edge,
$t_e$ = edge thickness of the lens,
$R_T$ = base radius of the spot block,
$R_L$ = lens radius.

The recesses should be at least 0.5 mm (0.020 in.) deep at the edge so that the lenses have a firm hold in the lens seats. The lens should also protrude from the block by at least 1.0 mm (0.040 in.), if at all possible. The residual

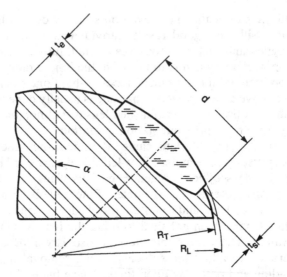

**Figure 4.40.**  Lens seat dimensions.

grinding or polishing slurry can this way be easily stripped off and will not foam. Both requirements can of course be met only when the lens has an edge thickness of at least 1.5 mm (0.060 in.). The recess diameters should be held 0.2 to 0.3 mm (0.008" to 0.012 in.) larger than the diameter of the lens blanks.

For second side blocks the thickness of the protective lacquer on the finished first side must also be considered. The lacquer layer is typically 0.01 to 0.02 mm (0.0004 to 0.0008 in.) thick. The center thickness tolerance must also be taken into account. This tolerance does not merely signify an attainable precision. Defective surfaces can be reworked within the thickness tolerance limits. Therefore the lens seats are recessed to a depth that ensures that

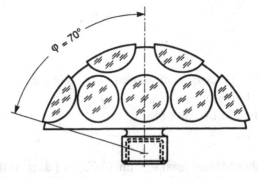

**Figure 4.41.**  Included angle = $2\varphi$.

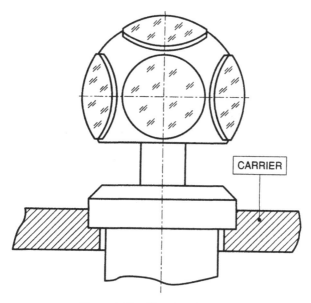

**Figure 4.42.** Hyperhemisphere.

the resultant lens center thickness lies just inside the top tolerance. A lens with a reworked surface will then lie nearer the lower tolerance limit. The depth of the lens seat can be accurately measured with a dummy disk inserted into the lens seat recess, as shown in Fig. 4.33.

Spot blocks are commonly covered with lenses to an included angle of $\varphi = 65°$ to $75°$ (Fig. 4.41). The lenses can be easily ground and polished with this pattern. Convex surfaces are not limited to this constraint because they can be worked as hemispheres or even as hyperhemispheres (Fig. 4.42). In most cases such strong curvatures are avoided because they are difficult to control. If hemispheric blocks must be polished, specially modified generating and polishing machines must be installed [7]. Most convex blocks with an included angle of $\varphi \leq 70°$ can be polished on standard equipment.

### 4.2.5. Number of Lenses per Block

The lens seats are arranged on the block in concentric circles. The number of lens seats for a given ring $r$ (Fig. 4.43) can be mathematically approximated by the circumference of the ring divided by the sum of the lens blank diameter and the separation between the lenses in the ring:

$$N_r = 2\pi \left( R_T \sin \frac{\alpha}{d_L + a} \right), \tag{4.9}$$

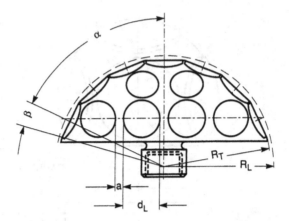

**Figure 4.43.**   Number of lens seats per row.

where

$N_r$ = number of lenses per ring,
$\alpha$ = included angle,
$R_T$ = tool radius,
$a$ = average lens separation,
$d_L$ = lens blank diameter.

The decimal values resulting from the use of Eq. (4.9) are not considered since the number of lenses can only be an integer, a whole number.

On the tool drawing the distance of the lenses within a lens circle is given by the angle $\gamma$ as measured from the center of one lens seat (recess) to the center of an adjacent one. The angle $\gamma$ can be calculated as

$$\gamma = \frac{360°}{N_r} \qquad (4.10)$$

The indexing head of the mill on which the spot block is mounted for boring of the lens seats is indexed by this angle $\gamma$ after completion of each lens seat (Fig. 4.44).

During the design layout, the first lens ring is placed close to the edge of the block. The ring is followed by the others toward the block center. Depending on the diameter of the remaining central zone, one, three, or four lens seats can be arranged there. The location of the bore axes for three or four lens seats in the center can be calculated as in Eqs. (4.11) and (4.12). Figure 4.45 shows this relationship.

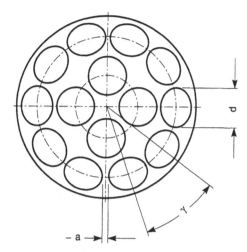

**Figure 4.44.** Typical pattern of lens seats.

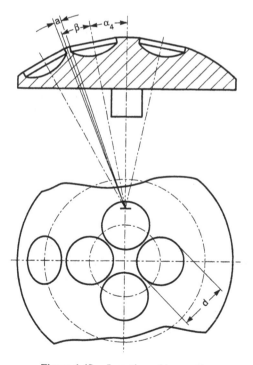

**Figure 4.45.** Location of bore axis.

For three lenses,

$$\sin \alpha_3 = \frac{d + a}{R_T \sqrt{3}},$$                                    (4.11)

and for four lenses,

$$\sin \alpha_4 = \frac{d + a}{R_T \sqrt{2}},$$                                    (4.12)

where

   $d$ = lens blank diameter,
   $a$ = lens separation,
   $R_T$ = tool radius,
   $\alpha$ = angle between block axis and lens seat axis.

The semidiameter of the lens and half the distance between the lenses form
the angle $\beta$:

$$\sin \frac{\beta}{2} = \frac{d + a}{2R_T}.$$                                    (4.13)

Equations (4.11) through (4.13) are derived later in this chapter.

### 4.2.6.  Spot Block Design Example

The design of spot blocks is best demonstrated by an example: A spot block
must be designed for a lens with the following dimensions (see also Fig.
4.46):

   $R_1$ = 36.0 mm (1.417 in.) concave
   $R_2$ = 94.0 mm (3.701 in.) convex
   $t_c$ = 3.0 mm ± 0.1 (0.118 ± 0.004 in.)
   $d$ = 30.0 mm (1.181 in.)

The dimensions are shown in both metric and English measure. Either can
be used for this example. It will result in the same tool.
   The raw lens blank is a pressing (Fig. 4.47) with the following dimensions:

   $R_1$ = 35.0 mm (1.378 in.) concave
   $R_2$ = 95.0 mm (3.740 in.) convex
   $t_c$ = 5.0 mm (0.197 in.)
   $t_e$ = 7.0 mm (0.276 in.)
   $d$ = 31.5 mm (1.240 in.)

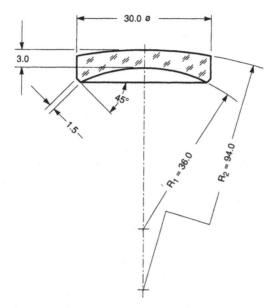

**Figure 4.46.** Lens specifications.

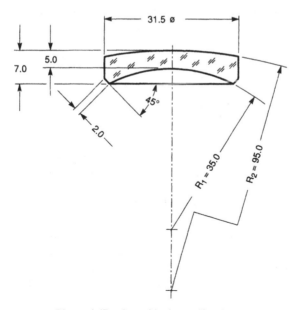

**Figure 4.47.** Lens blank specifications.

**Spot Block Design for the First Lens Surface (Concave).** The first surface worked is the concave side with the 36.0-mm radius. The block is made form cast-iron because it is small. Four lens seats have to be provided, although a block with six or even seven lenses would be geometrically possible. However, such strongly curved blocks would be too difficult to polish.

A lens seat that is 0.3 mm larger in diameter than the lens blank diameter will make it easier to block and deblock the lenses. Therefore 31.8-mm diameter lens seats must be provided for this example. In reality the diameter overage should be added to the maximum possible blank diameter to ensure that all blanks will fit the tool. This, however, may make the pocket too large for blanks that are at the bottom of the diameter tolerance. The formal practice is to make a thorough tolerance study and to choose the various spot tool dimensions accordingly.

Since the lens blank must contact the tool at the edge of the lens seat, the bottom of the lens seat must receive a concave radius that is shorter than the convex radius of the blank. With an assumed maximum blocking pitch layer of 0.2 mm, this radius $r_1$ calculates to be 82.8 mm concave.

When the first side has been ground and polished, the lens will still have an edge thickness $t_e = 6.0$ mm, since the thickness overage of 1.0 mm has been removed from that side. A recess depth of $t_s = 2.0$ mm has been chosen for this example. Using Eq. (4.8b), the base radius of the concave spot tool can now be calculated. Solving Eq. (4.8b) for $R_T$, we obtain

$$R_T = R_L + (t_e - t_s) = 36.0 + (6.0 - 2.0) = 40.0 \text{ mm}$$

Therefore the base radius of the spot tool is 40.0 mm. The separation between the lenses is chosen to be 3.5 mm, or about 10% of the lens blank diameter.

Since this tool will be a four spot, Eq. (4.12) is used to calculate the angle between the spot tool axis and the center of the lens seats:

$$\sin \alpha_4 = \frac{31.8 + 3.5}{36.0(\sqrt{2})} = 0.6934, \quad \alpha_4 = 43.9°.$$

The apex of angle $\alpha_4$ lies at the center of curvature of the block. The center of the lens seat lies on a radius $r_2$ around the tool axis that is defined as

$$r_2 = R_T \sin \alpha_4. \tag{4.14}$$

In this example this radius calculates to be 27.75 mm. Another way to express this is to say the lens seats are located on a circle concentric with the spot tool axis which has a diameter $D = 2r_2$, or $D = 55.5$ mm (Fig. 4.48). Note that if the centers of the lens seats at the calculated angle $\alpha_4$ are located on any other diameter $D$, then the resulting lenses will be wedged.

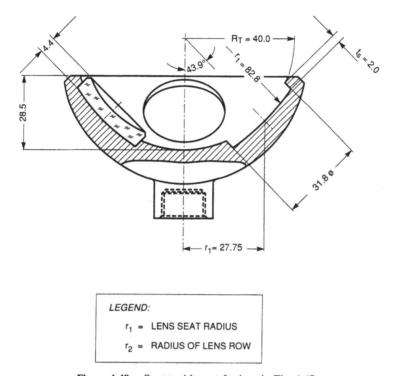

**Figure 4.48.** Spot tool layout for lens in Fig. 4.47.

In summary, the spot block to do the first side of the sample lens must have the following dimensions:

The base radius $R_T = 40$ mm concave.
Four lens seats are located on a radius $r_2 = 27.75$ mm.
Each lens seat has a concave bottom with $r_1 = 82.8$-mm radius.
The diameter of the lens seats is 31.8 mm.
The depth of the lens seats is $t_s = 2.0$ mm as measured at the edge.

**Spot Block Design for the Second Lens Surface (Convex).** The spot block for the convex lens side with radius $r = 94$ mm should be made from cast aluminum. To prevent unnecessary mounting thread wear, a threaded steel insert should be incorporated in the design.

The edge thickness of the lens after finishing both sides, but before final edging, is 5.3 mm. To allow for possible rework of one or both sides, about 0.05 mm of the center thickness tolerance is added to the nominal edge thickness to yield an edge thickness $t_e = 5.35$ mm. If we choose the depth of the lens seat to be $t_s = 2.35$ mm, then according to Eq. (4.8a), the convex base radius of the spot block is

$$R_T = R_L + (t_s - t_e) = 94.0 + (2.35 - 5.35) = 91.0 \text{ mm.}$$

The lens seats have also a diameter of 31.8 mm, just like those of the first side tool. Since the lenses must rest on the inner edge of the lens seat, the bottom surface must have a surface that is flatter than the first side concave radius. This means that the radius of the bottom surface of the lens seat must be longer and convex. This requires the use of a specially sized and shaped end mill. No more than an 0.2 mm-thick pitch layer should form under the lens center. To determine this radius, the sag of the concave lens surface must first be calculated by using the well-known sag formula:

$$s = R_L - \sqrt{R_L^2 - \left(\frac{d}{2}\right)^2}. \tag{4.15}$$

With a concave radius of $R_L = 36.0$ mm and a lens diameter of $d = 31.8$ mm, a sag of about 3.6 mm is calculated. The maximum pitch thickness, which is 0.2 mm for this example, is subtracted from this sag value, to yield a sag of 3.4 mm for the radius of the lens seat bottom. The radius can then be calculated by using an approximation of Eq. (4.15) solved for $r_1$:

$$r_1 = \frac{d^2}{8s} + \frac{s}{2}. \tag{4.16}$$

In this manner a radius of $r_1 = 38.9$ mm is calculated for the bearing surface of the lens seat, which is provided with an edge groove similar to the one shown in Fig. 4.28a. This arrangement ensures that the lens does not rest on its sharp edge but is supported evenly on the concave surface. Any surface damage occurring due to this contact will be removed when the oversized lens is centered and edged.

The correct design of the bottom surface of the lens seats is crucial to the successful implementation of spot tools. Even if the geometry of the spot tool is correct, an improperly designed bearing surface can cause unacceptable distortion of the polished lens surfaces, thus making the spot tool useless. In this example, a 0.2-mm maximum pitch thickness has been assumed. This choice is based on experience. Smaller lenses with thicker centers can tolerate deeper seats, whereas the more sensitive higher aspect ratio lenses can tolerate no more than 0.1-mm pitch thickness.

The reason for this is that blocking pitch has a much higher coefficient of thermal expansion than either the tool or the lenses. The thicker the pitch, the greater is the potential stress deformation on the lenses and the greater are the fringe irregularities of the polished lens surfaces. Because of the increased cost of boring recesses with spherical bottom surfaces, most designers think that they can save money by depending on straight or conical bores from standard end mills. The truth is that the one-time cost of making the tool right in the first place is far less than the impact a poorly designed tool can have on product yield and the need for constant rework.

Every lens seat, together with the lens separation distance, encloses an angle that can be determined by Eq. (4.13). The same angle $\beta$ is also valid as a measure from one recess angle $\alpha$ to the next. In this example the angle $\beta$ is calculated:

$$\sin\left(\frac{\beta}{2}\right) = \frac{31.8 + 3.5}{2(91.0)} = 0.194, \quad \frac{\beta}{2} = 11.2° \quad \text{or} \quad \beta = 22.4°.$$

Through the use of a test sketch, it was found that a pattern angle of 72° is favorable for the second side spot block of this example. This results in two lens rings with four lenses in the middle. Using Eq. (4.12), the recess angle $\alpha$ for the four central lenses can be calculated:

$$\sin \alpha = \frac{31.8 + 3.5}{\sqrt{2(91.0)}} = 0.2743, \quad \alpha = 15.9°.$$

On the drawing this angle is designated $\alpha_1$. To calculate the recess angle for the next row of lenses, $\beta = 22.4°$ is added to $\alpha_1$ to yield $\alpha_2 = 38.3°$. By adding another 22.4°, a recess angle $\alpha_3$ of 60.7° is calculated for the outermost row. When $\beta/2 = 11.2°$ is added to this for half of the outer lenses, we obtain a total included angle of $\varphi = 71.9°$.

The number of lenses in each row is determined by using Eq. (4.9):

Row 2    $N_2 = 2\pi[\sin 38.3° \ (91.0)]/(31.8 + 3.5)$
         $N_2 = 10$ lens seats

Row 3    $N_3 = 2\pi[\sin 60.7° \ (91.0)]/(31.8 + 3.5)$
         $N_3 = 14$ lens seats

Together with the four center lenses, the second side spot block will hold 28 lenses. Lift-out notches will not be needed in this case because these lenses have a sizable edge thickness.

The angle $\gamma$ by which the block must be indexed during the machining of the lens seats from one seat to the next is calculated by using Eq. (4.10):

Row 1    $\gamma_1 = 360°/4 \ = 90.0°$
Row 2    $\gamma_2 = 360°/10 = 36.0°$
Row 3    $\gamma_3 = 360°/14 = 25.7°$

In summary, the dimensions for the second side spot tool are

The base radius $R_T$ = 91.0 mm convex.
Four lens seats in the center at $\alpha_1$ = 15.9°.
10 lens seats in row 2 at $\alpha_2$ = 38.3°.
14 lens seats in row 3 at $\alpha_3$ = 60.7°.
Each lens seat has a convex bottom with $r_1$ = 38.9 mm.
The lens seat diameter is 31.8 mm.
The lens seat depth is $t_s$ = 2.35 mm as measured at the edge.

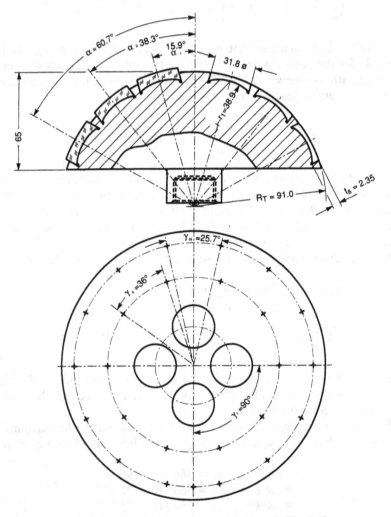

**Figure 4.49.**  Spot block seats layout for lens in Fig. 4.47.

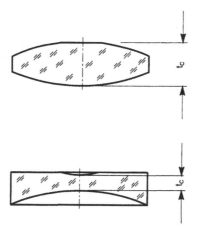

**Figure 4.50.** Pregrind center thickness.

These values are entered in the tool drawing, as shown in Fig. 4.49. The centers of the lens seats are indicated by crosses. With this information and a properly equipped mill, any qualified machinist can make such spot tools to the required precision.

One very important step that is often not done in order to save time is to check out the accuracy of a newly made spot tool with dummy lenses before the tool is released to production. These lenses, which can be made from any available scrap glass, must have the dimensions of the intended lens blanks. Flat spots must be ground on the convex side to the exact final center thickness (Fig. 4.50). When the second side of the lenses is ground on the spot block, the spots should be removed on the outer row and the inner row of lenses at the same time.

### 4.2.7. Lens Block Size Calculations

The more lenses that can be held on a block, the less expensive will be the grinding and polishing of each lens surface. Therefore it is always advisable to analyze all available options in terms of block geometry and the number of parts needed. This planning step should be done at the quotation stage to optimize from the beginning how many lenses can be processed at the same time.

For plano blocks or for blocks with long radii, each lens ring will hold $2\pi$ or approximately six more lenses than the next smaller ring (Fig. 4.51). Therefore the number of lenses on plano blocks equals the sum of an arithmetic series which is calculated for $N = n/2[2c + (n - 1)2\pi]$. If we choose as the starting member $c = 3$ (i.e., three lenses in the center) and use the approximate constant difference of $2\pi = 6$, while defining the number of lens rings on the block as $n = D/2d$, then the total number of lenses on a plano

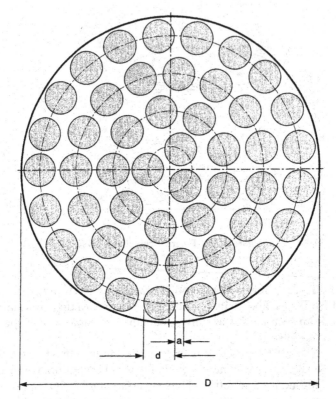

**Figure 4.51.** Typical layout for plano blocks.

block can be calculated:

$$N_p = \frac{3}{4}\left[\frac{D^2}{(d + a)^2}\right],$$  (4.17)

where

$N_p$ = number of lenses on plano block,
$D$ = diameter of plano block,
$d$ = lens blank diameter,
$a$ = separation between adjacent lenses.

The lens separation is a function of the lens blank diameter. It is most often considered as 10% of the diameter. This simplifies (4.17):

$$N_p = \frac{3}{4}\left[\frac{D^2}{(1.1d)^2}\right] = 0.62 \left(\frac{D}{d}\right)^2.$$  (4.18)

The number of parts $N_p$ is a theoretical value that cannot always be accommodated on a block. Another constraint is the blocking pattern which is determined by the arithmetic series. Therefore the usual practice is to calculate the maximum possible number of parts per block using Eq. (4.17) or Eq. (4.18), and then to calculate the next smaller number of parts that meets the requirements of the arithmetic series. This is a lot of calculating. Table 4.1 lists the possible block patterns for the three basic conditions, namely, one lens, three lenses, or four lenses in the center to simplify that choice. The more accurate constant of $2\pi$ has been used to generate the values for this table, which means that every fourth row will pick up one additional part per row.

Adherence to this pattern assures a well-distributed block surface that will be easy to polish. Although there is rarely a debate on the correct number of lenses per plano block for up to 19 lenses, differences of opinion exist for larger blocks. For instance, it can be shown that instead of 27 parts per block for the case of 3 lenses in the center with 2 lens rows, one can also make a 28-part block. That, however, is possible only if the radius, which defines the outermost row, is made larger to make room for the extra part. This necessitates a change in position for all other lenses on the block which could make this arrangement too open for the precise control of surface figure. Once the block size has gone beyond 38 parts, it hardly matters from a figure control standpoint if there is a difference of one to two parts per block either way. The importance of adhering to the lens pattern as defined by the arithmetic series lies in the predictability of the number of parts per block based on the lens diameter and chosen block diameter. This approach yields consistent results and does not leave room for guesswork. Such predictability is especially useful for the design of spot tools.

Sometimes it is necessary to know how large the diameter $D$ of a plano blocking tool must be for a given number of parts $N$ of a specific diameter

**Table 4.1.  Number of Parts per Block Based on Constant $2\pi$**

| Lens row | Lenses in center | | |
|:---:|:---:|:---:|:---:|
| | 1 | 3 | 4 |
| 1 | 7 | 12 | 14 |
| 2 | 19 | 27 | 30 |
| 3 | 38 | 49 | 53 |
| 4 | 63 | 77 | 82 |
| 5 | 95 | 112 | 118 |
| 6 | 132 | 152 | 159 |
| 7 | 176 | 199 | 207 |
| 8 | 227 | 253 | 262 |

and separation $(d + a)$. Solving Eq. (4.17) for $D$, we have

$$D = 2(d + a) \sqrt{\left(\frac{N}{3}\right)}. \qquad (4.19)$$

This calculation helps to determine if a block of the appropriate size is on hand or must be ordered, or if there is sufficient spindle capacity to run a block of that size. For instance, the block diameter $D$ of a 199 lens block for 12.7-mm (0.500-in.) diameter parts with an average lens spacing of 1.3 mm (0.050 in.) calculates to be 228 mm (about 9.0 in.). This is probably as big as one would want to make such a block, although larger blocks are possible.

While lens blocks with long radii follow the rules of Eq. (4.17), there comes a radius to diameter ratio where this equation will yield inaccurate results. Therefore another equation is needed to calculate the number of lenses for such spherical lens blocks (Fig. 4.52). This equation is derived from Eq. (4.17) by noting that a spherical block will accept the same number of lenses than a plano block with the same surface area. The surface area of a spherical segment is $A_s = 2\pi r h$, which is set equal to the area of a plano surface $A_p = D^2\pi/4$. From this we calculate the value $D^2 = 8rs$, which then replaces the value $D^2$ in Eq. (4.17). This leads to the following equation to calculate the number of lenses on a spherical radius block:

$$N_s = \frac{6R_L h}{(d + a)^2}, \qquad (4.20)$$

where

$N_s$ = number of lenses on a spherical tool,
$R_L$ = lens radius,
$d$ = lens diameter,
$h$ = height of outermost lens ring,
$a$ = lens separation.

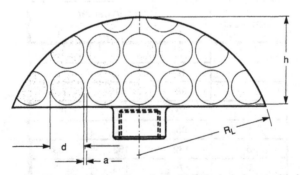

**Figure 4.52.** Number of parts for spherical blocks based on tool height $h$.

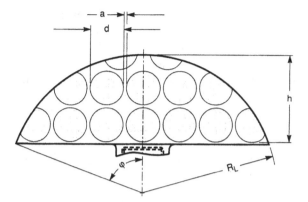

**Figure 4.53.** Number of parts per spherical block based on included angle $2\varphi$.

Another method has prevailed in recent years for which the effective block pattern height $h$ has been replaced by the effective included angle $\varphi$ of the lens block. This angle serves as a measure for the lens capacity of the block (Fig. 4.53). By replacing the value $h$ in (4.21) with the expression $r(1 - \cos \varphi)$, a new equation results:

$$N_s = \frac{6R_L^2(1 - \cos \varphi)}{(d + a)^2}. \tag{4.21}$$

**Example**

A block with lens radius $R_L = 88.1$ mm must be blocked to an effective included angle of $\varphi = 70°$ with lenses of 28 mm in diameter. The lens separation is to be $a = 2$ mm. What is the maximum number of lenses that the block can accept?

$$N_s = \frac{6(88.1)^2(1 - 0.342)}{(28 + 2)^2}, \quad N_s = 34 \text{ lenses.}$$

If the number of lenses for same lens block had been calculated with Eq. (4.15) by using $h = 88.1(1 - \cos 70°)$, the same result would have been obtained.

It should be noted here that the calculated lens capacity of the block is a maximum number only. The block will not accept more lenses because the ratio of surface areas will not permit this. The actual lens capacity is somewhat less, for the correct blocking pattern must also be considered. A block of 30 lenses appears to be the best choice here. For lenses with unusual geometries, such as very large edge thicknesses, an even smaller number of lenses would fit on the same block. The best approach is to lay out the pattern graphically, or calculate it as described for spot tools.

Lenses with steep curves may have to be blocked singly, or as three- or four-lens blocks. The previously derived equations do not give accurate results for these conditions. Since there is really nothing that needs to be calculated for single-surface blocks, more accurate equations are needed only for blocks with three or four lenses. To obtain an exact solution, this problem must be solved by geometric means. In a previous optics book [8] an equation was derived through analytic geometry to calculate the largest possible lens diameter that could be fit on hemispherical tools (Fig. 4.54). This led to the following result:

$$d = 2R_L \sin \frac{\sin \alpha}{\sqrt{\sin^2 \alpha + 1}}, \qquad (4.22)$$

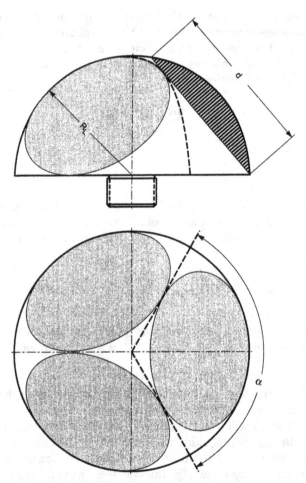

**Figure 4.54.** Hemispherical three-lens block.

where

$d$ = largest lens diameter,
$R_L$ = lens radius,
$\alpha$ = 180°/number of lenses.

With this equation the maximum possible lens diameter for three-lens and four-lens hemispherical blocks is

$$d_3 = 1.31R_L, \tag{4.23}$$

$$d_4 = 1.15R_L. \tag{4.24}$$

However, these type of blocks represent only an upper limit. Hemispherical three- and four-lens blocks are no longer fabricated today because it is not economical to do so. More general equations have been developed [9] for the determination of the effective blocking angle $\alpha$ for these kinds of blocks. The rationale for this derivation is based on a simple consideration: The centers of the lens surfaces of a three-lens block are the corners of an equilateral triangle. The sides of the triangle is the lens diameter $d$ plus the lens separation $a$. The circumscribed radius of this triangle is $r = (d + (a)/\sqrt{3}$ (Fig. 4.55). This radius $r$ forms a right triangle with the lens radius $R_L$ and the block axis. The central angle $\alpha$ can be calculated from this relationship. Combining it with the original expression, we obtain the following equation:

$$\sin \alpha_3 = \frac{d + a}{R_L\sqrt{3}}. \tag{4.25}$$

The corresponding logic for the four-lens block yields

$$\sin \alpha_4 = \frac{d + a}{R_L\sqrt{2}}. \tag{4.26}$$

The angle $\alpha$ is not only of interest as the block angle, but it is also the angle at which the lens seats are machined into spot blocks. In addition the edge angle $\beta$ is needed to calculate the effective block angle $\varphi$. The sine of $\beta$ is defined for both the three- and four-lens blocks as the quotient of half the lens diameter and the lens radius:

$$\sin \beta = \frac{d}{2R_L}. \tag{4.27}$$

The effective block angle $\varphi$ is then the sum of the two calculated angles:

$$\varphi = \alpha + \beta. \tag{4.28}$$

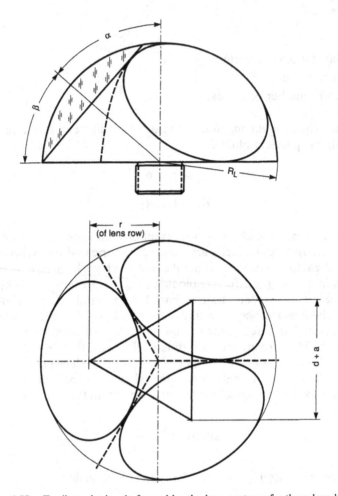

**Figure 4.55.** Equilateral triangle formed by the lens centers of a three-lens block.

## Example

The effective block angle for a three-lens block must be calculated. The lens radius is $R_L = 47.0$ mm, the lens diameter is $d = 52.0$ mm, and the lens separation is $a = 3.0$ mm. The use of Eq. (4.26) leads to

$$\sin \alpha_3 = \frac{52 + 3}{47\sqrt{3}} = 0.676, \ \alpha_3 = 42.5°.$$

According to Eq. (4.28), the value of angle $\beta$ is

$$\sin \beta = \frac{52}{2(47)} = 0.553, \ \beta = 33.6°.$$

Adding the two calculated angles as in Eq. (4.29), we obtain the effective block angle of $\varphi = 76.1°$.

### 4.2.8. Required Number of Tools

When considering the use of spot blocks, it is not only necessary to design the tools, it is equally important to accurately determine how many of these tools are needed. The customer's purchase order clearly defines the number of lenses that must be delivered and by what date. For larger orders, which can stretch out over months and even years, the rate of delivery becomes the factor that determines how many elements must be produced in a week, month, or quarter. It is this rate and the cycle times through the grinding and polishing operations and the associated blocking and deblocking steps that determine the number of tools that must be available.

Spot blocks are expensive precision fabrication tools that are of use only for those lens surfaces for which they were originally designed. Because of the dedicated nature and the high cost of this tooling, it is essential that there be an accurate estimate made to determine the number of required spot tools for each side before the milling machine is set up. Any misjudgment that leaves the supply of tools too low will require a new expensive setup on the mill to make the additional spot tools later. If more tools have been made than were needed, the efforts to produce the excess tools would have been totally wasted because the tools cannot be used for anything else. The primary determining factors are the daily production requirement for the lenses, the number of lenses that can be accommodated on each block, the daily fabrication time, and the total cycle time for the blocks.

The *cycle time* for the blocks is the time period during which the spot block is blocked with lenses, the surface is generated, fineground, inspected, polished, inspected again, and deblocked (Fig. 4.56). At the completion of

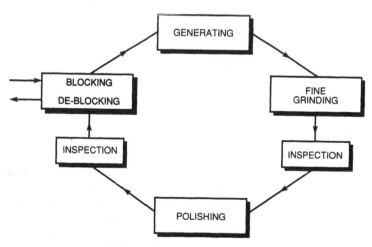

**Figure 4.56.** Typical use cycle for spot blocks.

that cycle, the spot blocks are cleaned and available to go into the next blocking cycle for another batch of lenses. The effective cycle times can be taken from the work flow plan. These planned times must be adjusted by an additional constant time factor that accounts for the return of defective parts, routine machine downtime, correction of grinding and polishing tools, heat-up and cool-down cycles during blocking and deblocking and similar unavoidable delays. Because of this correction factor, which differs from shop to shop, the cycle times are not a transferrable measure. In addition to the essential work-planning calculations, each shop should acquire some values for correction factors that are based on observation and experience at that shop.

The relationship of the values that influence the number of required spot blocks can be expressed by the following equation:

$$N_b \approx \frac{N_L T_b}{N_s T_d},\tag{4.29}$$

where

$N_b$ = number of required spot blocks,
$N_L$ = daily fab requirement of lenses,
$T_b$ = cycle time for spot blocks in hours,
$N_s$ = number of lenses per spot block,
$T_d$ = daily work time in hours.

## Example 1

Each spot block holds $N_s$ = 24 lenses. $N_L$ = 72 lens surfaces must be fabricated daily. The cycle time $T_b$ = 16 hr and the daily work time is $T_d$ = 8 hr. According to Eq. (4.29),

$$N_b \approx \frac{72(16)}{24(8)} \quad \text{or} \quad 6 \text{ blocks.}\tag{4.30}$$

## Example 2

In a large shop $N_L$ = 640 identical lens surfaces must be processed daily. The spot blocks hold $N_s$ = 4 lenses each. A conveyorized fabrication setup is used for this. The fine grinding is done with diamond pellets, and polishing is done with synthetic polishing foils on high-speed machines. The use of these fabrication approaches reduces the block cycle time to $T_b$ = 5 hr. The daily work time remains at $T_d$ = 8 hr. With this information the required number of spot block can be calculated using Eq. (4.29):

$$N_B \approx \frac{640(5)}{4(8)} \quad \text{or} \quad \text{about 100 blocks.}$$

The required number of grinding and polishing tools can be calculated in a similar manner. Much depends on the type of machines number of spindles in use for each machine. These relationships will be discussed in greater detail in Chapter 5.

**References for Section 4.2 (Spot Tools)**

[3] DIN 58 730 through 58 732.
[4] Threads per DIN 58 725.
[5] DAMA Optikmaschinen, GmbH, Germany.
[6] Pin seat standard (DIN 58 729), poker pin standard (DIN 59 727).
[7] For instance, Wilhelm Bothner, GmbH, Germany, model B-26-VA.
[8] Schade, Harry, *Arbeitverfahren der Feinoptik,* Deutscher Ingenieur. Verlag, Düsseldorf, Germany, 1955.
[9] Personal communication with H. Schade, 1984 (unpublished paper).

## 4.3. DIAMOND TOOL SPECIFICATIONS

Diamond tools have become such an important part of optical manufacture that by far the greatest majority of optics produced today have been shaped or ground with such tools. Diamond tools are used on a variety of optical fabrication machines. Some of these have been adapted for use with diamond tools, but most of them have been designed specifically for diamond tools. Because of this importance it is essential that a clear understanding exists of what is important when selecting diamond wheels and what options are available. Therefore the rest of this chapter focuses on some basic diamond wheel specifications and on detailed descriptions of the various diamond wheel types and how they are used.

### 4.3.1. Diamond Wheel Codes

The generally accepted diamond wheel code defines the type of diamond, the size of the diamond grains, the diamond concentration, the bond type and hardness, and the bond depth. In addition the wheel dimensions must also be specified.

**Diamond Wheel Code.** The code is in the form **A–B–C–D–E–F**, where

   **A** = diamond abrasive type—natural or synthetic, blocky or splintery;
   **B** = diamond grit size—in mesh numbers or in micron sizes;

**C** = bond hardness—in letter grades from J (soft) to T (hard);

**D** = diamond concentration—in percent from 25 (low) to 150 (high);

**E** = bond type—resinoid, metal, or electroplated;

**F** = bond depth—most commonly $\frac{1}{16}$, $\frac{1}{8}$, and $\frac{1}{4}$ in.

In the absense of an enforced standard, there are many departures from this format and individual diamond wheel manufacturers often change the order of this code and add other items such as the core type and ID and grades to describe modifications of the basic bond. A summary of these departures from the norm for the primary U.S. manufacturers and suppliers is shown in Table 4.2.

### 4.3.2. Diamond Grain Sizes

The various diamond types and typical grain sizes were already discussed in Chapter 3. Most of the discussion there applies also to diamond wheels, since the diamond powders are used both in free abrasive grinding and in bonded tools. Recall that the diamond particles used in diamond tools are called *bort*. They are carefully graded according to strict industry standards. The grading is either by mesh sizes or by micron sizes. Mesh-sizing standards permit a relatively wide particle distribution for each size. This is not a problem for the larger grit sizes, but such a wide range is unacceptable for the finer grit sizes of 50 microns or less that are used for critical finishing of critical surfaces. The majority of diamond tools used in optics falls into this category. To overcome this problem, the Micron Diameter Standard CS-261-63 was issued to define much narrower diamond size ranges. This standard

**Table 4.2. Diamond Tool Codes Used by Various Manufacturers and Suppliers**

| Manufacturer-Supplier | Ref. | Diamond tool code sequence | | | | | | |
|---|---|---|---|---|---|---|---|---|
| Amplex Corp. | (10) | A | B | C | D | E | BM | F |
| DP Inc. | (11) | A | B | C | D | E | BM | F |
| Do-All Co. | (12) | A | B | C | D | E | F | BM |
| EDP Co. | (13) | A | B | C | D | E | CT | |
| Fish-Schurman | (14) | A | B | C | D | E | F | BM |
| PDT Co. | (15) | A | B | C | D | E | F | BM |
| Super-Cut, Inc. | (16) | E | C | A | D | B | CT | F |
| Winter USA, Inc. | (17) | B | E | C | D | | | |

BM = bond modification    CT = core type

**Table 4.3.** **Relationship of Different Diamond-Sizing Methods and Their Effects on Stock Removal and Surface Finish [19, 20, 21, 22]**

| US Mesh | Micron | FEPA[18] | Stock removal | Finish |
|---|---|---|---|---|
| 40 | | | very heavy | very rough |
| 60 | | | very heavy | very rough |
| 80 | | D181 | very heavy | very rough |
| 100 | | D151 | very heavy | rough |
| 120 | | D126 | heavy | rough |
| 140 | | | heavy | rough |
| 150 | | D107 | heavy | medium rough |
| 160 | | | heavy | medium rough |
| 180 | | D91 | moderate | medium rough |
| 220 | | D76 | moderate | medium rough |
| 240 | | D64 | moderate | moderately fine |
| 270 | | | moderate | moderately fine |
| 320 | | D54 | moderate | moderately fine |
| 400 | | D46 | light | pre-fining |
| 500 | 45 | | light | pre-fining |
| 600 | 30 | | light | fining |
| 850 | (20) | | light | fining |
| 1 200 | 15 | | light | super-fining |
| 1 500 | (12) | | light | super-fining |
| 1 800 | 9 | | very light | pre-polish |
| 3 000 | 6 | | very light | pre-polish |

also limits the permissible percentages of particles that are larger and finer than the nominal size. In other words, the distribution curves for micron-size diamond powders are significantly steeper than those for mesh-size diamond powders. Table 4.3 compares the relationships between the various size designations, typical stock removal, and surface finishes. Table 4.4 shows typical applications of the different grit sizes in optics manufacture.

### 4.3.3. Diamond Concentration

The diamond concentration in bonded diamond tools is expressed in carats per cubic inch of bond volume. One carat equals 0.2 g. Table 4.5 shows this relationship. The first column lists the typical diamond concentration values used for diamond tools. The second and third column show the recom-

**Table 4.4.  Typical Diamond Size Ranges for Optics**

| Manufacturing operation | US Mesh range |
|---|---|
| Core drilling | 60 to 120 |
| Cut-off saw | 80 to 120 |
| Curve generating (rough) | 80 to 140 |
| Curve generating (fine) | 180 to 220 |
| Edging (rough) | 120 to 180 |
| Edging (fine) | 220 to 320 |
| Blanchard (rough) | 100 to 140 |
| Blanchard (fine) | 150 to 220 |
| Dicing saw | 180 to 320 |

mended concentrations for metal bonds and resinoid bonds, respectively. The fourth column lists the carats per cubic centimeter, while the fifth column expresses this value in carats per cubic inch. Finally, the last column lists the diamond content in percent of bond volume. For instance, a diamond concentration of 100 corresponds to a diamond quantity of 25% of the volume of the bond material. Another way to look at these data is by using the specific weight of diamond which is $3.52 \text{ g/cm}^3$. At 0.2 g per carat, one cubic centimeter of a 100-concentration diamond bond contains $(0.25)(3.52)/0.2 = 4.4$ carats.

The diamond concentration, however, is not the only determining factor of whether a diamond tool will work well for a given situation. The diamond size and the type of the bond material must also be carefully chosen and

**Table 4.5.  Range and Application of Diamond Concentration [23]**

| Diamond concentr. | Metal bond | Resinoid bond | Carats per $cm^3$ | Carats per $in^3$ | Diam. cont. in volume % |
|---|---|---|---|---|---|
| 25 | • | | 1.10 | 18.0 | 6.3 |
| 38 | • | | 1.65 | 27.4 | 9.4 |
| 50 | • | • | 2.20 | 36.0 | 12.5 |
| 75 | • | • | 3.30 | 54.0 | 18.8 |
| 100 | • | • | 4.40 | 72.0 | 25.0 |
| 125 | | • | 5.50 | 90.0 | 31.3 |
| 135 | | | 6.00 | 97.2 | 34.1 |
| 150 | | | 6.60 | 108.0 | 37.5 |

matched to the intended application. A properly chosen wheel will have optimum cutting efficiency, extended wheel life, and produce good surface quality and precise dimensional control.

In matching the diamond concentration to the intended application, there are no fast and hard rules of thumb that make the choice an easy one. The choice of diamond concentration cannot be made independent of grit size and bond material. The entire set of parameters that defines the wheel must be addressed. This includes the material to be machined and what is to be expected in terms of surface finish, removal rates, and dimensional accuracy. Only experience can provide the necessary answers.

There are, however, a few general facts that can help make it easier to determine the correct diamond concentration. The following list of facts were gleaned from the product literature of the leading diamond tool manufacturers and suppliers.

### Concentrations of 100 and Higher

- Best for heavy duty grinding
- Provides maximum wheel life
- Produces relatively coarse finishes
- Yields rapid stock removal
- For smaller contact areas only
- Requires the use of harder bonds

The use of very high diamond concentrations such as 150 or higher can reduce wheel life. The high percentage of diamond in the bond weakens the bond strength.

### Concentrations of 75 and Lower

- Best for standard to fine grinding
- Reduced stock removal
- Produces finer finishes
- Longer wheel life is possible
- Lower cost because less diamond is needed
- Typically better performance
- Allows for larger contact areas due to more clear areas for coolants.

### 4.3.4. Bond Matrix

Next to the diamond size and concentration, the type of bond and its hardness is a controlling factor that determines the usefulness of a diamond tool for a specific application. The exposed diamond grains that are held in the bond matrix make up the cutting edge of the tools. The bond must firmly hold the grains while they are cutting efficiently but readily release them when their edges become dull and inefficient. Thus the type and hardness of the bond determines to a large degree the stock removal efficiency of the tool and the surface finish produced with it. The choice of bond also affects the useful life of the diamond tool.

The most effective bond will hold the individual grains sufficiently long to provide optimum cutting action from each grain but will erode at such a rate to readily release those grains that have become dull and ineffective, thus exposing new diamond grains on a continuing basis. Figure 4.57 graphically represents the principle of this process. Figure 4.57a shows two grains A and B properly exposed for optimum cutting efficiency, while two other grains C and D are still embedded in the bond matrix. Fig. 4.57b shows the same grains after some time has elapsed. The cutting edges of grains A and B have worn flat, and the erosion of the bond has reduced their holding strength. They are just about ready to be released, while grains C and D are just being exposed. Figure 4.57c shows the condition a short time later when further bond erosion causes grains A and B to be released and exposes grains C and D to replace the diamonds just lost.

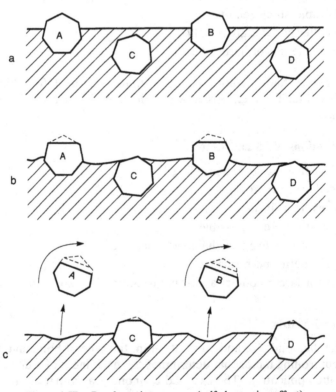

**Figure 4.57.**   Bond erosion process (self-sharpening effect).

Erosion of the bond and associated tool wear is not only normal but vitally necessary for optimum diamond tool performance. The difficulty lies in selecting the optimum combination of diamond tool parameters for a specific application and balance this with the economics of tool wear. The rate of bond erosion depends on many different factors. The choice of diamond tool parameters such as diamond size, concentration, bond type, and bond hardness are the more obvious factors, but the type of material that is abraded and the grinding pressures generated during that process also contribute greatly to bond wear.

For instance, when working hard, friable materials such as BK 7, coarse swarf (which is generated by a coarse grit diamond tool that is used at high infeeds) will erode the bond at much faster rates than fine swarf of the same material (which is generated with finer grit tools at lower infeed rates). Recall that swarf is a term generally used to describe the abraded material resulting from a fabrication process that uses diamond tools. For the same operating conditions and the same diamond tool, the swarf from a hard material will cause more rapid bond erosion than the swarf from a softer material.

The diamond grains must be held firmly by the bond until they have done their work. Grinding wear on the diamonds produces flat surfaces that increase surface area and thus friction between them and the workpiece. With the proper choice of bond, the erosion of the bond will permit this friction to overcome the holding force of the bond. This causes the dull grains to be pulled out just at the proper time, thus freeing new diamonds for action. If the bond is too soft (erodes too rapidly), the grains will be lost long before they have been used up. This reduces cutting efficiency and above all tool life. If the bond is too hard, the dull grains will not be pulled out. This means that no new diamond grains will be exposed, and the tool will eventually cease to remove material. When this happens, the increasing contact area between diamonds and workpiece drives up the working pressures, leading to frictional heating. This condition can destroy both the tool and the workpiece. The precision bearings in the machine spindles will also be adversely affected by such an overload. To prevent this from happening, such hard bond tools must be dressed frequently, and this will shorten the life of the tools appreciably. However, occasional dressing is essential for proper operation of diamond tools, and this topic is treated separately in this chapter.

In general the following can be said about bonds [24]: Hard bonds are used for hard and abrasive materials. Tools must be used with sufficient coolant supplies only. In choosing hard bonds, keep in mind that

- They have potentially long wheel life.
- They keep shape and profile very well.
- They perform best with small contact areas (narrow rim widths).
- They best perform with coarser grits and lower concentration.

Soft bonds are used on soft and heat-sensitive materials and on materials that tend to chip easily. Tools with soft bonds can be used with or without coolants. Additional considerations when choosing soft bonds are

- They have reduced wheel life.
- They produce finer surface finishes with same grit size.
- They allow for larger contact areas.
- They perform best with the finer grain diamonds.
- Their optimum performance is at lower concentrations.

### 4.3.5. Bond Types

There are three basic bond types for diamond tools, namely, resinoid, metal, and electroplated. A resinoid bond matrix is made up of cross-linked polymers into which the uniformly distributed diamond grains are firmly embedded. Metal bonds are typically composed of a bronze or steel matrix in which the diamond grains have been uniformly distributed. These bonds are made up of a uniform mixture of metal powder and metal-plated diamond grains that are sintered into a solid bond by using powder metallurgy technology.

Both of these bond types can be produced with bond depths of up to $\frac{1}{4}$ in. Diamond tools made with these bonds are used for light (resinoid) to heavy (metal) stock removal on large surfaces, or for general cutting and slicing. Electroplated diamond tools are made with very thin bonds of only about a hundred microns (a few mils) to capture and hold a single layer of individual diamond grains on a metal surface. Electroplated diamond tools are often preferred for operations that do not require large stock removal, such as beveling, profiling, and smoothing. They are generally much less expensive than either resinoid or metal-bonded tools because very little diamond is used in these tools. A fourth bond type, called a *vitrified bond,* is not used for diamond tools but is the bond material for abrasive wheels that are widely used in the metal and many other industries.

Resinoid bonds are best for applications where fine finishes are required and where only light grinding pressures can be exerted on fragile workpieces. Resinoid-bonded tools tend to cut freer than metal-bonded tools. They can also be operated at higher speeds, where they maintain efficient stock removal without generating much heat. Resinoid-bonded saw blades can be used for plunge grinding and can be operated equally well in the Z-cut or oscillating mode.

The characteristics of a resinoid-bonded tool are determined by the intended use. There are no easy rules of thumb, but there are basic limits that generally apply. The following lists some of these limits [25]:

| For Rough Cuts (wet grinding) | For Finish Grinding |
| --- | --- |
| Narrow rim width | Medium rim width |
| Coarse diamond sizes | Fine diamond sizes |
| High concentration | Medium concentration |

Resinoid wheels must be protected from excessive machine vibrations, mechanical shock, and thermal overload. The fragile bond can be easily damaged this way. Most resinoid-bonded tools can be used for low-pressure dry grinding, but they must always be used with plenty of coolant flow when working optical materials.

Metal bonds have a higher wear resistance relative to resinoid bonds. They are much less affected by mechanical shock and thermal buildup, which makes them a good choice for heavy-duty applications. Their wear resistance, however, makes them less efficient in terms of material removal rates compared with otherwise equivalent resinoid-bonded tools. The life of metal-bonded tools is many times greater than that of resinoids.

Metal bonds are typically made from either steel or bronze. Special alloy bonds are also available. They form intermediate bonds between the superior wear resistance of steel and the much softer action of bronze. The wear resistance of steel gives steel-bonded tools exceptionally long life. The erosion resistant bond must be used at high grinding pressures to maintain an acceptable self-sharpening effect. This limits the use of steel-bonded diamond tools to very sturdy, specially built production machines. These tools perform best on soft optical materials and for rapid stock removal on harder materials.

Bronze bonds are in widespread use in optical shops. They do well on optical materials, especially when used with light grinding pressures. Bronze-bonded diamond tools are the preferred choice when hard and brittle materials must be ground with a minimum of edge chipping.

Electroplated metal bonds exhibit high grinding efficiencies and excellent form retention. They are most often used for beveling and profiling of optical parts. Since these tools have only a single layer of diamonds, there can be no self-sharpening effect. Although these tools cut very aggressively at first, their grinding efficiency deteriorates with prolonged use. The metal plating is either copper or nickel. Nickel plating produces a tough and very durable metal layer that firmly holds the diamond grains in place. Copper plated bonds provide a softer alternative to the nickel plated tools. A chrome overplating is usually applied to provide corrosion resistance.

### 4.3.6. Bond Hardness

The hardness of the bond is another important parameter that determines the cutting efficiency and life of a diamond tool. Bond hardness is defined as the bond's ability to resist erosion and to hold the individual diamond grains firmly so that they can do their work most efficiently. Bond hardness is typically identified by a letter code. The softest bonds are identified with letters near the beginning of the alphabet. As the hardness increases, the letter code approaches the end of the alphabet (Table 4.6).

The information in Table 4.6 represents a fairly general consensus among several manufacturers and suppliers of bonded diamond tools. There are several cases where different hardness levels are assigned the same letter

**Table 4.6.  Bond Hardness Code**

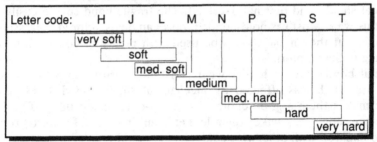

code. The user should verify this parameter when switching from one supplier to another to purchase an "equivalent" diamond tool. In general, it can be said that the softer bonds have the fastest cutting action but the shortest life. Medium bonds provide a somewhat reduced cutting efficiency at increased tool life. Hard bonds exhibit the least cutting efficiency but provide the longest tool life. These generalities must be further modified by many other factors such as bond type, diamond size and concentration, the workpiece material, the operating conditions, and the desired surface finish. Table 4.7 relates bond hardness to type of bond.

Because of widely diverging conventions in using the letter code and in describing the bond hardness, the information in Table 4.7 is also a general consensus from several different sources [26, 27, 28, 29, 30, 31]. There are numerous deviations from this summary. Typical bond depths for both metal and resinoid bonds are $\frac{1}{32}$, $\frac{1}{16}$, $\frac{1}{8}$, and $\frac{1}{4}$ in. The depth of bond affects wheel life and above all diamond tool cost.

### 4.3.7.  Tool Speeds

Diamond tools cut best when the diamond engages the work at the right speed. When the speed is too low, the diamond will not cut but rather tear the work. When it is too high, the diamond will wear more rapidly and will lose its cutting efficiency rather quickly. The spindle revolutions per minute (rpm) will not be sufficient to describe the tool speed because diamond tools

**Table 4.7.  Bond Hardness versus Bond Type**

| Hardness code | Bronce bond | Steel bond |
|---|---|---|
| L | soft | |
| N | medium | medium hard |
| P | medium hard | medium hard |
| R | very hard | hard |
| S | | very hard |

vary in diameter from less than an inch to several feet. The best way to express this value is to calculate how far one diamond grain on the periphery of the wheel travels in one minute relative to the stationary workpiece. By multiplying the wheel circumference by the spindle rpm, one obtains the peripheral tool speed. When the wheel diameter is in feet, this value is then expressed as surface feet per minute (SFM). Equation (4.31) gives a mathematical representation of this relationship:

$$V_p = \pi \left(\frac{d}{12}\right) V_s,\tag{4.31}$$

where

$V_p$ = peripheral velocity in SFM,
$d$ = diamond tool diameter in inches,
$V_s$ = spindle velocity in revolutions per minute.

The metric equivalent is expressed in meters per second (m/sec). A formula similar to Eq. (4.31) can be established for the metric case:

$$V_p = \frac{\pi d V_s}{60},\tag{4.32}$$

where

$V_p$ = peripheral velocity in meters per second,
$d$ = diamond tool diameter in meters,
$V_s$ = spindle velocity in revolutions per minute.

These equations can also be used to calculate the spindle rpm required to ensure that the peripheral speed is within the desired limits. For instance, Eq. (4.31) is solved for $V_s$ to yield

$$V_s = \frac{V_p}{\pi(d/12)}.\tag{4.33}$$

**Example 1**

A 6-in. diameter peripheral diamond wheel will cut best when run at a peripheral speed of 5500 SFM. What is the required spindle rpm?

$$V_s = \frac{5500}{\pi(6/12)} = 3501.$$

Therefore the optimum spindle speed is 3500 rpm.

**Example 2**

Equation (4.32) can be used to convert from SFM to m/sec. The diameter must be converted from inches to meters by dividing the 6-in. wheel diameter by the conversion factor of 39.37 in. per meter to yield 0.1524 m.
Entering this into Eq. (4.32),

$$V_p = \frac{\pi(0.1524)3500}{60} = 27.9 \text{ m/sec.}$$

A peripheral speed of 5500 SFM is equivalent to 27.9 m/sec.

Although there are divergent views on the optimum peripheral diamond wheel speeds, there is good agreement that it should not be less than 3000 SFM and no greater than 6000 SFM. The most often recommended optimum speed should lie somewhere between 4500 and 5500 SFM. The correct choice of the peripheral tool speed determines not only the stock removal efficiency and the surface finish but also how rapidly the wheel will wear out. A correctly chosen speed will optimize wheel wear; other speeds that are either too low or too high will actually shorten the useful life of the wheel.
Resinoid wheels are usually run faster than metal-bonded diamond wheels. They can be used for dry cutting without coolants or for wet cutting with oil-based or aqueous coolants. Only wet cutting is done in optics manufacture. The wheel speed must be reduced for large contact areas, but it can be greatly increased for smaller contact areas. Diamond wheels tend to cut softer at lower speeds, but they produce rougher surfaces. Conversely, at higher speeds they tend to be harder, but they produce better finishes.

### 4.3.8. Feed Rates

The opinions on the proper manner and rate of infeeds or downfeeds also vary greatly. Either the work is fed into the wheel, or the wheel is fed into the work. The actual mode is a function of the machine design, and it usually does not matter which mode is employed. What matters is that there is a relative motion between the diamond wheel and the workpiece. It is this relative motion that is called *infeed* or *downfeed*.
There are two basic feed modes to consider. First, the feed can proceed at a slow and constant rate, while the workpiece either rotates or oscillates back and forth. The material is removed in small increments with each revolution in a spiral manner, or even in a zigzag manner when the motion is linear as in sawing. Second, the initial feed can be rapid, and the material is removed from the workpiece as it makes at least one full revolution or traverses at least one full pass, depending on the type of operation performed. Figure 4.58 shows these conditions for a dicing saw. Some general

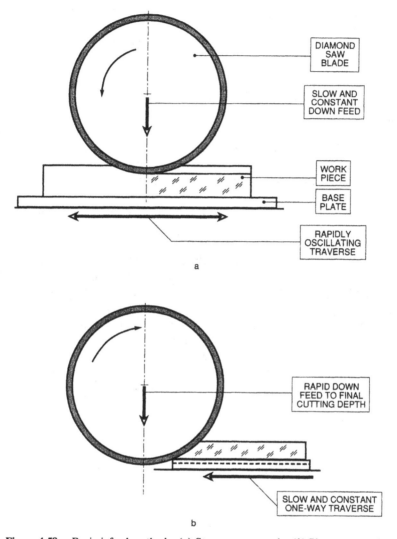

**Figure 4.58.** Basic infeed methods. (*a*) Step or z-cut mode. (*b*) Plunge cut mode.

recommendations for downfeeds in the step cut mode and traverse infeeds for the plunge cut mode are presented here:

### Downfeeds for Step Cut Mode

- 0.001 in./pass for 80 to 120 grit wheels
- 0.0005 in./pass for 120 to 220 grit wheels
- 0.00025 in./pass for 220 or finer grit wheels
- Traverse speeds up to the limit of machine capability

### Traverse Infeeds for Plunge Cut Mode

• 60 to 120 mm/min (0.040 to 0.080 in./sec) at production rates
• 13 to 25 mm/min (0.5 to 1.0 in./min) for precision dicing

These feeds and speeds are only very general guidelines. Higher or lower feeds and speeds can result in improved performance. Much depends on the specific application. The material, desired surface finish, required dimensional accuracy, stock removal rate, infeed method used, depth of cut, traverse travel speed, coolant used, even the machine, the operator, and the wheel type determine the best choice of feed rate [32]. Similar considerations apply for other stock removal operations that involve the use of diamond tools.

The best method to determine the optimum feed rates is to begin with an engineered first-order solution based on experience and the best available information. During the actual manufacturing stage the variables must be adjusted in a carefully controlled deliberate fashion while the process is carefully being monitored. One note of caution when doing this is that the wheel must be redressed after each test to provide for a common reference.

Besides evaluating surface finish and dimensional stability, one excellent way to monitor the performance of diamond tools and determine their cost effectiveness is to measure their removal efficiency in volume abraded per unit time (cm³/min), and the expected wheel life in terms of unit volume (cm³) per carat of diamond. Figures 4.59 and 4.60 were adapted from notes of a presentation on this topic [33]. These methods are useful only for high-

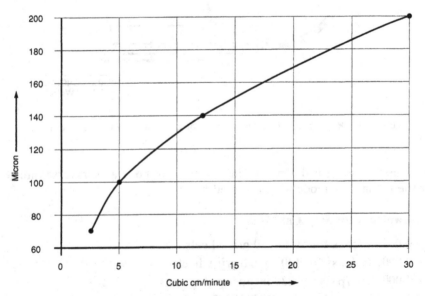

**Figure 4.59.** Stock removal efficiency of generating wheels.

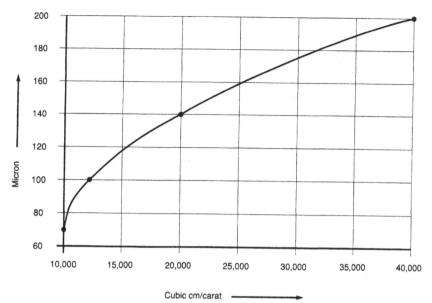

**Figure 4.60.** Average cutting life of generating wheels.

volume production. Therefore it should not be too surprising that these methods are highly developed among the volume optics producers in Germany and Japan.

### 4.3.9. Diamond Wheel Dressing and Trueing

Wheel dressing is a necessary process step that must periodically performed on any diamond tool when it shows signs of loading up, as would be evidenced by a slowly declining cutting efficiency. A suitable abrasive stone is held to the cutting edge of the tool, or the tool is allowed to cut into the stone. This step clears away compacted residue from the cutting edge and trims away a little bit of the bond material to expose new diamonds.

Wheel dressing is often not needed for a correctly specified diamond wheel when it is used as originally intended. As long as the grit size, diamond concentration, and bond type are suitably matched to its intended use, the tool will remain free-cutting due to the self-sharpening effect of proper bond wear. In addition good tool maintenance and care, such as cleaning after use and safe storage, will go a long way to keep the tool in top condition. As the tool nears the end of its life, or when a relatively minor change in the operating conditions causes the tool to load up, occasional light dressing may become necessary.

Resinoid wheels should be dressed as little as possible because their soft bond can be worn away rapidly. If dressing becomes necessary, it should be done lightly, and only with a white aluminum oxide dressing stick of 220 to

320 grit. Metal-bonded diamond tools can be dressed more often, but to reduce bond wear to a minimum, the dressing should be done more as an exception than as a rule. Gray silicon carbide sticks with grit sizes between 180 and 320 are the best choice for dressing metal bonds. Face wheels can be dressed by removing them from the spindle and grinding them lightly with loose abrasive on a plano-cast iron lap.

Frequent dressing is a sure sign that a serious mismatch exists between the diamond wheel and the workpiece or grinding method. Sometimes a change in the grinding method, such as changing feed rates and coolant flows, is all that is necessary to correct the problem. In severe cases, however, the diamond tool may prove to be unsuitable for the task, and the only sure way to solve the problem is to switch to a more compatible tool.

While light dressing is designed to wear a little of the bond material away to expose more diamond grains, the more aggressive process of trueing actually alters the shape of the wheel. Trueing should always be the initial step after a wheel or other diamond tool has been mounted on the spindle. This is especially important for finishing tools that establish critical dimensions, sharp edges, and fine surface finishes. Such tools must run absolutely true, or they will produce edge chips and poor surfaces. Improper trueing also leads to excessive and uneven wear of the cutting edge. However, it may not seem equally critical to take the time to true in a newly mounted rough diamond blade for a cutoff saw. There is usually little concern about some chipping and rough surfaces when cutting with this type of saw because subsequent fabrication steps will take care of any problems. Nevertheless, especially large diameter diamond wheels that do not run true can cause significant vibrations that can damage spindle bearings and lead to unacceptable wheel wear. Proper trueing of any newly mounted diamond tool is never a mistake.

Trueing must always be done with the tool mounted on the arbor or spindle on which it will be used in the machine. Single point diamond trueing dressers are available in many shapes, sizes, and configurations. These dressers are sometimes held by hand against the wheel, while they rest on a suitably adjusted post. More often, however, they are mounted on a cross slide, and the single-point diamond is passed across the cutting edge of the tool several times. Break-type trueing devices and independently driven grinding wheels are also used. Since trueing reduces wheel life even more than dressing, it should only be done when absolutely necessary and then only as little as needed to true the cutting edge.

## References for Section 4.3 (Diamond Tool Specifications)

[10] Amplex Corporation, Bloomfield, CT.

[11] Diamond Productions, Inc., Clifton, NJ.

[12] Do-All Company, Des Plaines, IL.

[13] Elgin Diamond Products Co., Elgin, IL.

[14] Fish-Shyurman Company, New Rochelle, NY.

[15] Precision Diamond Tool Co., Elgin, IL.

[16] Super-Cut, Inc., Chicago, IL.

[17] Ernst Winter & Sohn USA, Inc., Lynbrook, NY.

[18] FEPA = Féderation Européenne des Fabricants de Produits Abrasif.

[19] Do-All Company, catalog A-80-1.

[20] Rogers and Clarke Mfg. Co., catalog 8/80-1M-ST.

[21] Ernst Winter & Sohn USA, Inc., catalog ds 75e.

[22] Fish Surman Corp., catalog DG-395.

[23] Ernst Winter & Sohn USA, Inc., catalogs ds 75e and Lg Nr. 39/77c.

[24] Ibid. catalog ds 75e.

[25] Ibid. catalog ds 75e.

[26] Super-Cut, Inc., catalog SE 27.

[27] Rogers and Clarke Mfg. Co., catalog 8/80-1M-ST.

[28] Ernst Winter & Sohn USA, Inc., catalog ds 75e.

[29] National Diamond Laboratory, catalog 1969.

[30] Elgin Diamond Products Co., catalog No. 41.

[31] Precision Diamond Tool Co., catalog G-977.

## 4.4.  DIAMOND TOOLS

There are many different types of diamond tools made for use in optics fabrication. For each type there are numerous variations in size and configurations. There is a diamond tool available for nearly every process step. Table 4.8 lists the basic types of diamond tools and what they are typically used for. The tools on the list are described in detail in the following sections.

**Table 4.8.  Basic Diamond Tools and Their Applications**

| Diamond tool | Typical application |
| --- | --- |
| Core drills | coring and hole drilling |
| Saw blades | parting, slicing and dicing |
| *Blanchard* wheels | surface milling |
| Generator cups | curve generating |
| Peripheral wheels | edging, centering and surfacing |
| Beveling tools | beveling and chamfering |
| Diamond pellets | lapping and fining |

### 4.4.1.  Core Drills

Core drills are used on optical materials to drill holes, cut circular blanks from flat stock, or core rods from crystal boules. The outer diameter of the core drill must be sized accurately when drilling holes, and when rounds or rods are cored, the inner diameter is important.

Figure 4.61*a* shows a partial cutaway of typical core drill, while views *b* and *c* show a slotted core drill and a diamond drill, respectively. Most core drills consist of a thin-walled steel tubing that is silver soldered or brazed into a threaded brass collet or into a steel shank. Larger core drills are made usually from one piece with an integrated shank. The diamond bond is applied to the free end of the tubing. The bond is made of diamond bort embedded in a sintered metal matrix. The diamond grains can also be held at the cutting edge by an electroplated nickel layer. Resinoid-bonded core drills have been successfully used, but because the soft bond wears rapidly, these

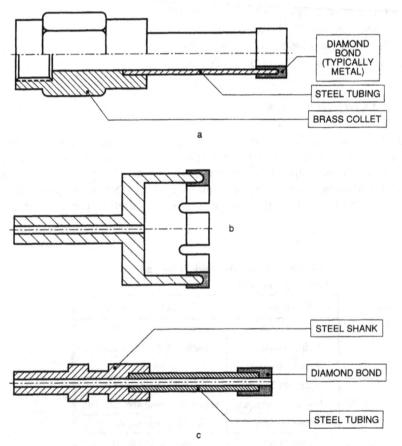

**Figure 4.61.**  Diamond core drills. (*a*) Typical core drill. (*b*) Slotted core drill with hollow integrated shank. (*c*) Hollow shank diamond drill.

drills are reserved only for use on soft and sensitive materials for which the other bonds would be too aggressive.

The majority of core drills have a continuous cutting edge. Only core drills that are larger than 75 mm (3 in.) in diameter must be slotted to ensure that the coolant can adequately penetrate to the contact area between the core drill and the workpiece.

The proper application of coolant is one of the primary factors that will determine the efficiency of the coring operation. Both surface finish and penetration rate are closely related to the effectiveness of the coolant supply. Core drill life can be greatly extended and workpiece damage can be almost totally avoided with sufficient coolant flow. The coolant must be liberally applied through the hollow center and also to the outside edge of the core drill to prevent uneven tool wear and maintain uniform stock removal. The coolant must be pumped to the cutting edge under sufficient pressure to provide effective cooling where the tool and the workpiece are in contact. It must also provide sufficient flushing action to carry away the swarf that is generated during the coring operation. The hollow center can easily be clogged if swarf is allowed to build up. A diminished or even temporarily interrupted coolant flow can lead to sudden tool and workpiece failure.

Any core drill that has been subjected to sudden stops or other mechanical shock, must be checked for run-out before it can be used again. Even a slight run-out will affect not only the surface finish and dimensional accuracy of the core or the hole, but it will lead to uneven wear of the bond. To prevent coolant flow interruptions, some core drills are provided with relief channels in the bond to ensure that a heavy flow of coolant can be maintained even under the most severe operating conditions. Submerged coring or core drilling with the workpiece completely covered by the coolant has proved useful for smaller core drills. However, it is a practice of questionable value for larger core drills because cavitation effects will prevent the coolant to reach the contact zone between the tool and the workpiece.

To get coolant from a fixed supply line into the hollow center of a rapidly rotating core drill, specially designed coolant heads or journals are used on the drill presses or milling machines most often used for this purpose. A typical coolant head [32] is shown in Fig. 4.62. A single unit brass housing has a fitting that accepts the coolant feed line. Spring-loaded Teflon washers provide the water seal. The Teflon washers are held in place with retaining screws. On some journals the lower washer accepts the core drill, whereas on others, the core drill shank is shaped to become an integral part of the journal. A hole in the shank allows the coolant to enter the hollow center of the core drill.

On occasion cores may break off during coring before the operation is complete. These cores can then become wedged inside the core drill. To remove them, straight holes for push rods or threaded holes for clean-out screws are often provided on core drills. Broken and wedged cores are a common problem on very long shank core drills with small diameters. These core drills are especially made to core solid state laser rods out of expensive

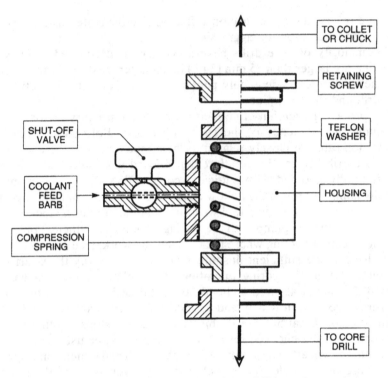

**Figure 4.62.** Exploded view of core drill journal.

laser crystal ingots. Some are as long as 300 mm (12 in.) with an inner diameter of about 6 mm (0.24 in.). The major problems encountered with this type of drill are the difficulty in bringing sufficient quantities of coolant to the cutting edge of the core drill and run-out associated with long but small drills. Both of these problems can lead to core breakage, which can plug up the hollow core drill. Therefore provisions to remove these cores must be made.

It may be difficult in most cases to attain the correct spindle speeds necessary for optimum cutting efficiency because most core drills are less than one inch in diameter. For instance, while the recommended rotational speed for a 1-in. diameter core drill is about 2000 rpm, the speed increases to 3000 rpm for a 12.5 mm (0.5 in.) diameter drill and to 4500 rpm for drills that are 6.3 mm (0.25 in.) in diameter. For very small drills, speeds in excess of 10,000 rpm are not uncommon. Figure 4.63 shows the recommended speed ranges for core drills [32]. Operating at or near the optimum operating speeds is another important consideration when working with core drills. Excessive speeds can lead to overheating because of insufficient coolant supply at the cutting edge. If the drill speed is too slow, the stock removal capacity of the core drill will decrease while the tool wear increases.

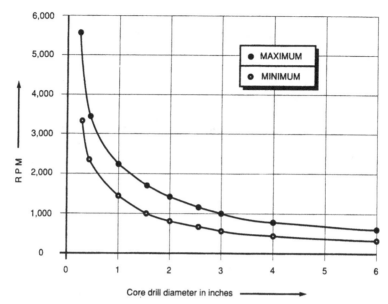

**Figure 4.63.** Recommended core drill speed ranges.

### Typical Core Drill Parameters

Coarse diamond grits are used most often at high concentrations in a medium-hard metal matrix.

Finer grits (<150 mesh) and low concentrations (<100) are not effective except in a resinoid bond formulated for special applications.

Typical grit sizes are the following:

Standard electroplated drills: 40, 60, 80 mesh
Smaller metal-bonded drills: 80, 100, 150 mesh
Finer electroplated drills: 80, 100, 150 mesh
Large metal-bonded core drills: 40, 60 mesh

Diamond impregnation depth is either 3 mm (0.12 in.) or 6 mm (0.25 in.).

Outer diameters range from about 3 mm (0.12 in.) up to 150 mm (6.0 in.).

Wall thickness ranges from 0.25 mm (0.010 in.) to 1.5 mm (0.060 in.).

Typical cutting depths range from 6 mm (0.24 in.) to 75 mm (3.0 in.).

The price for core drills can range from less than $50 for small drills to several hundred dollars for the largest sizes.

### 4.4.2. Diamond Saw Blades

Diamond saw blades are used on cutoff saws for rough cutting or parting, or on precision saws for sawing, dicing, slotting, grooving, and wafering of a wide variety of optical materials. Very thin ID saw blades are used on special machines to economically wafer semiconductor ingots. Paper-thin diamond saw blades are used on highly precise automatic microsaws to dice thin substrate to final dimensions. Difficult to cut materials can be sliced on a reciprocating wafering saw which uses a number of diamond-edged blades. Large pieces of glass have been successfully rough cut with a continuous diamond-edged blade on a modified bandsaw. These diamond tools will be discussed in this section.

Most diamond saw blades are thin peripheral wheels that have a diamond bond on the outer rim. The bond is continuous for wheels under 8 in. in diameter. Larger blades are often slotted or segmented to allow for the free

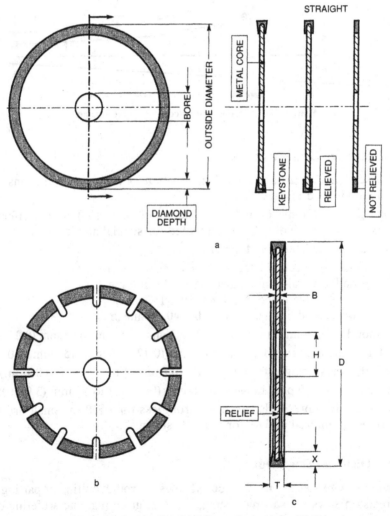

**Figure 4.64.** Diamond saw blades. (*a*) Standard diamond saw blade. (*b*) Slotted or segmented diamond saw blade. (*c*) Diamond saw blade dimensions.

flow of coolant to the area of contact between the blade and the workpiece. Figure 4.64 shows typical diamond saw blades. The bond can be metal, resinoid, or electroplated. The bond cross section is either straight or keystone shaped, as shown in Fig. 4.64a. Diamond saw blades have steel cores, whose core thickness are a function of the blade's diameter. The aspect ratio ($D:T$) ranges from 100 : 1 to 250 : 1. In other words, the core thickness can range from 0.25 mm (0.010 in.) for a 50-mm (2.0-in.) diameter blade to 2.0 mm (0.080 in.) for a 500-mm (20-in.) diameter blade.

High-aspect ratio saw blades are not very stiff, and they can cause unacceptable blade wander unless they are used with large flanges or hubs to support them properly on the machine spindle. Such blades have thin cores that can make narrow but not very deep cuts. Figure 4.65a shows a single

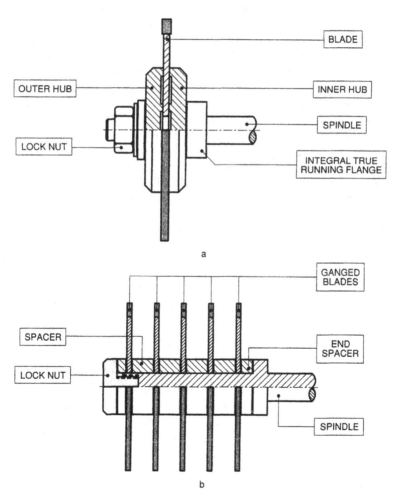

**Figure 4.65.** Saw blade mounting method. (a) Typical mounting of single saw blade. (b) Gang blade pack.

blade mounted with slinger hubs on a saw spindle, and Fig. 4.65$b$ shows a multiblade package designed for gang dicing or for grooving.

Diamond saw blades are designed to abrade large volumes of material, usually in deep but narrow cuts. To allow the coolant to get to the bottom of the cuts, the blades are relieved, which means that the core $B$ is thinner than the diamond cutting edge $T$ (see Fig 4.64$c$). The typical relief is 0.15 mm (0.006 in.) per side. It can be as small as 0.10 mm (0.004 in.) for small wheels and as much as 0.25 mm (0.010 in.) for large saw blades.

Diamond saw blades range from from 50 to 1200 mm (2 to 48 in.) in diameter with the 150-mm (6.0-in.) diameter blades being the most popular size used in optical shops. Cutoff wheels are typically in the 200- to 300-mm (8- to 12-in.) diameter range. The diamond depth $X$ (see Fig. 4.64$c$) ranges from 1.5 to 6.0 mm ($\frac{1}{16}$ to $\frac{1}{4}$ in.) in 1.5-mm ($\frac{1}{16}$-in.) increments. The diamond depth is 3.0 mm ($\frac{1}{8}$ in.) or less for blades of up to 300-mm (12-in.) diameter. Electroplated blades have diamond depth that rarely exceeds 1.5 mm ($\frac{1}{16}$ in.). Diamond grit sizes are 150 to 180 mesh for cutoff saws and 240 to 320 mesh for dicing blades. The diamond concentration is usually low, between 25 and 75. For very fine cuts the concentration can be as high as 100.

In addition to the standard diamond saw blades, there are a number of saw blades that are designed for special purposes. Microdicing blades and diamond ID saw blades have been specifically developed for low-kerf, high-efficiency slicing and dicing of semiconductor ingots and wafers. Figure 4.66 shows a typical ID blade and a microsaw blade assembly.

The core of an ID saw blade is a thin steel membrane that is stretched radially by the mounting screws in the rim until it is flat and the inner edge is running true. The diamond is electroplated to this inner edge. ID blades range in outer diameter from 200 to 400 mm (8 to 16 in.). The inner diameter allows for the slicing of 75 mm (3.0 in.) up to 150-mm (6.0-in.) round silicon boules.

Microsaw blades were originally developed to accurately cut silicon wafers that are patterned with integrated circuits (ICs) on automatic microsaws. The extremely thin blade was designed to cut along the so-called streets between the ICs on the wafer with a minimum kerf loss. The blade thickness can be less than 25 $\mu$m or one mil (0.001-in.) or as thick as 100 $\mu$m (0.004 in.). The blade exposure, which determines the maximum depth of cut, is a function of blade thickness and ranges from 0.50 mm (0.020 in.) for the thinnest blades to 1.5 mm (0.060 in.) for the thicker blades. The blade's outer diameter ranges from 50 to 52 mm (2.00 to 2.08 in.). The inner diameter is typically 40 mm (1.60 in.) [33]. The bond is typically a resinoid bond, but blades with a nickel alloy bond are also in use. Obviously the metal bond blade will have a longer life than any resinoid blade. Resinoid blades, however, will produce finer cuts under most conditions. Typical diamond sizes used are 5, 9, 20, and 70 $\mu$m. The finer diamond bonds are used for dicing of silicon wafers, the 20-$\mu$m blades have proved to be a good choice for dicing hard crystals such as garnet and sapphire, while the 70-$\mu$m blades are best for ceramics or glass substrate [33].

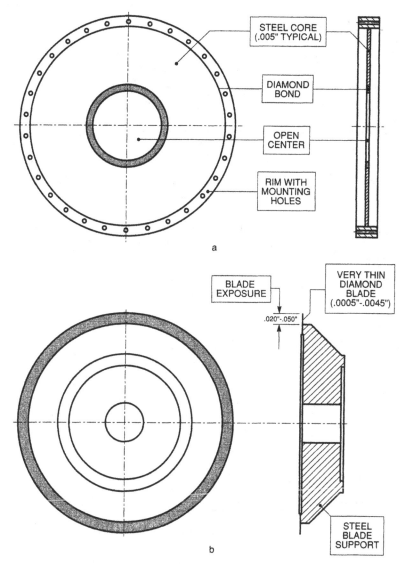

**Figure 4.66.**   Saw blades for special applications. (*a*) ID saw blade. (*b*) Microsaw dicing blade.

### 4.4.3.   Blanchard Wheels

Blanchard wheels are special purpose diamond face wheels for use on vertical surface grinders that operate on the Blanchard principle. Such machines, commonly known as Blanchards or mills, can be found in any optical shop where they are used to grind or mill a flat surface on multiple part blocks, stacks of parts or on individual pieces. Most optical materials can be milled this way although high density or high melting point materials can cause excessive wear or glaze over the wheel. All flat (or plano) surfaces on optical components, such as windows, filters, prisms and mirrors are milled on at least one side.

The diamond-impregnated bond is applied to an annular ring that is perpendicular to the axis of rotation. This ring is made of steel for all metal bonds and electroplated bonds and of aluminum for resinoid bonds. Figure 4.67 shows this arrangement. Typical Blanchard wheel dimensions are

Diameter    $D$    10 to 20 in. (10, 11, 12, 14, 18, 20 in.)
Thickness   $T$    $\frac{3}{4}$ to 2 in. ($\frac{3}{4}$, $\frac{7}{8}$, $\frac{1}{2}$ 1, 2 in.)
Width      $W$    $\frac{1}{4}$ to 1 in. ($\frac{1}{4}$ to 1 in. for up to 10-in. diameter
                    1 in. for larger than 11-in. diameter)
Depth      $X$    Typically $\frac{1}{16}$ or $\frac{1}{8}$ in.

Depending on the diamond grit size and the bond type, surfaces are either rough-milled for rapid stock removal or fine-milled to exact tolerances in preparation for final finegrinding. Intermediate finishes can be obtained with other diamond size and bond combinations.

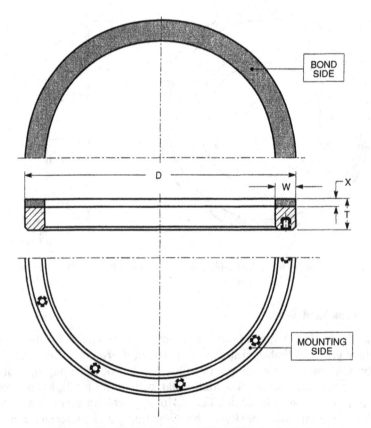

**Figure 4.67.** Typical surface grinding Blanchard wheel.

The bond for Blanchard wheels can be either resinoid or metal. The metal bonds can be steel, bronze, or an alloy. Electroplated wheels are also available for special applications. Although it is hard to generalize on the performance of specific diamond wheels with different materials, the following holds true for the best choice of bond in most cases [34]: A medium hard steel bond is recommended for heavy stock removal on soft glass. For heavy stock removal on hard glass, it is best to use a hard bronze bond of bond hardness R. Fine finishes on hard glass require the use of a hard alloy bond, while the same finish on soft and fragile materials can be best achieved with a soft bronze wheel. Resinoid wheels are preferred for milling of crystalline materials such as the most commonly used IR materials.

Of course the bond alone will not determine how good the finish will be. The size of the diamond and the diamond concentration also play a major role. The effect of the correct grit size and concentration, however, could be easily negated by an incorrect choice of bond material. The diamond grit sizes that are typically used for Blanchard wheels can range from 80 mesh for roughing wheels to as fine as 400 mesh for fining wheels. Intermediate sizes are also used, with the most common ranging from 240 to 320 mesh.

The most commonly used diamond concentrations for Blanchard wheels lie between 25 and 100. A concentration of 25 ensures fast stock removal for all grit sizes. It is the only acceptable choice for very fine wheels such as 400-mesh (30- to 45-$\mu$m) fining wheels. It is also recommended that these fine wheels be slotted to ensure adequate coolant flow to the contact area. A diamond concentration of 37.5 is the best choice for use on harder materials, with diamond sizes ranging from 80 to 150 mesh. It is also recommended for softer materials using 180- to 240-mesh diamonds. A 50 concentration is essential for milling soft glass with diamond grit sizes ranging from 50 to 180 [34]. Higher diamond concentrations are only used for special applications. It is best to discuss the necessity of higher diamond concentrations with a qualified diamond tool manufacturer or supplier.

Most Blanchard wheels have a continuous bond face as shown in Fig. 4.67. For heavy stock removal with wheels less than 18 in. in diameter, segmented diamond wheels exhibit improved removal efficiency over the continuous face wheels for the same bond, diamond size, and operating parameters. Segmented wheels are necessary for all Blanchard wheels larger than 18 in. There are two basic types of segmented wheels [34], as shown in Fig. 4.68. Individual segments that are composed of a diamond-impregnated bond layer on a steel backing (Fig. 4.68a) are either brazed or epoxied to a core ring. For the first type shown in Fig. 4.68b, the segments are mounted in a concentric arrangement, which is recommended for milling large surfaces. Two or more segments are always in contact with the workpiece. The second type (Fig. 4.68c) has the individual segments mounted radially so that they overlap to provide continuous contact with the workpiece. This arrangement is best for milling small parts.

The segments range from 0.75 to 1.50 in. in length (L) and from 0.25 to 1.00 in. wide (W); see Fig. 4.68a. The larger sizes are used on the concentric

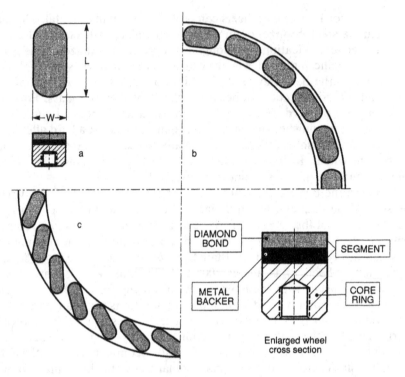

**Figure 4.68.** Segmented surface grinder Blanchard wheels [34]. (*a*) Wheel cross section. (*b*) Concentric mounting. (*c*) Radial mounting.

wheels. From 20 to 40 of these segments are typically mounted on a ring. The smaller segments are used for the radial-type wheels. Depending on wheel diameter, 50 to 100 segments are mounted on a ring. Almost all segments are made with metal bonds of varying hardness. Table 4.9 lists the various options.

The diamond grit sizes used for the segments are the same as those used for the standard continuous bond wheels. The diamond concentrations are,

**Table 4.9.  Bonds for Segmented Blanchard Wheels [34]**

| Diamond bond | Bond hardness | | Typical application |
|---|---|---|---|
| Steel | hard | (R) | lens blanks, photoflats |
| Steel | medium | (N) | lens blanks |
| Alloy | hard | (R) | large glass surfaces |
| Bronce | hard | (R) | general purpose |
| Bronce | medium | (N) | delicate glass |

however, somewhat higher. Specifically, the following diamond concentrations are recommended [34]: A concentration of 37.5 is used for fine-milling of surfaces with 400-mesh diamond grit (30 to 45 $\mu$m). A 50 concentration is recommended with coarser grits (46 to 80 mesh) to mill hard materials such as fused silica and ceramics. It can also be used with 150- to 320-mesh diamond to mill softer materials such as typical optical glass. Higher diamond concentrations of 75 to 100 are best for milling soft glass with 80 to 120 mesh grit.

The increased removal efficiency of segmented wheels stems from the fact that the liberally applied coolant can readily reach the contact area between the wheel and the workpiece and carry away the swarf as fast as it is generated. Reduced interference from swarf particles permits the diamond grains to penetrate the surface of the workpiece more effectively. This makes faster downfeeds possible, which in turn reduces milling time and improves the stock removal efficiency.

### 4.4.4. Generator Cup Wheels

Generator diamond wheels are specially shaped cup wheels that are used exclusively on curve generators to grind spherical surfaces on optical materials. Figure 4.69 shows several types of such wheels. The universal type wheel has a radiused bond edge (Fig. 4.69$a$). The straight cup wheel (Fig. 4.69$b$) is used to generate plano surfaces on prisms and other special shapes. These are similar to the Blanchard wheels that were already discussed in the preceding section, except that they are much smaller. The Bothner and Loh generators from Germany have been designed to work most efficiently with a wraparound-type radiused generator wheel (Fig. 4.69$c$). Other generating equipment requires the use of either convex- or concave-type cup wheels. The contact surface is either radiused or angled. Figures 4.69$d$ and 4.69$e$ show composite examples of such wheels.

Most generators operate on a constant infeed rate, for the wheel to remove an equal thickness of material with each revolution of the lens. The Bothner and Loh generators, however, operate in a plunge cut mode, whereby the cup wheel is fed into the workpiece to its final depth before the lens rotation is engaged and the spherical surface is generated in one slow revolution of the lens. This requires that the inner or outer edges of the wheel engage just as effectively as the radiused edge.

Generator wheels have been produced with outer diameters (OD) ranging from much less than 1 in. up to 10 in. Most of these are used in the OD range from 2 to 4 in. The larger diameters can only be used on large generators with powerful drive motors. The rim width $W$ is a function of the OD and ranges typically from 1.5 mm (0.060 in.) for the smaller diameters to more than 12.5 mm (0.500 in.) for the largest wheels. The overall height $T$ of the wheels ranges from a low of 12.5 mm (0.5 in.) to a high of 37.5 mm (1.5 in.). The rim radius $R$ is half the rim width $W$ for half-round bonds and ranges from 4.5 to 9.0 mm (about $\frac{3}{16}$ to $\frac{3}{8}$ in.) for the flat radius bonds.

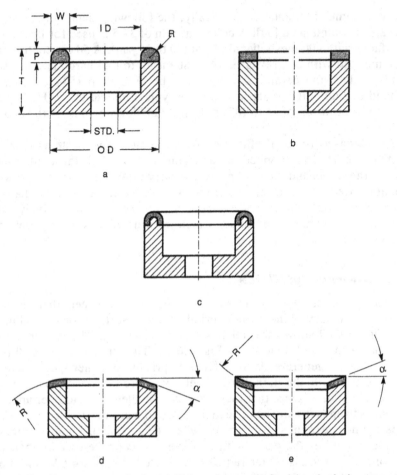

**Figure 4.69.** Generator cup wheels. (*a*) Universal type. (*b*) Straight wheel (ring type or cut type). (*c*) Bothner/Loh type. (*d*) Convex type (radius or angle). (*e*) Concave type (radius or angle).

The bond used for generator wheels is most often a medium to soft bronze bond. Steel bonds have been successfully used for the right type of work. Resinoid bonds would not stand up to the aggressive engagement required of curve generators. The diamond grit sizes range from 60 mesh for rough milling with large diameter wheels to 1000 mesh or finer for fine-milling with smaller wheels. In general 160 mesh diamond or finer is preferred for generator wheels up to 100-mm (4-in.) diameter, while 120 mesh is best for wheels that are larger than 100-mm (4-in.) diameter. Typically used diamond concentrations are 50, 75, 100, and 150. The diamond depth is most often 3.0 or 6.0 mm ($\frac{1}{8}$ or $\frac{1}{4}$ in.) but can be as much as 12.5 mm ($\frac{1}{2}$ in.) for some special wheels.

The minimum recommended operating speed for generator wheels is 3000 SFM. Since this speed is hard to achieve for small, less than 25-mm (1-in.) diameter wheels on standard curve generators, the diamond concentration must be increased to 150 to ensure a free cutting wheel. For wheels larger than 75-mm (3-in.) diameter, softer bonds should be used at a diamond concentration of 75 to reduce the problem of spindle stalling caused by increasing grinding pressures. The only other solution is to reduce the infeed rate, which would increase the generating time. A concentration of 50 with soft bronze bonds is best for generating curves on delicate parts that cannot tolerate high grinding pressures. Finer diamond wheels also work best with a low diamond concentration.

There are several different types of spindle adapters that permit the safe mounting of the wheels on curve generator spindles. Some of these are shown in Fig. 4.70. Each machine manufacturer and sometimes each machine model requires its own type of adapter.

Diamond tools are made by bonding a matrix that holds the diamonds to the edge of a metal core. This can be done by a simple casting method for the resinoid bonds or by a more involved sintering process for the metal-bonded tools. Figure 4.71 shows in a greatly simplified manner how metal-bonded cup wheels are made. A fine metal powder is thoroughly mixed with the appropriate diamond powder. This mix is then carefully distributed in a

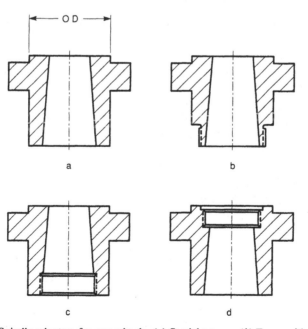

**Figure 4.70.** Spindle adapters for cup wheels. (*a*) Straight taper. (*b*) Taper with male thread. (*c*), (*d*) Typical tapers with female threads.

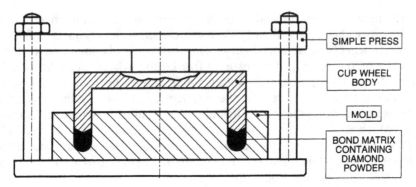

**Figure 4.71.**  Schematic of diamond tool manufacturing.

mold. The cup wheel core or body is placed into the mold. Compression bolts are tightened to force the body into the powder. The entire assembly is placed into an oven where it is heated until the metal powder begins to stick together. This process, called sintering, captures the individual diamond grains in a metal bond matrix. When the finished tool has cooled down, any excess bond material is ground off and the cutting edge of the tool is carefully dressed to run true to the axis of rotation.

### 4.4.5.  Peripheral Diamond Wheels

As the name implies, peripheral diamond wheels have the diamond-impregnated bond applied to their outer edge. Such wheels are typically used for edging, centering, and surface grinding of optics. Peripheral diamond wheels can be resinoid, electroplated, or metal bonded. The metal bonds are either bronze or steel. The choice of bond, diamond grit size and diamond concentration are determined by the intended application. The grit sizes typically range from 150 to 320 mesh, although other sizes are available for special requirements. The concentration can be 50, 75, or 100. Diamond-impregnation depth ranges from 1.5 to 6.0 mm ($\frac{1}{16}$ to $\frac{1}{4}$ in.) in 1.5 mm ($\frac{1}{16}$ in.) increments.

As it is true with all diamond tools, there are no definitive rules that apply to all situations. There are some basic rules of thumb, however, that can help establish a first order solution [34]. The final choice of diamond wheel specification may have to be refined through experimentation.

- For small, thin parts, a high diamond concentration of 100 and a fine diamond grit between 240 and 320 mesh in a medium hard steel bond may work best.
- A lower concentration of 75 and a coarser grit (150 to 180 mesh) in a medium to soft bronze bond has proved to be best for large, thick glass parts.

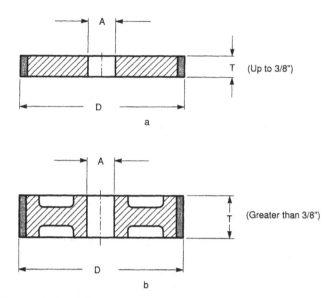

**Figure 4.72.**  Typical peripheral diamond wheels. (*a*) Straight wheel. (*b*) Weight relieved.

• For thin parts of hard materials such as fused quartz, a soft bronze bond with a 100 concentration and a 320 mesh diamond grit is a good choice.

The body or core of peripheral diamond wheels are made from either steel or aluminum. The arbor hole must be accurately specified so that the wheel can be mounted on the machine for which it is intended. Small-diameter or thin wheels of less than 10-mm (about ⅜-in.) width are usually straight walled, as shown in Fig. 4.72a. Larger and wider wheels must be weight relieved (Fig. 4.72b). Large wheels must be at least statically balanced; otherwise, there can be damage to the workpiece and excessive wear to the wheel. For the most critical applications, such as precision edging of small diameter parts and final centering of critical lenses, dynamic balancing at the operating speed is highly recommended.

The peripheral wheel width $T$ ranges from 1.5 to 25 mm ($\frac{1}{16}$ to 1 in.). Diameter ($D$) can be as small as 15 mm ($\frac{3}{4}$ in.) and as large as 750 mm (30 in.). However, the most common diameters are 150-mm (6-in.) diameter wheels with an aspect ratio ($D/T$) of about 25 : 1.

## References for Section 4.4 (Diamond Tools)

[32] Super-Cut, Inc., catalog SE-27.

[33] Polyohm Corporation, 1979.

[34] Precision Diamond Tool Co., catalog No. G-977.

# 5

# Optical Fabrication—Methods and Machines

This chapter is the focus of this book. The terms "optical fabrication" or "optical manufacturing" are used to describe a large number of specific shop activities that range from the shaping of the raw materials through grinding and polishing to finishing operations such as lens centering and the assembly of prisms. These and many other operations are described in this chapter, including the use of the necessary supplies, tools, and machines.

## 5.1. SHAPING OPERATIONS

Shaping is the first activity that must be performed in most cases. During this stage raw material or blanks must be cut, sliced, diced, cored, milled, and rounded into the basic shape of the finished part in preparation for the subsequent critical manufacturing stages. For this reason shaping is also often referred to as "preparation."

### 5.1.1. Cutting of Optical Materials

Sheet glass and blocks, ingots, and plates of a wide variety of optical materials must first be cut up into manageable sizes. Sheet glass is typically cut by the scribe and break method. Blocks and plates of glass and other optical materials, as well as ingots of crystalline materials, are parted with abrasive wheels on a glass saw.

**Scribe and Break Method.** Sheet glass is most economically cut with a glass cutter. By scribing one surface with the cutter, the surface tension is interrupted along the scribe line. Since the surface tension on the opposite side is unchanged, the sheet wants to break along the scribe line. When light pressure is applied to the side opposite the scribe, a fracture will develop that propagates along the scribe line until it has progressed along the entire length of the part. At that point the two pieces will separate and the break will be complete. The glass should be separated or broken immediately after the scribe operation since the stress difference in the sheet tends to rapidly equalize which results in rough breaks.

The glass cutter can be equipped with a diamond point or with a hardened metal wheel (Fig. 5.1). The diamond requires less scribe pressure than the wheel. This makes a diamond cutter a better choice to scribe thin sheet glass. The hardened metal wheel cutters are more effective for sheet glass thicker than 1 mm (0.040 in.).

The glass cutter is usually guided along a ruler or some other guide as it is drawn across the glass. The distance between the ruler and the scribe line is

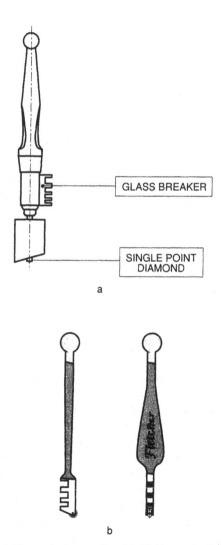

**Figure 5.1.** (*a*) Diamond glass scribe. (*b*) Single steel wheel glass scribe.

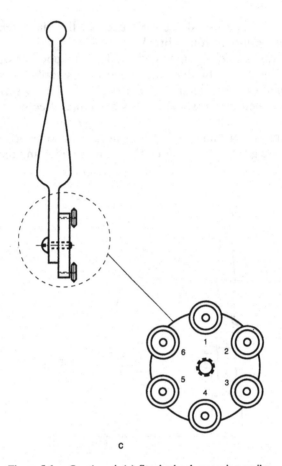

**Figure 5.1.** *Continued.* (*c*) Steel wheel magazine scribe.

2.5 mm (0.100 in.) for the most common glass cutters, but other designs may require a greater or smaller offset, as shown in Fig. 5.2.

Experience is required to optimize the proper holding of the glass cutter and to provide an adequate and uniform scribe force. The surface on which the glass sheet rests should be quite flat and free of debris such as glass splinters. It should also be obvious that the glass surface must be free of oils and other contaminants. The break should look smooth and shiny when viewed from the side. It must not be fractured or dull in appearance. It is always a mistake to repeat a cut along the same scribe line.

The glass scribed with the glass cutter is easily broken by light finger pressure. Especially clean breaks result when the glass is not separated by bending but by pulling the two halves apart. For thick glass the break is initiated by lightly tapping the beginning of a scribe line from the opposite side. When a large sheet is cut into many small pieces, the break sequence

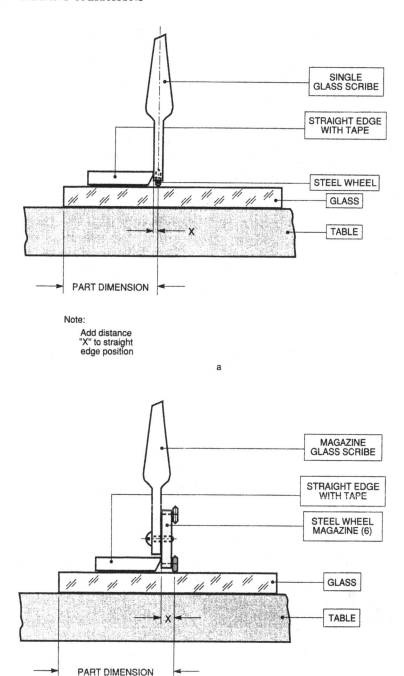

**Figure 5.2.** (*a*) Setup for guided scribe line (single glass scribe). (*b*) Setup for guided scribe line (magazine glass scribe).

must be chosen so that after each break two approximately equal-sized parts result.

Ground glass is scribed with a large wheel cutter. When cutting pieces for lens blanks from thick ground glass, a large glass cutter with a 10- to 15-mm (0.400- to 0.600-in.) diameter hardened steel wheel is preferred. Such a glass cutter is guided with both hands along a premarked scribe line. Since surface tension is greatly reduced or entirely eliminated on ground surfaces, it is much more difficult to break ground glass than fire-polished sheet. Single pieces can be separated through a light blow with the point of a hammer against the side opposite the scribe line. A wooden dowel can also be used to separate the piece (Fig. 5.3). A small press can be useful for larger quantities.

Circular pieces can be scribed with a special cutter (Fig. 5.4). The glass is scribed with the cutter, and the disk is broken out with finger pressure against the side opposite the scribe line. Afterward the surrounding glass is parted. Disks with large diameter are scribed with a compass glass cutter

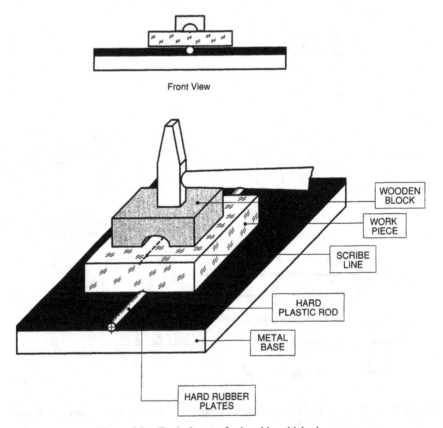

Front View

WOODEN BLOCK

WORK PIECE

SCRIBE LINE

HARD PLASTIC ROD

METAL BASE

HARD RUBBER PLATES

**Figure 5.3.**  Typical setup for breaking thick pieces.

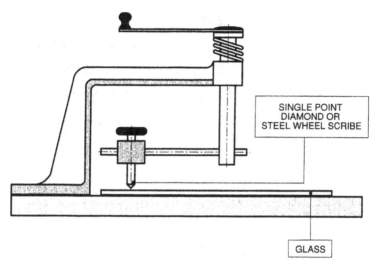

**Figure 5.4.** Circle scribe machine.

(Fig. 5.5). The compass has a suction cup in the center that adheres to the glass. The wheel is adjustable on the rotating arm so that various disk radii can be set.

If such circle cutters are not available, large circular pieces can be cut by the scribe and break method from sheet glass with a succession of straight cuts that are tangent to the desired circle. The resultant polygon must then be rounded by other means as described in Section 5.1.4.

**Scribe and Break Machines.** Manual scribe and break methods are perfectly adequate for a low to moderate volume of a few hundred parts. If, however, a large number of parts have to be cut with the scribe and break method on a sustained basis, it is necessary to mechanize the process as much as possible.

A glass-cutting machine is often used when a large number of equal-sized sheets must be cut. The position and scribe force for each glass cutter can be set accurately on such a machine so that even lower-skilled workers can produce good parts. The attainable accuracy for pieces cut from 2-mm (0.080-in.) thick sheet glass is approximately $\pm 0.2$ mm ($\pm 0.008$ in.). Thicker sheets tend to flare, which means that they tend to break at a slight angle. The resulting dimensional accuracy of the cut parts is then reduced. It is advisable to scribe crossing cuts from both sides of the sheet.

A number of commercially available semiautomatic scribe and break machines have been developed just for that reason [1, 2]. These commercial machines, which are designed for cutting parts from large sheets of flat glass, use several hardened steel roller cutters. Each wheel cutter is independently adjustable to an exact position and scribing pressure. Once properly set up,

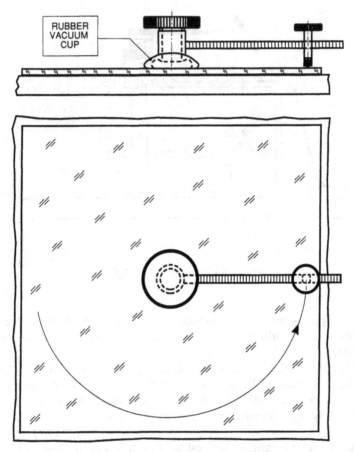

**Figure 5.5.** Compass glass cutter.

the machine drags all cutters across the glass sheet in one pass. The resulting long strips can then be easily broken off by hand or by mechanical means. Similar machines that use a series of diamond points have also been custom built for special applications.

More precise machines have been developed for scribe and break small geometries on thin plane-parallel substrate to very closely held tolerances. These machines utilize a single precision diamond point that is mechanically dragged across the surface line by line. An indexing feature rotates the workpiece by 90°, and the automatic S&B sequence is repeated. After breaking the scribe and separating them, square or rectangular parts result with tight dimensional control.

For very small parts cut from thin substrate, a die separation technique has proved useful, whereby the uncut substrate is held by friction on a sticky plastic sheet [3] that is tautly stretched over a sheet metal frame. When the

S&B sequence is complete, the plastic sheet is uniformly stretched in all directions which cleanly separates the majority of the resulting parts. The individual parts can remain adhered to the plastic sheet for shipment to the customer.

One such machine [4] looks very much like a precision microsaw. It is specifically designed for using a single-point diamond scribe rather than a diamond wheel to scribe patterned semiconductor subtrate. This machine has also been successfully used on thin glass substrate for optical applications. It can hold dimensional tolerances to ±0.05 mm (±0.002 in.) for 1 mm (0.040 in.) thick glass substrate. Thicker substrate will exhibit a greater tolerance range because of the unavoidable flare that results from the break stage. The scribe point is mounted on a pivoted arm which is well supported to give it excellent stiffness. A sliding weight is used to vary the pressure on the scribe point. The workpiece table is mounted on a linear slide which permits motion only in and out as seen from the front of the machine. After each scribe the point is raised slightly to lift it above the surface and the workpiece table is stepped over a preset distance by a servomotor. After the scribe position has been located to within 0.025 mm (0.001 in.), the point is engaged and the next scribe is made. When all programmed scribes have been made in one direction, the work table rotates 90° to repeat the process in the other to complete the dicing of the substrate.

One of the most important considerations for diamond point scribe and break machines is the proper selection and use of the single diamond point. This is a topic that goes beyond the scope of this book. Diamond suppliers are often the best sources for useful information to determine the optimum approach.

### 5.1.2. Glass Saws

When the glass sheets become too thick for scribe and break, or when the resulting dimensions become too inaccurate, or when the scribe and break method is no longer cost-effective, then the glass or optical material must be cut on special glass saws.

Glass saws that use diamond cutting wheels play an important role in optical manufacturing. They are used to cut or slice a broad variety of optical materials from ingots or other forms of raw material (see Chapter 2). But more important, precision saws that utilize diamond blades are used to cut, slice, or dice optical parts and often to final dimensions.

There are several basic types of saws used for a variety of different applications that are described in this section:

- Cutoff glass saws
- Diamond edged band saws
- Reciprocating slurry saws

- ID diamond saws
- Crystal wire saws
- Precision slicing and dicing saws
- Precision microsaws

**Cutoff Saws.** Cutoff saws are most commonly used for rough cutting of glass and quartz. Manageable pieces of the raw material are cut off from large blocks or from ingots. These pieces are in the form of slabs or plates and suitably sized for further processing into lenses, windows, and prisms. There are a number of different machines in use.

Cutoff saws are very simple machines that consist of a diamond wheel that is mounted on a horizontal spindle and a suitable work table for supporting and guiding the workpiece. The spindle is belt driven by a constant speed motor. Beneath the spindle is the workpiece table that typically runs on rails to permit the workpiece that is mounted on the table to be fed into the rotating diamond wheel. Appropriately designed splash guards and a recirculating coolant supply system complete the machine. A schematic of such a machine is shown in Fig. 5.6.

The saw blade can be cooled with either a water and emulsion mix or with coolant oils. A coolant reservoir is built into the machine base from which the coolant is pumped up to the blade. The saw blade should run at approximately 25 m/sec (4900 SFPM) circumferential speed.

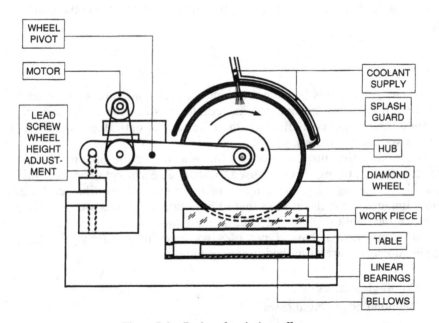

**Figure 5.6.** Basics of typical cutoff saw.

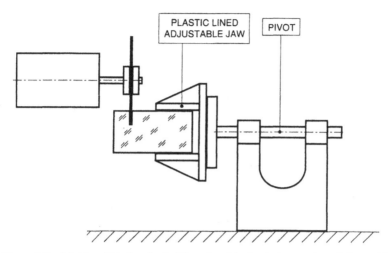

**Figure 5.7.**   Multiposition jaw for holding glass blocks during cutting of plates. [5].

The spindle support is mounted on a pivoted mount that makes it possible to raise or lower the diamond blade relative to the table surface. This provides for a simple depth of cut adjustment. Many of the older cutoff saws have workpiece tables that are pushed by hand, others are fed into the blade by a simple weight, while the more modern saws have lead screw driven workpiece tables. The more sophisticated cutoff saws have not only a linear infeed but a crossfeed as well to facilitate the cutting of successive slices.

Figure 5.7 shows a saw with a jaw-type clamping arrangement [5] that holds blocks of glass securely in place while slices are cut off manually. This machine is particularly useful for cutting uniform slices off large blocks. It is used in the following paragraphs to describe typical glass sawing operations.

In preparation for sawing a glass block on the saw, cut lines are marked on the glass. The width of the cut must be considered for this step. It consists of the saw blade thickness plus 0.2 mm (0.008 in.) for run-out. This width is often called *kerf*, and the volume of the cut in the material is the *kerf loss*. During actual use some cutting losses are inevitable. But since optical materials are valuable, it pays to carefully think through how the block can be laid out and cut up most favorably. Figure 5.8 shows an example.

After the cuts have been clearly marked, the glass is clamped into the workpiece holder, which is similar to a vise (Fig. 5.7) except that the jaws are lined with a plastic material to prevent damage to the clamped block. By rotating the workpiece holder around the pivot, the saw operator makes sure that the marked cut line touches the saw blade in every attitude of the workpiece. If this condition is not met, the workpiece is repositioned in the holder. When several plates are to be cut from the same block, the workpiece can be advanced by one plate thickness with a turn of the handwheel that is typically found on the right-hand side of the machine.

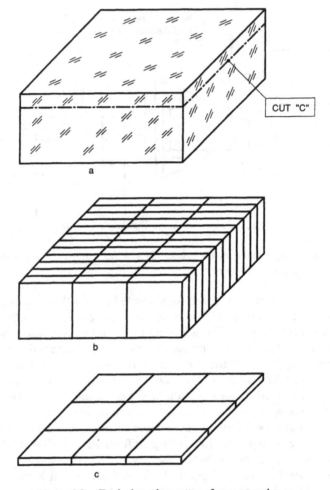

**Figure 5.8.** Typical cutting pattern for square pieces.

For short cuts the swing and the clamped workpiece can be guided with the right hand, while the left hand holds the glass and removes the cut piece. The left hand also serves as support when heavy pieces are cut. Such slices would tend to break off under their own weight after the cut has sufficiently progressed. This would result in unacceptable corner fractures. The work-piece advance can be provided by a pulley and weight for large cuts. This self-acting advance of the swing can be limited by an adjustable positive stop. The operator needs only be present during the final stages of the cut to receive the sawed-off slices by hand.

When a piece is so large that it cannot be sawed apart with one pass, the workpiece is rotated by 180°, and a second cut is made from the opposite side. To do this, the locking screw must be released. This then frees the pivot of the workpiece holder. If many equally spaced cuts must be made, as is sometimes required for the preparation of prisms, a suitably designed holding jig can be built into the workpiece holder. This eliminates the need for the time-consuming marking of the cut lines. Such a jig should be made from wood or plastic. One glass manufacturer [6] has stated that a glass saw with a standard 240-grit metal-bonded diamond blade will cut through 15 cm$^2$ (2.3 in.$^2$) of BK 7 optical glass in about one minute.

On occasion, one will find a specially modified band saw for cutting large sheets of thick glass. A diamond-edged band saw blade is used with such a saw. Liberal amounts of water coolant must be supplied to the blade and the workpiece. In addition to cutting through thick sheet glass, this type of band saw can also be used quite effectively to rough cut slices off ingots and large blocks of optical bulk material.

**Slurry Saws.** The slurry saw is designed to cut thin slices from optical, crystal, and semiconductor materials with a multiblade assembly that uses an oil-based abrasive slurry as the cutting medium [7, 8]. A heavy-duty blade head that holds a tensioned blade pack is driven by an eccentric drive assembly to oscillate back and forth on accurate linear ways that are supported by a sturdy machine base. The workpiece is mounted on the lift which is hydraulically raised at light but constant pressure into the reciprocating blades while the recirculating abrasive slurry is continually and liberally applied. This arrangement is shown in Fig. 5.9.

A variation of this approach has the blades in a fixed position while the workpiece oscillates [9]. It is claimed by the machine builder that much

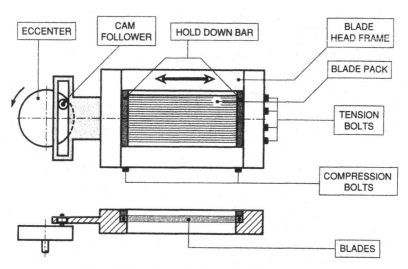

**Figure 5.9.** Schematic of reciprocating slurry saw.

higher stroke frequencies are possible with this machine as compared with the standard version because the workpiece is much lighter than the blade frame. Higher cutting rates should be attainable with this configuration.

The key to good results with slurry saws is the proper assembly and tensioning of the multiblade packs. These blade packs can be either assembled by the user, or they can be purchased from the machine manufacturer already assembled to the user's specifications. Studies have shown that there is no clear-cut advantage to using one method over the other.

There are three ways in which blade packs can be assembled. They can be built up from individual spacers and blades prior to use and taped together for installation in the blade frame. This type of assembly is usually done by the machine operator. The preassembled blade packs are assembled by the supplier in a special fixture, and the alternating spacers and blades are either epoxied or pinned together while the blades are stretched straight and the spacer stacks are compressed.

Blade packs are built up from blue spring steel blades that are separated by accurately lapped plane-parallel steel spacers (Fig. 5.10). The blade thickness ranges from 0.006 to 0.010 in. The blade height can be $\frac{1}{4}$, $\frac{3}{8}$, or $\frac{1}{2}$ in. The blade length is typically $16\frac{1}{4}$ in. The spacers must be about 0.2 mm (0.008 in.) thicker than the desired thickness of the slices when a 320-grit SiC abrasive is used to allow for abrasive wear on the slices. The required spacer thickness over and above the thickness of the slice is a function of abrasive size. Table 5.1 shows this relationship.

The minimum center-to-center spacing between blades is determined by the minimum part thickness, including sufficient fab allowance, the blade

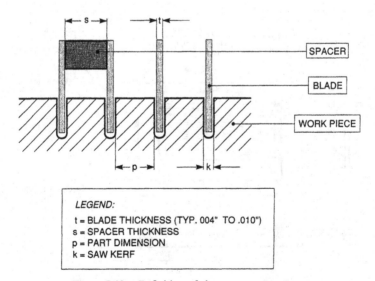

LEGEND:

t = BLADE THICKNESS (TYP. 004" TO .010")
s = SPACER THICKNESS
p = PART DIMENSION
k = SAW KERF

**Figure 5.10.** Definition of slurry saw parameters.

**Table 5.1. Minimum Spacer Thickness Overage Depends on Abrassive Size**

| Abrasive size Mesh | μm | Wear/side (inch) | Blade walk (inch) | Minimum spacer thickness overage (inch) |
|---|---|---|---|---|
| #300 | 35 | .0035 | .001 | .008 |
| #320 | 32 | .0028 | .001 | .0075 |
| #400 | 25 | .0025 | .001 | .006 |
| #600 | 15 | .0015 | .001 | .004 |
| #800 | 12 | .0011 | .001 | .003 |
| #1000 | 9 | .0007 | .001 | .0025 |

thickness, and the actual spacer thickness plus the overage. This is best explained by example:

What is the required spacer thickness when 0.2-mm (0.008-in.) blades are used to cut 0.5-mm (0.020-in.) thick slices with 300-grit slurry? From Table 5.1 we find that the minimum spacer thickness overage is 0.2 mm or 0.008 in. The blade-to-blade spacing which is 0.7 mm (0.028 in.) must be added to this value to yield a spacer thickness of 0.9 mm (0.036 in.). In an actual case, spacers used to cut 0.46-mm (0.018-in.) thick slices with 0.15-mm (0.006-in.) thick blades using a 320-SiC slurry were 0.89 mm (0.035 in.) thick.

Once the blade pack has been assembled for installation, it is inserted in the blade head frame. The pack is visually centered in the frame, while compression bolts are tightened at each end to apply a strong compressive force where the spacers are so that the blades cannot slip when the pack is tensioned. Only moderate force is needed for epoxied or pinned blade packs. Once locked into position, the blades are tensioned by alternating the tightening of four tensioning bolts. A torque wrench is often used for this critical step to ensure that all four bolts are tightened uniformly. The required torque is a function of blade cross section and the number of blades in the pack. For instance, for a 0.15-mm (0.006-in.) thick blade that is 12.7 mm (0.500 in.) wide, a minimum torque in inch-pound is 1.8 times the number of blades in the pack. For a 100-blade pack, a minimum bolt torque of 1800 inch-pound is required for proper tensioning.

This recommended tensioning is intended to physically stretch the blades to about 80% of their elastic limit. It results in an appreciable elongation of the blades that can be measured with a simple linear scale. This type of measurement is often used to determine the point at which proper blade tensioning has been achieved. For instance, a 1.5-mm (0.060-in.) elongation is required to properly tension a 0.2-mm (0.008-in.) thick by 12.7-mm (0.500-in.) wide spring steel blade pack. Greater elongation must be measured for thinner blades and smaller elongation for thicker blades. The exact values are best established empirically, since much depends on blade material, blade and spacer condition, compression forces, and other factors.

When the blade pack has been correctly tensioned, the blades can be properly aligned to run parallel to the direction of oscillation by adjusting their position with the aid of two draw-over bolts. If this parallel adjustment is not done, or if it is done improperly, any misalignment of the blades would cause them to cut with their sides and not just with the cutting edge. This would not only lead to thickness problems for the slices, but it would also cause excessive wear on the blades which can lead to premature blade failure. If the blades fail before the workpiece has been cut through, many slices will also be broken. The prevalent adjustment method is to attach a dial indicator to the machine frame and to measure the out-of-parallel condition of the outer blade by contacting the blade while the blade head is manually moved back and forth. Any run-out is taken out by adjusting the draw-over bolts. Both sides of the blade pack should run well within 0.025 mm (0.001 in.) parallel from end to end.

The properly tensioned and aligned blade pack oscillates back and forth as an oil-based SiC abrasive slurry is liberally applied to the top of the blades. The workpiece is firmly waxed or epoxied to a glass baseplate into which the blades can cut. This assembly is then mounted on the lift which gently but firmly forces the top of the workpiece against the bottom of the oscillating slurry covered blades. The blades cut very slowly but relentlessly into the workpiece with feed rates that are measured in millimeters per hour.

There are two different lift mechanisms. The older model slurry saws have a single-post lift which tends to stick, causing an erratic upward motion. This behavior is reflected in the typically longer cycle times on single-post saws as compared to the cycle times achieved on the more modern saws that are equipped with improved four-post hydraulic lifts. For instance, one comparative study conducted by the author showed that the cycle time for the one-post machine was on average 18 hours compared to an 11-hour cycle on a four-post machine to cut the same part under otherwise identical conditions. Typical cycle time variations from cycle to cycle are on the order of about ±3 hours. The economic advantage of slurry saws lies in the fact that, once set up, one operator can easily service several saws while building blade packs and setting up machines for new cycles.

The width of the blade packs is a function of the machine design, and it can range from 190 mm (7.5 in.) to 250 mm (10 in.). The maximum number of blades that can be safely tensioned also depends on the type of machine and ranges from about 225 up to 325. These values correspond to a minimum blade-to-blade spacing of 0.6 mm (0.024 in.). According to Table 5.1 a setup with 0.15-mm (0.006-in.) thick blades and a #600 slurry will yield slices that are 0.5 mm (0.020 in.) thick. The stroke length is about 200 mm (8 in.) which limits the size of the workpiece to about 150 mm (6 in.) to be on the safe side.

The maximum stroke speed ranges from 70 to 150 cycles per minute (cpm), with typical operating speeds at about half those values.

The blades will wear from the abrasive action, and most blade packs are good only for a total cutting depth of 25 to 50 mm (1 to 2 in.). Much depends on the type of material cut. The wear is not uniform as the center of the blade, which is nearly always engaged, wears more rapidly than the ends. This wear pattern causes the blades to eventually ride up on the workpiece at each end of the stroke. As a result the blade frame tends to slightly lift off the ways. The unmistakable sound that this motion emits signals that a stroke length reduction is required. To maintain an efficient cutting speed, any stroke length reduction should be compensated for by a corresponding increase in stroke speed. Since speed is expressed in terms of length traveled per unit time, it means when the stroke length is decreased, the cycle speed must be increased by a corresponding amount to keep the blade speed relative to the workpiece constant.

The optimum cutting speed depends on many factors, and it must be empirically established for each case. Some of the factors are the abrasion resistance of the material that must be sliced, the maximum safe lift pressure that does not cause blade flexure, the number and thickness of the blades, the abrasive type and size, the slurry distribution, and the stroke frequency.

The slurry system operates in a continually recirculating mode. The slurry is usually a mix of SiC abrasive powder with an oil or a similar vehicle that has desirable suspension properties. The slurry is pumped from a slurry tank to a reciprocating manifold from which it is uniformly distributed across all blades. The typical slice that results is slightly tapered and ridged. The taper results from the breakdown and settling out of the abrasive grain in the slurry and the considerable wear on the blades. The ridging is the result of blade flexure due to erratic lift pressures and decreasing slurry cutting efficiencies. Both problems are fairly typical and are a problem only as a matter of degree.

Slices as thin as 0.3 mm (0.012 in.) have been routinely cut from fused silica substrates as large as 50 by 50 mm (2 by 2 in.). Germanium (Ge), silicon (Si), crystal quartz, most glasses, and many ceramics have all been successfully sliced on this type of saw. Round rods can also be sliced, but care must be taken to support the rod sufficiently to avoid breakout where the blades cut through the bottom third of the rod. Dicing of flat substrate is also possible if the workpiece support is designed for 90° indexing after the first cut has been completed.

To calculate the slice yield per workpiece, we can use the blade kerf (see Fig. 5.10) and the part thickness. Kerf is the blade thickness plus the sum of allowances. The following simple equation can be used:

$$N = \frac{L - K}{P + K}, \tag{5.1}$$

where

$L$ = workpiece length (width of blade pack),
$K$ = kerf width,
$P$ = part thickness.

**ID Saws.** ID saws are special types of saws that are designed for making very fine cuts through large silicon ingots. These are automatic machines that produce very thin slices of silicon (wafers) and slices of materials with similar hardness such as GaAs and fused silica with a minimal kerf loss that is as low as $\frac{1}{10}$ of the kerf produced by standard blade slicing. The obvious advantage of this approach is to maximize the yield per ingot and thus reduce slicing cost per part [10, 11].

A very thin 0.10 to 0.15 mm (0.004 to .006 in.) stainless steel blade ranging in diameter from a little more than 400 mm (16 in.) to nearly 700 mm (27 in.) [12, 13] has a concentric hole that allows for slicing ingots from 150 to 200 mm (6 to 8 in.) in diameter and is stretched over a tensioning ring until it is very stiff and resists flexure. The inner edge of the ID hole has diamond abrasive electroplated to it to serve as cutting edge. The tensioned blade is mounted in a blade head assembly that is driven at a spindle rpm to give a peripheral speed of about 3300 SFPM at the cutting edge. Obviously the cutting edge must be running true and the head assembly must be properly balanced to avoid damaging vibrations. A schematic of a tensioned blade is shown in Fig. 5.11.

The ingot is typically a round silicon boule, but ingots with square or rectangular cross sections and of different hard materials such as fused quartz or crystal quartz can also be sliced on such a saw. The ingot is epoxy mounted on a graphite support base that is fastened to the infeed bar. The infeed bar is stepper motor driven to an accuracy of 2.5 $\mu$m (0.0001 in.). Such an infeed step occurs after the completion of each cut to set up for the next one.

Liberal amounts of coolant must be supplied to the cutting edge, to the blade exit, and to both sides of the blade. This is essential to keep the blade free cutting and to prevent the buildup of abraded material (swarf) that can lead to blade wander and unacceptable cuts, as shown in Fig. 5.12. Improper blade tensioning will add appreciably to this problem.

The cutting blade head assembly typically moves downward into the stationary workpiece at feed rates ranging from less than 25 mm (1 in.) per minute up to 250 mm (10 in.) per minute. Upon completing the cut a small distance into the graphite support, the blade head rises until the blade clears the work and the workpiece feed steps over to position the ingot for the next slice. The slices can remain standing on edge, as shown in Fig. 5.11, or they safely drop into a chute, or they are picked off one by one by a vacuum pickup device and automatically loaded into a cassette.

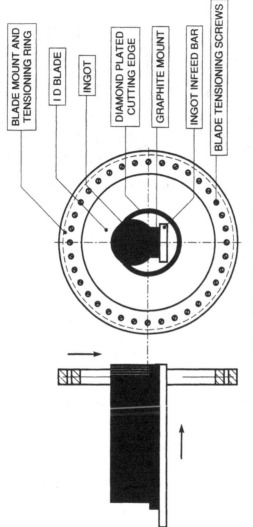

BLADE MOUNT AND
TENSIONING RING

I D BLADE

INGOT

DIAMOND PLATED
CUTTING EDGE

GRAPHITE MOUNT

INGOT INFEED BAR

BLADE TENSIONING SCREWS

**Figure 5.11.** Schematic of ID saw.

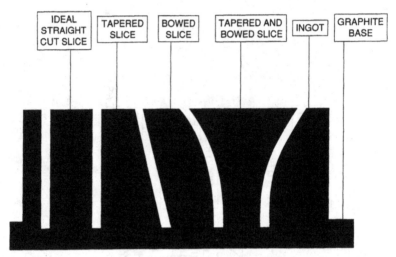

**Figure 5.12.**  ID saw silicing problems (greatly exaggerated for clarity).

**Wire Saws.**  Wire saws are often preferred for cutting sensitive crystals that are too valuable to be cut with too much kerf loss or for crystals that cannot tolerate the cutting stresses produced by standard diamond sawing.

A diamond-impregnated stainless steel or beryllium copper wire is used with these saws, or a high tensile strength steel wire to which diamond grains are held by an electroplated copper layer. The diameter of the wires range from a low of 0.075 mm (0.003 in.) with 9 $\mu$m diamond up to 0.25 mm (0.010-in.) wire impregnated with 45 $\mu$m diamond.

The wire is pulled taut over a pulley system, as shown in Fig. 5.13, as it is reeled off a shuttling capstan very similar to those used in weaving machines. While one loop of the wire reels off to engage the crystal, another loop is wound up on the capstan. As the capstan shuttles back and forth, the wire remains aligned with the idling and tensioning pulleys.

Slices are cut from sensitive crystals at slow speeds and light pressures. Since there is little or no heat generation and cutting stresses that can induce twinning are all but nonexistent, even the most difficult crystals can be cut on such a saw. The typically thin wires used will guarantee a minimal kerf loss, which is a very important consideration for expensive materials. In addition the diamond grit sizes used for the wire saw are much smaller than those used in standard diamond saw blades, so cut surfaces require only a few mils overage for subsequent fine-grinding and polishing operations. These machines [14, 15], however, are very slow and only well suited for prototype lab work. They are not a useful alternative for even low-volume production.

**Slicing and Dicing Saws.**  Slicing and dicing saws are the workhorses in many optical shops. They are either modified standard surface grinders or preci-

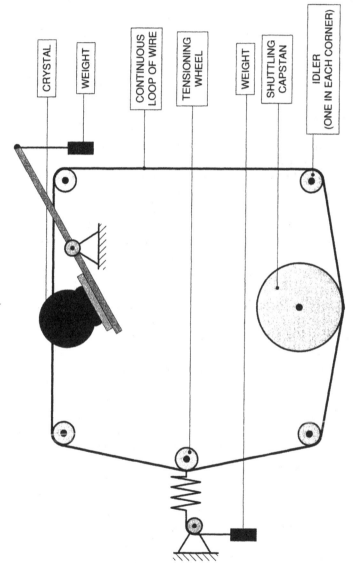

CRYSTAL

WEIGHT

CONTINUOUS
LOOP OF WIRE

TENSIONING
WHEEL

WEIGHT

SHUTTLING
CAPSTAN

IDLER
(ONE IN EACH CORNER)

**Figure 5.13.** Schematic of simple wire saw.

sion machines that have been specifically designed for making precise cuts in optical materials.

Low-cost saws used in optics manufacture are based on the surface grinder that is most often found in machine shops. They are equipped with a very short spindle that is supported at only one end. This cantilevered spindle arrangement is quite acceptable when only a single diamond blade is used for slicing and dicing. Two blades can be mounted if the separation between them is not too great. However, gang dicing methods where many blades are mounted on the same arbor have proved to be very economical. This is especially true when a large number of cuts have to be made.

If multiblade gang dicing is to be done with these machines, the spindle arbor must be made longer to accept more blades. This requires that the spindle bearing be strengthened. It also means the installation of a more powerful motor to overcome the significant grinding forces that the simultaneously engaged blades exert on the spindle. These are very expensive modifications to an otherwise inexpensive machine. Saws that are specifically designed for multiblade slicing and dicing will be a more cost-effective choice in the long run. They will be described separately in this chapter.

But even if the modification is done properly, success cannot be guaranteed because the long cantilevered spindles tend to flex under normal cutting loads. This condition produces blade walk which results in dimensional inaccuracies, excessive blade wear, poor cuts and finish, perpendicularity problems, or even breakage and damage of both workpiece and diamond blades.

Figure 5.14 shows a typical slicing and dicing saw in a slicing mode. The workpiece, which is either a thick plate or a waxed stack of thinner plates, is wax blocked to a glass baseplate which, in turn, is waxed or cemented to a mild steel blocking plate. The disposable glass baseplate serves a very important function by allowing the diamond blade to cut all the way through the workpiece and well into the baseplate. This way there will be no breakout along the bottom edge of the workpiece due to an insuffiently deep cut or an improperly supported surface. The plane-parallel metal blocking plate allows the workpiece to be firmly held to the magnetic chuck on the table.

When the workpiece is waxed to the base and blocking plates, it is customary to align the workpiece edges with two adjacent edges of the blocking plates. This is especially important when the workpiece needs to be diced as well. This arrangement is shown in the top view (bottom half of Fig. 5.14) where the first cut has already been made and the blade has just started the second cut after a stepover of the saw table. The stepover distance is determined by the part thickness, the blade width, and the kerf width. More is said about this later. The method described so far assumes noncritical dimensional tolerances.

A definition is in order here. *Slicing* is cutting through the workpiece longitudinally from end to end, which yields long strips. *Dicing* is the same process after the workpiece has been rotated 90° to yield either square or rectangular parts. The square option is shown in Fig. 5.15.

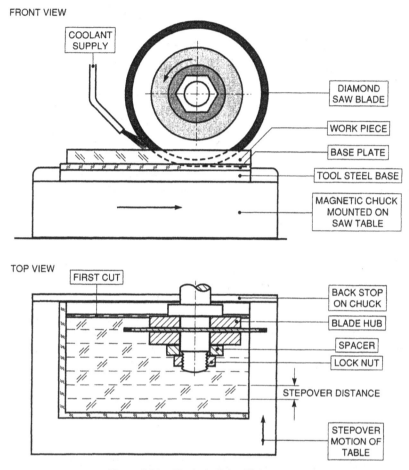

**Figure 5.14.**   Typical slicing/dicing saw.

For more critical requirements it is best to make a waste cut first and then step over for the second cut to yield an accurate part dimension. This is necessary for several reasons. A waste cut trims off any out of square condition and any irregularities of the typically scribe and break edge. It also establishes a clearly defined reference cut from which to begin the required step over. Figure 5.16 shows this and the rectangular option in greatly exaggerated detail.

The first and last cuts leave a small strip of waste material. The minimum width of this waste should not be less than about $\frac{1}{10}$ of the depth of cut; otherwise, the thin waste remainder may break off before the cut is complete which could lead to blade walk-off and thus dimensional inaccuracies.

Some saw operators prefer the practice of lowering the cutting edge of the diamond wheel below the top surface of the workpiece or stack with the

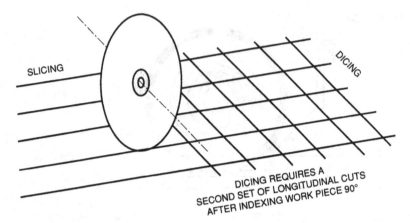

**Figure 5.15.** Difference between slicing and dicing.

blade behind the stack. The operator then moves the table back very slowly while the wheel is slowly turned back and forth by hand until a faint scraping sound is heard, which indicates that the wheel has touched the back of the workpiece. This method is inherently less accurate, at least for the first row of parts, than the waste cut approach for the following reasons:

First, the back edge of the workpiece is rarely a good reference because of the uncertain condition of the sides of the stack. Second, the saw operator

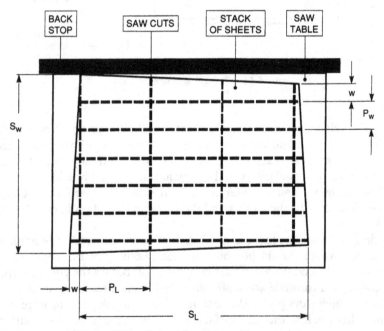

**Figure 5.16.** Pattern for slicing with waste cuts.

cannot be certain that the contact between blade and stack is light enough so that the relatively thin blade is not bent by some small, yet measurable amount. And finally, the reference could easily be lost when the wheel must be raised to clear the top of the workpiece to set the blade for the first cut. As a result of these concerns it is best to reserve this particular approach for cutting parts oversize for later shaping to final dimensions by other methods such as milling or grinding.

The depth of cut is a function of blade thickness. The thicker the blade, the deeper the cut that can be made. There are no hard and fast rules that go beyond this general statement. The optimum depth of cut must be established separately for each case. Another important consideration is hub clearance. The hubs of a mounted diamond blade (Fig. 5.14) physically limit the depth of cut. Hub diameters are a function of blade thickness. The thinner the blade, the larger the hub diameter should be. The reason for that is that the hubs provide stiffness to the blades. This prevents blade flexure under typical cutting loads. Any blade flexure can lead to blade wander that affects part dimension or even lead to blade failure, which can be a real problem especially for resinoid-bonded blades.

The safe optimal feed rate is determined by the type of blade, the depth and mode of cut, the material type, and the coolant efficiency. For instance, 12.5-mm (0.500-in.) high stacks of 1.5-mm (0.060-in.) thick glass sheets can be safely cut with a 320-grit resinoid blade at a feed rate of 12.5 mm (0.500 in.) per minute. A copious supply of properly directed coolant is necessary for this. Faster feed rates are possible, but they are not as safe.

**Precision Dicing Saws.** Precision dicing saws have been specifically designed to dice optical and semiconductor materials to final size. These machines use diamond blades ranging in thickness from 0.15 mm (0.006 in.) up to 1.5 mm (0.060 in.). The blade thickness is a function of grit size for the thinner blades, with the thinnest blades requiring the finer diamond sizes.

A single blade automatic precision saw [16] with a cantilevered but rigid arbor is shown in Fig. 5.17. This type of saw can be programmed to automatically slice and dice large stacks of optical substrate.

The following operating limits are composite values for number of precision dicing saws: The magnetic work tables have infinitely variable longitudinal feed rates ranging from 1 mm (0.040 in.) per minute up to over 6000 mm (240 in.) per minute. The stepper motor driven crossfeed for blade stepover provides accuracies down to 2.5 $\mu$m (0.0001 in.). The longitudinal travel is usually limited to 500 mm (20 in.), and the maximum stepover distance lies between 150 mm (6 in.) to 200 mm (8 in.). The typical blade diameter is 150 mm or 6 in.

Some saws feed the work through the blades longitudinally to the full cutting depth. At the end of the cut, the blade is raised and returned to the starting position. The table automatically steps over the preset distance (part dimension + kerf) for the next cut, and the process is repeated until the workpiece has been cut.

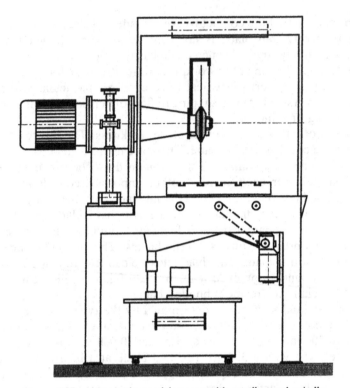

**Figure 5.17.** Automatic precision saw with cantilevered spindle.

When the blade cuts into the work in the direction of the cut, it is called a *plunge cut*. When the rotation of the blade is reversed, it is called a *climb cut*. In the square cut mode the spindle rises after each cut, the table indexes over, while the spindle returns to the start position for the next cut. Some saws have instant spindle rotation reversals at the end of each pass so that the blade will cut in both directions either in the plunge mode or the climb mode. For a zigzag cutting mode the spindle does not rise after each cut. The blade travel continues until the blade disengages fully from the workpiece, the blade rotation will reverse while the table indexes over, and the table will feed the workpiece into the blade for the next cut.

The depth of cut can be set to full depth, or the cut can proceed in small incremental steps with each pass. Another mode is the Z-cut for which the blade is lowered at a slow but constant rate as it oscillates longitudinally through the workpiece. The lowering of the blade can proceed at a constant rate, or it can occur only in one direction. In the latter case the blade returns to the starting position for the just completed pass without any downfeed. The blade downfeed commences again when the longitudinal feed in initiated.

Two simple equations based on the conditions shown in Fig. 5.18 can be set up to determine not only how many parts a certain plate will yield but

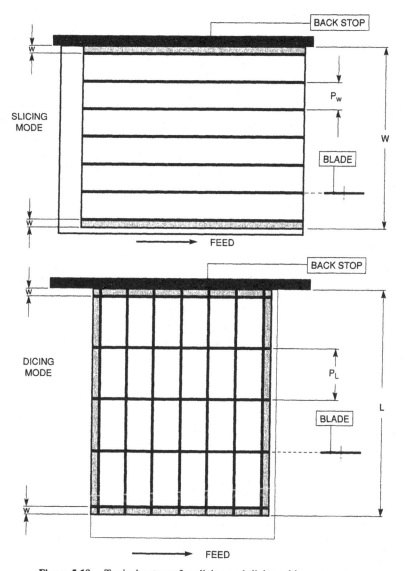

**Figure 5.18.** Typical pattern for slicing and dicing with waste cuts.

also what the optimum dicing pattern is when both the parts and the plates are rectangular. The same equations can be used when one is square and the other rectangular, or when both are square.

The number $N_w$ of rows across the sheet width $S_w$ for parts with side $P$ and blade kerf $K$ is

$$N_w = \frac{S_w - K}{P + K}.\tag{5.2}$$

For the sheet length SL, the equation is:

$$N_L = \frac{S_L - K}{P + K}. \qquad (5.3)$$

In most cases $N_L$ and $N_w$ will be a decimal fraction, but for practical reasons these values must be integers or whole numbers because only complete parts are of interest. The decimal fraction represents the sum of the two waste cuts ($W_1$ and $W_2$). If this remainder is multiplied by the sum of part dimension and kerf $(P + K)$, the material available for waste cuts can be determined. This is best explained by example.

**Example 1:**

A large number of 25-mm (1-in.) square parts must be cut to size from 1.5-mm (0.060-in.) thick sheets of 125 by 150 mm (5 by 6 in.) glass. A 0.9-mm (0.036-in.) diamond saw blade is used that produces a kerf width of 1.0 mm (0.040 in.). The stack height must be limited to 15 mm (0.6 in.).

For the sake of clarity, the calculations for this example are made with the inch values. Metric values can be used as well, and they will produce the same results when the correct conversion factors are applied.

$$\text{Summary: } S_L = 6.0 \text{ in.} \quad P = 1.0 \text{ in.}$$
$$S_w = 5.0 \text{ in.} \quad K = 0.04 \text{ in.}$$

Using Eq. (5.2),

$$N_L = \frac{6.0 - 0.04}{1.0 + 0.04} = 5.73.$$

This result means that five rows of parts can be cut across the 6.0-in. length by making six cuts. The remaining waste cuts are 0.73 (1.04) = 0.76 or on 0.38 in. per side.

$$N_w = \frac{5.0 - 0.04}{1.0 + 0.04} = 4.77.$$

Across the 5.0-in. width, four rows can be yielded with five cuts. The waste cuts are 0.77 (1.04) = 0.80 or 0.40 in. per side.

Since the depth of cut is no more than 0.6 in., the minimum acceptable waste cut per side should be 10% of that value, or 0.060 in.. The waste cuts in this example are more than adequate. If the material is valuable, it may be better to use smaller sheets to avoid unneccessary waste.

The total number of 1-in. square pieces that can be cut from each sheet is $N_S = N_L + N_w = 5(4) = 20$. The stack height in this example was 0.6 in. This limits the number of sheets per stack to $0.6/0.06 = 10$, which means that the stack can be only ten sheets high. Therefore the stack in this example will yield $10 \times 20$ or 200 parts.

In most cases, however, rectangular parts must be cut from rectangular sheets. There are two possible ways to do that. Either the long edge of the parts ($P_L$) runs parallel to the long edge of the sheet ($S_L$), or $P_L$ runs parallel to the short edge ($S_w$) of the sheet.

Equations (5.2) and (5.3) must be modified to represent these conditions. For the first case where $P_L$ is parallel to $S_L$,

$$N_L = \frac{S_L - K}{P_L + K}, \tag{5.4}$$

$$N_w = \frac{S_w - K}{P_w + K}. \tag{5.5}$$

For the second case where $P_L$ is perpendicular to $S_L$,

$$N_L = \frac{S_L - K}{P_w + K}, \tag{5.6}$$

$$N_w = \frac{S_w - K}{P_L + K}. \tag{5.7}$$

These equations yield two solutions depending on how the cuts are aligned. By calculating both cases, it is easy to determine which solution will yield the greater number of parts.

**Example 2:**

Using the same values from the preceding example but cutting parts that are 0.5 by 1.0 in. (i.e., $P_L = 1.0$ in. and $P_w = 0.5$ in.), we have the following:

Case 1. Using Eqs. (5.4) and (5.5),

$$N_L = \frac{6 - 0.04}{1 + 0.04} = 5.73 \quad (5 \text{ rows}),$$

$$N_w = \frac{5 - 0.04}{0.5 + 0.04} = 9.19 \quad (9 \text{ rows}).$$

Then, $N_S = 5 \times 9 = 45$ parts per sheet.

Case 2. Using Eqs. (5.6) and (5.7),

$$N_L = \frac{6 - 0.04}{0.5 + 0.04} = 11.04 \qquad (11 \text{ rows}),$$

$$N_w = \frac{5 - 0.04}{1 + 0.04} = 4.77 \qquad (4 \text{ rows}),$$

And, $N_s = 4 \times 11 = 44$ parts per sheet.

At first glance it seems that it does not matter which approach is used, since the difference is only one part per sheet or ten parts per stack. But the analysis must be carried further to include the waste cuts. When this is done, it becomes clear that only ten rows can be yielded across the sheet length for case 2. This reduces the stack yield for case 2 to 400 parts. Therefore case 1 is clearly the better choice as $45 \times 10$ or 450 parts will be yielded by each stack.

A difference of ten parts per stack does not seem significant when only a few stacks must be cut up. But when 1 million parts must be cut, the extra ten parts per stack will save over 50 stacks that will not have to be cut up. This is a rather significant difference, in terms of both materials and supply savings and savings in labor hours. It is always a good idea to check all available options.

**Multiblade Dicing Machine.** When a large number of parts must be cut up, multiblade dicing can greatly speed up the process and reduce unit cost. Multiblade dicing is also referred to as *gang dicing*.

One special saw [17] has been designed for precision dicing of thin semi-conductor and glass substrate using a multiblade arbor. It has a 100-mm (4.0-in.) long cantilevered air-bearing spindle that is belt driven over a three-step pulley to drive the spindle at either 4000, 6000, or 8000 rpm (Fig. 5.19). Either thin single blades or easily interchangeable preassembled multiblade packages can be used.

For the most precise cutting to finished size with tolerances down to ±0.050 mm (±0.002 in.), a saw must be used that has a spindle that is firmly supported on both ends. Such a saw is a Swiss made machine [18] that is schematically represented in Fig. 5.20. Blade wander caused by spindle flexure is completely eliminated with such a rigid spindle and vibration-free cutting yields edges that are practically free of chips, even when cutting brittle materials.

The precision spindle is firmly mounted between rigid arbors that allow free rotation but do not give even under heavy cutting loads. This design makes these machines ideally suited for multiblade gang dicing. These saws are semiautomatic in that they can be set up with one or more diamond blades in a preprogrammed operating cycle that engages the blade, com-

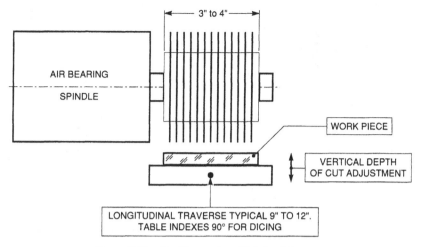

**Figure 5.19.** Cantilevered multiblade dicing saw.

pletes the cut, raises the blade, returns the workpiece to its starting position while it simultaneously steps over a predetermined distance and initiates the next cut. This process continues without operator intervention until the workpiece has been cut and the stepover has run against a preset limit switch. The accuracy of the stepover distance is claimed to be 0.01 mm (0.0004 in.), and this claim is supported by outstanding results. The machine can be programmed in either the plunge cut or the climb cut mode, and at the same time, it can be operated in either straight cuts or the Z-cut mode.

Special preassembled precision multiblade packs [19] are available for this type of machine (see Fig. 5.21). The blade pack can be designed in such a

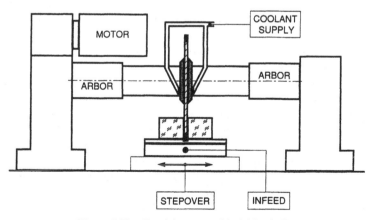

**Figure 5.20.** Precision saw with rigid spindle.

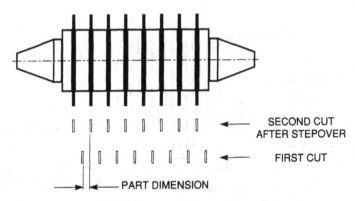

**Figure 5.21.**   Preassembled gang dice package.

way that only every second cut is made with each pass. Only one stepover is then required to cut all parts to size in one direction. For small parts which require many blades, the blade pack can be designed to cut every third or even every fourth cut, requiring two or three stepovers, respectively. The cost of blades is often an economic constraint on how many blades can be mounted in a pack. There is, however, another limit on the number of blades per pack that is determined by the power capabilities of the spindle motor.

The blades in the pack are carefully ground in to precise dimensions in both blade thickness and separation between blades. The blades are separated by spacers. By carefully trueing in each blade, the blade run-out is reduced to a minimum. This limits the maximum kerf loss. The blade-to-blade spacing can be ground in in such a way so that the part dimensions that are cut with the new blade pack will be initially near the bottom of the tolerance and will increase toward the top of the part tolerance as the blades wear. When the top tolerance on the parts has been reached, the pack is retired for refurbishment. Parts can be sliced and diced on a precision saw with such multiblade pack to an accuracy of ±0.025 mm (±0.001 in.), although ±0.050 mm (±0.002 in.) is more typical. Multiblade gang dicing only makes economic sense when a large number of the same size parts must be cut to close tolerances on a sustained basis.

Diamond blade wear is an important condition that must be constantly monitored when cutting parts to final size. This is especially true when cutting with resinoid-bonded diamond blades. As dicussed in Chapter 4 in the section on diamond tools, proper bond wear is essential for keeping a wheel or blade cut efficiently. As a result a certain degree of blade wear must be expected, even under the most ideal conditions. Excessive blade wear, however, is usually due to an incorrect choice of diamond size, bond type, and concentration for the intended application. Inadequate coolant supply and distribution and gross misuse of the blade, such as feeding it faster than it could be expected to cut, are other common reasons for premature blade failure. Among those insufficient cooling of the cutting region where the blade engages the work is one that is most often found to be responsible for the problem.

For instance, if the sides of a resinoid-bonded blade are adequately cooled but the center of the blade is not, the center will wear at a faster rate than the sides. This can lead to the formation of a groove in the center of the cutting edge, and thus to inaccurate cutting and ultimately blade failure. A more likely condition is that one side of the blade is more efficiently cooled than the other. In time a taper will develop on the cutting edge of the blade that is slanted toward the side that is inadequately cooled. Such a wedge-shaped cutting edge will cause the blade to wander, and this results in dimensional inaccuracies on the parts and ultimately in early blade failure.

It can never be stressed enough that ample coolant must be supplied at all times during the cutting cycle to the point where the blade engages the work. The distribution of coolant must be such that the center of the blade's cutting edge and both sides receive sufficient volume and a roughly equal flow of coolant. One can never have too much coolant, but not enough coolant always causes serious problems.

To overcome these difficulties, it has been suggested (and often tried) to submerge the workpiece in coolant and cut through in this way. Although seemingly logical, this approach may the worst way to distribute coolant because cavitation caused by the rapidly revolving cutting edge of the blade will almost guarantee that no coolant gets to where it is needed the most. Therefore submerged diamond blade cutting is not recommended.

The optician is sometimes called upon to estimate how many diamond saw blades are needed for a specific job. If it is known from previous experience with similar work how much material volume can be cut with a certain diamond blade before it must be replaced, then it is not difficult to calculate how many blades are needed. Using the previous example of the rectangular $5 \times 6 \times 0.6$ in. high stack that is cut into $0.5 \times 1$ in. rectangular parts, the following can be calculated:

For case 1, in the long dimension six cuts by 6 in. long by 0.6 in. deep are required, and in the short dimension the corresponding values are ten cuts by 5 in. long to the same depth. The blade kerf remains at 0.040 in. All the information needed to calculate the volume of material that must be abraded for each stack is now available. It has been already established that each stack will yield 450 parts.

$$V = 0.04 \ [(6 \times 6 \times 0.6) + (10 \times 5 \times 0.6)]$$
$$= 2.06 \ \text{in.}^3 \ \text{per stack.}$$

Assuming that 100,000 parts must be produced at a 95% yield, at least 234 stacks must be cut up. A typical 6-in. diameter 320-grit resinoid blade can abrade about 200 in.$^3$ of glass or 150 in.$^3$ of fused quartz before it must be replaced.

By using this information, it can be shown that for a glass material (234 × 2.06)/200 ≈ 2.4 or at least 3 diamond blades are needed. If the parts were made from fused silica, then (234 × 2.06)/150 ≈ 3.2 or 4 blades would be required to cut up the same number of parts.

**Microsaws.** Microsaws were originally developed for the precise dicing and separation of patterned IC wafers. A number of very fine and highly precise microsaws are commercially available [20, 21, 22]. They all have the same basic function in common, although they may differ in some of the individual features.

These machines are automatic to a high degree. Many machine operating parameters are preprogrammed. Most commonly used substrate materials such as silicon (Si), germanium (Ge), gallium arsenide (GaAs), gallium phosphide (GaP), sapphire and alumina ($Al_2O_3$), and a wide variety of glass substrate can be sliced and diced on such a machine. The wafer size capacity is typically limited to 6-in. diameter. The workpiece (wafer) is held by vacuum to a worktable that moves through the blade at fairly high cutting rates. Most microsaws use a single blade, but some saws have been set up for multiblade dicing.

The diamond blades for the microsaws are of a unique design as shown in Figs. 4.66b. The resinoid-bonded or electroplated metal-bonded diamond blades are only 50 to 100 mm (2 to 4 in.) in diameter and are very thin. Diamond blades as thin as 20 $\mu$m (<0.001 in.) can be used to produce cuts with a maximum saw kerf of 35 $\mu$m (0.0015 in.). Even the thickest blades rarely exceed 0.25 mm (0.010 in.). One of the constraints imposed by such thin blades is that the maximum depth of cut is limited.

A safe blade exposure ratio rarely exceeds 1:20. This means that the maximum blade exposure is about 0.5 mm (0.020 in.) for a 0.025 mm (0.001 in.) thick blade and 5.0 mm (0.200 in.) for a 0.25-mm (0.010-in.) thick blade. The safe depth of cut is usually about 0.005 to 0.010 in. less than that. These are merely basic rules of thumb, and much depends on the material cut and the choice of operating parameters.

Because of the small blade diameters, the spindle must rotate at a high rate to provide the necessary peripheral speed for optimum cutting efficiency of the blade. The spindles are typically direct drive, water-cooled, nearly frictionless air-bearing spindles that rotate at speeds ranging from 10,000 to 50,000 rpm. The speeds are infinitely variable over that range. Air-bearing spindles provide such high rotational speeds with minimal vibration. Cutting speeds (feed rates) are very high, ranging from a low of 150 mm/sec (6 in./sec) to a high of 400 mm/sec (16 in./sec.). The optimum speed must be experimentally determined because it is dependent on other operating parameters, the type and size of blade and the properties of the substrate materials.

After each pass, the workpiece table is stepped over by a preprogrammed increment that represents the part dimension. The claimed stepover accuracies are within ±5 $\mu$m (0.0002 in.) for each cut, and the nonaccumulative indexing accuracy is within ±10 μm (±0.0004 in.). When all cuts have been made in one direction, the workpiece table is rotated (indexed) by 90°, and the process is repeated for final dicing. Some machine have also 60° and 120° index positions for special dicing patterns.

As is true for all diamond dicing, an efficient coolant supply is essential for keeping the blades at their optimum cutting potential. The machines incorporate a number of innovative features to simplify the process and to optimize the quality of the cuts. A magnified image of the cut is seen on a TV monitor which alerts the operator to cutting problems such as blade wander and chipping. The blade diameter is optically sensed for the automatic correction of the depth of cut to compensate for blade wear. Programmed automatic blade dressing at specified time intervals keeps the blades at optimum cutting efficiency. An integral wafer cleaning and drying system and an integrated automated wafer loading and unloading feature make the machines nearly fully automatic.

As an example [23] of typical dicing times, it takes about 5 minutes to completely dice a 4-in. diameter silicon wafer to a depth of 0.010 in. at a cutting speed of 4 in./sec. into 1-mm (0.040-in.) square parts.

### 5.1.3. Boring and Coring in Glass

Holes and recesses can be machined into glass using a variety of methods. The chosen method depends on the nature and size of the hole or recess, as well as on the condition of the workpiece. A modern CNC mill that has been modified for use with diamond tools can be programmed to greatly facilitate the shaping of optical parts to complex geometries at a very high degree of precision. This includes the boring of holes and the cutting of recesses.

**Diamond Drill.** Holes up to 3 mm (0.120 in.) in diameter are bored with a diamond drill that uses a single diamond as the cutting point. Such diamond drills can be obtained from any manufacturer of diamond tools by specifying the desired diameter. The glass is bored halfway through one side and finished by boring from the opposite side. This approach requires accurate part and tool alignment. The coolant is turpentine oil which is applied with a brush. The eyeglass opticians use diamond drills to bore the holes for the mounting screws for rimless glasses.

**Hardened Metal Drill.** Holes of up to 12-mm (about 0.5-in.) diameter can be bored into thick glass with a hardened metal drill. These tools are available as fluted drill bits with hardened metal cutting edges or as pointed drills. They yield equal results, but the fluted drill can be better used in a drill jig. Here too the hole is started on one side and finished from the other (Fig. 5.22).

Since the location of the hole cannot be set, the drill tends to wander easily. It is best to do this type of work with a drill jig. This will ensure that the opposing holes will line up accurately to yield a clean hole. The coolant

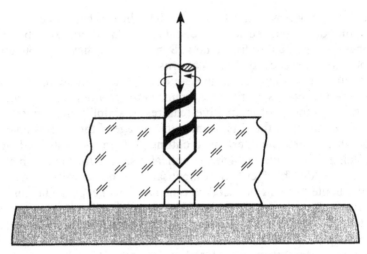

**Figure 5.22.**   Hole drilling in glass using fluted drill.

can be cutting oil, petroleum, or turpentine oil. For deep holes the drill must be raised frequently so that the abraded glass can be flushed away by the coolant.

This approach is mentioned here only as an alternate method when there is a need to drill a hole and a suitable diamond drill is not available. The accepted method, of course, is to drill or core holes with with diamond tools that are designed for that purpose.

**Diamond Core Drills.** Larger diameter holes are cored with diamond core drills. These tool are used in most optical shops. Their shape and basic dimensions have been standardized [24].

Only a 1- to 1.5-mm (0.040- to 0.060-in.) wide zone is abraded with these drills. This makes it possible to feed the drill quite rapidly. A coring rate of 30 mm/min (1.2 in./min) can be achieved for a 50-mm (2.0-in.) diameter BK 7 part. Often the solid core and not the bore is the point of interest, for round disks and round rods are fabricated in this way (Fig. 5.23).

The bond material for the diamond in the cutting edge is usually a sintered metal matrix. Diamond sizes ranging from 90 to 150 $\mu$m (D90 to D150) are those most often used [25]. During the coring operation a liberal amount of coolant is supplied through the hollow spindle of the drill press. The most effective core drill edge has grooves through which the coolant can escape. But even with this feature the core drill should be lifted repeatedly during the coring of deep holes so that the appreciable amount of glass swarf that is generated can be carried away more effectively by the coolant flow.

The parts that are set up for coring must be cemented to a disposable glass baseplate that supports the coring pressure when the drill exits the glass. This prevents chips on the underside of the glass. The core drill is allowed to cut into the baseplate only a small distance so that the glass plug remains

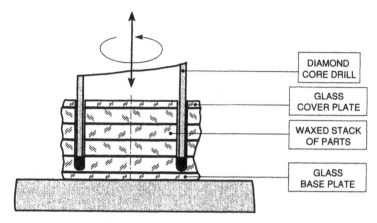

**Figure 5.23.** Coring of round glass disks.

cemented to the plate and does not jam up in the hollow core drill. Figure 5.23 shows a stack of parts with a baseplate and a cover plate. Cover plates are usually not necessary, but they are an effective and necessary protection when polished parts must be cored. Thin plates must be stacked up for coring so that the cored edges will be quite clean. The plates must be quite flat, and there should be no voids between the surfaces.

The diameters of successively cored holes or plugs can correspond to within 0.050 mm (0.002 in.) from part to part. But in time the diamond bond wears not only on the cutting edge but also on the sides. When the bond is exhausted, the resulting holes will have become smaller in diameter by as much as 0.2 mm (0.008 in.), while the core plug diameter will have increased by the same amount. Adjustable core drills are available on special order for the larger diameters to compensate for such wear.

When core drilling is only an occasional requirement in an optical shop, a rotary union can be placed between spindle and core drill, and the coolant is fed through it from the side. Rotary unions are offered by most suppliers of diamond products as an assessory to their line of diamond core drills.

It is much better when the drill press or coring machine is provided with a hollow spindle. In addition, the work table must be provided with stops for locating the workpiece, and the machine must be equipped with a spray guard. One special coring and boring machine [26] executes a series of programmed boring and coring steps automatically.

For critical materials that cannot be machined with diamond tools, holes or plugs can be cut with a rotating length of brass tubing while an appropriate free abrasive slurry is applied. The tubing must run true to ensure a clean cut.

**Ultrasonic Coring and Shaping.** Unusual or otherwise impossible geometries can be cut into a variety of optical materials with ultrasonic abrasive machining. A specially shaped anvil oscillates at a rate of 20 KHz along its axis.

Since the anvil does not rotate, it is possible to cut odd-shaped holes or recesses. The cutting tool, which resembles a cookie cutter, is silver soldered to this anvil. Each change in shape and size requires a new cutter.

The arrangement of an ultrasonic drill, as shown in Fig. 5.24, is similar to that of a conventional bench drill press. It has a manual downfeed which brings the tool into contact with the workpiece. Instead of an electric motor, it has an ultrasonically driven head which is energized by high frequency ac current from a generator. The cutting tool that is silver soldered to the anvil is made of nonhardened but wear-resistant steel. It must have the opposite shape of the desired bore or recess. Together with an abrasive, the axially oscillating tool does the actual shaping. Boron carbide with a 280-grit size has been proved to be especially effective. But silicon carbide and alumina can also be used. The abrasive is suspended in water, and this slurry must be pumped to the tool in sufficient quantities for cutting depths greater than 5 mm (0.200 in.).

The cutting tool wears at a rate of about 2% of the abrasion rate of the glass, which means that it must be replaced quite frequently. The penetration time in plate glass to a depth of 1 mm (0.040 in.) is approximately 0.5 minute. The rate is correspondingly slower for deep holes because it is more difficult to supply a sufficient amount of slurry.

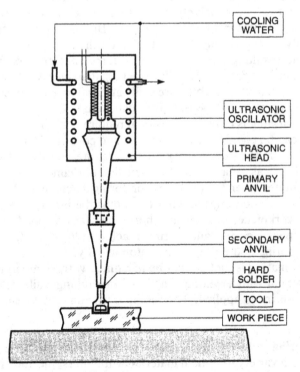

**Figure 5.24.** Ultrasonic drill.

**Sandblast Cutting.** A very gentle process that is a refined method of sand-blasting can be used to drill and shape glass and crystals. A fine-grain abrasive powder is mixed with a carrier gas in a mixing chamber. The abrasive is blown under a pressure onto the workpiece surface through a nozzle that has the shape of the desired hole or recess. The material is removed nearly damage free. This allows for clean cuts on sensitive crystals. The minimum kerf can be as small as 0.1 mm (0.004 in.).

### 5.1.4. Rounding

Most lenses and many other optical components are made round to simplify mounting them into instruments. This requires a rounding step. Different rounding methods are used depending on the number of parts and the required precision. Coring with diamond core drills has already been described in Section 5.1.3 as an effective rounding method. Centering is a related operation that differs from rounding in that the optical axis must be brought coincident with the mechanical axis. This requirement does not exist for rounding. Lens centering is described in detail in Section 5.6.

**Manual Rounding.** Manual rounding methods can be used when only a few prototype parts must be made. The inclusion of these methods in this chapter does not suggest that manual rounding methods are a cost-effective alternative to machine rounding. They are included mainly for the sake of completion and also to decribe alternate methods when machine rounding is not possible.

The blanks are first cut into squares in preparation to manual rounding. For larger parts they are cut into octagonal shape by carefully "breaking" off the four corners. Especially formed glass pliers are used for this task (Fig. 5.25). It is also possible to successfully perform this operation with any other similarly shaped pair of pliers. Small pieces of the glass edge are grasped by the jaws of the pliers and broken away sideways. For plate thicknesses of 4 mm (0.160 in.) or more, sizable pieces can be broken off in this way. Thin glasses will break, so only fine, dustlike glass chips should be broken off. A diameter overage of 2 mm (0.080 in.) must be provided for the subsequent abrasive rounding. For this and all similar operations, safety glasses must be worn.

For rounding large diameters several plates are stacked with wax, and the stack is mounted to a small plano tool. The tool is then screwed to a spindle that runs at about 1000 rpm (Fig. 5.26). A ring-shaped steel sheet with its ends bent outward to serve as handles is used as a rounding tool (Fig. 5.27). The sheet metal ring is placed around the rotating stack, and with the application of water and an abrasive slurry, the ring is moved axially up and down. This vertical movement must be carefully controlled or the edge will take on a barrel shape if this is not done correctly, and the rounds in the stack will have different diameters.

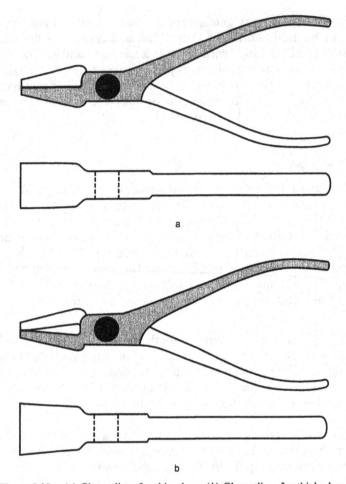

**Figure 5.25.**   (*a*) Glass pliers for thin glass. (*b*) Glass pliers for thick glass.

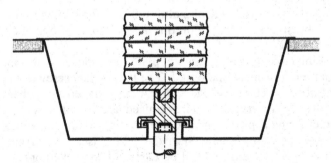

**Figure 5.26.**   Mounting of glass plates for manual rounding.

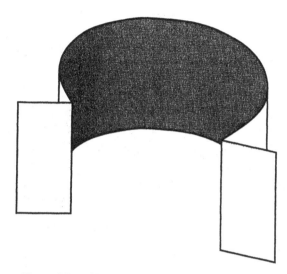

**Figure 5.27.** Sheet metal tool for manual rounding.

For smaller diameters the pieces are stacked 100 to 150 mm (4 to 6 in.) high in a stacking fixture. The individual parts are aligned such that the ends are perpendicular to the stack axis (Fig 5.28). If it is a metal fixture, it should be prewarmed or the glasses may fracture. The long stacks resulting from this step are ground round on a cast-iron plano tool with a coarse abrasive. A wooden strip fastened above the plano tool inside the slurry pan of the grinding spindle will simplify this task (Fig. 5.29).

The stack rests against the strip and is turned by hand; while some pressure is applied with one hand, the other feeds slurry to it. The stack can be transformed into a round roll in this manner, but it will still be out of round by about 0.050 mm (0.002 in.). A sufficiently large diameter overage must be provided to perform this step. The stack or roll is marked with a pencil and

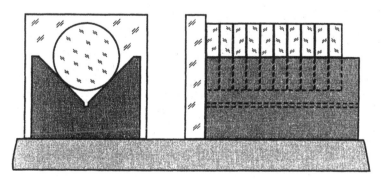

**Figure 5.28.** Stacking of glass rounds in vee-block.

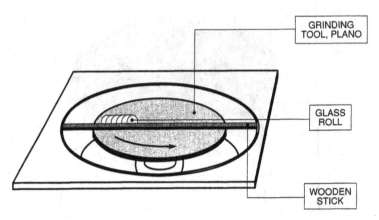

**Figure 5.29.** Manual rounding on a tub grinder.

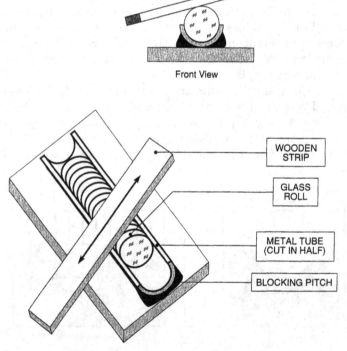

Front View

**Figure 5.30a.** Manual rounding in cut tubing.

ruler along the edge parallel to its mechanical axis. It is then heated to just barely soften the wax, and the individual disks are rotated relative to one another by a few degrees. The pencil mark is used as a guide to verify that the disks have been twisted uniformly and that the total twist is 360° from one end of the roll to the other. After the roll is realigned and allowed to cool in a vee-block, it is rounded as before to the desired diameter. These steps reduce the out-of-round condition of the resulting parts to about 0.025 mm (0.001 in.).

When the diameter of the rounded glass disks must be accurate to within 0.01 mm (0.0004 in.), an added step is necessary. A diameter overage allowance of 0.2 mm (0.008 in.) must be made. It is then rounded off in a length of a tubing half section or between two pieces of steel round stock fastened to a baseplate (Fig. 5.30). With the application of fine abrasive slurry, the roll is

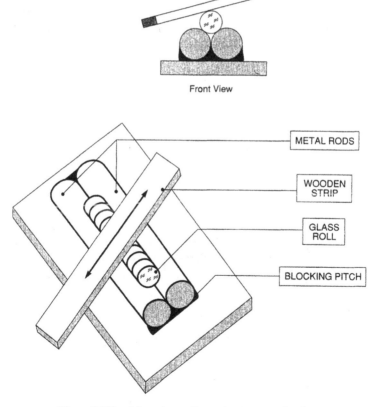

**Figure 5.30***b*. Manual rounding using two metal rods.

rotated with a wooden strip in a diagonal motion until it has reached the exact diameter. The strip engages the roll especially well when it is lined with a thin sheet of rubber.

**Machine Rounding.** Machine rounding is the only economical method to round optics, even when production volumes are low. Manual rounding requires the direct engagement of an optician or a skilled operator, but machine rounding allows for a much more efficient utilization of their labor, giving them time to perform other tasks while the machine is running. One operator can run two or more machines in support of volume production.

**Rounding Machines.** Rounding machines used in optical shops are designed to grind the outer diameter (OD) of individual parts or stacks of parts that are waxed together. They basically consist of a workpiece spindle that runs parallel to a diamond wheel spindle. Either the workpiece spindle oscillates axially past the rotating but otherwise stationary diamond wheel or the wheel oscillates past the stationary workpiece spindle. The separation between the spindle axes can be reduced at a slow but steady rate to feed the rotating diamond wheel into the counterrotating workpiece to grind it into a circular shape. This arrangement is shown in Fig. 5.31.

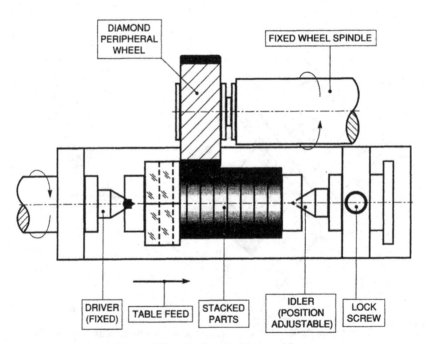

**Figure 5.31.** Basics of edging machine.

The workpiece spindle usually drives one end of a tail stock, while the opposite live center rotates freely. The workpiece (often a stack of plane-parallel parts) is held between the driven center and the idler so that it rotates around its axis. Other machines are set up to accept an exchange spindle to which the workpiece is fastened. With this particular arrangement, an edging machine can also function as a basic centering machine, but the infeed controls are often not fine enough for critical work. The two ends of the tailstock are synchronously geared to rotate at the same speed. This arrangement is best for large and solid workpieces. For instance, such edge-grinding machines (OD grinder) have been developed to grind large silicon ingots (boules) to accurate diameters prior to slicing them into wafers.

Commercially available edging machines suitable for the optics shop tend to be fairly large. The maximum diameters that these machines will handle range from 150 mm (6.0 in.) to 400 mm (16.0 in.). The maximum thickness of the workpiece or stack ranges from 100 mm (4.0 in.) to 200 mm (8.0 in.). The exact limits depend on the design of the machine.

These edgers can also edge very small parts down to 3 mm (0.120 in.) or less. The length of the rod or stack must then be reduced appropriately since the grinding pressures excerted by the grinding wheel can easily break a slender rod or cause a stack of parts to separate.

A large edger [27] which has been specifically designed for OD grinding of silicon boules (ingots) can grind 150-mm (6.0 in.) diameters to a length of nearly 600 mm (24.0 in.). It uses either one or two diamond face wheels. In the single-wheel mode it operates like any other edger. When operating in the two-wheel mode, the first wheel makes the rough cut to remove the bulk of the material and the second wheel smoothes the surface and grinds the workpiece to the required diameter (Fig. 5.32). The face wheel approach is thought to produce much less work damage because its grinding forces are directed along the boule axis rather than at right angles to it as is the case with peripheral diamond wheels.

Workpieces can be rods of solid material, individual rounds, or stacks of plane-parallel parts. These are rounded as they turn between the tail stock and are fed past the fixed diamond wheel which is incremented after each pass by a preset amount. The oscillation can be set to go in either direction. Metal runners are waxed to each end of the workpiece. The runners have tapered seats that fit the tailstock centers. One of the two runners has a coupling groove that engages the driver.

While the corners of large squares must be trimmed off prior to stacking them for edging, this preshaping step is not needed for diameters of under 30 mm (1.2 in.). The square cut pieces must have an overage of 2 mm (0.080 in.) over the final diameter.

There are many different edging machines each with their own characteristics. To list them here would be far too confusing. It is best to limit the discussion of machine operating parameters only to typical values. This will be true throughout this chapter whenever machines are discussed.

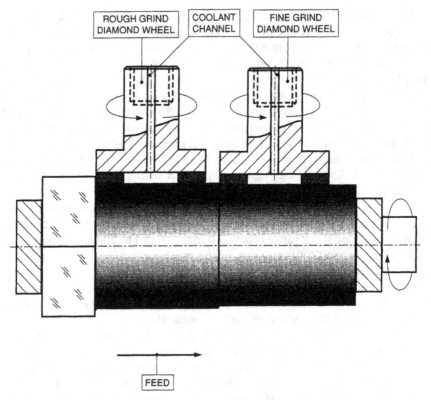

**Figure 5.32.**   Two-stage edger using face wheels.

The operating parameters for edging machines are

Rounding accuracies of ±0.050 mm (±0.002 in.) are common, and precision edgers can hold diameter tolerances to better than ±0.025 mm (0.001 in.), even under volume production conditions.

The workspindle speeds are infinitely variable from a few rpm to several hundred rpm to accommodate different diameters and materials. The larger machines have typically the slower spindle speeds.

The longitudinal feed of the workpiece table or the feed of the diamond wheel stage can be varied infinitely from about 5 mm (0.200 in.) per minute to 200 mm (8.0 in.) per minute. The longitudinal feed can go in either direction.

After completion of each pass the diamond wheel infeed can be controlled manually or it can be an automatic preset advance. Infeeds can be set in increments of 25 $\mu$m (0.001 in.) over a maximum range of 10 mm (0.400 in.).

Modern edgers are equipped with integral recirculating coolant systems. As is true for all diamond tool operations, copious supplies of coolant must be applied to the region where the wheel contacts the work. The most commonly used coolant is water to which a water soluble emulsion has been added.

Peripheral diamond wheels for edgers or rounding machines range from 150-mm (6.0-in.) diameter to 200-mm (8.0-in.) diameter. They are usually 25 mm (1.0 in.) wide. Because a significant amount of material must often be removed during rounding, the diamond is quite coarse at 120-grit (D120). The removed material is turned into swarf, which must be carried away by the coolant to maintain the optimum cutting efficiency of the diamond wheel. A low diamond concentration between 25 and 50 is used for free cutting and to facilitate the removal of the swarf. The diamond wheels rotate at about 3000 rpm, which gives them a peripheral speed between 24 and 30 m/sec (4500 to 6000 SFPM). These wheels are available with rounded edges or with lead tapers on one or both edges. The diamond face or cup wheels used on some machines (see Fig. 5.32) are little more than half as large, and they must run proportionately faster to attain a similar peripheral speeds.

Figure 5.33 shows an edger with a vertical workpiece spindle and horizontal edging face wheel. This type of machine [28, 29] was designed for edging large and heavy workpieces. A separate spindle to accept a diamond face beveling wheel can be adjusted from a horizontal to a vertical position.

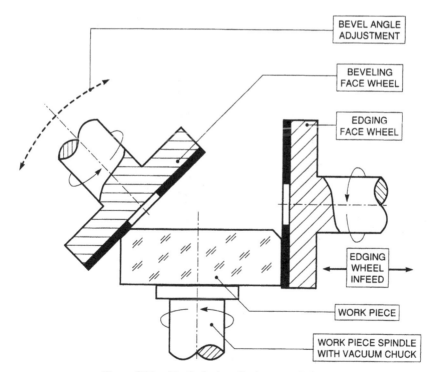

**Figure 5.33.** Vertical edger for large workpieces.

Most machines used for rounding are specifically designed for this type of work. There is, however, a plano surface generator [30] that can be converted to produce round parts to a diameter of up to 50 mm (2.0 in.). This may be a useful alternative for small optical shops which only on occasion require a rounding capability.

Most precision rounding machines or edgers are of the horizontal design. Some of these [31, 32] are found in many American optical shops. Figure 5.34 shows such a rounding machine schematically. These machines operate best when transfer spindles are used. The workpiece is waxed to a chuck which is then mounted on a fixed workpiece spindle. This could be a precision transfer spindle [33] that can be used directly on a precision centering machine. The eccenter drive motor imparts a lateral oscillation to the wheel spindle table. The edge of the diamond wheel moves across the edge of the workpiece as the wheel is fed in. A powered lead screw infeed moves against a preset micrometer stop that shuts off the machine. The infeed accuracy is within 5 μm or (0.0002 in.).

With suitable cams and cam followers, edging machines can be adapted to edge noncircular shapes such as squares, rectangles, ovals, ellipses, and polygons. Some vertical curve generators can also be set up for rounding and coring of large workpieces. Curve generators are discussed in Section 5.2.2.

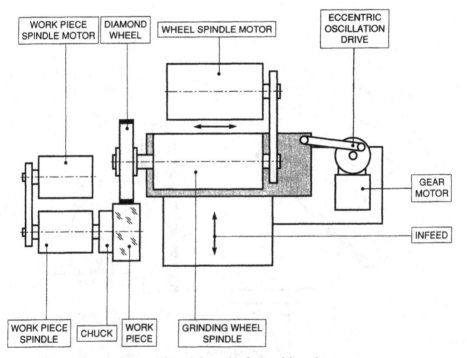

**Figure 5.34.** Schematic of a precision edger.

**References for Section 5.1 (Shaping Operations)**

[1] Bilco.

[2] Sommer & Maca, Chicago, IL.

[3] Nitto Blue, Nitto Optical Co. Inc., Japan.

[4] Karl Süss, Germany, 3″ Automatic Scriber.

[5] Wilhelm Bothner, GmbH, Germany.

[6] Schott Glasswerke, Germany.

[7] Varian/Norton, Lexington, MA.

[8] Meyer Burger, AG, Switzerland.

[9] PR Hoffman, Carlisle, PA.

[10] Kachajian, G. S., and P. Aharonian, "Cutting costs with improved wafering technology," *Microelectric Manufacturing and Testing,* January 1982.

[11] Kachajian G. S., "High precision slicing for solid state devices," *Industrial Diamond Review,* February 1982.

[12] Silicon Technology Corp., Oakland, NJ.

[13] Meyer Burger, Model TS 23.

[14] South Bay Technology, Inc., Temple City, CA.

[15] Laser Technology, Inc., North Hollywood, CA.

[16] Leica, Lohnberg/Lahn, Germany.

[17] Semitron Limited, Montvale, NJ.

[18] Meyer Burger, Model TS-3.

[19] Industrial Tools, Inc. (ITI), Ojai, CA.

[20] Disco Abrasive Systems, Japan.

[21] Micro-Automation, Inc. Sunnyvale, CA.

[22] Tempress, Inc., Los Gatos, CA.

[23] Die Separation, P. S. Berggraf, Semiconductor International, July/August 1980.

[24] DIN 58 744.

[25] DIN 848.

[26] Meyer Burger, Boring Automat, Model ABA-2.

[27] Siltec, Menlo Park, CA, Model 580.

[28] R. Howard Strasbaugh, Inc, San Luis Obispo, CA.

[29] DAMA Darmstadt, Germany, Model RMD-500.

[30] Wilhelm Loh, Wetzlar, Germany, Universal Surface Generator.

[31] Strasbaugh, Model 7H.

[32] Rogers & Clarke Manufacturing Co. Rockford, IL, Model E-150.

[33] Georg Müller Kugellagerfabrik KG, Germany, GNM Spindle.

## 5.2. MILLING OPERATIONS

The machining of optical materials with diamond tools for preparing surfaces for grinding and polishing is called *milling*. This is a collective term that is used to describe plano milling (Blanchard grinding), curve generating, and more recently also the fine-milling done prior to high-speed polishing.

Fine-milling operations have become important for the volume fabrication of optics. To a large degree they have replaced the grinding with loose abrasives. Although more expensive tools and machines are required for fine-milling, these operations can be more economical than abrasive grinding because of shorter work cycles and because several machines can be operated by the same person.

Depending on the type of milling done, only a limited amount of diamond is used up during these operations. As a rule of thumb, about 3000 to 4000 cm$^3$ (180 to 250 in.$^3$) are ground up by one carat of diamond. During an optimal milling cycle, 0.2 to 0.6 cm$^3$ (0.01 to 0.07 in.$^3$)) of glass is ground away each minute. This optimum abrasive potential is almost never fully exploited because of the high grinding forces that would be required. If these forces become too high, the last part of the workpiece would not be able to resist the pressure where the tool exits the surface and it would break out. This tendency can be reduced, however, by rounding the cutting edges of the diamond tools (Fig. 5.35).

### 5.2.1. Surface Mills

Surface mills or surface grinders that are based on the Blanchard configuration have found widespread application in optical shops for the generation of plano surfaces on individual parts and on blocks of parts.

One of the smallest of these [34] can be found in virtually any optical shop. It has a 400-mm (16-in.) diameter electromagnetic workpiece table and it uses a 280-mm (11-in.) diameter abrasive or diamond face wheel. This machine, and others like it, generate or grind plano surfaces on a wide variety of optical materials. The workpieces can be single parts or numerous smaller ones that are held on the workpiece table either individually with suitable hold down fixtures or wax blocked to a common blocking tool. The massive contruction sets this type of machine apart from other similar machines that have become available in the recent past. The rigidity of the machine ensures repeatable accuracy in the low micro inch range.

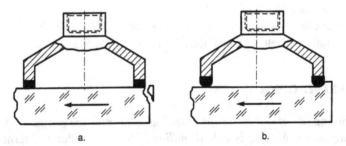

**Figure 5.35.** Plano milling with diamond face wheels.

The loaded workpiece table is manually moved underneath the raised up diamond wheel by a handwheel or a power feed. While the table rotates, the diamond tool spindle is then lowered slowly until the wheel just begins to contact the work. After contact has been made, the diamond tool spindle is switched on and an aqueous coolant is supplied by a submerged pump from the base of the machine and liberally applied to the wheel and workpieces at a rate of 75 liter/min (16 GPM). The tool spindle is then automatically lowered at a preselected feed rate to remove the desired amount of material. A preset limit switch is triggered to shut off the downfeed when the correct thickness has been reached.

Best surface finishes are achieved when the diamond wheel is permitted to dwell (spark out) for several revolutions of the workpiece table; the wheel is then manually raised very slowly by a few mils. After this disengagement the powered head can be activated to return the tool spindle to the load position.

The squareness of the massive vertical tool spindle column relative to the workpiece table determines the flatness of the milled surface. This alignment must be periodically checked and corrected if necessary by adjusting one or two of the three leveling screws. Figure 5.36 schematically shows a top view of the machine.

By slightly tilting or biasing the tool spindle a small amount, the milled surface can be made to be a little concave or convex. It is often desirable to grind a block from the outside in, and a slight concave bias would help do

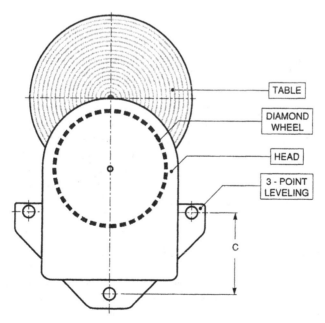

**Figure 5.36.**   Three-point head leveling feature on Blanchard.

that by raising or lowering the rear screw. For instance, the amount of rear screw height adjustment on a No. 11 Blanchard to effect a deliberate amount of concavity on the block can be calculated by the following approximation:

$$\Delta = \left(\frac{4sDc}{d^2}\right), \tag{5.8}$$

where

$s$ = desired sag over the block diameter,

$D$ = diamond wheel diameter (typically 10 or 11 in.),

$d$ = block diameter,

$c$ = axial distance from the rear screw to the axis that connects the other two screws (see Fig. 5.36),

$c$ = 17 in. for a No. 11 Blanchard,

$c$ = 23.25 in. for a No. 22 Blanchard.

### Example

By how much must the rear screw be raised on a No. 11 to produce a concave sag of 0.0001 in. (about 8 fringes) over a 12-in. diameter block?

$$\Delta = \frac{4(0.0001)(10)(17)}{(12)^2}$$

$$= 0.0004.$$

A 0.0004-in. adjustment of the rear screw is required to effect the desired concavity on the block.

The magnetic workpiece table must be kept very flat and perfectly square to the wheel spindle axis, unless of course a deliberate tilt has been introduced. The abrasive wheel or even the diamond wheel of the mill can be used to grind the table flat and square to the tool spindle. This is the usual procedure, but only minimal grinding should be done with a diamond wheel to avoid loading it up with metal removed from the table surface. Copious amounts of coolant must always be supplied to the contact area between the tool and table.

Optical parts, which are almost always made of nonmagnetic materials, are generally wax blocked to a magnetic tool plate or to a nonmagnetic Pyrex blocking body held on the table by magnetic wedges. Stacked square or rectangular parts are usually waxed into metal jigs that are then magnetically held to the table. Blocks of waxed parts should have an open center (i.e., no parts in center of block) to avoid rough surfaces caused by the minimal circular velocity near the center of the table. This problem can also be overcome by mounting the block slightly off-center.

A typical surface mill [34] has the following operating limits:

The 250- or 275-mm (10- or 11-in.) diameter diamond ring face wheel (see Chapter 4) rotates at 1200 rpm to produce a peripheral speed that ranges from 15 to 18 m/sec or 3000 to 3500 SFPM.

The 20-in. diameter workpiece table, also called the *chuck,* can be set to rotate at speeds of 15, 24, 41, and 64 rpm. The different speeds are obtained from a variable speed gear reducer.

The vertical downfeed of the tool head that brings the diamond tool into contact with the work can be activated either manually or by a powered feed. In the power mode the downfeed is infinitely variable between 0.075 and 1.75 mm/min (0.003 and 0.070 in./min). It has also a fine feed range from 2.5 to 250 $\mu$m/min (0.0001 to 0.010 in./min). The effect of the downfeed rate is monitored with a load meter. The tool head downfeed has a built-in stop that shuts off the machine after a downfeed of 1.2 mm (0.048 in.). Upon completion of the dwell cycle, the head rise rate is quite rapid.

Another surface mill that can be found in many optical shops has been specifically designed and built for optical use [35]. It has the same basic operating parameters as just described, except that this machine is useful only for relatively light work and it works best with slow downfeed rates. This limitation is due to the light tool head design. In contrast to the massive head structure of a Blanchard, the vertical tool spindle head of this machine is supported by a 150-mm (6.0-in.) diameter hollow column that can yield to high grinding back pressures. But when used as recommended by the machine manufacturer, it will also produce very good surfaces that will require only a minimum of fine-grinding to remove all mill marks.

There are larger machines [36, 37] of a very similar design and operating characteristics. These have a 900-mm (36-in.) diameter chuck and use a 450-mm (18-in.) diameter diamond wheel. These large machines are occasionally found in optical shops.

Electromagnets exert a powerful force once they have been magnetized. This is especially true, the larger the contact surface is. Therefore tool and jigs must have very flat bearing surfaces that contact the chuck. They must also be free from any damage such as dents and burrs along the edges and in the corners. It is best to put liberal chamfers on the jig edges to minimize this problem. Precision jigs that are designed to transfer a precise angle, parallelism, or dimension to the parts are typically lapped in by free abrasive grinding. They must be treated with particular care.

The estimated life of a 180-grit resinoid-bonded diamond wheel for a standard vertical surface mill [34, 35] when used with an emulsion-fortified aqueous coolant [38] is about 3000 in.[3] for fused silica and more than 4500 in.[3] on optical glass (BK 7).

The coolant must be changed quite regularly to take ground up swarf out of recirculation. Significant amounts of swarf are generated in a regularly used surface mill. For a volume production operation, this can amount to more than 100 in.$^3$ of swarf per day for each machine. Therefore daily coolant changes are recommended for these type of operations.

The reduction in the coolant's effectiveness due to the accumulation of swarf causes the generation of irregular surfaces on blocks and other large surfaces. This can result in grinding, polishing, and thickness problems later on. More important, excessive swarf accumulation in the recirculating coolant will erode the diamond wheel bonds and thus greatly reduce the life of the wheel.

**Prism Milling.** The plano surfaces of prisms are inclined to one another by specific angles. The prism blanks are either rough cut blanks or prism pressings. The prism faces must be milled to near-net shape prior to grinding and polishing. Several special milling machines have been developed specifically for the shaping of prisms [39, 40]. These machines are designated "universal" because they can also be used for rounding and sawing.

The mounting fixtures used for the milling of prisms are mounted on the machine table which is connected to a longitudinal stage. The speed of the longitudinal feed is adjustable by a gear train. One or two mill head units are mounted on the machine for milling the individual prism faces. These units are available with preset or with adjustable angle settings of the tool spindles so that every prism angle can be milled (Fig. 5.37).

Several prism surfaces can be milled at the same time on these machines. To accomplish that, the tool spindles are mounted on opposite sides of the prism stage (Fig. 5.38). If they must be mounted on the same side, an offset must be provided so that the diamond tools cannot interfere with each other. The angle settings are adjustable in order to mill prism angles to about one arc minute accuracy.

When sharp edges on roof prisms have to be milled, the resulting roof edge should be free of chips. To achieve that every time, the direction of the spindle rotation must be chosen so that the cutting edge of the diamond face wheel turns into the prism edge. It then exits the surface where the prism has an obtuse angle (Fig. 5.39). The spindle tilt must be adjusted in such a way that the trailing edge of the diamond tool, which runs toward the sharp prism edge, does not contact the glass. A separation of 0.01 mm (0.0004 in.) is sufficient to do that. If the head tilt adjustment is more than that, then the generated surface would become unacceptably low (concave).

Since considerable amounts of glass must be removed when milling prisms, a correspondingly liberal supply of coolant must be provided. Although water-based emulsion coolants will perform adequately, much better results are obtained in finish and speed when oil coolants are used. The critical machine components are also much less likely to be damaged by corrosion with oil coolants. The use of oil coolants, however, requires the

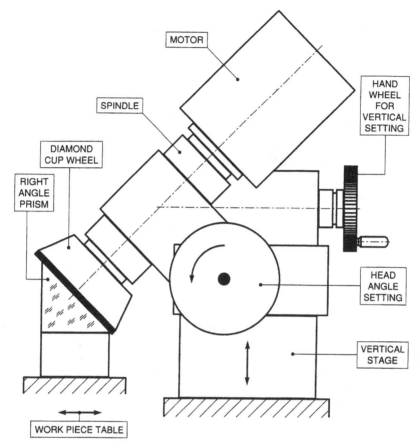

**Figure 5.37.**   Mill head set for milling the hypotenuse of a right-angle prism.

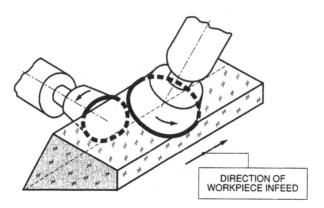

**Figure 5.38.**   Prism profile milling with two mill heads.

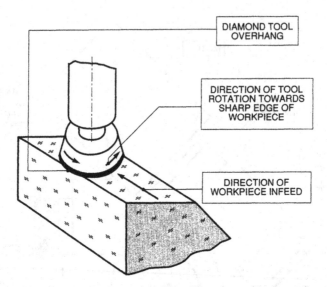

**Figure 5.39.** Optimum setup for milling parts with sharp edges.

installation of expensive ventilation and oil vapor precipitators. To ensure that the coolant is not misdirected, it is contained by a circular brush that surrounds the diamond wheel (Fig. 5.40). Despite this arrangement the oil stream that emerges from the spindle center can squirt into space when irregularly shaped prisms are being milled. In such a case an additional oil nozzle should be installed through which the contact point can be supplied from the outside. The amply supplied oil and the resulting oil spray and oil fog are contained in a vented spray enclosure. The oil flows through a return into a tank where the swarf settles out, and it is pumped back from there for reuse.

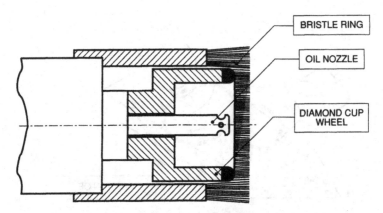

**Figure 5.40.** Circular brush for limiting coolant dispersal.

Because of legitimate environmental and health concerns, the use of oil coolants may be sufficiently restricted in the near future that water-based coolants will become the only safe alternative. This, however, will be at the price of cutting efficiency, surface finish, machine corrosion, and diamond tool life. The price of safety can be substantial.

The prisms are fastened to a fixture that is mounted on the table of the machine. For large prisms, which are typically shaped individually, this fixture can be a simple steel plate. The prism is ground plano with loose abrasives on its largest side and waxed to the steel plate (Fig. 5.41). All other dimensions are then referenced to the waxed-down prism face.

Smaller prisms are preferrably mounted in fixtures. The fixture remains on the machine table, and the prisms are mounted individually and held down with strong spring pressure. A spring-loaded mounting is required because glass cannot be rigidly clamped like metal parts in a vise. The spring pressure must be strong enough not to yield to the grinding pressures that the glass is subjected to. Leaf springs made of steel covered with plastic on the contact surfaces have proved to be the best for mounting prisms.

A sphere generator can also be used to generate plano surfaces [41]. Fixtures that hold prisms can also be milled on such a machine just like plano blocks of lenses. The mounting elements for the prisms are fastened to a plano plate, which can hold a number of them so that many prisms can be processed at the same time. After completing each surface, the prisms are flipped in the fixture for milling the next side. In this manner finished milled prism become available for each cycle of the machine. It is best for optimum efficiency to use two fixtures in rotation. While the prisms in one fixture are remounted, the other prism fixture runs in the machine. This results in minimum downtime for the machine.

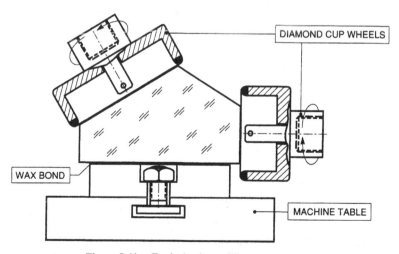

**Figure 5.41.** Typical prism milling arrangement.

## 5.2.2.  Curve Generating

Spherical surfaces are milled, or more specifically generated, on a special machine called a *curve generator*. It consists of two spindles with one of them tiltable with respect to the other. The fixed spindle accepts the workpiece, while the tilt spindle supports the diamond tool.

A curve generator consists in its basic form of an upper spindle that holds the diamond tool and a lower spindle that holds the workpiece. The tool is a diamond cup wheel and the workpiece is either a single surface or a block of parts. The two spindles are inclined at an angle to one another. Since the workpiece spindle is fixed in a vertical arrangement, the tool spindle is tilted relative to the workpiece spindle. This tilt angle in conjunction with the diameter of the diamond cup wheel determines the radius of curvature that is generated on the workpiece. The tool spindle rotates at a high rate to achieve the necessary cutting speed, while the workpiece spindle rotates at a much slower rate. By slowly feeding the workpiece into the rotating diamond wheel, the optician generates a spherical surface. Another method allows the tool to plunge into the workpiece to the final depth of cut, and a slow rotation of the workpiece then generates the spherical surface.

The diamond cup wheel must run over the center of the workpiece and should extend over the edge by at least 1 mm (0.040 in.). The resulting curvature with radius $R$ on the workpiece is determined by the diameter of the diamond cup wheel $d$, the radius of the cutting edge $r$, and the tilt angle $\alpha$ of the tool spindle (Fig. 5.42). The diameter $d$ is the arithmetic mean of the inner and outer diameters of the cup wheel.

Generator cup wheels are described in Section 4.4.4. In the absence of clearly defined standards for diamond tools, the user specifies the tool dimensions to suit a particular need. The German optics industry has found it advantageous to standardize diamond tool dimensions, including those for generator cup wheels [42]. The choice of the cup wheel size is determined by the size of the workpiece (Fig. 5.43). It can also be calculated.

$$d > \frac{2[R \pm r]\sin(\varphi)}{2}, \qquad (5.9)$$

where

$d$ = mean diameter of cup wheel in millimeters,
$R$ = radius of curvature in millimeters,
$r$ = radius of cutting edge (+ for convex, − for concave),
$\varphi$ = effective block angle.

The tilt angle $\alpha$ of the tool spindle is calculated according to Eq. (5.10) for convex (Fig. 5.44) and Eq. (5.11) for concave surfaces (Fig. 5.45):

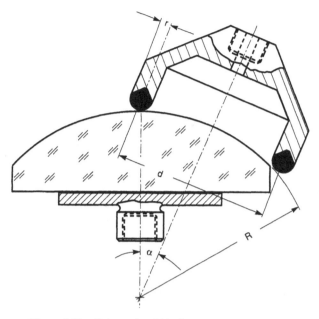

**Figure 5.42.** Schematic of ideal curve generator setup.

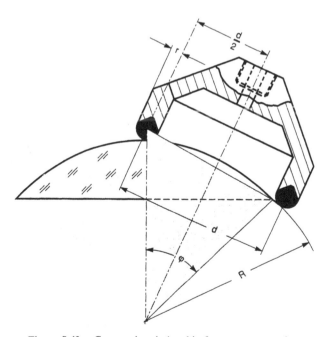

**Figure 5.43.** Geometric relationship for curve generating.

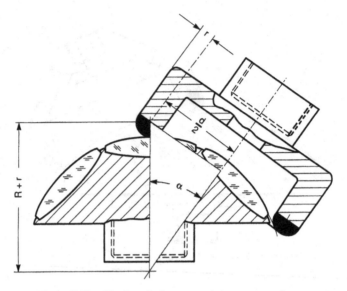

**Figure 5.44.** Head angle for generating convex surfaces.

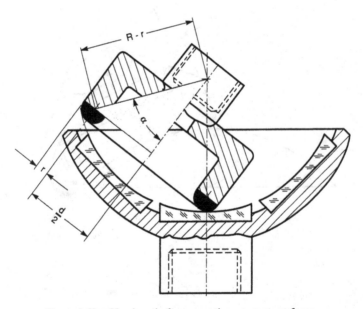

**Figure 5.45.** Head angle for generating concave surfaces.

For convex surfaces

$$\sin \alpha = \frac{d}{2(R + r)}, \tag{5.10}$$

and for concave surfaces

$$\sin \alpha = \frac{d}{2(R - r)}, \tag{5.11}$$

where

$\alpha$ = tilt angle of the tool spindle,
$R$ = radius of curvature,
$r$ = radius of the cutting edge,
$d$ = mean diameter of the cup wheel.

## Example 1

A convex lens block of radius $R = 46$ mm must be generated. The effective block angle is $\varphi = 72°$. Assume that the tool edge radius $r = 2$ mm. This is all the information needed to calculate the required cup wheel diameter per Eq. (5.9):

$$d = 2(46 \text{ mm} + 2 \text{ mm})\sin \left(\frac{72°}{2}\right)$$

$$= 55 \text{ mm.}$$

From the DIN list [42] of standard cup wheels, the one with $d = 56$ mm is chosen. The radius of the cutting edge is specified as $r = 2$ mm. The tilt angle $\alpha$ of the tool spindle is determined by using Eq. (5.10):

$$\sin \alpha = \frac{d}{2(R + r)}$$

$$= \frac{56}{2(46 + 2)} = 0.583.$$

The tilt angle $\alpha$ is then 35.7°.

## Example 2

A concave lens block of radius $r = 52$ mm must be generated. The effective block angle $\varphi$ is 68°. Again, assume the tool edge radius $r$ to be 2 mm.

According to Eq. (5.9) the required cup wheel diameter is

$$d = 2(52 - 2) \sin \left(\frac{68°}{2}\right)$$

$$= 57.4 \text{ mm}.$$

By referring to the DIN listing [42], a cup wheel with a diameter $d = 63$ mm is selected. The listed cutting edge radius is $r = 2$ mm. With this information the tilt angle of the grinding spindle is calculated per Eq. (5.11):

$$\sin \alpha = \frac{63}{2(52 - 2)} = 0.630.$$

This results in a tilt angle of 39.1°.

Many of the modern curve generators operate on a fully automatic pre-programmed sequence. All generators can be set up to generate both convex and concave surfaces either as single surface lens or on lens blocks, which are usually spot tools. While, at least in theory, most generators can be set up to generate plano surfaces, it is best to use a plano surface mill that has been specifically designed for that purpose. Two or three spindle production curve generators have been built in the past, but almost all curve generators in use today are single-spindle units. Some curve generators are specifically designed to serve the needs of the opthalmic industry. The head angle settings for such curve generators are calibrated in diopters.

Consistently high accuracy of curve-generated surfaces both in terms of radius of curvature (form) and surface roughness (finish) is very important for automated lapping on pel-grinders and for subsequent high-speed polishing operations. To do that on a consistent basis, curve generators must be very rugged, repeatably and predictably precise, and operate in a stable manner ideally without any vibration. This requires that the machine base be massive and that all spindles are of the highest quality.

Curve generators have been built with horizontal as well as with vertical workpiece spindles. The machines with horizontal spindles are very rugged and also less expensive than those with vertical spindles. However, large workpieces, such as large spot blocks, are more easily mounted when the spindles are vertically arranged. These two options are shown in Fig. 5.46.

About 75% of all curve generators have vertical spindles. The remainder have the spindles in a horizontal arrangement. Although it is often assumed that horizontal generators are designed for small workpieces and the vertical variety only for larger workpieces, the advertised workpiece ranges do not support that conclusion. Both spindle arrangements are used for very small lenses down to 3 mm (0.120 in.) in diameter. The largest workpiece capacity generators, however, are limited to the vertical spindle arrangement.

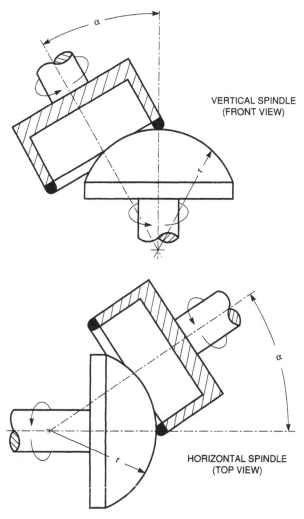

**VERTICAL SPINDLE
(FRONT VIEW)**

**HORIZONTAL SPINDLE
(TOP VIEW)**

**Figure 5.46.**    Basic spindle arrangements for curve generators.

More than half of all available models are designed for generating lenses or lens blocks of 150-mm (6.0-in.) diameter or less. Almost all of the commonly found curve generators fall into this category. The remainder are for special applications up to a workpiece capacity of 900-mm (36.0-in.) diameter.

**Tool Spindle.** The upper tool spindle can be tilted relative to the vertical workpiece spindle by means of a worm gear (Fig. 5.47). This tilt angle is called the *head angle;* it is typically no greater than 50°. Most curve genera-

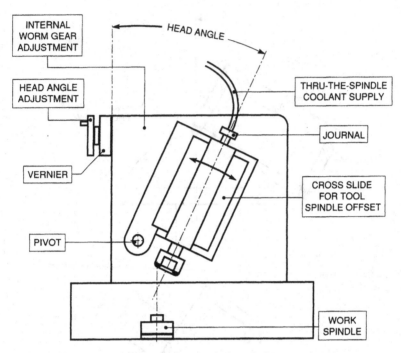

**Figure 5.47.** Schematic for a typical vertical curve generator.

tors can be adjusted to a head angle of 45°, while for others this range is extended to 60°. A 45° head angle will theoretically generate a hemispherical surface. Larger angles are needed to generate convex hyperhemispheres.

Modern machines have a digital head angle readout to an accuracy of 0.1° or 6 arc minutes. Some more advanced machines have motorized head angle adjustment and a digital angle encoder readout to a 0.01° or 36 arc sec accuracy. The spindle arrangement can be either vertical or horizontal. Vertical tool spindles can be tilted either to the right or the left of the operator, or they can be tilted away from the operator. The tool spindle is always tilted away from the operator for horizontally aligned tool spindles.

Not only can the tool axis be tilted relative to the fixed workpiece spindle, it can also be laterally offset to ensure that the cutting edge of the diamond tool is positioned exactly over the axis of rotation of the workpiece spindle. This adjustment usually extends to both sides of the axis, ranging from 75 to 125 mm (3.0 to 5.0 in.) per side. Some generators have provisions for small axial adjustments to compensate for diamond wheel wear and to allow very fine radius adjustments. One manufacturer [43] utilizes an eccentric tool spindle design that permits a small 0.3-mm (0.012-in.) adjustment to bring tool spindle and work spindle axes into one plane (Fig. 5.48). While most generators feed the workpiece into the rotating diamond cup wheel, a few

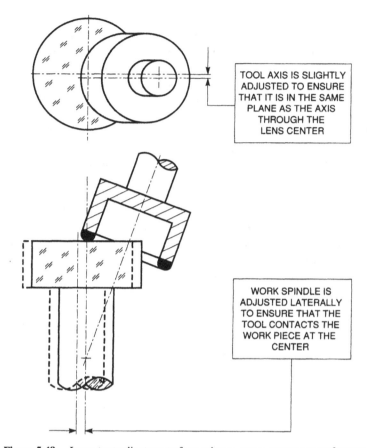

**Figure 5.48.** Important adjustments for optimum curve generator performance.

designs depend on the downfeed of the tool spindle into the slowly rotating but otherwise fixed workpiece.

**Coolants for Curve Generators.** An efficient coolant supply is critical for any high-speed diamond-machining process. Therefore all curve generators supply sufficient amounts of coolant through the hollow bore of precision spindles [44] with additional feeds from above and the sides. The coolant system consists of a pump, coolant supply lines, spray nozzles, and a reservoir, often called a *sump*.

Low viscosity oils [45, 46] are preferred by many optics manufacturers as coolants. These oils effectively cool the diamond wheel and the workpiece, and they efficiently carry off abraded glass material. Other shops prefer to use water-based coolants to which emulsions have been added. These water emulsions have the advantage that on machines with fast-running diamond wheels, no oil fog but only water fog is generated. It is also easier to dispose of swarf residue because the entrapped emulsion traces are biodegradable. Several emulsions have proved to be useful. The abrasive efficiency of the diamond wheels drops by about one-third as compared to oil coolants. That is particularly noticeable on workpieces with large surface area.

The coolant is fed through the hollow tool spindle. A brass tubing is passed through the spindle for this purpose. The tubing is equipped with a nozzle at its end that sprays the oil sideways (Fig. 5.49). After cooling the abrasion process, the coolant runs off into a reservoir. The suspended coarse glass swarf is filtered out with simple paper filters before the coolant returns to the reservoir, which has typically three settling compartments. The coarse swarf particles that the filters did not catch settle out in the first compartment. After that the coolant runs over a weir into the second compartment where the finer swarf settles out and the coolant becomes much cleaner. From there it continues over a second weir into the third compartment from which the coolant is pumped up for reuse in the machine. Additional coolant is frequently supplied from the outside through a movable metal feed tube to make up for unavoidable losses.

According to a proven rule of thumb, the amount of coolant in liters that should be supplied per minute is equal to the diameter of the cup wheel as measured in millimeters. This requires that the volume of a cooling sump is circulated once every minute. The abraded swarf can only settle out partially under these conditions. That is why central cooling systems are preferred in which larger quantities of coolant can circulate.

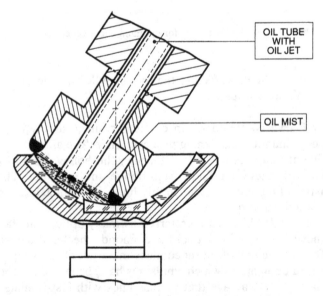

**Figure 5.49.**   Through the spindle coolant supply.

When oil coolants are used, the swarf that accumulates during the clean-
out of the coolant sumps contains a quantity of oil although the residue may
appear relatively dry. For this reason the swarf cannot be discarded without
prior purification.

**Cup Wheels.** Standardized diamond cup wheels [42] are tapered for mounting them on the tapered tool spindle, as shown in Fig. 5.50. To facilitate that,
the corresponding tool spindles are standardized [47]. The tapered spindle
terminates in a fine thread that holds the cup wheel.

There are many deviations from this standard. For instance, the cup
wheels on many curve generators are held to the tool spindle by other
means. A holddown nut holds the wheel firmly mounted on the tool spindle
for some of the more popular curve generators [48].

When blocks of small lenses up to about 20-mm (0.800-in.) diameter must
be curve generated, cup wheels with a fine diamond size of about 70 $\mu$m
(D70) can be used. Coarser sizes such as 90 $\mu$m (D90) or 110 $\mu$m (D110) are
chosen if the lenses are larger. For large uninterrrupted surfaces, as would
be encountered when milling large mirrors, wheels with grain sizes of up to
22 mm (D220) are used. A diamond content of 50 to 75 concentration is
recommended. Bronze bonds have proved to be the best choice.

Accurate data on the best use of diamond sizes, concentration, and bond
materials are not available. The optimum parameters are determined by trial
and error during which experience points the way. To use an available
diamond tool universally, the optimum abrasion potential may have to be
sacrificed and longer generating or milling times have to be expected.

For optimum cutting of diamond tools, a fairly narrow range of circumferential speeds must be maintained. This means that the tool spindle speeds
must be matched to tool diameters. The smaller the tool diameter, the higher
the necessary revolutions per minute. Circumferential speeds are expressed
either as meter per second (m/sec) or surface feet per minute (SFPM). One
meter per second equals approximately 200 SFPM.

The peripheral speeds for tool spindles based on possible machine speed
settings and diamond tool size capacities for commercially available curve

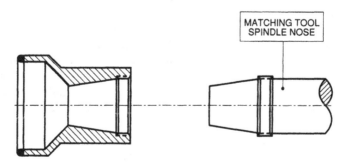

**Figure 5.50.** Diamond cup wheel with conical seat.

generators range from a high of 32.5 m/sec (6500 SFPM) to a low of 13 m/sec (2600 SFPM) with an average of 21 m/sec (4100 SFPM). The ideal cutting speed for diamond tools used on optical glass is 27.5 m/sec or 5500 SFPM . These limits should be carefully considered, depending on the intended application, when selecting a curve generator. Otherwise, expensive modifications may have to be made on the machine after purchase.

The tool spindle speed is adjustable on some curve generators by a stepped pulley drive to compensate for different cup wheel diameters. But even for machines with fixed spindle speeds, the deviation of the circumferential speed of the cup wheel remains within acceptable limits, since large cup wheels are typically used on large machines and small cup wheels on small machines. The tool spindles on the small machines run at higher rpm than on the larger machines so that the circumferential speed of the cup wheels remains approximately the same.

**Workpiece Spindle.** Two different infeed approaches have been developed for the workpiece spindle on curve generators. For the first approach the infeed is primarily controlled by a pneumatic or hydraulic cylinder. The infeed is uniformly distributed over the generating cycle, while the workpiece makes several hundred revolutions. Therefore the material is removed in thin spiraling layers. Such an infeed system can be equipped with one cylinder for rapid feed and another for precision feed (Fig. 5.51). The workpiece is brought into initial contact with the rapid feed. The precision feed acts during the generating phase, and it runs against an end stop that terminates the infeed. This triggers a timer that allows the wheel to dwell a preset time and gives the machine a predetermined period to return to the starting position. Smooth surfaces can be generated in this manner while maintaining good accuracy.

The second approach limits the workpiece during the generating step to only three to ten revolutions. With this approach the diamond cup wheel is immediately advanced for the full depth of cut, which is about 1 mm (0.040 in.) deep, so it is occasionally referred to as the plunge method. The infeed of the tool spindle is controlled by a cam that runs against an extension spring which in turn attempts to pull the workpiece away from the diamond cup wheel (Fig. 5.52).

Initially the cam advances the workpiece, which is mounted on the spindle until it contacts the diamond cup wheel. Immediately following this, the full cutting depth is realized as the diamond tool plunges to the preset depth. The spindle remains in this position for at least one revolution so that the generated surface will not develop noticeable steps. Finally, the cam releases the workpiece spindle which is retracted to the starting position by the extension spring.

In the more common mode the workpiece mounted on the spindle rises in a smooth motion by means of a hydraulic-assisted pneumatic lift from a retracted load position into the tilted and rotating diamond cup wheel while the workpiece spindle slowly rotates. The workpiece spindle moves upward

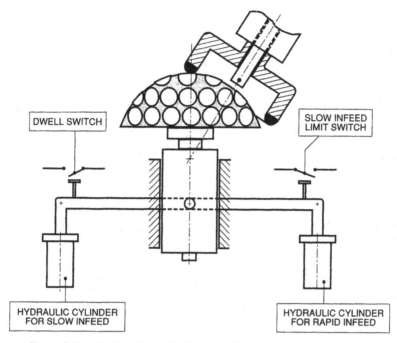

**Figure 5.51.** Work spindle infeed with rapid and slow feet rate control.

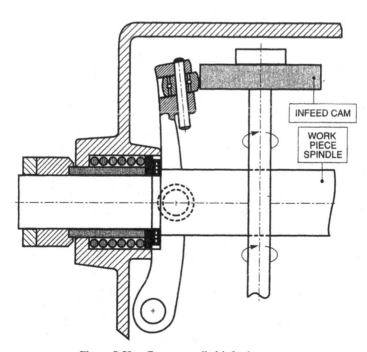

**Figure 5.52.** Cam-controlled infeed system.

for vertical spindles, forward or lateral for horizontal spindles, against a solid preset, micrometer-controlled stop for precise control of center thickness of the generated parts. Modern curve generators have digital spindle height readout to monitor the precise up or forward position of the workpiece spindle. Production machines have a two-stage workpiece spindle feed rate with about 80% of the spindle travel in a fast feed mode to bring the workpiece from the load position to close proximity of the diamond wheel. The remaining distance of the total spindle travel is fed in at a precisely controlled slow feed rate for the actual surface generation. The feed rates are infinitely variable from 0.2 mm/min (0.008 in./min) to 9 mm/min. (0.360 in./min.)

For plunge mode curve generators the horizontal workpiece spindle is cam driven into the rotating diamond cup wheel against a preset fixed stop for center thickness control to a repeatable accuracy of less than 0.025 mm (0.001 in.). With careful operational procedures, thickness tolerances can be held to 0.01 mm (0.0004-in.). The feed is smooth thanks to the hydraulic-assisted pneumatic infeed mechanism which is essential to produce chip-free edges and surfaces free from deep generating marks. Slow feeds to the finished thickness setting are used as the surface is generated. Machines can usually be set for a timed dwell (spark-out) ranging from a few seconds up to several minutes, depending on application and size. When the preset infeed limit has been reached, the workpiece retracts to the load/unload position.

In addition to the two infeed modes there are two different ways how the tool and workpiece can be engaged. Either the diamond tool feeds downward into the workpiece, or the workpiece rises into the fixed tool. These two infeed options are shown in Fig. 5.53. Both options can be used with vertical or horizontal spindle arrangements. Only one machine was identified as feeding the tool spindle downward into the workpiece [49]. Several other specialty curve generators [50, 51] operate on a fully automated sequence during which both spindles are moved toward each other at preprogrammed feed rates and distances.

The workpiece spindle rotational speed depends on three modes of operation. The spindle feeds in at a preset constant rate with variable spindle rotation ranging from moderate to high speeds. The spindle feeds in at a preset constant rate with the spindle speed at a fixed rate. The spindle feeds in with the plunge mode with preset spindle speeds ranging from a one or two revolutions per cycle to several per minute. Some of the more advanced machines permit changing the workpiece spindle rotation during the generating cycle. This feature can be programmed.

In general workpiece spindle speeds decrease as workpiece sizes increase. The diamond wheels can only abrade a certain volume of material per unit time, depending on diamond wheel design and part material. Figure 5.54 shows fitted maximum and minimum work spindle speeds in rpm as function of workpiece diameter for the constant infeed mode (Fig 5.54a) and the plunge mode (Fig. 5.54b). The numbers in the graphs are average values

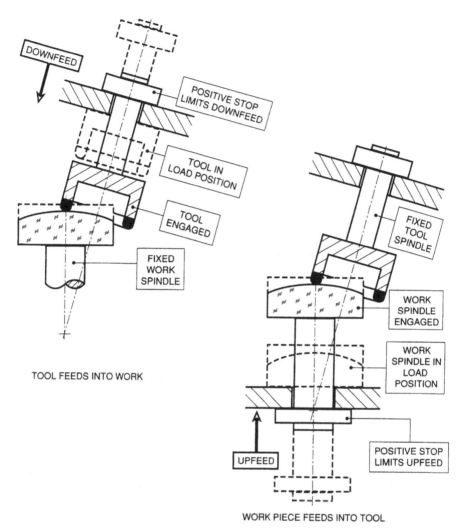

**Figure 5.53.** Basic tool engagement mechanics for curve generators.

that have been fitted for clarity. There is considerable spread in the data. The workpiece diameters are represented in metric and English measure.

Almost all curve generators depend on hydraulic-assisted pneumatic cylinders to smoothly lift up or push horizontally the workpiece spindle into the rotating diamond cup wheel. Compressed air must be available at the machine for the pneumatic lift. It can also be tapped for possible compressed air chucking of parts. Some machines are also equipped with hollow workspindles to permit vacuum chucking of single lenses. A vacuum system for this application consists of a vacuum pump, a ballast tank, and sufficiently flexible but noncollapsible vacuum hoses.

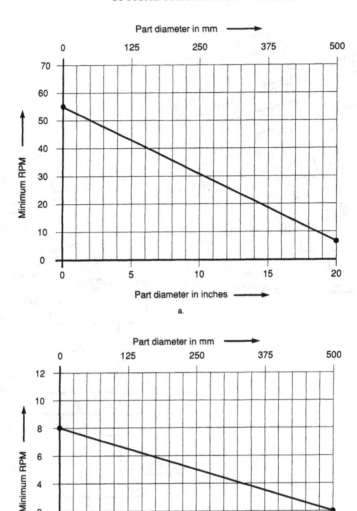

**Figure 5.54.** (a) Workpiece spindle speeds for curve generators: constant rate infeed mode. (b) Workpiece spindle speeds for curve generators: plunge infeed mode.

**Automatic Load and Unload Features.** For speeding up the curve genera-tion of lenses that have to be done in large numbers, automatic caroussel load features have proved to be very effective. A circular part carrier plate is indexed into position for part loading or unloading. Depending on the part diameter, up to 48 lenses can be held by one carrier plate. The parts are

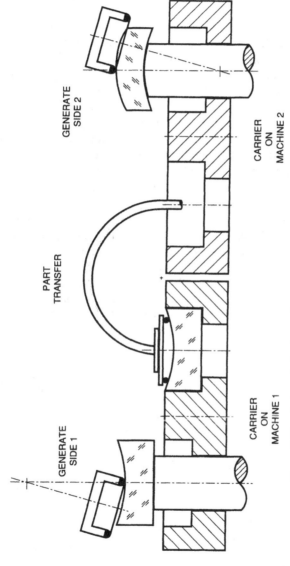

**Figure 5.55.** Automatic load-transfer-unload system for production curve generators.

loaded and unloaded via a hydraulically operated vacuum pickup arm. A load/unload sequence typically takes only 5 to 6 seconds. Two machines can be paired to permit the transfer of parts that have been generated on side 1 on the first machine for generating side 2 on the second machine. Figure 5.55 shows this schematically.

When a large number of spot blocks or single lenses with the same radius must be generated, the use of a fully automatic curve generator [52] is the most economical solution. The workpieces are loaded onto a caroussel on such machines. The ascending workpiece spindle lifts one workpiece from the caroussel and positions it into the starting mode. The workpiece is held on the spindle with a clamp chuck or by pneumatic means. Depending on the type of machine, the workpiece is fed into the cup wheel or the cup wheel sinks into the workpiece. The diamond cup wheel grinds until a stop ends the infeed. Upon rundown of the generating step, the spindle is lowered, and the generated workpiece is returned to the caroussel. A switch mechanism is activated that advances the caroussel by one station, which then brings a new workpiece into position.

There are several Japanese production generators that are equipped with slanted automatic loading magazines that will hold up to 100 lens blanks. Generated lenses will then be automatically loaded into a finished part carrier. Some curve generators are equipped with auxiliary beveling spindles. These beveling heads automatically cut bevels or face flats (Fig. 5.56). A few of the larger generators that have a tool spindle that can be set up in a vertical position can be used for other operations, as shown in Fig. 5.57.

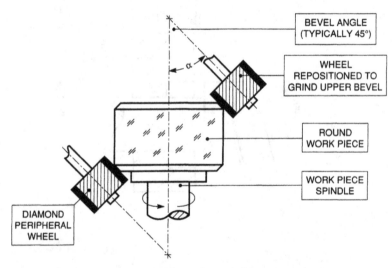

**Figure 5.56.**  Beveling of round parts with a curve generator.

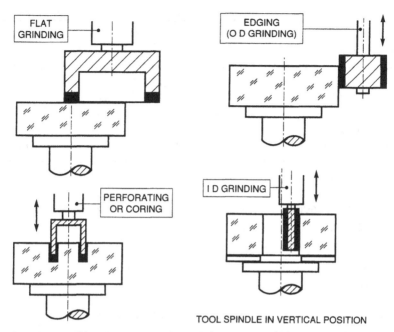

**Figure 5.57.** Alternate setups are possible with some curve generators.

**CNC Curve Generators.** CNC curve generators made their debut during the late 1980s. To date, there are two U.S. [53, 54] and one German manufacturer [55] with a marketable product. Other optical fab machine builders will undoubtedly follow suit. The first CNC curve generators [53] were introduced in 1989. These were two standard curve generators that were converted to CNC operation. The generators [56] were modified by the addition of dc servomotors and optical encoders for each axis. The position of either axis can be independently monitored and controlled by a programmable CNC controller. Another manufacturer [54] quickly followed in 1990 with a similarly modified CNC version of their standard curve generator. In the meantime a German company [55] has also come out with a CNC curve generator for automatic generation of single lenses or lens blocks ranging in diameter from 100 to 350 mm (4 to 14 in.).

These machines are available in either a three-axis or four-axis mode. All current CNC generators are of the vertical configuration. Some of the existing curve generators can be retrofitted for CNC operation by adding a separate CNC controller module in addition to the dc motors and rotary encoders.

The three-axis machine has one axis for laterally offsetting the tool spindle so that the edge of the diamond cup wheel can be precisely positioned over the center of the workpiece axis, a tilt axis that is used to set the head angle needed to generate a specific radius of curvature with the chosen diamond cup wheel, and the vertical upfeed of the workpiece that defines the center thickness of the generated parts. The four-axis version, in addition, allows for the positioning of the tool spindle along its axis to compensate for tool wear and to make adjustments for changing cup wheel dimensions. The three-axis CNC generators costs more than 50% more and the four-axis machines nearly two times what the equivalent standard curve generators would cost today.

The CNC machines can be programmed to automatically set the head angle and position the tool edge based on the following data which are input by the operator:

- The desired radius of curvature either convex or concave
- The selected cup wheel ID and OD
- The required center thickness
- The curve generation cycle (speeds, feeds, and dwell)
- The ring diameter of the integral spherometer

The programming is done by a simple set of instructions in the popular menu format that relies on commonly used generator terminology so that any optician familiar with curve generator setup but unfamiliar with machine programming can quickly learn how to perform this task.

After entering the data, the controller calculates the correct head angle setting and the required tool edge offset. A closed loop servo control system sets the head angle and tool position based on the calculation. The actual positions are monitored by the operator on a video monitor. The accuracy of these settings is determined after the first surface is generated by making a radius measurement with a built-in digital spherometer that has first been zeroed against a master radius. The CNC controller then detects any departures from the programmed positions, recalculates the settings, and automatically makes the necessary corrections in head angle and tool edge position if instructed to do so. Any differences can be the result of inaccurate input data, or they can be the result of wear of the diamond cup wheel's cutting edge. This feature gives the machine a unique self-correcting capability that can lead to a very consistent output.

**Micro Generator.** One unique curve generator [58] needs to be mentioned here because it has no rival. It has been specifically designed for generating small diameter lenses with steep curves from round plane-parallel blanks. This programmable, fully automatic machine can handle blanks up to 15-mm diameter (0.6-in.), and it can generate hyperhemispheres up to 240°. Because of the small diamond tools used, the tool spindle is air driven and rotates at more than 40,000 rpm.

The horizontally arranged spindles are first inclined to one another by the head angle, which was first calculated based on the radius of curvature and the effective diameter of the tool. Both the workpiece spindle and the diamond tool spindle are initially in a retracted position to facilitate loading. The lens blanks are held on the workpiece spindle by either a vacuum chuck or a mechanical clamping device. The rotating workpiece spindle moves forward against a preset stop. The high-speed tool spindle advances rapidly at first, then proceeds at a slower, predetermined feed rate to generate the radius on the lens blank. The infeed of the tool spindle stops when a preset micrometer position that determines the center thickness of the generated part is reached. It is advisable to allow a dwell cycle of at least 30 seconds to smooth out any surface roughness caused by spindle vibrations. When the dwell cycle is completed, both spindles retract back to their load position.

**Use of a Curve Generator.** The setup of a standard curve generator involves the following sequence of steps: The tool spindle is initially set to the previously calculated head angle. Since the diamond wheels tend to wear at the contact edge, this setting is typically inaccurate, and a setup tool that has the dimensions of the generated spot block or that of a single lens must be used to make the final adjustment. The setup tool is mounted on the workpiece spindle for this purpose instead of the workpiece, and the spindle is moved to the farthest point of the infeed. By loosening a lock screw on the machine, the diamond cup wheel can be brought into contact with the setup tool, and the head angle can be corrected to make the cutting edge of the diamond cup wheel contact the setup tool all around. The lock screw is then tightened in this position, the setup tool is removed, and the machine is returned to the starting position. The diamond tool must be retracted during these steps so that it cannot interfere with a newly mounted spot block or single lens. The coolant oil control valve is closed during this setup phase. After completion of the adjustment, it must be opened again, and the workpiece is mounted on the spindle. The support of the machine is moved with a lever to the end stop, and the machine is started by pushing a button. The setup data and the setup tool, and any other setup aids used, are recorded in a log book or an index file.

**Inspection of Generated Surfaces.** Generated parts are inspected for dimensions, surface shape, and surface roughness. Edge chips are also noted at that time. For single lenses the dimensional accuracy is limited to center thickness. This is measured with a dial indicator setup (Fig. 5.58). An overage of 0.1 mm (0.004 in.) per surface relative to the finished center thickness must remain after curve generation. This means that the generated lens must be 0.2 mm (0.008 in.) thicker in the center than the finished lens. The finished lens thickness is in most cases the maximum allowable center thickness tolerance.

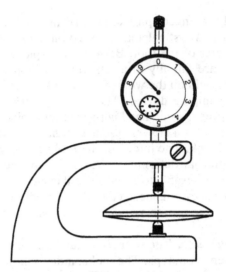

**Figure 5.58.** Typical center thickness gauge.

The thickness is measured indirectly on generated spot blocks. Corresponding to the design of the spot block, the on-the-block dimension is predetermined as measured between lens surface and block surface. This dimension serves as indirect measure of the lens center thickness. It is measured with a depth gauge (Fig. 5.59). This dimension must be very accurately determined after generating the second side.

The radius of curvature is verified with a handspherometer that has been previously calibrated on a radius standard (Fig. 5.60). In the past the generated lens was tested for correct radius by rubbing it in the next following grinding tool. A correct radius would have rubbing marks extending from the edge to about half of the surface toward the center. This test method is still useful when only a few parts have to be made.

The depth of generating marks serves as a measure of surface roughness of the generated surface. This characteristic is generally not measured, however, but is visually judged, relative to a comparison standard. The roughness is primarily determined by the diamond grain size used. Changes in that can occur when the highly utilized tool spindle of the machine has become defective. This can set up vibrations that increase the depth of generating marks. An insufficient coolant supply can also cause rough surfaces.

### 5.2.3. Milling with Diamond Pellets

Blanchard grinding or curve generating, which can be described as coarse milling operations, use diamond face wheels or diamond cup wheels, respectively, to establish a geometrically correct surface for further processing.

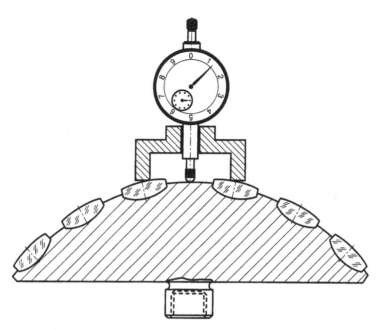

**Figure 5.59.** Lens thickness measurement on spot blocks.

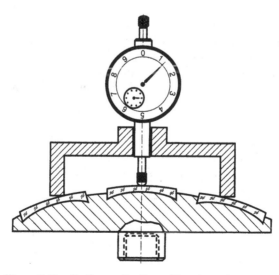

**Figure 5.60.** Radius verification with a ring spherometer.

The subsequent operations of premilling and fine milling use sphere laps that have been studded with small diamond pellets that are now being used effectively as a replacement for the once widely used loose abrasive grinding. Premilling is an intermediate step similar to the traditional pregrinding step to prepare a generated surface for fine milling.

**Premilling.** Single lenses, for instance, eyeglass lenses, are frequently premilled with diamond pellets after they are generated. The tool is made up of metal-bonded pellets with diamond grain sizes 90 $\mu$m (D90) down to 70 $\mu$m (D70). The diamond concentration is low ranging from 35 to 50 (C35 to C50). The surfaces are premilled on the same machine with which they are later fine milled. When working the second side with this method, the lens thickness cannot be accurately controlled.

**Fine Milling.** The precision working of optical glasses with metal-bonded diamond is called *fine milling* or *pellet lapping*. This is a relatively new process that has proved to be very cost-effective under certain operating conditions. Fine milling replaces fine-grinding with loose abrasives. The remaining pores in the fine-milled glass surface are only 2 to 3 $\mu$m deep, so they tend to polish out faster than those that were fine-ground with loose abrasives. Fine milling requires expensive preparations, and the method should be preferred over loose abrasive fine-grinding only when the diamond tool can be used at least for 1200 runs.

**Fine-milling Tools.** The fine-milling tool, often referred to as *pellet lap*, is covered with metal-bonded diamond buttons. The word pellet has become the accepted term in the shops. The shape and size of the pellets are standardized [59], as listed in Table 5.2, although in many cases the pellet dimensions and composition is often decided by supplier and user.

The pellets are made up of diamond powder that is captured in a bronze bond. They are available with a variety of diamond grain sizes. Small tools up to about 20 cm$^2$ (1.2 in.$^2$) are mounted with pellets of diamond grain size

Table 5.2.  **Dimensions of Metal-Bonded Diamond Pellets per DIN 58 745**

| Diameter d | | Thickness t | |
|---|---|---|---|
| mm | inch | mm | inch |
| 4 | .160 | 2 | .080 |
| 5 | .200 | 2 | .080 |
| 6 | .240 | 3 | .120 |
| 8 | .320 | 3 | .120 |
| 10 | .400 | 3 | .120 |

D7 to D10 (7 to 10 $\mu$m). For larger tools pellets with D15 to D25 (15 to 25 $\mu$m) are preferred. The most effective diamond concentrations lie between 35 and 50 (C35 to C50).

Small diameter pellets are used for tools with short radii, while the larger ones are used for longer radii and for plano tools. Pellets with a diameter in excess of 10 mm (0.400 in.) are rarely used. If the pellets were larger, then the abraded glass cannot be completely flushed away, which can lead to glazing of the pellet surfaces. When that happens, the pellet will not remain in contact with the workpiece, and the resulting surface will not be smooth.

The pellets are mounted with a hard curing adhesive on properly ground in spherical or plano tools. The adhesive should be thermosetting and remain quite viscous throughout the bonding cycle. Two-component epoxies that set up in about 1 hour at 200°C (392°F) are recommended as being well suited for this task. The pellets are first slightly deformed between two dies so that they will conform to the curvature of the tool (Fig. 5.61). The bond material should not be stressed too much during this operation or it will become brittle.

The radius of the prepared concave or convex tool differs from the final radius by the thickness of the pellets. The pellets should be uniformly distributed to avoid tool imbalance. They should be set closer together near the edge of the tool as compared to those in the center. The distance between pellets should be 25% to 33% of the pellet diameter at the tool edge and 50% to 67% near the tool center. It has been occasionally recommended to arrange the pellets not in rings but rather in a spiral pattern to avoid zones with diminished abrasion effectiveness. However, for a sufficiently dense pellet pattern, the oscillation of the workpiece will tend to average out such zones.

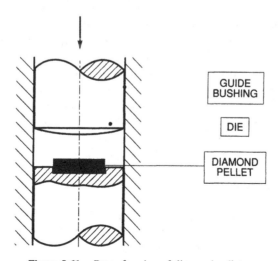

**Figure 5.61.** Press forming of diamond pellets.

The diameter of the fine-milling tools must be 20% to 30% larger than the workpieces. For single surfaces, such as eyeglass lenses, an even larger pellet tool diameter is often selected. This diameter ratio, in conjunction with the workpiece oscillation over the tool, should ensure that the curvature of the workpiece remains unchanged for a long time.

The fine-milling tool must have an axial hole for the coolant supply. The tools cannot be purchased but must be made in the optics shop for each lens radius and block size combination. A drop of the thoroughly mixed epoxy is applied to the back surface of the pellets, and they are then firmly pressed into the concave carrier tool. The pellets adhere much better when the carrier tool is not only turned but also lightly lapped. The pellets are arranged either in a ring or in a spiral pattern. It is no longer customary to accurately arrange the pellets according to premilled seats in the carrier tool. To prevent the shifting of the pellets during the curing of the adhesive, a soft cloth is stretched over the tool, which is then firmly pressed against the mating tool. In this fashion the mounted tool is placed into an oven for the thermal curing of the adhesive according to the instructions of the adhesive manufacturer.

Before the fine-milling tool can be used, it must be ground in with loose abrasives against the mating grinding tool. Initially an abrasive of 45 mm (OS45) is used, which is followed by a grind with 10 $\mu$m (OS10) abrasive. The curvature is verified on the fine-milled test piece with a hand spherometer and a radius reference. A more accurate method is to grind the pellet tool against a matching cast-iron tool with water only. This makes the surface of the cast-iron tool sufficiently reflective so that its curvature can be tested with a test plate. Fine-milling tools that have changed shape after prolonged use are corrected in the same manner.

**Cooling Liquids.** Cooling for fine milling is accomplished with an emulsion that is a mix of water and a water-soluble oily liquid. Such emulsions, which also protect the tools from rust, are commercially available from chemical companies [60, 61]. Many of the emulsion coolants used by optical shops are proprietary formulations. Some shops have had success with a simple mixture of water and glycerin. The surfaces resulting from fine milling with emulsion coolants are somewhat shiny and occasionally exhibit light scratches.These scratches, however, are not a problem, since they are not very deep and are easily polished out. A very fine but matte surface typically results when coolant oils are used during the fine-milling stage.

**Pel Grinders.** Machines used for fine milling are usually called *pel grinders*. They are special high-speed machines that are designed to prepare the generated surfaces of single spherical lenses or spherical lens blocks for polishing by using diamond tools instead of the traditional free abrasive method.

It has been often demonstrated that spherical lens surfaces can be fine-ground on a pel grinder in a fraction of the time than that required for free abrasive grinding. Typical cycle times are measured in tens of seconds rather than in minutes. For instance, a typical pel grinder cycle for a medium size BK 7 lens block is about 30 seconds. During that short time 0.001 to 0.002 in. of glass will have been removed when a pellet lap is used with 15-$\mu$m diamond in a soft bronze bond. Smaller workpieces can be completed in 15 to 20 seconds, whereas larger ones may take 1 minute.

A staged two-step fining method is often employed that uses two pel grinders in tandem. The first is set up for fining; the other uses a finer diamond tool for a prepolishing cycle. If the staging of the tools and machine settings is done properly, subsequent high-speed pad polishing cycles can often be cut in half because of the much reduced subsurface damage and the improved consistency of the radius of curvature.

While the majority of grinding tools used on pel grinders are diamond pellet laps, some operators have found that solid face diamond impregnated radius laps can offer advantages in efficiency and surface finish. Diamond tools are discussed in Chapter 4. There are a great variety of pel grinders commercially available that cover a wide range of sizes and lens radii. For instance, on some units hemispheric laps for radii as small as 3 mm (0.12 in.) can be used, while larger machines accept pellet laps with long radii and diameters exceeding 200 mm (8 in.). Even larger capacity machines exist (up to 20-in. diameter) for very long radii and for plano surfaces.

Whereas most pel grinders are designed to mount the pellet tool on the bottom spindle and run the workpiece on top under the pin, others are set up to run the tool on top. Some pel grinders have hollow spindles and hollow pins for the coolant so that they can be run either way. The arrangement where the tool runs on the bottom and the lens block on top is the preferred option for high-volume production because the block can be quickly exchanged after completion of the cycle. This option is shown in Fig. 5.62.

Similar to standard grinding and polishing machines, two different overarm drives are used to accommodate such a large range of curvatures and sizes. A radial stroke is used for short radius work. It pivots around the center of curvature of the lens radius. With this arrangement the pressure that is transmitted through the pin seat remains constant during the cycle. The pin, and frequently also an extender rod, must be set up in such a way that the center of the workpiece radius coincides with the swing axis of the overarm (Fig. 5.63). Such machines are limited to radii of curvature of 150 mm (6.0 in.) or less. The extent of the pin travel is measured in degrees with some machines operating at pivot angles of 80° or more. These machines are well suited for grinding hemispheres and also hyperhemispheres. The majority of radial stroke machines have pivot angles ranging from 20° to 45° to either side of vertical. An additional offset bias can further extend the angular sweep to one or the other side.

These adjustments allow the workpiece to be moved meridionally by the oscillating overarm at a rate of 20 to 30 cycles per minute (cpm). The workpiece travel is biased over to one side or toward the rear of the pellet tool. In

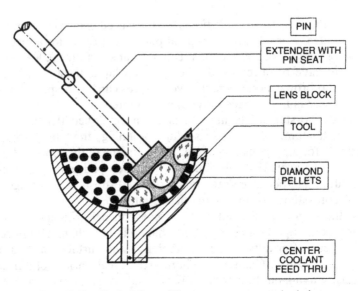

**Figure 5.62.**  Typical-fine-milling arrangement on pel grinder.

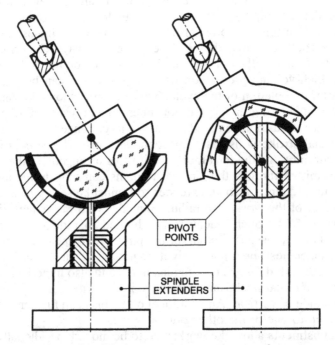

**Figure 5.63.**  Lens blocks must pivot around center of curvature.

other words, the workpiece is moved across the tool asymmetrically. When these motions are well tuned relative to the size of the tool and the workpiece, the resultant surface shape can then be maintained for a long time. Although it should be possible to express the motion of the overarm swing in mathematical terms, no one has as yet succeeded in predicting the results by using a practical formula.

A horizontal sweep motion of the overarm is used for longer radii work. Similar to standard grinding and polishing machines, the stroke length and overarm offset is a function of workpiece and the size and shape of the tool. The workpiece is held firmly by the pin around which it can turn freely. When the pin can engage near the pellet surface but at a distance from the center of curvature of the lens surface, as is the case for plano parts, then the pin can be set close to the workpiece (Fig. 5.64). This condition is not possible for steep convex surfaces. It therefore is necessary to attach an extender rod which has a pin seat at its top end. The ball of the pin engages the seat, and the pin acts far from the center of curvature of the surface (Fig.

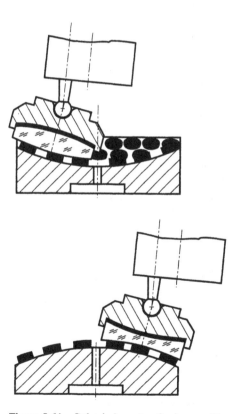

**Figure 5.64.** Pel grinder setup for long radii.

5.63). While the majority of pel grinders have been designed for spherical work, some are specifically intended for pellet lapping of large plano surfaces [62, 63, 64].

There is also a wide divergence of work spindle speeds ranging from a low of 20 rpm for the largest workpieces to a high of 3600 rpm for the smallest diameters. This translates to an average peripheral speed of 3 m/sec (600 SFPM). The speeds typically used for polishing are about half of these values. Almost all pel grinders use infinitely variable work spindle drives to give the operator the opportunity to optimize the process for the specific task. The tool pin overarm oscillates at rates ranging from 20 to 100 cpm, with 30 cpm the most common speed. Most pel grinders have a fixed speed oscillating drive but there are a few machines with stepless variable overarm drives.

The pressure exerted by the pin on the workpiece must be quite high in order to optimize rapid stock removal. Only high stock removal rates can ensure short cycle times. This pressure is derived from a pneumatic pressure cylinder that is fed by a 60.8 N/cm² (90 psi) line, and it is infinitely variable from full line pressure down to zero pressure. A line pressure of 90 psi translates to a 3 psi pressure at the pin. For radial-type pel grinders the pressure is always directed toward the center of curvature of the lens block. This ensures that a uniform pressure is exerted on the workpiece regardless of stroke position. The optimum grinding pressure is about 10 N/cm² (14.8 psi). This translates to a pressure of 0.5 to 1.0 kp/cm² at the workpiece surface.

In the mks system of units 1 Newton (N) is a unit of force that gives a standard mass of 1 kilogram (kg) an acceleration of 1 m/sec². As a measure of pressure, as used here, 10.13 N/cm² equals 14.7 psi. Both of these values express normal atmospheric pressure.

Some machines have added features either for special applications or to make them more versatile. For instance, the pin can be driven to provide a high-speed counterrotation to the diamond tool. This feature increases stock removal efficiency. The tool pin can also be offset to impart an additional circularly eccentric motion to the tool, which tends to randomize the tool action. Another special feature is a tiltable workpiece spindle that helps to eliminate the bothersome "dead" spot in the center of strongly curved lenses.

One very important feature is the coolant supply. Since pelgrinding is a high-speed diamond-lapping process that is done with considerable down pressure, sufficient amounts of coolant must be continually and liberally applied to the area where the tool contacts the workpiece. To facilitate this, most machines use hollow tool spindles and perforated pellet laps to feed the coolant directly through the hollow spindle to the center of the tool. Additional coolant must also be applied to the periphery of the tool.

The coolant pump is located in the base of the machine. It is also possible to provide a central cooling supply system that serves several machines. The milling swarf is separated from the returning coolant by using disposable filters. Also centrifuges that deposit the swarf on the inner walls are well suited. For one type of centrifuge the swarf is removed by means of an interchangeable rubber sleeve.

Aqueous emulsion coolants tend to become alkaline in time because the very fine glass swarf partially dissolves in water. This affects the fine-milling process unfavorably as it appears that the pellets have become dull. Through the addition of alum powder, $Al_2(SO_4)_3$, the coolant can be neutralized. Because of this problem the pH value of the coolant must be regularly monitored with litmus indicator paper. The resultant pH value should lie close to a near-neutral 7.

Pel grinders are machines intended for volume production. They operate typically in a fully automatic mode once they have been set up and programmed. All cycles run off timed controllers that are preset to regulate speeds, pressures, and cycle times. They are sometimes equipped with automatic carousel workpiece feeds. Figure 5.65 shows such an arrangement [65] where parts are raised out of the carousel by a small distance right into the oscillating lap.

Most pel grinders are small single-spindle machines because it would be difficult to effectively service a greater number of spindles since the fine-milling cycles are typically short. Despite this constraint, dual spindle machines are not uncommon, and some pel grinders have been built with four

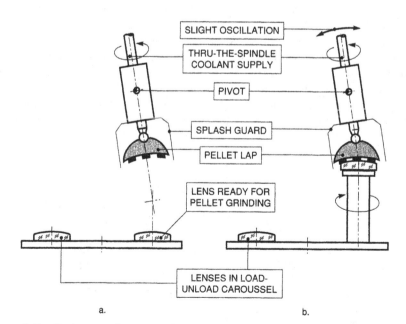

**Figure 5.65.** Basic setup of an automatic pel-grind machine. (*a*) Machine at start of cycle. (*b*) Machine during pel-grinding cycle.

spindles. These machines require a minimum of floor space but provide a potentially high throughput rate because of the typically short cycles. They can also be operated by semiskilled labor because once set up, they are almost totally automatic. This makes them a very cost-effective mode of operation for the right job. The special nature of tooling (spot tools for blocks) and the high cost of making the pellet laps make these machines useful only for ongoing, high-volume production.

A fine-milling method using diamond pellets for which the overarm oscillation has been eliminated was probably first applied by an American optician. The axes of workpiece and tool are firmly fixed with this method under an empirically determined angle. The relative motion between workpiece and tool surfaces is provided by only the spindle rotation. When this type of machine [66, 67] is carefully set up, the resultant surfaces are sufficiently accurate for most applications.

**Typical Operation of a Pel Grinder [68].** Before starting the operation, several important preparatory matters must be taken care of. First, the workpiece, which can be a single lens or a lens block, must be firmly held with only a minimum of run-out. The use of carefully designed spot blocks ensures true running lens blocks. Second, a properly broken in pellet lap for which the radius has been verified must be on hand. Finally, all speed and pressure settings must be made according to previously established procedures. The following machine adjustments must be made (see Fig. 5.66): stroke length or arc length of stroke, centering of the pin or pin offset, and overarm stop. Timers must be set and the recirculating slurry system must be checked to make sure that it is fully charged.

A typical pel grinder operation for a convex lens block running on the vertical workpiece spindle proceeds as follows: A pneumatic cylinder keeps the pin in a raised position to allow the lens block to be safely mounted on the spindle. Once the coolant is activated and the tool pin is engaged, the pneumatic pressure is applied. The preprogrammed cycle is then started.

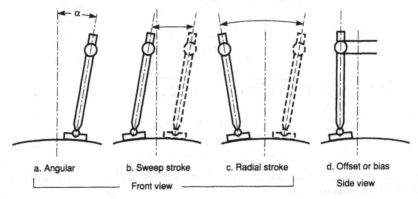

a. Angular    b. Sweep stroke    c. Radial stroke    d. Offset or bias

Front view ————————————————    Side view

**Figure 5.66.**  Basic pin adjustments on pel grinder.

The load pressure is infinitely variable from normal pressure to full line pressure. During a standard operation three pressure cycles are typically used. A low start-up pressure assists the tool from slowly breaking in the surface of the lens block. Since the tool should always engage the block on the edge first, the immediate application of full pressure would cause binding of the tool, which can lead to chipping of the parts and the premature wear of the tool. After the initial break-in cycle, the full working pressure is applied to bring the machine to optimum grinding efficiency. At the end of the cycle the pressure is gradually reduced, while coolant continues to be liberally applied. This cycle helps to prevent the formation of scratches during the finishing phase.

A strong, free-running, vertically mounted workpiece spindle holds the lenses or lens blocks. The spindle is hollow on some pel grinders to supply coolant to the lens/tool interface for the most efficient distribution of coolant. This only works, however, with lens blocks. The overarm pin is also hollow for the same reason.

**Application of Fine Milling.** After the ground in pellet tool is mounted on the machine spindle, the most advantageous overarm swing and the required grinding time is determined through the use of a test piece. A production workpiece is then put under the pin, and the start button is pushed. Upon completion of the cycle the machine stops automatically, and the completed workpiece is exchanged for a new one. The finished workpiece should be washed immediately in warm water so that the adhering emulsion does not attack the blocking wax that holds the lenses on the block.

The diamond pellets engage aggressively at first, but when enough glass has been removed so that tool and workpiece have the same surface area, the stock removal seems to cease. This effect is most certainly the result of reduced grinding pressures. It can be used advantageously for the precise control of lens center thickness.

When pellet tools become dull or glaze over, it is necessary to grind them against a mating cast-iron tool with a suitable loose abrasive. The abrasive size and type must be matched to the bond type and diamond concentration. Experience is the best guide to determine this. The abrasive must erode away a small amount of the bond to remove glazed over or dull diamond points and to expose new and sharp diamonds.

For fine-milling tools to retain an accurate surface shape for a long time, fine milling should be done with two pellet tools. The pellets of the first tool should have a diamond grain size of 35 $\mu$m (D35), and those of the second tool approximately 10 $\mu$m (D10).

**Testing of Fine-milled Surfaces.** The fine-milled surfaces must be tested for surface quality and curvature. The visual inspection consists of the use of an eye loupe to determine if the generating marks have been entirely removed and if the surface is scratch free.

The curvature can be verified with a test plate, since the fine-milled surfaces are sufficiently shiny. In most cases, however, the sag of the curvature is measured with a hand spherometer relative to a reference curve. Spherometers are discussed in detail in Chapter 6. Since the pit depth on fine-milled surfaces is minimal, only a small amount needs to be polished off the surface and the resulting radius change is only minor. This means that the sag as measured with the hand spherometer should deviate from the target value by no more than one micron when the diameter of the spherometer ring is at least as large as the radius of curvature of the surface. The lens center thickness need to be tested only infrequently because the stock removal during fine milling is for the most part constant.

### References for Section 5.2 (Milling Operations)

[34]  Cone-Blanchard Machine Co, Windsor, VT, No. 11 Blanchard.

[35]  Mildex, Pittsford, NY, Yoshikawa, Model YGS-16.

[36]  No. 20 Blanchard.

[37]  Yoshikawa, Model YGS-20, and Loh, Model Spheromatic Plano.

[38]  Quaker 101, 1 part to 50 parts water.

[39]  Wilhem Loh, Wetzlar, Germany, Loh Universal Milling Machine, Model UFM.

[40]  Loh Universal Milling Machine, Model UFMS.

[41]  Loh curve generator, Model RF-2.

[42]  DIN 58 741.

[43]  Rogers & Clarke, (R&C).

[44]  GNM Müller, Germany.

[45]  Esso Sommentor 43.

[46]  Shell 5585.

[47]  DIN 58 740.

[48]  Loh, Model RF-1 (Spheromatic).

[49]  Coburn Industries, Inc., Model VG-1.

[50]  R&C Micro curve generator.

[51]  CNC generator (Strasbaugh and R&C) and Loh, Prism Grinding Machine CNC.

[52]  Wilhelm Bothner, GmbH, Germany, Model B26VA.

[53]  Rogers & Clarke.

[54]  R. Howard Strasbaugh, San Luis Obispo, CA.

[55]  W. Bothner, Germany.

[56]  R&C, Models G-150 and G-300.

[57]  Strasbaugh, Model 7N.

[58]  R&C Microgenerator G-15.

[59]  DIN 58 745.

[60]  Castrol Super Edge.

[61]  Houghton Grinding Emulsion.

[62] Loh, Model PLM-400.

[63] Mildex, Udagawa, Model 51.

[64] Strasbaugh, Model 6AQ-SP.

[65] W. Bothner, Germany.

[66] Leico, Germany.

[67] Loh Spheromatic.

[68] per Rogers & Clarke.

## 5.3. LOOSE ABRASIVE GRINDING

Grinding is the process of shaping optical materials with free or loose abrasives. The abrasive powder is mixed with water and the grinding is done on grinding tools made of cast iron. Free abrasives are discussed in Chapter 3.

The fabrication process now known as *grinding* may be called *lapping* in the future. In this way the designation will be in harmony with the term customarily used in the metal industry for this process. But the time-honored term "grinding" continues to be widely used in most optical shops.

The pregrinding and fine-grinding steps are now frequently replaced by finishing the glasses with diamond tools. This is the fine-milling process described in the preceding section. But for prototype lenses and low-volume production, free abrasive grinding continues to be widely used.

### 5.3.1. Theoretical Considerations

The loose abrasive is mixed with water to form a slurry that is applied between the grinding tool and the glass workpiece. The individual grains of this slurry roll back and forth between the glass and tool during the grinding process. This causes the sharp edges of the abrasive grains to penetrate into the surface. The process leads to local fracturing of the outer layers of the glass and thus to the removal of tiny glass splinters. As seen in the microscope, numerous small concave clam shell fractures, called *grinding pits*, cover the ground surface. Maksutov [69] determined that the fractures that extend into the material under the visible surface are about as deep as the remaining grinding pits (Fig. 5.67). More recently Hed [70] determined that individual fractures go much deeper than the earlier model predicted. As expected, coarse abrasives produce deeper pits than finer abrasives.

To investigate the effect of the abrasive grain size on the resultant surface roughness, a series of controlled grinding experiments were conducted [71]. A number of different grit sizes of the same type of grinding compound were used on BK1 optical glass samples. Table 5.3 lists a series of such grinding test results. The depth of the grinding pits resulting from this test is compared to the grit size of the abrasive used. The depth of the pits was measured on the edge of optically contacted test pieces.

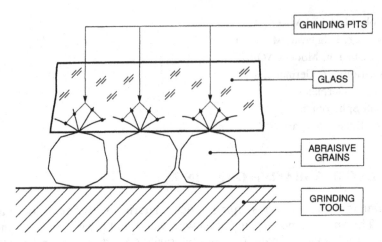

**Figure 5.67.**   Grinding model per Maksutov.

By measuring along an edge, one cannot be certain that either the deepest or the shallowest grinding pits are included. The results, however, clearly show that the finer abrasive grains are more effective than the larger grains. While the large abrasive grains produce pits that have about $\frac{1}{10}$ the depth of the grain size used, the finer grains produce pits that are $\frac{1}{2}$ as deep as their size. This effect can be explained that during the grinding process the sharp edges and corners of the grains are primarily active, and these are more numerous on finer grains.

Also the hardness of the abrasive and the hardness of the optical material that is being ground influence the grinding pit depth. Corundum produces deeper pits at a specific grain size than, for instance, the softer garnet of the same size. Likewise the pits in softer glass are deeper than those in harder glass. This fact can be easily demonstrated when an attempt is made to grind and polish crown and flint glasses on the same block, which is not a customary or recommended practice.

**Table 5.3.   Depth of Gridning Pits as Function of Abrasive Size**

| Abrasive grain size (µm) | Grinding pit depth (µm) |
|---|---|
| 150 to 300 | 15 to 35 |
| 30 to 60 | 6 to 15 |
| 9 to 30 | 4 to 9 |
| 1 to 10 | 2 to 6 |

The crown glass parts would polish out faster than the softer flint glasses, since the abrasive will have left deeper pits in the flint glass. Theoretically, softer glasses should polish faster than harder glasses, but this effect is often not enough to overcome the longer polishing times required to remove the deepest grinding pits.

As described above, the pits in glass surfaces that have been ground with loose abrasives have some range of varying depth. This is especially true for fine-ground surfaces. When such a surface is first polished, it takes on a shiny appearance in a relatively short time. But a few scattered pits, or *digs* as the opticians call them, will remain for an appreciable time. One can assume that the edges of the freshly applied abrasive penetrate the surface at once, thus producing deep grinding pits. Since these points and edges are very fragile and tend to be in the minority, the grains will break down and be smoothed out soon. What then remains is a mass of slurry with reduced grinding power with which the deeper pits can no longer be ground out. A new application of fresh slurry will simply repeat this process.

It is very disturbing when the grain sizes vary over a broad range. The coarser grains attack the surface strongly on the edge and on prisms, for instance, they will leave deep grinding pits behind in the corners. Most quality abrasives in use today are closely sized, so this is not as much of a problem as it used to be.

Through the use of fine milling with diamond pellets, the fine-ground surfaces are almost free from pits. But because of the elaborate preparations needed for pellet lapping, loose abrasive fine-grinding will retain its usefulness for many applications.

## 5.3.2. Grinding Tools

Spherical or plano grinding tools are needed against which the lenses or lens blocks are ground with an abrasive slurry. Grinding tools are specifically discussed in Chapter 4. Since the grinding process typically covers several abrasive grain sizes, one tool must be available for each grain size. Therefore one complete set of grinding tools must be available for each specific radius. Table 5.4 lists the optimum number of tools per set.

The lay-in tools are not needed when spot tools are used. When the surface is also generated on the spot block, then only the prefine and the fine-grinding tools are required. The curvature of the grinding tools must be corrected such that the next finer abrasive will initially grind from the edge. This means that the concave grinding tool set has a progressively shorter radius for the next finer grind, while the tool radius becomes progressively longer for the convex tool set.

The reason for this is the necessity to grind off a uniformly deep layer to remove the grinding pits from the preceding grind (Fig. 5.68). After the grind the radius of a convex surface will be slightly shorter and that of a concave

**Table 5.4.   Typical Set of Spherical Grinding Tools and Their Application**

| Type of tool | Abrasive grain sizes (µm) |
|---|---|
| Roughing | 115 to 230 |
| Pre-grinding | 45 to 75 |
| Lay-in | Only used for pitch blocking |
| Pre-fine | 17 to 29 |
| Finegrinding | 3 to 13 |

surface will be a little bit longer. A grinding tool that is designed to remove a concentric layer of material at least to the depth of the deepest grinding pits must have a shorter radius when it is a concave tool and a longer radius when the tool is convex. As a result both tools will grind from the edge toward the center. No such compensation required for plano grinding tools.

In the ideal case, the radius difference between consecutive grinding tools would be equal to the grinding pit depth plus the abrasive grain size. But to avoid having to maintain the grinding tools for coarser grains absolutely accurate, one typically deviates in practice from this theoretically possible case. Through measurements done on a series grinding tools [72], practical radius differences were determined that correspond to the data shown in Fig. 5.69.

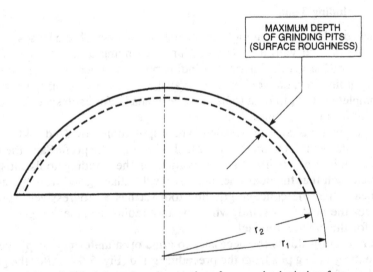

**Figure 5.68.**   Schematic representation of a ground spherical surface.

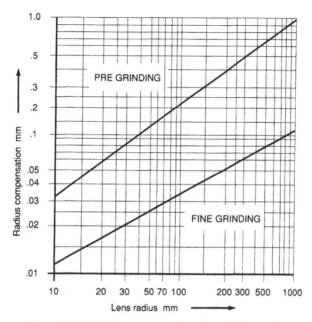

**Figure 5.69.** Radius compensation for grinding tools.

It is customary to provide concentric grooves for large diameter grinding tools. These grooves are usually spaced every 30 mm (1.2 in.) from the center. They have a typical width of 2 mm (0.080 in.) and a depth about 3 mm (0.120 in.). The water-based slurry settles in the grooves and is slowly released from there during the grinding process. This promotes even tool wear, which makes it possible to maintain the shape of the tool constant for long periods of time.

The turned or milled raw tools are accurately ground in by the optician. The fine-grinding tool is prepared first, since its resultant ground surface can be measured against a test plate for determining the accuracy of the radius of curvature. The curvature of the raw tool is judged first with the use of a radius gauge or a hand spherometer. After that is done, the surface of the tool is smoothed out with a reject lens and fine abrasive. Radius gauges and spherometers are described in Chapter 6.

This future fine-grinding tool will then serve as a lay-in tool for a pitch button block made of reject or dummy lenses. Prior to hardening of the pitch, the block must be firmly pressed against the lay-in tool once more so that the lens surfaces assume its curvature as accurately as possible. The resulting lens block is fineground in the fine-grinding tool and prepolished with a suitable pitch polisher. At this point the surface should be 3 to 5 fringes concave for large blocks and 5 to 10 fringes concave for shorter radii when measured against the test plate. This measurement can also be made with a 1-$\mu$m accuracy hand spherometer that must first be calibrated against the test plate. The measurement must show a deviation that corresponds to a radius difference of about 0.15 to 0.20 mm (0.006 to 0.008 in.).

When the curvature of the fine-ground block deviates from the required value, then the fine-grinding tool must be corrected with a hand-held abrasive stone either near the edge or over the center, as shown Fig. 5.70. The radius of a concave fine-grinding tool will either become longer when the edge zone is corrected or shorter when the correction is made near the center of the tool. For a convex tool the radius change would be in reverse order.

When the fine-grinding tool has been properly corrected, the prefine tool must be ground in. A lens block ground in on such a tool should contact the fine-grinding tool lightly at the edge. This requires that a concave prefine tool must have a longer radius and a convex tool a shorter radius than the fine-grinding tool.

The pitch button block of reject lenses is again carefully heated, pressed out against the prefine tool, and later ground in it with a finer abrasive slurry. The block should grab lightly from the edge when rubbed against the fine-grinding tool if this step has been done correctly.

The lay-in tool for pitch button blocks is ground by using the identical method. A lay-in tool is used to temporarily adhere lenses to its surface in

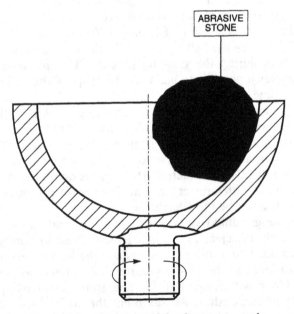

**Figure 5.70.** Radius correction for concave tools.

preparation to pitch button blocking. Some opticians make the distinction between a concave lay-in tool and a convex set-up tool. The surface of these tools must be especially well ground in so that the lenses adhere firmly to the tool surface.

The pregrind tools are also prepared in the same manner so that the lenses ground in them contact the next finer tool from the edge. The next tool in this case is the lay-in tool. Pregrind tools in which single lenses are ground should be no larger than 180 mm (7.2 in.) in diameter. They would otherwise develop zones, and it would be difficult to maintain their shape.

### 5.3.3. Pregrinding

Instead of the more economical curve generating, lens surfaces can also be ground in by hand from a round blank. This is often the best course of action when only a few lenses have to be made and the curvatures are not too steep. This pregrinding is done typically in two, and occasionally in three, steps.

During the initial coarse grinding stage, also called *roughing*, an abrasive with a grain size of at least 120 $\mu$m is used. Lenses and prisms receive this way their basic shape. Roughing also removes the skin on glass pressings and prepares them for further processing. This stage must be avoided on many commonly processed optical materials such as polycrystalline materials because the surface damage caused during roughing would be unacceptably deep.

This initial step is followed by the actual pregrinding which begins with abrasives ranging in grain sizes from 45 to 80 $\mu$m. The surface roughness and surface shape are improved during this step, and the center thickness of lenses or the lengths of prism sides are brought to exact dimensions. High-quality optics are adiitionally ground with abrasives of grain sizes between 30 and 40 $\mu$m. These are usually the coarsest abrasives used on brittle infrared (IR) materials such as germanium (Ge), silicon (Si), and many of the other optical crystals used in IR and laser systems.

For the just described grinding steps a thickness overage of about 0.3 mm (0.012 in.) must be left after rough grinding and 0.1 mm (0.004 in.) after pregrinding per side. This requires that a lens that can have a finished maximum thickness of 10.00 mm (0.400 in.) is rough ground to a thickness of 10.60 mm (0.424 in.) and preground to 10.20 mm (0.408 in.). These guidelines were developed for standard quality optics made from BK 7 or similar optical glass. The thickness overages for components made from other optical materials must be established by trial and error.

It is customary to use the maximum allowable center thickness of the finished component as the target. The reason for that is that the still available thickness tolerance allows for one or two regrinds in case scratches are picked up during fine-grinding or polishing that can only be removed by grinding.

The rotational speed of the grinding spindle is about 1500 to 2000 rpm for tools up to 60 mm (2.4 in.) in diameter. This speed is stepped down to about 1000 rpm when larger tools are used. If the speed is too high, the abrasive grains are thrown off the tool before they can become engaged. This greatly reduces their abrasive action.

The pregrind spindle should run only half as fast as the rough grind spindle. When the pregrind tool runs at a sufficiently slow rotational speed, the ground surfaces become more accurately spherical or more plano. The glass surface is also easily sucked to the tool surface when finer abrasives are used, and the lens can slip out of the optician's hand at higher spindle speeds.

The rough grinding and pregrinding steps are typically done by hand. The lens is most often held by the left hand, while the right hand supplies the slurry. The preferred use of the left hand during manual operations on optics machines is in part the result of the spindle direction that runs counterclockwise because of the right hand threads used. Some opticians, however, prefer to grind with their right hand, and there is really no good reason to assume that that is incorrect.

The goal during rough grinding is fast material removal. This requires high grinding pressures. For instance, when a convex surface is roughed in, the lens blank is pressed against the inner surface of the concave tool with the thumb of the left hand (Fig. 5.71).The lens blank must be moved between tool edge and tool center during this operation so that the cast iron tool, which also experiences wear, does so uniformly. When the lens blank reaches the tool edge, it is turned slightly about its own axis with the index finger of the same hand. This is necessary to avoid grinding a wedge on the lens. A lens is wedged when the edge is not of equal thickness around its circumference. When roughing out large lenses in shallow concave tools, the blank can be pressed with the ball of the left hand against the tool and moved

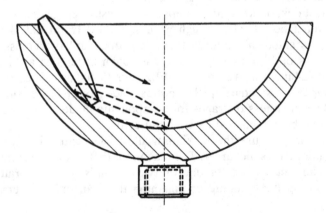

**Figure 5.71.**  Manual grinding of a lens.

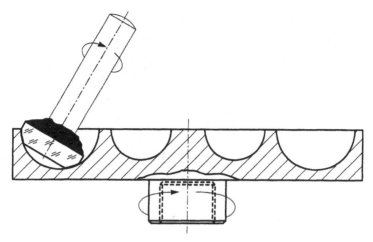

**Figure 5.72.** Grooved grinding tool to grind a small but steeply curved convex lens.

from center to edge. The right hand supplies the slurry and turns the lens blank.

Steeply curved convex surfaces can be ground in a groove tool (Fig. 5.72). The lens is blocked to a wooden handle by which it is held into a matching groove, while it is turned around its axis under constant flow of sufficient amounts of slurry. A curved surface can be ground much faster in a grooved tool than in a spherical tool. The resulting surfaces, however, are not quite spherical, and the lens must be finished in a spherical tool. To accurately maintain the center thickness of such lenses, the blanks are first ground to thickness on a plano tool. A spot of the plano surface remains in the center of the lens when it is ground in the grooved tool. This spot is removed later during fine-grinding.

When manually roughing concave surfaces of manageable size, the lens is usually held between the middle and/or index finger and thumb of the left hand and is moved from the edge across the center of the grinding tool. After each back and forth motion, the lens is turned slightly. The right hand supplies the slurry as previously described. Experienced grinders can work with both hands interchangeably.

The lens grinder holds a lens that has convex surfaces by pulling it against the rotation of the pregrind tool while turning the lens as described before (Fig. 5.73). When a lens has a strongly curved surface on one side and a surface with a longer radius on the other, then the shorter radius is ground first. The correction of the center thickness is done on the longer radius because the typically larger grinding tool for the longer radius removes the glass faster.

The grinding pits from the roughing operation are ground out with finer abrasives during pregrinding, and the resulting surfaces must be accurately

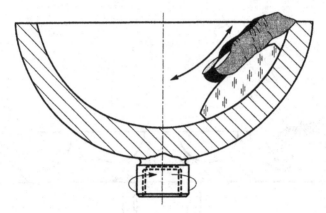

**Figure 5.73.** Manual grinding of a convex lens.

spherical to the exact radius of curvature. The center thickness of the lenses are brought to a precise dimension for which deviations in the range of 0.02 to 0.05 mm (0.001 in. to 0.002 in.) are acceptable.

Spot blocks are also preground on occasion. Smaller blocks are screwed to a grinding tool handle for that purpose. Larger spot blocks or plano blocks are preground by hand only so that they will end up with a regular surface. This type of grinding is generally done on a tub grinder under a hand lever. The hand lever runs on the left side and is moved between tool center and tool edge. It is possible to influence the curvature of the surface with this movement or to maintain the proper curvature once it is obtained. If the workpiece is moved more over the edge, then it will be attacked more in the center. That means that convex surfaces are made flatter and that concave surfaces are made steeper, assuming that they run on the spindle and the tool runs under the pin. When one grinds more near the tool center, the workpiece is ground more from the edge. This makes convex surfaces more convex and concave surfaces less concave. The relative size of the grinding tool plays an important role in this process. While the hand lever is guided with the left hand, the right hand applies the slurry with a brush. The parallelism of plano plates or correction of edge thickness variations on spot blocks is achieved by manually pressing down on the high side of the block. A short grind under the hand lever is necessary after such a correction to grind the surface regular again.

**Inspecting Preground Surfaces.** Preground surfaces are inspected for the quality of grind. The center thickness, surface shape, and edge parallelism must be verified for lenses. Angles and the lengths of the sides must be measured for prisms. The surfaces are visually inspected with the unaided eye to determine if the grinding pits from the preceding grinding steps have been ground out. Also obvious scratches are noted. Deep grinding pits are more easily seen when the lens surface is moistened and the lens is held in front of a bright lamp. While inspecting the surface, any edge chips are noted as well.

The thickness of the rough ground lenses is measured with a simple gauge, which is most often a dial indicator mounted on a stand that has a typical accuracy of 0.01 mm (0.0004 in.). The plunger of such dial indicator is activated by a lever, since it is often used with wet hands (Fig. 5.74). During the measurement the approximate lens center is visually located. The lens can be turned between anvil and plunger to facilitate that measurement. By attaching a centered lens seat to the table of a dial indicator, it is possible to accurately measure the center thickness at the exact lens center. Only departures from a preset absolute dimension are measured with the dial indicator. This requires that the dial indicator be zeroed to a set of gauge blocks that add up to the required dimension.

The surface shape is verified by rubbing the lens against the next finer grinding tool. During the final pregrind stages it is rubbed against the lay-in tool. After rubbing against the pregrind tool, the rough ground surface should show rub marks to about 25% of its diameter. The preground surface should show rub marks that are strongest at the edge and gradually weaker toward the center. The central half of the lens diameter should show no marks when it is rubbed against the lay-in tool. The evaluation of these rub marks can be made easier by marking the lens or block surface with a soft pencil first. While this seems like a method from the dark ages, it can be a very handy way to quickly and quite accurately determine the fit of the curvature relative to a known reference.

The modern way is to check the curvature with a hand spherometer. But unless a reference glass exists for the exact curvature for each tool, this

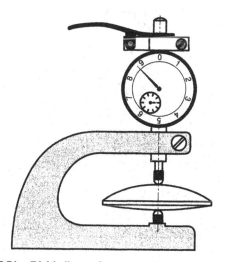

**Figure 5.74.**   Dial indicator for center thickness measurement.

approach would involve a number of calculations to gain the same insight that the simple rubbing method provides to the experienced optician. In addition it is not possible with the spherometer to determine if the surface is truly spherical.

The edge parallelism of preground lenses is commonly verified by eye, but a measurement may be necessary for optically weak lenses. A dial indicator setup that locates the edge of the lens underneath the indicator point will be required to make this measurement. As the lens is rotated, variations in the edge thickness are a measure of the wedge in the lens. This method is usually referred to as the true indicator reading (TIR) method of measuring the edge thickness variations (ETV).

The thickness of lenses on spot blocks is measured with a dial depth gauge. Such a depth gauge can be a modified hand spherometer (Fig. 5.75). The resultant dimension is a reference measurement only. When the spot block includes test lenses during the initial run, this thickness reference measurement can be easily determined.

### 5.3.4. Fine-Grinding

Fine-grinding with an aqueous abrasive slurry is usually the last process step before polishing. All grinding pits of preceding grinds are removed, and the surface is ground to an accurate shape. This process step is typically subdivided into prefine and fine-grinding. Abrasives with grain sizes of 15 to 35 $\mu$m are used for prefine grinding, while fine-grinding is done with 12 $\mu$m

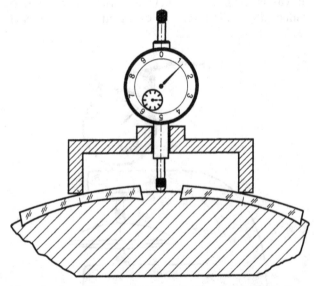

**Figure 5.75.** Measurement of lens center thickness on spot blocks using a reference dimension.

down to 3 $\mu$m. The finer abrasives within these ranges are used for small lenses, the coarser ones for larger lenses.

Small lenses or single surfaces are often fine-ground by hand on the pedal bench or on the motor-driven bench. Larger lenses and spot blocks are processed on a grinding machine. These machines are described in detail in Section 5.5. Lenses must be blocked to wooden handles when they are fine-ground by hand (Fig. 5.76a). When the lenses have to be finished later on the machine, they are first blocked to a threaded plate and afterward screwed to the handle(Fig. 5.76b). Small blocks are also mounted on a threaded handle.

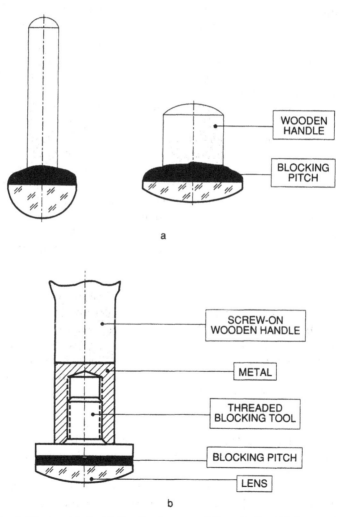

**Figure 5.76.** (a) Lenses blocked to handles for manual fine-grinding. (b) Screw-on handle for blocked lenses.

The fine-grinding tool is always mounted on the work spindle of the machine for manual work. The workpiece is guided by hand from the top as the grinding tool rotates underneath. This is true whether concave or convex surfaces have to be worked.

Abrasive slurry is applied with a brush to the fine-grinding tool and the workpiece is set on or into the tool, as the case may be. The tool spindle rotates at moderate speed of about 200 to 300 rpm. The handle holding the workpiece is moved back and forth so that it moves between the center and edge of the tool. The workpiece is allowed to rotate with the tool from time to time so that it ends up in another position to the direction of motion. The surface will grind uniformly if this is done correctly. After an abrasive slurry has been applied several times, fine-grinding continues without further addition of slurry until the slurry on the tool turns dark. After that the workpiece is removed and cleaned. It serves no beneficial purpose to grind "dark" or "black" for too long. Tests have shown that dark ground abrasive slurry has no longer any abrasive powers. It merely smears over the grinding pits without grinding the surface any finer.

The curvature of the surface can be influenced by the manner in which the handle is moved. If the stroke is short and the workpiece remains near the center of the tool, then the surface of the workpiece is attacked preferentially at the edge. Convex surfaces fit more convex, concave surfaces become less strongly curved. Both surfaces therefore will become more convex when measured against a test plate. With longer strokes away from the tool center, the workpiece is ground more in the middle. The tool is also wearing at the same time. In the first case it wears more in the center; in the other case it wears more at the edge. With a suitable application of the workpiece motion, the tool can wear uniformly and thus maintain its shape for a long time.

When a tool is so severely worn that the required surface curvature can no longer be obtained through a change in the grinding motion, it must then be corrected. A piece of grinding stone is typically all that is needed to do this (Fig. 5.70). A fine file is useful for correcting convex tools (Fig. 5.77). Also grinding the tool in against a mating tool is quite common. However, when more material must be removed, a file that has been ground sharp can be used to serve as scraper. The tool material should be removed uniformly without creating grooves to ensure that the corrected zones of the tool contribute to the grind without delay.

It sometimes happens that single surfaces for rework or for fabricating prototype lenses must be ground on large grinding tools that were designed to grind large blocks. Such lenses must be fine-ground on a stationary tool. To do that, the lens is held by hand, preferably close to the surface that must be ground and is moved in a circular motion over the tool. In short intervals the lens is then held at another part of the circumference to make sure that it grinds uniformly. If one would allow the tool to rotate, the lens

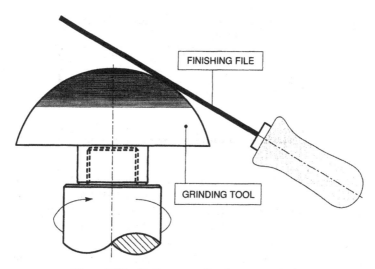

**Figure 5.77.** Radius correction for convex tools.

surface would be strongly abraded at the edge, and it would then be very convex if measured against a test plate.

When fine-grinding on a regular grinding machine, the overarm with the pin executes the side motion over the rotating spindle. The most common practice is that the convex workpiece or the convex grinding tool is mounted to the spindle. The concave part runs on top and is guided by the pin. Only very flat convex surfaces are preferrably run on top. Plano surfaces are typically run on top as well unless they are large and heavy. With this arrangement the loading and unloading of the workpiece, and the larger grinding tools used make it possible to reduce grinding times.

The water and abrasive slurry is applied with a brush. It is more desirable to apply only small amounts of slurry but to apply it more often. The surface of the tool can only retain a limited amount of slurry. The excess drips unused into the catch pan. Abrasive slurries should not be ground down too much on the overarm machine as they tend to scratch this way. Also the workpiece surface should still be slightly wet at the end of the grinding cycle. When the surface becomes too dry, the workpiece can suck down onto the grinding tool from which it can be separated only with considerable force. This condition is especially dangerous for large concave blocks. The tightly adhering block will suddenly separate from the grinding tool when firmly pulled, and the lenses can strike against the tool and be damaged.

The drive belt of the fine-grinding machine is laid on the stepped pulley so that the work spindle runs faster than the overarm spindle. The rotational speed of the overarm spindle is chosen on the basis of the size of the workpiece. This relationship is shown in Table 5.5.

**Table 5.5. Eccenter Spindle Speed as Function of Workpiece Diameter**

| Workpiece diameter mm | inch | Eccenter spindle speed (approximate RPM) |
|---|---|---|
| Up to 50 | 2.00 | 100 |
| Up to 100 | 4.00 | 70 |
| Up to 250 | 10.00 | 50 |

When a multispindle machine is loaded with fine-grinding work, the optician will set it up in such a way that all spindles run for an equally long time span. He or she works according to process cycles. To accomplish this, all spindles are loaded at the same time, and the blocks are removed from them at the same time. The workpieces must be approximately of equal size. The fine-grinding cycles on the machine lie between 5 and about 25 minutes for each abrasive grain size used.

Scratches are the most common defects found during fine-grinding. They can have a variety of causes. The primary cause is contaminated abrasive slurry. The commercially manufactured abrasives of today are for the most part of excellent quality from the raw material, preparation, and sizing standpoint. Contamination is in most cases caused by the user.

Scratches can also result when the slurry is too wet. In such a case the abrasive film between the cast-iron surface of the grinding tool and the workpiece surface is pushed aside, and the glass comes in contact with the tool surface. The same effect can be noticed at high grinding pressures, especially on soft glasses. With sharp-edged lenses there is the additional danger that small glass chips flake off when the workpiece is placed on the tool, which can also lead to scratches. When grinding too dry, a surface roughness similar to scratches can result which can locally penetrate 0.1 mm (0.004 in.) deep into the surface of the glass.

The overarm motion of a fine-grinding machine also influences the surface shape of the lenses. When the overarm is set for a wide swing, the part running on top will grind more from the center and the lower part more from the edge. In other words, the upper and the lower part are becoming more strongly curved. If compared against a test plate, we find that the upper part has become more concave and the lower part more convex. When, on the other hand, the pin crosses the center and small excentric oscillations are used, the upper part will become more convex as it is attacked more from the edge. The part running on the bottom is attacked more in the center, and this will make it more concave. Through such action, the surface shape of the grinding tool changes little by little.

It is possible to find a machine setting at which the surface shape can be maintained over longer periods of time. In addition the ratio of workpiece size to the size of the grinding tool influences the surface shape. It is customary to give the part running on top an included angle that is 10% to 15% smaller than the lower part. The part that runs on the bottom, which is mostly convex, has an included angle that is proportionately larger. Also the diameter of a plano grinding tool that runs on the spindle must be no less than 15% larger than the diameter of the workpiece.

An automatic slurry supply that recirculates slurry for repeated use is not recommended for precision fine-grinding. The abrasive tends to break down between glass and tool, and only the slurry that had been pushed aside, not the slurry that was actually engaged, would still be useful. The abrasive slurry could only be effectively transported through the supply pipes if it is greatly diluted with water, and this would not result in a clean fine-grind. Recirculating fine-grinding slurries can be used quite effectively for non-critical applications. The dilution ratio of the slurry should be 10% or less suspension-treated abrasive powder in water to prevent premature settling of the compound in the lines. This, however, reduces the abrasive effectiveness of the slurry.

**Inspection of Fine-Ground Surfaces.** Fine-ground surfaces are inspected for surface quality and surface shape.The center thickness of lenses must be measured as well.

The surface quality is inspected with a loupe in the light of a test lamp. Defects are grinding pits from preceding grinds and grinding scratches. Defective parts are returned for rework. When reworking these parts, one must consider how much of the lens center thickness will change during regrinding. The center thickness measurements are made as was described earlier in the section on pregrinding.

The surface shape can be tested with a test plate after a flash polish, or it can also be checked with a hand spherometer. For both tests one must consider that for glass lenses 0.01 to 0.02 mm (0.0004 to 0.0008 in.) will be polished off from the ground surface. Different, empirically developed values apply to other optical materials. The radius of a fine-ground convex surface must therefore be larger and that of a concave surface smaller than the test plate radius.

In both cases the test plate will contact the surface at the edge, and it will show a number of interference fringes. The number of fringes can be easily determined with an approximate equation. It depends on the diameter of the test surface, which is generally the diameter of a lens on the block, the surface radius, the wavelength of the light source used, and the thickness of the layer that is to be polished off (typically 0.02 mm or 0.0008 in.). When the blue interference fringes across the test plate are counted during the white light interference test, the required number of fringes can be calculated using the approximation per Eq. (5.12):

$$Z \approx 600 \left(\frac{d^2}{r^2}\right) \Delta r, \qquad (5.12)$$

where

$Z$ = number of blue interference fringes that should be seen on a pre-polished fine-ground surface,

$d$ = diameter of the test surface in millimeters,

$r$ = radius of the test surface in millimeters,

$\Delta r$ = layer thickness in millimeters that must be polished off to remove all grinding pits.

Equation (5.12) is valid only for metric dimensions. For dimensions in inches, the same results are obtained when the constant 600 is multiplied by 25 mm/in. This leads to (5.13):

$$Z \approx 15\,000 \left(\frac{d^2}{r^2}\right) \Delta r. \tag{5.13}$$

**EXAMPLE:**

A fine-ground and prepolished surface has a radius of 46.5 mm (1.860 in.). The surface is tested with a 30-mm (1.200-in.) diameter test plate and 0.02 mm (0.0008 in.) must be removed during polishing. How many blue fringes should the test plate show when the test is done in white light illumination?

From Eq. (5.12),

$$Z \approx 600 \left(\frac{30^2}{46.5^2}\right) 0.02$$

$$\approx 5 \text{ blue fringes,}$$

and from Eq. (5.13)

$$Z \approx 15\,000 \left(\frac{1.2^2}{1.86^2}\right) 0.0008$$

$$\approx 5 \text{ blue fringes.}$$

When a uniform thickness of 0.02 mm (0.0008 in.) were polished off in this example, the test plate would fit "white." In other words, it would be a perfect test plate match. To avoid drifting into the convex range during polishing where the test plate would contact in the center, two additional fringes are added. This example is intended to show that the number of fringes can be calculated. The skilled optician, however, relies on his or her experience.

The deviation in the curvature can also be found with a hand spherometer. As a calculation will show, the sag deviation $\Delta s$ becomes constant when the measurement is made with a spherometer ring that has a diameter approximately as large as the radius of the tested surface (Fig. 5.78). The sagittas of the two surfaces are then

$$s_1 = r_1 - \sqrt{(r_1)^2 - \frac{(r_1)^2}{4}}$$

$$s_2 = r_2 - \sqrt{(r_2)^2 - \frac{(r_2)^2}{4}}$$

It follows that

$$s_1 = r_1 \left(1 - \sqrt{\frac{3}{2}}\right),$$

$$s_2 = r_2 \left(1 - \sqrt{\frac{3}{2}}\right),$$

since $s_1 - s_2 = \Delta s$, $r_1 - r_2 = \Delta r$, and $\Delta s = \Delta r(1 - \sqrt{3/2})$. This leads to

$$\Delta s = 0.134 \, \Delta r. \tag{5.14}$$

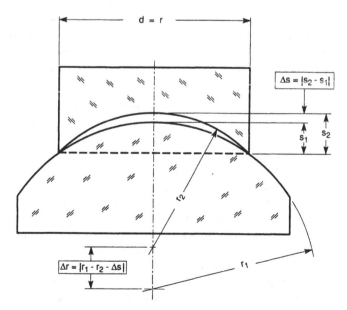

**Figure 5.78.** Spherometer measurement when $d = r$.

The condition that has to be met for this approximation is that the two radii are close to one another. This is the case here where they differ only by the thin layer that must be polished off which is a $\Delta r$ of about 0.01 to 0.02 mm (0.0004 to 0.0008 in.).

Therefore, when the diameter of the spherometer ring is almost as large as the surface radius and the required removal in polishing is assumed to be $\Delta r = 0.02$ mm (0.0008 in.), the hand spherometer should show a reading of $\Delta s = 0.134 (0.02) = 0.0027$ mm or $\Delta s = 0.134 (0.0008) = 0.0001$ in. The radius of the surface does not influence the result. For surfaces that were fine milled with diamond pellets, the grinding pit depth $\Delta r$ is less than 0.01 mm ($>0.0004$ in.). For this case a $\Delta s$ of only 1 $\mu$m results when the ring diameter equals the surface radius.

Convex surfaces have a longer radius prior to polishing. They therefore show a smaller $\Delta s$ relative to the test plate. Concave surfaces have a shorter radius prior to polishing and thus show a larger $\Delta s$.

### 5.3.5.  Double-Sided Grinders

Two-sided lapping, which is also called *planetary lapping*, is not a new technology because it has been used in the metal industry for quite some time. Its application in optics fabrication has been tried from time to time with varying results. In recent years, however, the double lapping method has been successfully extended into two-sided polishing of high-quality optical substrate. This makes double-lapping and two-sided polishing attractive and cost-effective alternatives to the conventional grinding and polishing of plane-parallel parts. Uniform stock removal from both sides of the workpiece during the same operation all but eliminates the typical bowing and springing of lapped and polished surfaces due to the so-called Twyman effect which is caused by differential surface stresses in fine-ground substrate.

**Operation.** A planetary rotation of parts, which are sandwiched between two counterrotating lapping plates, is produced by a gear system. The parts are captured in perforations cut into thin carriers that are part of the gear system. The gear system consists of an outer ring gear that surrounds the bottom plate. The bottom plate has a perforated center that accepts the inner sun gear. The ring gear and the sun gear can rotate in the same direction but at different rates of rotation, or they can counterrotate. The geared carriers run between these two gears, as shown in Fig. 5.79.

In the first case only a slight amount of precession or recession is introduced, but in the case of counterrotating gears the carriers rotate about their own axis as they move in the direction of the ring gear rotation, which is usually the same as that for the lower lapping plate. The top plate floats on the parts but is prevented from rotating by friction or by a fixed arm. The top

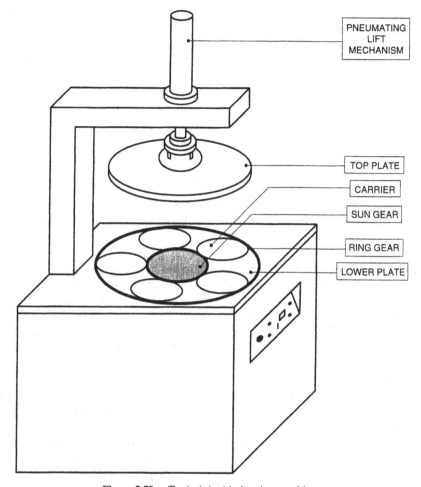

**Figure 5.79.** Typical double lapping machine.

plate can also be driven to rotate in the opposite direction to that of the bottom plate. These arrangements result in a four-way planetary motion:

- Counterclockwise rotation of the bottom plate
- Clockwise axial rotation of the carriers
- Counterclockwise orbital precession of the carriers
- Clockwise rotation of the upper plate

The gear ratio, expressed in number of teeth, is selected so as to prevent periodic synchronicity which would be detrimental to the process and would cause uneven wear of the tool. In the ideal case a workpiece should have to make up to 1000 revolutions before it returns to its original position relative to the lapping plates.

The method of asynchronous four-way motion produces excellent flatness and parallelism to very precisely held thickness while yielding high-quality surface finishes on both sides at the same time. The following results have been achieved at volume production rates:

- Thickness control to less than 5 $\mu$m (0.0002 in.)
- Parallelism to within 2.5 $\mu$m (0.0001 in.)
- Surface flatness to within 1 fringe
- Surface finish to good optical quality (80–50)

Although this method was originally designed for lapping, it has since been very effectively adapted for double-sided pad polishing of plane-parallel parts.

**Typical Machine Dimensions.** The lap diameter can range from a 100-mm (4-in.) diameter tabletop machine for precision lapping of quartz frequency-control crystals to a giant 1.9-m (75-in.) diameter production machine that is capable of lapping and polishing large sheets up to 600 mm (24 in.) in diameter or with a 600-mm (24-in.) diagonal if rectangular or square. The most common production machine sizes have a table diameter of 600 mm (24 in.) and 900 mm (36 in.).

The carrier diameters are approximately 35%, and the maximum part circle is about 30% of the plate diameter. These average values are for the standard five-carrier machine. Figure 5.80 shows the ratios of other types of carrier arrangements.

The annulus of the lapping plate must be smaller than the diameter of the part circle because the parts must extend a small distance over the edge of the lapping plate to keep it flat. A value of 25% of the plate diameter is the average for a five-carrier machine. This means that the part circle is between 10% to 15% larger than the track width of the annulus.

The part circle is a function of the carrier diameter, and it defines either the largest possible part diameter or, more likely, the diameter of the outermost row of part pockets. The part circle ranges from 70% of the carrier diameter for small diameter plates to up to 95% of the carrier diameter for the largest machines.

Part sizes determine the size of the ring gear, sun gear, plate diameters, and the size and number of carriers. This dependency results from the relationship between part diameter and gear ratio (number of teeth). The number of gear teeth is in a general way a function of the lapping plate diameter. The smaller teeth numbers are associated with the smaller machines, which are best suited for smaller parts. Many exceptions exist, however. Typical gear teeth numbers are 15, 21, 32, 50, 56, 66, 96, 106, 118, and 150, but other values also exist. Each size requires a different set of gears and laps. The

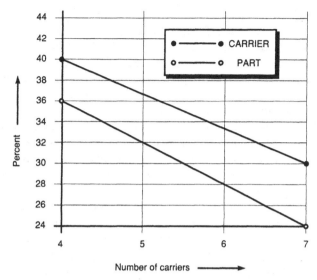

**Figure 5.80.** Carrier or maximum part diameter as percent of plate diameter.

plate speeds are usually preset, and the carrier rotation is determined by the gear ratio or the number of gear teeth. The gears can be cast or made up of a circular row of pins. The latter variety can cause warpage of thin carriers.

**Special Applications.** Nearly any double lapper can be set up for two-sided pad polishing, except that the lap drive motor may have to be replaced with a more powerful motor to overcome the appreciable frictional forces generated during this type of polishing. There are basically four types of double-lapping machines:

1. Standard double-lapping machine
2. Standard two-sided polishing machine
3. Double lapper for thin parts (smaller annulus)
4. Two-sided polishing machine for thin parts

Double-sided polishing can be done either with standard slurries or with chemo-mechanical pad polishing methods. Polishing can be done with or without pneumatic polishing down pressure.

Cylindrical parts, such as laser rods or steel pins, can be accurately lapped round on double-sided lapping machines. A specially designed carrier is needed for this approach to keep the rods at an angle of 15° inclined to the direction of the bottom plate rotation. The counterrotating plates cause the rods to roll. This makes them very round if the rod material is sufficiently hard and uniform.

**Lapping Plates.** Most lapping plates are made from a good quality cast iron. The better plates are made from a high-grade finely structured cast iron (mehanite). The better grade plates are preferred because they wear more uniformly and thus hold their flatness longer. They also last nearly twice as long as standard quality cast-iron plates before they have to be replaced. The lapping surface of the plates usually remains smooth, or the plates can be serrated with custom radial or crosscut grooves.

By attaching synthetic polishing pads such as polyurethane or poromerics (see Chapter 3) to both laps of a set, a double lapper can be readily converted into a two-sided polisher. The slurry supply system must be replaced or thoroughly cleaned out, and the plate drive motor may have to be replaced with a more powerful one.

For special applications involving lapping very hard materials such as ceramics and ferrites, ceramic plates have proved to hold up better. Plates made from stainless steel, copper, and tin have also been used for special applications. On small lab-type machines, lapping plates made from Pyrex and crown glass have proved to be useful for lapping soft and delicate materials. Although glass plates tend to lose their flatness more rapidly than the other lap materials, they are easily reconditioned by regrinding them on a plano lap using standard optical fab practice.

**Top Plate Lift.** While the top plate can be handled fairly easily for machines up to 300 mm (12 in.) in diameter, the weight of the top plate rapidly increases with increasing plate diameter. The top plate of a 1900-mm (75-in.) diameter double-lapping machine weighs a whopping 900 Kg or 2 tons [73]!

For this reason machines larger than 300 mm (12 in.) in diameter are equipped with top plate lifting mechanisms. The same feature is used to lower the top plate onto the work at the beginning of the lapping cycle. As the plates get larger and heavier, an additional tilt feature is added to the lift that tilts the plate 90°. This simplifies the inspection of the lap condition. For the largest machines separate movable lifting and tilting mechanisms are used. These units are also used to transport the heavy top plates for reconditioning.

A more important function of the lift mechanism is to counterbalance the often significant weight of the larger top plates. If the weight of the plate were not limited, it would cause unpredictable removal rates and wear of the plates, and thus breakage of the workpieces and damage to the drive mechanism.

The weight of top plates is counterbalanced through stepping motors that automatically respond to load conditions to control the pressure on the parts by slightly raising the plate. The heavy top plate can be raised and lowered in this way to an accuracy of 0.025 mm (0.001 in.). This feature is also used at the end of the cycle to raise the top plate by several mils in order to make it easier to load and unload parts.

The automatic sequencing of the pressure by a pneumatic or hydraulic plate elevation system is another important feature of top plate lifters. Severe chipping and breakage of fragile parts could result if the full working pressure were applied from the beginning to the end of the lapping cycle. The best way to control lap pressure is through a preprogrammed lo–hi–lo sequence. The cycle is started with near-zero pressure, which is then ramped up to full operating pressure where it is kept for a prescribed time interval of several minutes, to be reduced to near-zero again during the run-down phase. The start-up mode for fragile parts should include the use of a soft-start motor to avoid damage to the parts at the start of the cycle. Once the full lapping pressure has been reached, the machine will run for the preset time, after which the pressure will be reduced. The decrease in pressure at the end of the cycle will occur over a shorter time span than the pressure ramp-up.

**Slurry System.** The abrasive slurry must be supplied through a series of holes in the top plate. This is done through a slurry supply arrangement that rotates with the top plate and supplies a carefully metered amount of slurry to recesses in the top plate from which the slurry is permitted to seep through the holes to the workpieces. The metering is done by a multichannel peristaltic pump system. The slurry supply system can be one directional or it can be a closed-loop recirculating system. Some machines have an optional water rinse feature to flush off most of the slurry at the end of the cycle.

**Thickness Control.** The minimum part thickness is a function of lapping plate diameter. The larger the lapping plate, the greater will be the minimum part thickness that can be controlled. Figure 5.81 shows this relationship. Because of the much greater forces acting on the parts during double-sided polishing, the minimum part thickness is usually twice that of the minimum grinding thicknesses. In addition to the control of the lapping pressure, there is another very important feature that helps control accurate part thickness.

There are several ways to monitor part thickness during the lapping stage: The simplest of these is based on a timed cycle shutoff or on a lap counter. It operates on the assumption of accurately knowing the removal rate which usually varies with material, part configurations, load conditions, type of slurry, and the slurry supply system. There must be good empirical data and very good process control to make this method yield predictable results. Table 5.6 lists typical removal rates for fused silica parts and typical cycle times.

The second method is to monitor the lap position with an accurate electronic height gauge that measures against a reference surface on the top plate. The gauge mount is fixed with respect to the plate. The lapping cycle is stopped when the preset thickness position has been reached. The electronic

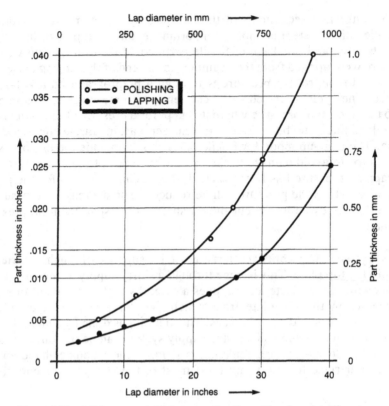

**Figure 5.81.** Minimum part thickness as function of lapping plate diameter.

indicator can be replaced by a capacitance probe that remains accurate to within 0.012 mm (0.0005 in.) over a wide range of thicknesses. Its primary advantage lies in the noncontact nature of the measurement.

A third approach uses air sensors to detect the separation between the top and bottom plates. This approach is best for parts thicker than 1 mm (0.040 in.). Repeatable thickness control accuracies within a few tenths have

**Table 5.6. Removal Rates and Cycle Times for Lapping Fused Silica**

| Abrasive (µm) | Max. relief layer per Side (µm) | Removal (µm) | Average rate (µm/min.) | Time (min.) |
|---|---|---|---|---|
| 35 | 175 | 350 | 25 | 7 |
| 25 | 125 | 250 | 10 | 12 |
| 12 | 60 | 120 | 6 | 10 |
| 5 | 25 | 50 | 2 | 12 |

been achieved with this method. It can also be preset to shut down the machine when the proper thickness has been achieved.

A separate method of thickness control that monitors the frequency of witness crystals is useful only for lapping piezoelectric quartz crystals of a thickness of 1 mm (0.040 in.) or less. The accuracy of this method is very good, usually within 5 $\mu$m (0.0002 in.).

**Machine Drives.** It is important to maintain and accurately control the rotational speeds of the various gears and plates. For this reason most double lappers use infinitely variable speed dc motors to drive the ring gear and the lower plate, the center sun gear, and the top plate independently. These motors should have a soft start feature with an adjustable time delay. The speeds of the various drives is controlled by the adjustment of knobs on the control panel and monitored by analog or digital electronic tachometers. The maximum plate speed for small double lappers is 60 rpm. For larger machines, it is less than 40 rpm, with only a few rpm for the largest machines.

**Special Features.** With a variety of machines available for a wide range of applications, it is unavoidable to find a number of special features. Some of these are:

- A programmable machine controller that regulates critical operating parameters such as speeds, pressures, and cycle times for near fully automatic operation.
- Maintenance of lap flatness during the cycle by controlling the slurry temperature or by utilizing a recirculating liquid cooling system in the lower plate. This feature carries off excess heat generated during the lapping and especially during the polishing cycle that otherwise would lead to distortion of the plate surface.
- Some machines can be programmed to maintain the rotation of the upper plate after the machine has been shut off. This tends to break the often significant suction that makes it difficult to remove the parts during the unload cycle.
- Larger machines have typically retractable ring gears to make it easier to load and unload parts. Also a loading table that is flush with the surface of the lower plate facilitates the load and unload operation. Loaded carriers are simply slid onto the plate during loading and slid off when the lapping cycle is complete.

**Plate Flatness Control and Correction.** Even with the most careful process control, uneven plate wear will inevitably occur. For this reason the flatness of the plates must be regularly checked and monitored, and must be periodically reestablished when there has been unacceptable wear.

Plate flatness is measured with a three-point bar spherometer that has at least a 2.5 $\mu$m (0.0001 in.) resolution. The flatness is measured at several positions on the plate. When the reading exceeds 5 $\mu$m (0.0002 in.) across the annulus, it is time to correct the plate flatness. This measurement should be done on a regular basis.

Flatness correction, also called *plate reconditioning,* should be done frequently when operating in a volume production mode. For small- and medium-size machines, it is best when two sets of lapping plates are available. One set can be reconditioned while the other set is on the machine lapping parts. Laps should be routinely reconditioned once each shift. It is a lot easier and faster to maintain lap flatness in this way than to wait until the lap is so badly worn that a complete resurfacing is necessary.

Lap flatness can be maintained and slightly out-of-flat laps can be reconditioned with the periodic use of bronze or cast-iron dressing gears. This operation is greatly facilitated when the carrier rotation is reversed. Bronze gears are also used to reestablish lapping plate surfaces that had to be refaced on a lathe or on a surface grinder. This method should only be used as the last resort because it is a labor-intensive operation that also reduces the life of expensive lapping plates, and the replacement of bronze correcting gears is a costly proposition. The most cost-effective way to correct lapping plates and to maintain their flatness is to lap them on an annular machine from time to time with a medium-fine (25-$\mu$m) abrasive for about one-half hour.

**Carriers.** Carriers are essential to make double lapping possible. These are thin, circular pieces of metal or plastic that have a specific hole pattern to accept parts for double lapping. The periphery of the carrier is geared to engage the sun and the ring gears of the machine. This arrangement imparts a cyclical motion to the parts while the carrier is rotating in a planetary motion because of the different speeds of the two gears, as shown in Fig. 5.82.

Carrier size and thickness is a function of part size and part thickness. The smaller the part, the smaller is the carrier. The thickness of carriers must be as thick as the thickness of the finished part will permit. It should not be more than 0.025 mm (0.001 in.) less in thickness than the minimum allowable part thickness when lapping parts that are 0.25-mm (0.010-in.) thick or less. For thicker parts the carrier thickness should be about 15% less than the minimum part thickness. A greater allowance is required for double-sided polishing to allow for the compressibility of the pad.

The hole pattern in the carriers must be symmetrically laid out around the carrier center. There is usually no hole in the center of the carrier to prevent preferential (zonal) plate wear in the central zone. Enough material must be left between the holes and the geared edge of the carrier so that they will not fail under normal operating conditions. This webbing should not be less than 3 mm (0.120 in.) for thick carriers and 6 mm (0.240 in.) for thinner ones. The pattern must permit the edges of the part to sweep over the edge of the lapping plate by about 20% of the part diameter. The holes that accept the parts should have the same geometric shape as the parts that go into them

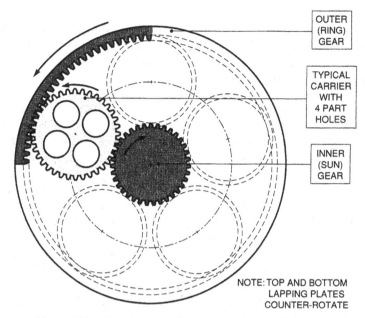

**Figure 5.82.** Basic configuration during double sided lapping.

but they must be larger by about 1 to 1.5 mm (0.040 in. to 0.060 in.) to allow the parts to move in the pockets. The danger of edge damage to thin parts can be reduced if the holes are only 0.25 mm (0.010 in.) larger.

Carriers can be made from a number of metals and plastics. The best choice depends on the specific application and the thickness of the parts. For instance, blue spring steel carriers can be used as thin as 0.05 mm (0.002 in.) for lapping very thin parts such as frequency control crystals. Stainless steel is most durable. Mild steel can also be used. Zinc alloy carriers are corrosion resistant. These are also a good choice for fragile parts where edge chipping would be a serious problem with other carrier materials. Reinforced fiberglass is a popular choice for thicker parts down to 0.4 mm (0.016 in.) thick. These fabric-reinforced resins do not work well with aqueous slurries because they tend to absorb water, which causes them to swell up. Vynil carriers are an economic choice for many noncritical applications. They can be used for part thicknesses down to 0.25 mm (0.010 in.). Lexan and Lucite (optical acrylics) are the best choices for polishing, since they don't cause scratches. These plastics are not very strong, however, and they can be used only to a thickness down to 0.5 mm (0.020 in.).

### 5.3.6. Annular Machines

There are many different types of annular machines ranging from 300-mm (12-in.) diameter lab models to giant machines with tables of more than 3 m (10 ft) in diameter. These are used for free-float lapping and polishing of plano surfaces, and most can be converted from lapper to pad polisher. An increasing number have been specifically built for free-float pitch polishing.

Annular machines used in optical shops for plano fine-grinding are referred to as *continuous grinders* (CG) and when they are set up for pitch or pad polishing they are called *continuous polishers* (CP). These terms and abbreviations will be used in this section.

Annular machines suitable for prototype, and R&D work are usually small tabletop machines with table diameters ranging from 300 to 450 mm (12 to 18 in.). For small shops and light production, 600-mm (24-in.) models have proved to be quite adequate. The best sizes for production are 900-mm (36-in.) and 1200-mm (48-in.) machines. Larger machines ranging from 1.5 m (60 in.) to 3 m (120 in.) have been built for special applications. Such large machines, however, are difficult to operate and to keep flat.

Despite a wide range of table sizes all annular machines share certain dimensional relationships, which are expressed in terms of table diameter. The width of the annulus lies between 30% and 40% of the table diameter, with 36% as the average value. The table ID is about 25% of the OD.

The parts are held in septums that are constrained by driven rings. The typical annular machine has four rings, although other arrangements are also possible. When the machine is set up for pitch polishing, there are only three rings available to accept parts; the fourth station is reserved for the conditioning plate which keeps the polishing surface flat. This conditioner is often referred to as the *bruiser*. The ring ID, which is another way of saying *maximum part diameter*, averages 35% of the table OD with a range of 30% to 40%. As a rule of thumb one can say that the maximum part circle is about 33% of the annular table diameter.

The table speed of fixed speed drive machines is roughly a function of table diameter with the smaller machines running faster and the larger running slower. It is difficult to be more specific because there are so many departures from this basic observation. Much depends on the specific application. Most modern machines, however, are equipped with stepless variable speed drives. Figure 5.83 shows the typical speed ranges for annular lapping (*a*) and polishing machines (*b*).

The basic annular machine consists of a plano lapping plate that has a hole or relief in the center to create an annulus-shaped lapping surface. Typically four workpiece rings that are constrained by a centrally mounted yoke rest on the lapping surface or are suspended just slightly above it. The workpieces (parts) are placed face down into the workpiece rings, and they float under their own weight on the abrasive slurry that is supplied to the lapping surface. If the parts are too light, they can be individually weighted down, or a pneumatic pressure plate could be lowered to provide the required grinding pressure. As the table (lap) slowly rotates, the rings rotate under friction in the yokes and the parts are ground flat.

While most annular machines have fixed yokes, some smaller machines are equipped with yokes that impart a small 6 to 12 mm (0.24 in. to 0.48 in.)

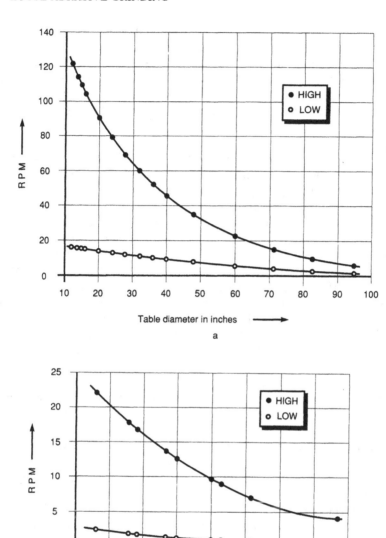

**Figure 5.83.** Speed ranges versus table diameter. (*a*) Annular lapping machines. (*b*) Annular polishing machines.

back and forth motion to the rings. This added motion is thought to make it easier to keep the lap flat. Oscillating rings cannot be used with pneumatic downpressure plates. In addition to the ring oscillation, most machines use caliper-type ring positioning which can be adjusted a small distance by turning a knob to precisely change the position of the rings relative to the lap. This feature is also designed to control the flatness of the lap and thus the parts.

**Rings.** When the annular machine is set up for lapping, the rings are usually free floating on the lapping surface. The rings can be powered, however, so that they maintain proper synchronization of the ring and table rotations. This prevents slipping, which can lead to uneven plate wear in time. Powered rings are essential for pad or pitch polishing or for fine lapping on ceramic laps. The rings for annular pitch polishers must be driven for precise synchronization. When the rings ride on the pitch lap, they must have runners of a suitable material. For instance, glass runners must be cemented to the contact surface of the rings when polishing glass parts. When polishing other materials, the material used for the runners must be chemically and thermally compatible with the part material and it must be of sufficient hardness to take on a polish at the same rate as the workpiece material. In most cases this means that the same material is used for both parts and runners.

To avoid this extra expense and reduce the load on the pitch polisher, the rings can be held slightly above the pitch surface by three rollers that run in a groove on the outside diameter of the rings. The rings are held in position by a yoke that has two rollers, one at each end, one of which can be driven to rotate the ring by friction drive. The yoke can also be adjusted to reposition the ring and slightly oscillated by a small adjustable distance to prevent the formation of "tracks" on the pitch surface. This arrangement also imparts a random motion to the workpieces.

Optical parts can be polished singly as unblocked parts, or they can be blocked on a plano blocking tool. They float within the constraints of cutouts in an aluminum or plastic plate, called a *septum*, which is supported by the ring just above the pitch surface. The synchronization of table, ring, and bruiser plate rotation is even more important for annular pitch polishing than it is for either lapping or pad polishing on annular machines. Closed-loop synchronization of ring rotation with the rotation of table and conditioner is an important option to consider when converting an annular machine from pad to pitch polishing.

**Annular Table.** Annular lapping plates or tables are typically made from fine-grain cast alloy that must have uniform density and hardness to ensure uniform wear. High-density cast iron and select grade steel are the materials most often used for these tables. Tables of up to 1.2 m (48 in.) in diameter are usually cast in one piece and ground flat on a suitable surface mill. Larger tables are cast in pie-shaped sections and assembled before the lap surface is established. The lapping surface is almost always serrated by a cross grid goove pattern or a radial groove pattern. This facilitates the uniform distribution of abrasive slurry. Proper grooving can increase the lapping efficiency by as much as 50%.

Table flatness during lapping is maintained by trueing rings that ride on the lap. The workpieces are held within these rings. A true flat surface will tend to go low (concave) if the rings are rotating by only frictional forces in the same direction as the table. Technically this occurs because the relative velocity between the ring and the table is a minimum at the outside edge of the table and a maximum at the inside. This situation can be changed by reversing the ring rotation relative to the table so that the maximum velocity difference is at the outside edge of the table. This arrangement will preferentially wear the outer zone of the table and restore table flatness.

For critical applications where very good surface flatness must be maintained, the lapping plate (or table as it is often called) can be equipped with an integrally mounted water cooling system. This limits thermal differences and can lead to the deformation of the lapping table. The temperature and pressure of the infeed and outflow of the water cooling system are constantly monitored and automatically controlled by adjusting the flow of water.

The smaller machines of up to 450 mm (18 in.) are often equipped with interchangeable laps made from a variety of different materials. Glass and Pyrex plates are often preferred for lab and prototype machines. Many different materials can be lapped on glass and Pyrex, mainly because they are noncontaminating. Lapping plates made from these materials can also be easily resurfaced by conventional grinding. When very hard materials such as ceramics, ferrite, sapphire, and tungsten must be lapped, it is best to use diamond abrasives. Special laps with a porous structure are required to provide small voids for holding the typically octahedral diamond grains. Laps with such properties are most often of composite materials made with powdered metals and resins or ceramics. The most common metals used are steel, copper, tin, and a tin-lead alloy. There are also nonmetallic composite materials used for diamond lapping. The annular table drive motor power must be increased when diamond abrasives are used as the lapping medium, since it must be able to overcome the significantly higher resistant forces that are generated during diamond lapping. An annular machine not specifically modified for diamond lapping may not be suitable unless the drive is changed.

One of the most important applications of annular machines is in their conversion to continuous polishers (CP). This is a relatively recent development for large machines, although pad and pitch polishing on this type of equipment has been done for some time on a smaller scale. The use of CPs started in the 1960s and by the 1980s nearly half of all plano precision optics were polished on CPs [74]. The advantages provided by carefully designed and properly maintained continuous polishers account for their wide acceptance in the precision optical manufacturing industry. Semiskilled operators can achieve excellent flatness and good surface finishes even on irregular-shaped plano surfaces as long as the machine is used properly. Most optical shops are now relying on custom-made or commercial CPs for most of their plano-polishing work.

Most CP's are specifically designed for pitch polishing. The conversion to a pad polisher is as easy as cleaning the table and applying a precut synthetic polishing pad to it. A semihard perforated pad has proved to be the best choice. Because the pad should ideally be in one piece, this approach is generally limited to machines with less than 1.2-m (48-in.) diameter tables. Pads have been successfully applied to machines as large as 2.4 m (96 in.). The oversized pads are constructed of three or four separate pieces sectioned in such a way as to form one polishing surface with minimal seams. The sections are curved away from the direction of rotation and their trailing edge is cut on an angle so that it slightly overlaps the leading edge of the next section. The idea is to prevent sharp edges from getting caught on the machine parts and coming loose during the polishing. This would lead to a sudden catastrophic pad failure, which can cause serious damage to the parts and even to the machine.

Continuous polishers made specifically for pitch polishing have a massive table with a 4 : 1 diameter-to-thickness ratio. The table is either water cooled if it is made from cast iron or used without additional cooling if it is made from granite. It must be thermally insulated from heat generated from within the machine by drive motors, gear drives, and bearings. Such measures to ensure proper thermal control of the pitch lap tables are important because uncontrolled thermal changes adversely affect flatness. An optional temperature control system typically consists of a chiller, a heater, a thermostat, and a recirculating pump system with suitable plumbing connected by a journal to the rotating table. The cooling water recirculates in a closed loop through channels inside the table, carrying away excess heat generated during the polishing cycle. Temperature control to less than ±3°C (about ±6°F) can be achieved with a commercial system. For critical applications even better thermal control could be required.

Mechanical vibrations must also be closely controlled and kept to an absolute minimum. This is one of the reasons why most annular lappers (CG) cannot be readily converted to a annular pitch polishing machine (CP). Several modern CP utilize an air-bearing table to provide vibration-free rotation. The table rides on a thin cushion of compressed air, which tends to dampen out any machine vibrations. A commercial machine [75] is vee-belt driven, as shown in Fig. 5.84. This greatly reduces the transfer of vibrations of the motors and drives to the lap table. These features also greatly improve flatness and surface quality of parts.

After the table surface is heated, preferably with a battery of heat lamps, the pitch lap is cast by pouring melted pitch onto the table as the table slowly turns. Removable dams made from strips of aluminum or other suitable

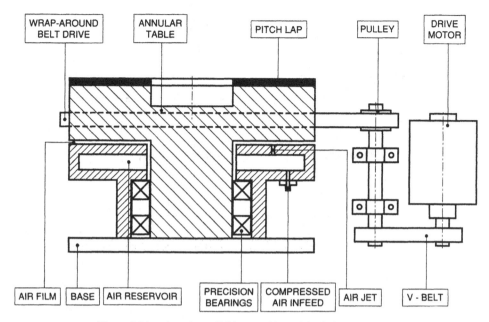

**Figure 5.84.** Annular polishing machine with air-bearing table.

material must first be attached to the OD and ID of the annular table so that the pitch cannot run off the table. The pitch thickness is a function of table diameter, with small 300-mm (12-in.) diameter machines requiring a pitch thickness of about 13 mm (about 0.5 in.) and larger machines requiring proportionately greater pitch thicknesses. The pitch lap for a 1.2-m (48-in.) CP is often cast to a thickness of 50 mm (2 in.). This is shown in Fig. 5.85a.

The freshly cast pitch lap must be made flat after it has been allowed to cool to room temperature. This is done with a pitch cutting mechanism that is mounted across the annulus from center to edge. Figure 5.85b shows this configuration. A manually operated or motor driven leadscrew moves a specially shaped pitch cutter across the lap surface while the table turns. In this mode the CP acts like a giant lathe. Cooled water should be applied to the point of contact to keep the pitch sufficiently brittle for efficient cutting. The cutter can be lowered after each pass so that the entire annular surface can be faced off flat.

The large flat pitch surface must be grooved to permit the slurry to uniformly cover every part of it. The choice of grooves depends a great deal on the operator's preference and to some degree on the specific application. Some prefer either concentric or spiral grooves, while others insist on radial grooves. The pitch cutter can be used to cut these grooves into the pitch surface. Circular and spiral grooves are cut with the table rotating slowly. Circular grooves can be cut with a manual lead screw drive. Spiral grooves can only be cut accurately with a motor-driven lead screw because the pitch

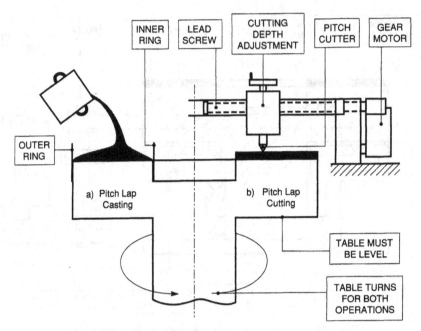

**Figure 5.85.** Preparation of annular pitch laps.

cutter traverse speed must be constant. Radial grooves are cut with the table locked in a fixed position for each cut.

**Conditioner.** For most annular lapping machines, the workpiece rings also act as conditioners to prevent zoning of the lapping plate and to keep it flat. This arrangement is not sufficient for annular pitch polishers (CP). The control of pitch lap flatness depends on a large conditioning plate, called the *conditioner* or *bruiser*. Most annular precision polishing machines have four stations but only three accept rings to hold the parts. The fourth station, usually the one in the inaccessible back position, is set aside for the conditioner. Figure 5.86 shows this arrangement.

The lap flatness is controlled by the precise radial positioning of the conditioner relative to the pitch annulus. This adjustment can be made while the machine is running. A witness flat rides along in one of the rings. It is periodically checked for flatness in an interferometer. Any change in flatness alerts the operator to a developing out of flat condition of the pitch annulus. For instance, if a 150-mm (6-in.) diameter flat shows one fringe low (concave), then the flatness across the 400-mm (16-in.) annulus is likely to be $400^2/150^2$ (1 fr) or 7 fringes high (convex). The position of the bruiser is then moved in a small distance toward the center of the machine to reduce the convexity of the annular surface.

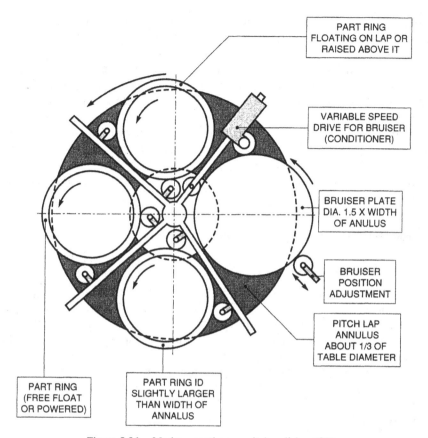

PART RING
FLOATING ON LAP OR
RAISED ABOVE IT

VARIABLE SPEED
DRIVE FOR BRUISER
(CONDITIONER)

BRUISER PLATE
DIA. 1.5 X WIDTH
OF ANULUS

BRUISER
POSITION
ADJUSTMENT

PITCH LAP
ANNULUS
ABOUT 1/3 OF
TABLE DIAMETER

PART RING
(FREE FLOAT
OR POWERED)

PART RING ID
SLIGHTLY LARGER
THAN WIDTH OF
ANNALUS

**Figure 5.86.** Modern continuous pitch polisher (CP).

Although there are many departures to this rule, it is generally true that a conditioning ring for an annular lapper is about 35% larger than the width of the annulus. For continuous polishers the diameter of the bruiser plate is about 50% larger than the annulus is wide. Bruisers are often made from uniform grain, high-density granite. Thin glass rounds, which are arranged in a uniform but nonconcentric pattern, are usually cemented to the bruiser plate to form the surface in contact with the pitch lap. Conditioning plates made from very low expansion materials such as ULE or Zerodur can run on the lap directly.

The conditioners for annular pitch polishers are independently driven to ensure that the optimum synchronization between table and conditioner rotation is kept constant. The variable speed conditioner drive motor is controlled by a closed-loop servo system to maintain the synchronization throughout the polishing cycle. If this is not done, then the synchronization would be lost, and the conditioner will spin or lag relative to the table as the lap surface flatness changes or as the rate of slurry feed becomes erratic.

Large conditioners are heavy, and a hoist is often needed to counteract their weight during polishing and to lift them off the machine for inspection. A manually operated hydraulic lift mechanism raises the conditioner off the pitch lap when the CP is not in use. If this were not done, the weight of the conditioner would cause it to settle into the lap surface, thus destroying the polisher. The hydraulic lift can also be used to remove the conditioner entirely from the machine. This is necessary for inspection of the bruiser surface and when a new lap must be cast. Special weights can be added to the conditioner of smaller machines to affect the correction of out-of-flat lap conditions.

## Special Features for Annular Machines

### Speed Control and Timers

• Modern machines are equipped with infinitely variable speed controls for most machine functions. These are no longer a luxury but an essential requirement for cost-effective operation. The operating speed is monitored by a *tachometer*, which displays the speed either on a digital display or on an analog gauge. Soft materials are lapped at slower speeds than harder materials. The requirements for pad or pitch polishing are somewhat different but they serve the same purpose.

### Load Gauges

• The potentially high lapping and polishing forces acting on the table and drive train require the use of torque or load meters to avoid overloads, which can lead to stalling of the drive motor. Since these are usually heavy-duty motors, it is more likely that the parts get damaged first. High overloads can occur during lapping and pad polishing, especially when high down pressure is applied to the parts by pneumatic pressure plates to increase the stock removal and speed up the polishing cycle. Since the parts float on annular pitch polishers under their own weight or with only light additional pressure, overload conditions are unlikely and load meters may not be needed. The load meters are basically ammeters that sense the increase in current drawn by the drive motor to overcome the increased torque. For this reason, they are often referred to as torque meters.

### Vacuum Chucks

• Vacuum chucks have been developed for use on annular lapping and polishing machines. While they have found widespread use in semiconductor substrate preparation, they are rarely used in optics manufacture. Optical parts are either free floating or wax blocked to a common blocking tool.

### Load and Unload Features

· Many annular lapping machines have a loading table that is at the exact height of the lapping plate. This permits the loading of one set of rings while another set is still running on the machine. This arrangement reduces downtime to a minimum. This feature is not useful for pad and pitch polishers since polished surfaces are lifted off the machine and never slid off.

### Machine Construction

· The machine base for an annular machine must provide a high degree of rigidity to ensure a vibration-free operation that is essential for precision lapping and polishing. This is an especially important consideration for annular pitch polishers (CP). In all cases the table drive train should be very quiet and should generate as little heat as is practical. This usually requires a belt drive, heavy-duty thrust bearings and a massive machine base. Much, however, depends on the specific application and the size of the machine.

### Down Pressure

· The pressure provided by the weight of the parts is usually inadequate for rapid stock removal during lapping and high-rate polishing. Only very large and heavy parts would provide sufficient pressure but such parts are rarely, if ever, processed in an optical shop. Therefore it is necessary to provide additional pressure. This can be done by placing weights on the parts or blocks. This is the usual method for annular lapping machines (CG) or for annular pitch polishers (CP). Most modern annular lapping and polishing machines come now equipped with pneumatic down pressure to ensure lapping and pad polishing pressures between 0.20 to 0.35 Kg/cm$^2$ (3 to 5 psi) on the parts. To ensure that the workpieces are kept in intimate contact with the lap or pad surface, the pressure plates must be self-aligning by a pivoting joint that connects it to the pneumatic ram.

### Slurry System

· Almost all, except the smallest annular machines, are set up with a closed-loop recirculating slurry system. The continually agitated slurry is supplied to the lap surface at a carefully measured flow rate to ensure uniform stock removal. For pad polishing the slurry is often supplied in copious amounts to flood the pad surface. This provides sufficient polishing slurry for rapid polishing, and the excess flow serves a thermal control function by carrying away heat generated by the significant friction between the pad and parts. An external heat exchanger is often used for critical polishing operations to carefully control the slurry temperature and keep it in a narrow range. The flow rate of slurry fed to a CP is usually not very high to avoid undue buildup of slurry on the pitch lap. Only a sufficient amount of slurry should be applied to maintain a constant polishing rate and to avoid the formation of dry spots on the lap. An exception to this is submerged polishing (bowl polishing) which has been successfully done on CPs. This approach does not rely on a recirculating slurry flow, but on a specific volume of slurry that is contained on the lap by suitable dams on the inside and outside edges of the table. This approach can be used for pitch polishing low-scatter surfaces.

**Results Obtained with CG's and CP's.** The efficiency of abrasive lapping on annular machines depends on the type of material and the configuration of the workpieces, the size and type of the abrasive slurry, machine parameters such as relative velocity between lap and workpiece, as well as pressure and cycle time. According to Preston's equation that describes the abrasive and polishing processes, the relative velocity is even more important than the pressure applied to the parts. Surface roughness is also a function of these fundamental parameters and any discussion of typical values is not very meaningful unless the material is specified. For instance, there have been reports of crystal quartz and fused silica parts lapped to a finish of 0.1 $\mu$m rms (2 $\mu$in. rms) [76] or 0.05-$\mu$m $R_a$ (1-$\mu$in. $R_a$) [77]. $R_a$ is the designation for average surface roughness as measured with a stylus profilometer. Surface flatness attainable during lapping is a direct function of the flatness of the lapping plate. If the machine is used properly, flatness of less than one fringe per inch is attainable, but two fringes per inch is more common. The part thickness can be controlled to within $\pm 2.5$ $\mu$m ($\pm 0.0001$ in.) and parallelism of 50 $\mu$in./in. (10 arc secs) have been measured on large parts. Workpieces as small as 3 mm (0.120 in.) in diameter and as large as 900 mm (36 in.) in diameter have been successfully lapped on annular machines.

The results of annular polishers are typical for precision optical manufacture. Pad polishing operations are used primarily for the rapid polishing of commercial quality optics, and any statements on typical flatness and surface finish may be misleading. The results that can be achieved with a properly controlled annular pitch polisher or CP can be spectacular. Flatness of $\frac{1}{8}$ wave are routine on most parts, and $\frac{1}{20}$ wave or better have been done on many occasions. Surface roughness is typically somewhat more than can be achieved on a conventional polishing machine but comporable finishes have been produced with a well-controlled machine using the submerged slurry polishing technique.

**References for Section 5.3 (Loose Abrasive Grinding)**

[69] Maksutov, Generation of Optical Surfaces.

[70] Hed, P. P., D. F. Edwards, and J. B. Davis, "Surface damage in optical materials," *LLNL, Optical Fabrication and Testing, Technical Digest,* Series Vol. 13, 1988.

[71] Harry Schade, Franke & Heidecke, Germany, private communication.

[72] J. Steppi.

[73] Saga, 1975.

[74] Strasbaugh, from 6C9 series.

[75] Lapmaster International, Morton Grove, IL, 48 in.

[76] P. R. Hoffman.

[77] Lapmaster.

## 5.4. POLISHING

The fine-ground surfaces become reflective and transparent during the fabrication step called *polishing*. The optical surfaces attain their final shape at the same time. Metallic polishing tools are covered with a viscoelastic material to which an aqueous slurry of a mildly abrasive polishing compound is applied.

### 5.4.1. Theoretical considerations

It is difficult to find a satisfactory explanation for the polishing process, since one cannot see what goes on between the glass and the polishing tool. As a result only the effects of these processes can be interpreted, or the assumed processes can be reconstructed through parallel experiments.

The polishing process was initially seen as a flow phenomenon during which the glass flows from the asperities of the ground surface toward the pores, thus filling them in. One researcher [78] drew steel pins under pressure across a polished glass surface. The microscopic examination revealed that the glass was deformed in a trenchlike fashion without being torn, leaving the edges of the trench raised like a levee. This fracture-free displacement of surface material, however, did not explain the verifiable material removal.

Another researcher [79], among others, determined on the basis of his investigations that polishing is a purely mechanical removal process comparable to a very fine grinding process. This supposition coincides with the observations made by the optician during polishing. It is commonly known that coarse, and therefore aggressive, polishing compounds polish faster than finer compounds. That fact can lead one to infer the mechanical effect of polishing compounds. Also a thickness and weight reduction can be established after polishing. For instance, a 5-cm² large piece of plateglass became thinner by 6 $\mu$m after 1 hour of polishing time and a disk of 85-mm diameter made of optical glass BaSF 7 became 0.2 g lighter after 3 hours of polishing. Surfaces of sensitive glasses that were polished with coarse abrasive tended to form stains. That observation permits the assumption that a visually no longer discernible surface roughness exists. This is only explainable through an abrasive effect of the polishing compound.

The existence of such a residual microstructure on an ultra-low-scatter surface for ring laser gyros has been visually verified by very high (1000×) magnification microscopic examination with a specially fine-tuned Nomarski microscope [80]. In addition scatter measurements with a laser scatterometer showed evidence of numerous fine, parallel microsleeks on the super polished surface, which showed up as identical scatter peaks nearly 180° apart. The peaks are thought to be the result of a grating effect from the multitude of the parallel microsleeks. On several samples a number of such matching pairs of scatter peaks were detected but each with a different angular separation. This was interpreted as the result of the final few sweeps of the block over the rotating lap at the conclusion of the polishing cycle. The final polishing was done only with water. No new slurry was applied. Therefore the microsleeks could have only been caused by polishing compound particles that were embedded in the pitch surface.

From other sources, polishing was explained as a chemical process. Through the friction of the polishing compound and the action of the water used in polishing slurries, the glass surface is supposed to be chemically altered so that the abraded material is dissolved in the water and is carried off by it. In fact the water in automatic slurry systems used on polishing machines becomes alkaline after prolonged use. That leads to the conclusion that glass is dissolved in the water. Also locally high temperatures are required for polishing. The heat tends to promote chemical reactions and a dissolution processes at the same time. In response to this theory, shorter polishing cycles are supposed to have been achieved by adding salts of hydrofluoric acid to the polishing compounds.

A combination of these various theories has been proposed [81] to develop a comprehensive polishing theory. It proposes that chemical as well as physical processes take place. Chemical processes predominate during fine polishing, whereas for coarse polishing the mechanical processes take over. Other studies [82] have shown that the most likely polishing mechanism is a complex chemo-mechanical process that is dependent on the nature of the optical material, the type of polishing compound, the characteristics of the pitch lap, and other specific operating parameters.

It is a basic characteristic of polishing compounds that only certain rare earth oxides [83] are effective in producing optically polished surfaces. While metals and certain hard crystals can be flawlessly polished with diamond powder, optical glass and many other optical materials cannot be polished with it. A unique paper [84] on the nature and behavior of optical polishing pitch goes into considerable theoretical detail. This important paper has been paraphrased in the section on optical polishing pitch in Chapter 3 of this book.

### 5.4.2. Manual Polishing

Manual or hand polishing is done on the pedal machine or on the motorized bench. The workpieces are typically prototype lenses, repair of finished lenses or prisms, and smaller test plates. The parts should not be larger than 60 mm (2.4 in.) in diameter. Otherwise, they could be no longer properly guided by hand, and the force required on the pedal bench to overcome the resulting friction would become too great.

The polishing tool is mounted on the spindle. The workpiece is typically blocked to a wooden handle (Fig. 5.87). Strongly curved convex surfaces are blocked on a long handle, while less strongly curved surfaces are blocked on a correspondingly shorter handle. The handle is held close to the surface on longer radii, while it is held beyond the center of curvature for shorter radii. The polishing tool should be about as large as the lens for strongly curved surfaces, and it should be larger in diameter by about 25% for longer radii,. The edge of the tool is neatly trimmed. A hole is cut into the center for pitch tools so that the viscoelastic polishing pitch can flow toward the center.

For larger polishing tools it is customary to cut rings into the pitch with a knife (Fig. 5.88). These rings accept some of the slurry and release it again in a predictable manner. When the rings are cut near the center, the polishing tool will contact the workpiece more at the edge of the surface. Rings that are cut near the edge permit the tool to polish more from the center. This effect can be further enhanced by the way in which the workpiece is moved

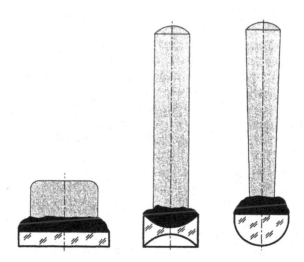

**Figure 5.87.** Lenses blocked to handles for manual polishing.

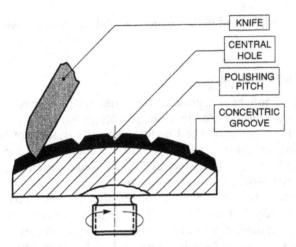

**Figure 5.88.** Cutting grooves into a pitch polisher.

across the tool. When the workpiece is moved across the tool center with short strokes, the workpiece is preferentially attacked at the edge. When the strokes are long and the part is moved over the edge, the center of the workpiece surface is favored. These effects result from the viscoelastic nature of polishing pitch. The pitch will slowly fill in the grooves, and this pitch flow changes the shape of the tool.

Strongly curved surfaces are hand polished on a rapidly rotating polisher with a brisk motion. For longer radii and for plano surfaces, these movements are guided primarily in linear fashion, while the polishing tool, which is mounted on the spindle, rotates at a much slower speed. Only in this way can an accurate surface be achieved.

The radius of curvature of the concave cast iron polishing tools is 1 to 2 mm (0.040 to 0.080 in.) larger than the radius of the polished surface. For convex polishers the tool radius is correspondingly shorter (Fig. 5.89). Many opticians prefer much thicker pitch layers, but it is doubtful that the same degree of control can be achieved with such a polisher. The applied pitch layer ought to be somewhat thicker in the middle than at the edge to make allowance for the pitch that will eventually flow to fill the grooves cut into the surface. The polishing pitch is poured into the preheated cast-iron tool and pressed out with the lens that needs to be polished.

Cerium oxide and zirconium oxide serve as the preferred polishing compounds today. The use of the once preferred iron oxide is now very rare. The polishing compound is mixed with water and is applied with a suitable brush. Only a small amount of slurry is applied because the polishing effect is greatest when the polisher becomes nearly dry. When polishing very soft glasses or crystals, the polishing slurry is taken with a brush from a rub-in tool. Such a tool consists of two fine-ground glass plates between which the polishing compound is rubbed before it is put to use. The lower baseplate is

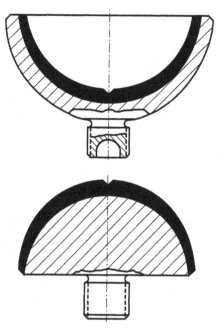

**Figure 5.89.** Polishing tools in cross section.

somewhat larger and thicker than the upper plate, which is typically blocked to a wooden handle (Fig. 5.90). The greatly improved quality of currently used polishing compounds may make this practice unneccessary.

**Hand Correcting.** Hand correcting is done on plano surfaces using stationary polishers. This method is primarily used to remove stains and sleeks from polished optics. The surface flatness is maintained or even improved.

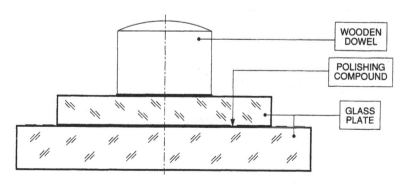

**Figure 5.90.** Rubbing tool for polishing compound.

The polishing tool is covered with a layer of soft polishing pitch which is no more than 2 mm (0.080 in.) thick.This layer must be applied uniformly. The cast-iron tool should be ground very flat. If the tool were slightly concave or convex, then the finished pitch lap would unquestionably polish from the edge or from the center, respectively.

The hand-correcting tool is pressed out on a flat glass plate which has been polished flat and tested against a plano test plate or a plano interferometer.The polisher is placed face down on the glass plate and weighted down as shown in Fig. 5.91. To prevent the polishing pitch from sticking to the plate, it is first dusted with dry polishing compound. When the polisher is needed, it is removed from the plate by a light tap from a small soft-tipped mallet. It should never be forcibly slid or lifted off the plate because the exact plano polishing surface could easily deform.

To offer the optician the greatest latitude in hand-polishing plano surfaces with varying flatness conditions, three pitch polishers are sometimes used. Each polisher is pressed out against its own pressing plate prior to use. One of the pressing plates is low (concave) by a fringe or two, one is high (convex) by the same amount, while the third is between 1/2 to 1/4 fringe low. This nearly flat polisher is used for the final correction. The other two are used during the early stages of hand polishing to correct for the initial out-of-flat condition or to correct for overcompensation. While one polisher is in use, the other two are being pressed out.

The polishing tool is firmly mounted in a fixture that has three adjustable jaws to hold large and small polishers. Two jaws are fixed while the third is used to lock in the polisher (Fig. 5.92). The polishing slurry is applied to the polisher on which it is uniformly distributed with a plano glass plate. In this way the pitch layer remains plano and coarse grains or agglomerates of the compound are pushed aside. Since rouge (i.e., iron oxide) tends to charge the polisher quickly, it is still used for hand polishing along with the contemporary compounds.

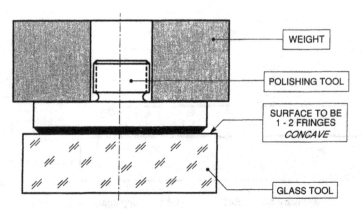

**Figure 5.91.**  Pressing out a pitch hand polisher.

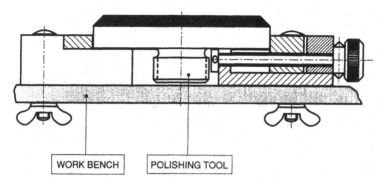

WORK BENCH      POLISHING TOOL

**Figure 5.92.** Tool clamp for hand polishing.

The glass part that must be polished is held close to the surface between thumb and middle finger of the right hand. Parts that tend to distort easily are blocked stress free on a piece of glass tubing (Fig. 5.93). For asymmetric surfaces, such as found on prisms, dummy pieces are fastened to them that are ideally made from the same glass as the prisms. These pieces even out the otherwise uneven polishing pressures so that the surface can be uniformly polished. The polished backside of the part is covered by a protective lacquer. Pitch buttons are then applied to the well-dried lacquer. The prewarmed dummy piece is blocked to these buttons. The buttons are pressed out just enough so that a separation of no less than 2 mm (0.080 in.) remains between the glass parts.

The workpiece is now moved in desirably straight strokes across the polisher. After every few strokes the workpiece is turned by a few degrees. During the first strokes some pressure is advantageous. Later the surface will tend to suck down. The strokes are randomly directed in such a way that little by little the entire surface of the polisher is used. After 5 to 10 minutes the polisher is most likely used up and is again pressed out against the plano glass plate. As a result several polishers must be available to be used in rotation.

The workpiece must not dig into the pitch because that would destroy the polishing surface. The polisher must always be slightly moist, never dry. Skilled opticians can repolish the surfaces of roof edge prisms without damaging the sharp roof edge or round it. This method can only be mastered after considerable practice.

### 5.4.3. Machine Polishing

Machine polishing can replace the motions of the workpiece and the polisher; grinding and polishing machines are discussed in Section 5.5. The optician sets up the machine according to the job at hand, prepares the polishing tool, and monitors the process. The polishing compound is applied by hand with a brush or by a recirculating slurry supply. Several spindles can

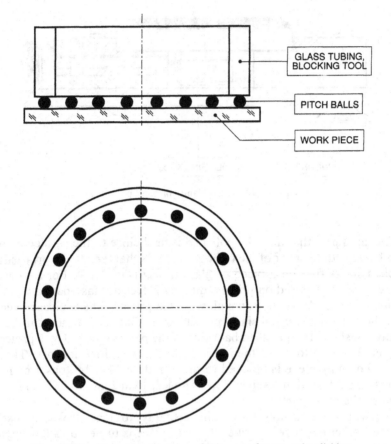

**Figure 5.93.** Stress-free blocking of thin parts for manual polishing.

be serviced in this way, and large surfaces can be polished that can no longer be managed by hand. Grinding and polishing machines are set up according to the following rules listed in Table 5.7, or the machine settings are corrected on the basis of this list.

The tool size mentioned in the table requires further explanation. A tool or workpiece that is mounted on the machine spindle must be 10% to 20% larger than the one that runs on top. The included angle serves as measure for strongly curved surfaces. The actual tool diameter is used for long radii or for plano tools (Fig. 5.94).

Large numbers of the same part are typically polished on these machines. It is best when the surfaces are polished only as accurately as is required by the print specification. Higher quality requires more time. This makes the parts much more expensive to fabricate. Particularly the accuracy of the surface shape is specified in many different ways. Less accurate surfaces can be polished on high-speed machines with pads or plastic polishers. A polishing time of only a few minutes is often all that is required with this

**Table 5.7. Guidelines for Setting-up Grinding and Polishing Machines**

| Workpiece on Bottom, TOOL on TOP | | Workpiece on Top, TOOL on BOTTOM | |
|---|---|---|---|
| Polish from **EDGE** | Polish from **CENTER** | Polish from **EDGE** | Polish from **CENTER** |
| *Large swing* of Eccenter | *Small swing* of Eccenter | *Small swing* of Eccenter | *Large swing* of Eccenter |
| *Pin forward* of Center | *Pin over* Center | *Pin over* Center | *Pin forward* of Center |
| *Swing* bias | *No swing* bias | *No swing* bias | *Swing* bias |
| *Larger* tool | *Smaller* tool | *Larger* tool | *Smaller* tool |
| *Coarse* polishing compound | *Fine* polishing compound | *Coarse* polishing compound | *Fine* polishing compound |
| Radius of polishing tool nearly the *same* as lens radius | Radius of concave polishing tool *larger* than lens radius | Radius of polishing tool nearly the *same* as lens radius | Radius of convex polishing tool *smaller* than lens radius |

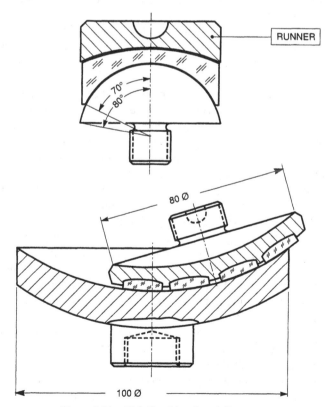

**Figure 5.94.** Relationship of tool diameters.

approach. Very high surface accuracy is achieved with pitch or wax polishers. The machines run slowly for pitch polishing, and several hours of polishing time are typically required for each side.

**Felt and Cloth Polishing.** Wool felt is only rarely used for polishing today.
The process will be described here only for the sake of completeness. Felt
and other polishing pads are described in greater detail in Chapter 3.

The type of felt that is used for polishing is white wool felt about 3 to
10 mm (0.120 to 0.400 in.) thick. The types with a medium high [85] to high
density are best suited. A circular piece that corresponds to the surface of
the polishing tool is cut out of a felt sheet. The felt is first softened in water
and then formed between the heated polishing tool and an equally heated
matching tool. The matching tool has the desired radius of the polisher. Both
tools with the felt between them are mounted in a press or vise until they
have cooled down. The felt form is now attached to the polishing tool with
black stick wax. The radius of the cast-iron polishing tool must differ from
the polishing radius by 2/3 of the felt thickness. This means that for a new
tool, the felt is thicker in the center and compressed at the edge. The cooled
felt tool is broken in for 1 hour without polishing compound by using the
matching tool or a suitable workpiece until the felt surface is smoothed out
and has conformed to the desired radius.

Coarse polishing compounds are used on felt polishers. The recommended rotational speed of the spindle for a tool diameter of 100 mm (4.0 in.)
is about 800 to 1000 rpm. At the same time the machine executes about 50
overarm swings per minute.

If one follows the progress of polishing with felt by periodically observing
the condition of the polished surface, it appears that the grinding pits continuously spread out until they blend together and the felt polish is completed.
This is quite different than the polishing behavior when pitch or plastic
polishing tools are used. The grinding pits continue to become smaller until
they disappear. Felt polishing can produce very clean surfaces that are free
from pits, scratches, and sleeks, unless insufficient fine-grinding leaves too
many deep pits that can result in an unacceptable surface condition called
*orange peel*. Even though the finish can be very clean, the surfaces are not
exactly spherical or plano, and the edges of the parts are typically rounded.
The workpieces, which are most often glass, become quite warm when
polished on felt polishers. Temperatures of 70°C (160°F) are not uncommon.
Therefore the glasses must be blocked with a hard blocking pitch. This is
especially important when lens blocks are felt polished, since the lenses can
shift during the polishing cycle.

The felt polisher changes its shape during use. Especially steeply curved
tools tend to polish hard from the edge at first. As time goes on they begin to
polish from the center. Therefore, when polishing large surfaces, it is customary to change polishers during the polishing cycle. If a polishing operator
should begin to polish the blocks with old polishing tools, the center would

be polished out in about 15 minutes. The edge zone would remain grey, and the surface would polish out with the same polisher only after an extended polishing time. If the operator, however, switches to new polishers that still polish well from the edge, the surface will be polished out after a few minutes. The experienced operator knows the polishing tools and will avoid extreme curvature changes.

Felt polishers also tend to glaze over with a combination of compacted polishing compound and abraded glass. This condition can form large smooth patches on the polisher, and this greatly reduces its polishing effectiveness and can also lead to polishing scratches. The usual remedy is to rough up the felt surface with a stiff bristle brush or a wire brush.

Condensor lenses that don't have to match a test plate are typically polished on felt. In the past corrective eye lenses were also worked in this way. However, the thin synthetic polishing pads used for these today are much better.

Cloth polishing is very similar to felt polishing. Cloth polishers are used for hand polishing on the pedal or motor bench. The cloth has to be a firm material such as that often used for men's clothing, billiard cloth, or synthetic fabric. The radius of the cast-iron tool that carries the cloth deviates from the polishing radius by about 0.5 mm (0.020 in.). If possible, the metal surface should be ground in. The cloth is bonded to the tool with sealing wax, and it will conform easily to the curved surface in most cases. Occasionally the optician uses polishing pitch to mount the cloth to the tool and allows the pitch to penetrate the material. This yields a polishing tool that has the characeristics of a pitch polisher and at the same time stands firm like a cloth polisher.

**Polishing with Plastic Polishers.** Attempts have been made for some time to polish with plastic polishing tools. The early attempts failed because the plastic absorbed water and changed shape as a result. Later it was shown that the plastic materials failed during the then-customary warm and dry polishing. Only wet polishing was possible with plastic polishers, and polishing compounds had to be tailored to this process. The machines had to be equipped with automatic slurry feed systems. After these requirements had been recognized and incorporated, several other plastics suitable for polishing were found. They are now available as polishing pads, foils, or thermoplastics.

When there are only modest demands on surface shape, as for eyeglass lenses, it is possible to polish with precut polishing pads. They are only 0.8 mm (0.032 in.) thick and have self-adhesive backing for easy mounting. The polishing surface is a very fine synthetic polishing felt that holds the polishing compound very well.

Polishers made from hard and porous polyurethane are now commonly used as a polishing surface for precision optics. Polishing tools made with this material produce clean surfaces and test plate fits of 3 to 5 fringes can be readily maintained. The edge roll on the surface is typically so small that it is easily removed during centering and edging of the lenses. Therefore a surface quality is achieved which suffices for most optical components. Polyurethane typically comes in 1.3-mm (0.052 in.) thick sheets.

Also thermoplastics that are stiffened with fillers are available. However, the making of a polishing tool is more difficult with this material than with pads, but thermoplastic tools last much longer.

Highly accurate surfaces almost without any edge roll and with a test plate fit of 1 to 3 fringes can be polished with a special plastic foil [86]. The typically circular foils are finely grooved on the polishing surface. Other thermoplastic materials have also been developed in the meantime with very similar polishing properties.

**Mounting of Plastic Pads.** Depending on the pad thickness, the radius of the cast-iron polishing tool must deviate from the finished radius. Since the pad thickness is not deformable, the pad forms a parallel layer on the cast-iron tool. Therefore, when a 1.3-mm (0.052 in.) thick pad is used, a concave tool must have a radius that is longer by that amount and a convex tool must have a radius that is correspondingly shorter. The tool radius should be held to an accuracy within 0.01 mm (0.0004 in.) and the cast-iron surface should preferably be ground in.

A round piece that corresponds to the polishing tool surface is cut from the pad sheet. It is cut in a crosslike pattern so that it can be applied to the curved tool surface without folds and creases (Fig. 5.95). An adhesive is applied to both the foil and the tool surface. The adhesive is typically diluted and must be applied repeatedly. In the supplied consistency the adhesive would be unevenly applied, and lumps would form. That would be very damaging for the very flexible plastic foils, but even the much stiffer polyure-

**Figure 5.95.**   Polishing pad cut to conform to spherical tool surface.

thane pads would pattern these irregularities to the polishing surface. Using this mounting method, the pads or foils are cemented to the cast-iron surface and firmly pressed out. A matching tool with a compressible interleaf, such as several layers of a soft material, can uniformly press down the pad.

The polishing surface of the tool must be ground in with an exactly fitting diamond pellet grinding tool. The shape of this grinding tool corresponds to the radius of the lenses that are to be polished later. To make the tool agressive, coarse diamond pellets are used with a grain size of about 50 $\mu$m (D50). The polishing tools that have deformed in time are corrected with the same pellet tool.It is important to maintain the typical groove structure when polishing tools lined with Desmopan foil are ground in.

**Thermoplastic Polishers.** Polishing tools that are lined with thermoplastics can be made in such a way that the plastic layer is initially thicker in the center than at the edge. These tools last longer and require reconditioning less often.

To prepare a thermoplastic-lined tool, a cast-iron radius tool is chosen that deviates from the radius of the polished surface by only 1 mm (0.040 in.) for concave tools and by 2 mm (0.080 in.) for convex tools. A matching radius tool that corresponds to the shape of the polished surface is required to press out the plastic. The thermoplastic material is heated in the concave tool of the two matching halves until it is formable. It is then pressed into the desired shape with the convex tool. After all parts have cooled down, the shell-like plastic part can be removed and cemented into the concave polishing tool. In most cases a special cement is used that is compatible with the plastic material. The spherical polishing surface is then quite accurately shaped on a lathe or on a milling machine. The groove structure on the surface that is typically generated in this way has a positive influence on the subsequent polishing characteristics of the tool. These grooves should not be entirely lost during the necessary smoothing operation with the diamond pellet tool.

The manufacture of plastic polishing tools as briefly described here is actually quite involved. The polishing times that are achievable, however, are so short relative to pitch polishing that in most cases an advantage can be demonstrated after polishing only 50 surfaces on average.

To achieve good polishing results, the part that is to be polished must be prepared in fine-grinding or in fine milling with a well-fitting and uniform curvature. A suitable pad, foil, or thermoplastic type must be chosen for the specific polishing tool. It should have the best-fitting shape and size in relation to the workpiece. For instance, the lower running tool is typically 30% larger than the workpiece since it is preferable to run the overarm on the left and away from center. This setup makes it easier to apply the slurry. The swing angle of a radial stroke or the swing distance of a sweep motion overarm is adjusted to meet this condition. The slurry is a mix of 1 kg (2.2 lb) of polishing compound with 10 liters (8.8 qt) of water. The slurry supply nozzle is adjusted in such a way that it strikes the workpiece or tool most effectively.

An engineered instruction or direction for the adjustment of the swing angle or overarm swing distance cannot be provided. We are still dependent for this on the experience of the optician who knows the effect of the adjustment functions from practical trial and error. Each type of machine has its own characteristics in this regard.

The goal for all adjustments for plastic polishers is that the part is polished out after a predetermined time and that it then fits the test plate to meet the specification. It is best that the polishing cycle not be interrupted by an intermediate test. It follows from these requirements that the optician is needed only during the difficult setup phase and for the subsequent supervision. The machines can then be operated by production workers.

Good polishing results with plastic pads and foils can be expected only with a compatible polishing compound [87]. The thermoplastic foils polish very well with pure cerium oxide, whereas the polishing effect with zirconium oxide is quite small. No such difference in the effectiveness of both of these compounds has been found during pitch polishing. Physically the two compounds differ in weight: 100 cm³ (6 in.³) of cerium oxide weighs about 128 g (0.28 lb), whereas the same amount of zirconium oxide weighs about 101 g (0.22 lb). This difference can be used to determine which polishing compound is on hand if the identification has been lost.

For chemically sensitive glasses the ideal polishing slurry should have a neutral pH of about 7.0. The glass surfaces can be attacked already during polishing when the water in the slurry becomes mildly acidic or alkaline. Chemically insensitive glasses polish best with slightly acidic slurries. The polishing compound remains better suspended in a slightly acidic slurry and is thus supplied to the polishing process in the desired concentration. Polishing compound mixed in an alkaline slurry settles out quickly, and it tends to build up and stay in the supply pipes of a recirculating slurry system. This results in the undesirable condition that more water than polishing slurry is supplied to the workpiece and diminishes the polishing effectiveness proportionately.

One way to control this problem, and any contamination of the slurry, is to supply each spindle with its own slurry. A slurry system that uses a peristaltic pump recirculates slurry through a compressible plastic tubing directly from the pan back to the spindle. This way each spindle has its own slurry supply. This approach is especially useful when several dissimilar materials have to be polished on the same machine.

**Pitch Polishing.** The best optically polished surfaces are achieved with pitch polishing. Polishing pitch is a mixture of wood tar pitch and colophonium. See Chapter 3 for a detailed description of polishing pitch. Occasionally natural resins with similar characteristics are also used to make polishers, and certain synthetic resins are used as additives. Polishing pitch is used in

**Table 5.8. Guidelines for Using Different Polishing Pitches**

| Hard<br>Polishing pitch for: | Soft<br>Polishing pitch for: |
|---|---|
| *Smaller* workpieces | *Larger* workpieces |
| *Small lenses* on larger blocking tools | *Large lenses* on smaller blocking tools |
| *Harder* optical materials | *Softer* optical materials |
| *Standard quality* optics | *High precision* optics |
| *High pressure* and high speed polishing | *Low pressure* and slow speed polishing |
| *Convex* surfaces | *Concave* surfaces |

the optical shops in different hardnesses. The optician chooses the most suitable type for the task. The guidelines listed in Table 5.8 can be useful for this choice.

The layers of polishing pitch on the tools are not very durable. They get warm and soften under pressure, and they tend to flow toward the center and edge of the tool. The essential grooves that are cut into the pitch contribute to the limited life of a pitch polishing tool. In addition the pitch tool can be easily damaged when it is allowed to run dry. When the highest precision is not required, fillers can be added to the pitch to make it more rigid. Pitch-impregnated felt is used for many polishing jobs, and cotton fabric soaked in pitch has also been used with success. In recent times a number of synthetics are being used, and water-soluble materials are added to the pitch to make it porous. A number of shop tests have confirmed, that the polishing action is much less effective when the polishing compound is mixed in with the melted pitch.

The polishing compound should form a crust on the pitch polisher. This is called *charging the surface*. Only after the formation of this crust can a good polish be expected. The rouge (or iron oxide) that was used extensively in the past charged the surface rather effectively. Cerium oxide or zirconium oxide can on occasion fail to form a crust. Then the polishing compound and the polishing pitch are said to be incompatible. The reason for this problem is not yet known. When a crust fails to materalize, the optician should select another polishing compound or make a more compatible pitch mixture.

The spherical polishing tool is covered with a pitch layer in such a way that its center is initially thicker than the edge (Fig. 5.96). The pitch will be used up in time through wear and groove cutting. The polisher is used up when the pitch layer becomes thinner in the center than at the edge. For

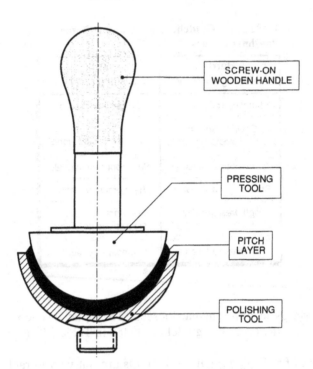

**Figure 5.96.**   Making a concave pitch polisher.

concave polishers the radius of the cast-iron tool is about 2 to 3 mm (0.080 to 0.120 in.) larger than the polishing radius. For convex tools, which are always larger than the workpiece, the radius difference is 2 to 4 mm (0.080 to 0.160 in.). Some opticians prefer even thicker pitch layers, but that practice is not based on a clear understanding of the viscoelastic nature of pitch. A thick layer of pitch can make the pitch flow in unpredictable ways, and controlling the surface figure with such a polisher is a matter of luck. The pitch is melted or poured into the concave polishing tool and then pressed out with a pressing tool or with the workpiece itself. The pressing tool is covered with polishing compound to prevent the pitch from sticking to it. Immersion of the pressing tool in soapy water is also a customary practice.

The cast-iron tool for a convex polisher is screwed to a wooden handle and heated up (Fig. 5.97). In this condition the tool is immersed partially in thoroughly blended liquid pitch, which then adheres to the tool. The pitch is at first formed with a wet hand and then pressed out in the pressing tool.

Plano polishers are cast. For that purpose the cast-iron plano lap is leveled with a bubble level, and the edge is surrounded with a strip of vellum paper that does not let the pitch flow out. Masking tape is also used quite often. The pitch tool is then neatly trimmed around the edge after the pitch has cooled to room temperature, and a hole is cut into the center of the

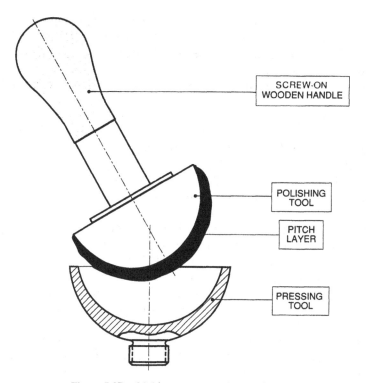

**Figure 5.97.** Making a convex pitch polisher.

surface. During the polishing process, the pitch will flow toward the edge and also toward the center. If the center hole were omitted, the pitch would rise there, and the resulting surface flatness on the workpiece would become quite irregular.

Several circular grooves are cut into the pitch in addition to the center hole. They take up the slurry and slowly release it to the surface during polishing. These grooves make the polisher more elastic. When the grooves are cut preferrentially in the center, the pitch layer becomes more elastic in the center, and the tool polishes more from the edge. With grooves near the edge, it polishes more from the center.

The polishing of a lens block with a pitch lap proceeds as follows: Uniform grooves are cut into the pitch lap, which is then slightly warmed by immersion in warm water. Spindle speeds and the overarm swing are preadjusted on the machine. The convex part, whether workpiece or polisher, is screwed to the machine spindle. The pitch polisher is then painted with a polishing slurry, and the concave part is brought into contact. The poker pin in the overarm is inserted in the pin seat, the machine is started, and the workpiece is prepolished until fringes can be seen with a test plate. On the basis of the test results, further grooves may have to be cut into the pitch, and the overarm swing may have to be adjusted. Depending on the size of the workpiece, the polishing cycle will proceed for 1/2 to 1 hour until another test and stroke corrections are made. Small blocks tend to polish out in 3 to 4 hours of polishing time; larger blocks can take twice as long. During this time polishing slurry is applied with a brush, and pure water is intermittently sprinkled on.

The polishing process with brush application should be done preferably nearly dry because it is then most effective. The polishing tool emits under these conditions a whistling sound. For pitch polishing with automatic slurry supply, the 10:1 (10 parts water and 1 part compound) aqueous slurry is prewarmed to about 40°C (100°F) before it is applied. Since less friction is generated during automatic slurry feed, only little heat is produced. Temperature is always needed during pitch polishing.

With brush feeding, water is applied for a few minutes prior to completion of the polishing cycle. Medium-sized workpieces are frequently finished off by hand in that the almost dry part is briefly held until it is completely dry at which point it is pulled off the polisher. The surfaces tend to be so clean then that they can be inspected with a loupe and a test plate. The value of this technique has been challenged, however, because sensitive glasses can pick up sleeks this way.

When the planned polishing cycle between tests has run its course, the workpiece is removed from the machine, rinsed off in warm water, and inspected with loupe and test plate. Surface flaws and errors in the surface shape can result during the polishing stage. When the test plate is used and the interference fringes run from the edge toward the center, this indicates a concave departure from the required sphere. For instance, one would say that the test plate fit is 4 fringes or rings "low." A test plate fit for which the fringes run from the center toward the edge is convex. Correspondingly the fit is called 4 fringes or rings "high."

When the test plate fit is "low," grooves are added to the center of the polisher, and the overarm swing is adjusted according to the guidelines noted in Table 5.7. The polishing tool will polish more from the edge, and the test plate fit will become less concave. When the surface is "high," interference fringes will run from the center toward the edge. Rings are cut into the polisher near the edge, and the overarm swing is also corrected according to Table 5.7.

If the grinding pits are uniformly polished out upon completion of the polishing cycle but the test plate fit is not correct, it is likely that the workpiece received the wrong curvature during fine-grinding. The test plate fit could also be irregular. When the irregularity is in the center of the block, the central hole is either too large or too small. When the test plate shows an irregularity near the edge, the polish was too much from the edge or the polishing pitch was too soft.

Surface flaws are remnants of grinding pits near the edge or in the middle of the polished surface. When the surface fit permits, one can polish more from the middle or more from the edge. But if the surface fits the test plate

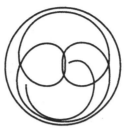

**Figure 5.98.** Path of the overarm pin relative to the machine spindle when the overarm spindle runs slower than the machine spindle.

well, the polishing cycle must continue uniformly until these flaws are polished out. As explained above, the radius of curvature of the fine-ground surface was probably incorrect.

Sleeks are flaws that appear frequently on surfaces. They are fine, fracture-free scratches. They are created when a contaminated polishing slurry has been used or when the surface contact between glass and pitch has been interrupted near the end of the polishing cycle. Such interruptions can occur when polishing too much from the edge or from the center. Sleeks are also formed when the polisher is too cold, and therefore the pitch too hard, so the contact is lost. These sleeks can be polished away with a warmed polisher and a little polishing compound. When too much compound is used, stains can form on glass surfaces. These stains tend to disappear quickly when a little vinegar or a few drops of acetic acid is added to the slurry.

The rotational speed of the work spindle relative to the number of overarm swings can be varied by adjusting the speed controls of modern machines. On older machines this would be done by changing the belt on stepped pulleys. When the spindle runs faster than the overarm motion, the pin will trace a sweeping motion over the work spindle, as shown in Fig. 5.98. This type of motion is useful for strongly curved workpieces.

When the speed is changed so that the work spindle runs slower than the overarm motion, the pin describes a path that approaches a straight line (Fig. 5.99). As is already understood from the discussion on hand polishing, such motion is good for flat radii and plano surfaces.

**Figure 5.99.** Path when the overarm spindle runs faster than the machine spindle.

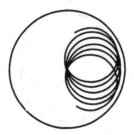

**Figure 5.100.** Path when the overarm and machine spindle are running at approximately the same speed.

The most unfavorable condition is achieved when spindle and overarm run at the same speed (Fig. 5.100). In this case the polisher is used in an irregular manner, and the results are totally unpredictable.

The speed of a polishing machine can be best expressed in overarm swings per minute. The average overarm swings vary with the size of the workpiece. The speeds listed in Table 5.9 are typically used for pitch polishing with brush feeding. The rpm of the work spindle is adjusted relative to the values in the table and the guidelines as described in Table 5.8.

Workpieces that must fit a test plate accurately must run slower on the polishing machine than those for which a test plate fit of several fringes is sufficient. The test plate fit tends to become irregular at faster speeds of the machine. The polishing effect is so strong at these higher speeds that the point in time at which the correct test plate fit has been achieved cannot be determined with any degree of certainty.

The workshop temperature should be kept between 22°C and 26°C (72°F to 79°F) so that the polishing pitch will conform easily to the lens surfaces. Once the polishing process has begun, it will generate its own heat through the resulting friction. This heat must not be allowed to dissipate; this means that cold outside air or cool drafts must be kept away from the polishing

**Table 5.9.  Eccenter Swings per Minute for Pitch Polishing**

| Workpiece diameter mm | inches | Approx. eccenter swings per minute |
|---|---|---|
| 30 | 1.20 | 60 |
| 50 | 2.00 | 50 |
| 100 | 4.00 | 40 |
| 150 | 6.00 | 35 |
| 200 | 8.00 | 30 |
| 300 | 12.00 | 25 |

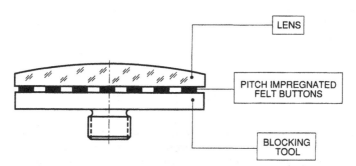

**Figure 5.101.**   Stress-free blocking of sensitive components.

machine. Sleeks and scratches are unavoidable when polishing on pitch in cold rooms.

This temperature rise during polishing introduces an uncertainty in the test plate fit. Under higher temperatures especially thin lenses will tend to deform because the glass and the blocking pitch have different coefficients of expansion. The fit of the lenses will differ when their surface temperature is appreciably warmer than room temperature. Only when the lens is de-blocked and cleaned, the true surface shape will become apparent. It should be remembered in such a case by how many fringes and toward which side the fringes have changed. One will typically find that the on the block test plate fit in the warm polishing condition most closely approximates the off-the-block surface fit at ambient temperature.

Very stress sensitive lenses are worked as single surfaces. They are blocked on felt buttons to minimize stress deformation (Fig. 5.101). Felt buttons are small soft felt discs, 5 to 10 mm (0.200 to 0.400 in.) in diameter, which have been soaked in a soft pitch. The lenses are then polished always wet with especially fine polishing compounds so that only little heat develops.

### References for Section 5.4 (Polishing)

[78] Klemm, W. "Zum Poliervorgang an Spröden Körpern" (On the polishing process on brittle materials).

[79] Karow, H. H., Notes on DIC microscope.

[80] DIC microscope, Karow.

[81] Prof. Dr. Kaller, Zeiss, Jena.

[82] Brown, N. J., and L. M. Cook, "The role of abrasion in the optical polishing of metals and glasses," *The Science of Polishing Topical Meeting, Technical Digest,* 1984.

[83] Silvernail, W. L. "The role of cerium oxide in glass polishing." *The Science of Polishing Topical Meeting, Technical Digest,* 1984.

[84] Brown, N. J. "Optical polishing pitch," *LLNL,* course notes, 1987.

[85] h3 per DIN 612 000.

[86] Carl Zeiss, Desmopan.

[87] M. Eichhorn, Dissertation, Technische Universität, Berlin, 1977.

## 5.5. GRINDING AND POLISHING MACHINES

A survey of currently available grinding and polishing machines identified 160 different models that are commercially available from ten different manufacturers. Although extensive, this survey was nevertheless not exhaustive; the actual numbers are probably higher. Why is there so much diversity for such a limited market? Perhaps the answer lies in the simplicity of the machine and the constant demands by opticians and optical shops to incorporate customized features. The large number of available models results in purchase decisions that are often based on nothing more than the likes and dislikes of the opticians rather than on the technical merits of the product. This situation causes confusion and uncertainty.

The grinding and polishing machines are the backbone of any optical shop. The currently used machines, although equipped with many modern features, incorporate the basic principles that governed the machines used during the early days of optics manufacture. Fig. 5.102 a,b

There are two basic overarm motions represented in Fig. 5.102. One is a nearly horizontal sweep-motion, and the other is a radial motion, whereby the overarm pivots around an adjustable point on the work spindle axis which corresponds to the center of curvature of the lens radius. The sweep-type motion is preferred for all plano surfaces and for radius work where the overarm rise is small compared to the block diameter. The radial motion is best for steeper curves and is essential for the controlled polishing of domes, hemispheres, and convex hyperhemispheres. All grinding and polishing machines come with one of these two overarm motions.

Grinding and polishing machines are available in many sizes and configurations. Most are the standard sweep-motion single or multispindle machines and a good percentage are of the radial-motion type. Yet there are a number of special purpose G&P machines that merit a closer look. Among these are tub grinder, pedal bench, high-speed multispindle, high-speed ganged radial-motion, stick polisher, toric and cylinder, and Draper machines.

### 5.5.1. Tub Grinders and Pedal Bench Machines

The tub grinders and pedal bench machines are designed for manual grinding and polishing of prototype optics and for making test plates. Tub grinders have a single spindle that accepts the workpiece or the tool. Figure 5.103 shows such a machine as seen from above. In this particular view the grinding tool is mounted on the spindle and the workpiece runs on top. To prevent slurry from getting into the spindle bearing, the slurry pan has a raised lip

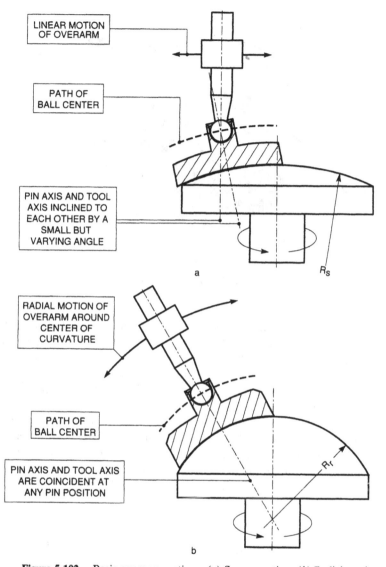

**Figure 5.102.** Basic overarm motions. (*a*) Sweep motion. (*b*) Radial motion.

and the spindle has a slinger made from rubber or plastic. This arrangement, shown in Fig. 5.104, is typical for most grinding and polishing machines.

A foot switch activates the spindle and permits speed variations of 10 : 1. An adjustable overarm holds the tool pin in a firm clamp that can be offset to either side of center. Once the workpiece is placed on the tool, the pin is engaged in the pin seat, slurry is manually applied, and the motor is started. The pivoting overarm is then moved back and forth by hand, and this guides the tool across the rotating workpiece. The opposite arrangement where the

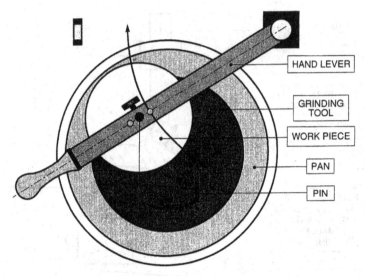

**Figure 5.103.**   Hand lever in top view.

tool runs on the bottom and the workpiece on top is equally valid. Tub grinders are useful for radius and plano tool correction, for breaking in pellet tools, and for conditioning pad polishing tools. The machines can also be used for manual roughing and fine-grinding of prototypes and low-volume production. Tub grinders have also been used for limited prepolishing of noncritical optics and for buffing of test plate backs.

Pedal benches are more useful for more precise work done by hand. They can be driven by pedal power like an old sewing machine, or they can be motorized. Most pedal machines incorporate both options. Pedal benches that were once quite common in optical shops have become a very rare sight. The decline in use has been a direct result of the inherent labor-intensive

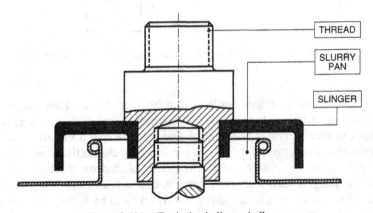

**Figure 5.104.**   Typical grinding spindle.

aspects of this type of operation, which usually requires the most skilled and thus highest-paid opticians. These machines are no longer economically viable for production. They are, however, ideal for making small test plates and for salvage, prototype, or development work. With the foot pedal option, the optician has both hands free to firmly hold and properly guide the workpiece over the tool. Because of the total involvement of the optician in this operation, spindle speed, pressure, and the stroke pattern can be changed easily and quickly. The experienced optician can "feel" the surface develop toward the desired shape and surface quality. Some of the finest work was once done on these pedal benches. It is regrettable that these skills have almost totally vanished.

### 5.5.2. Sweep Motion

The majority of grinding and polishing (G&P) machines use the sweep-type overarm motion for precision grinding and polishing. This motion is equally useful for slow-speed and high-precision polishing or for high-speed and commercial quality production polishing. The slurry can be hand-fed or supplied to the spindles by a recirculating slurry system. In general hand-fed machines are run relatively slow because they are usually set up for pitch polishing. Recirculating slurry systems are typically incorporated on production machines for volume production of single lenses or lens blocks of medium to small lenses.

The basic sweep-motion machine, shown in Fig. 5.105, is constructed as follows: A vertical spindle accepts the workpiece which is often a lens block and sometimes a single lens. A second vertical spindle terminates in an eccentric plate that is connected to a horizontally pivoting overarm by a connecting rod. The eccentric drive causes the overarm to oscillate laterally over the center of the workpiece spindle. The overarm accepts a vertically held pin that either drives a grinding or polishing tool for convex radii or the lens block for concave surfaces. For certain conditions, the order of workpiece and tool can be reversed.

Moderate speed multispindle sweep-motion machines with central recirculating slurry system are the most widely used machines in the precision optics industry because they provide high precision at production rates. The sweep-type grinding and polishing machines are characterized by versatility and simplicity. They are easy to operate and have a large radius range as well as a wide range of adjustability for nearly any condition. Some multispindle sweep-type motion machines are specifically designed for high-speed production grinding with either abrasive slurries or with diamond pellet laps. The same machines can also be used for high-speed polishing with hard pitch or synthetic pad polishers.

Double eccenter machines [88] were developed for the grinding and polishing of large surfaces. Two eccentric drives harmonized in both revolution and throw distance impart a randomized motion on the overarm with which zone formation on the surfaces can be avoided. A crank is sometimes used in

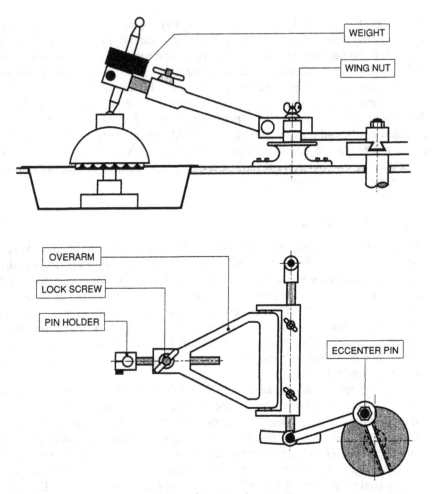

**Figure 5.105.** Sweep-motion overarm.

addition to the overarm on simpler machines to provide the polishing motion. The polisher is moved by the crank from the top, while the large workpiece mounted on the spindle remains stationary (Fig 5.106).

Sweep-motion machines are rarely useful for steep curves. By adding a tilt feature to the normally vertical work piece spindle, the usual radius range can be greatly extended. The tilt angle usually lies between 0° and 30°

On a system designed for small parts, the strongly curved workpiece is screwed to a handle. This handle has a seat for the pin, and it is moved in a circular motion by a crank on the driven pin. The polishing tool is fastened to the spindle which rotates in the opposite direction against the crank rotation. This arrangement is very practical because it allows the polished parts to be quickly changed (Fig. 5.107). Another machine [89] functions in a similar

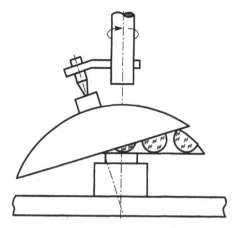

**Figure 5.106.** Machine for felt polishing with stationary lens block and crank-driven polisher.

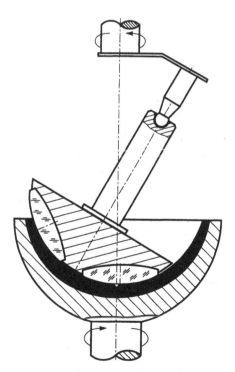

**Figure 5.107.** Polishing arrangement for a convex lens block on a crank-driven polishing pin.

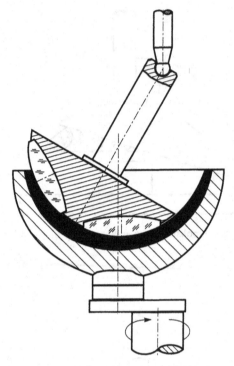

**Figure 5.108.** Polishing arrangement for a convex lens block on a crank-driven polishing tool and a stationary pin.

manner but the crank is built into the work spindle. This is shown in Fig. 5.108.

Figure 5.109 shows a typical sweep-type overarm with a number of common adjustment options and their effect on the location of the pin relative to the workpiece spindle. The sweep angle $\alpha$ is a function of the eccenter radius $R$. When the eccenter is out as far as it will safely go, then $\alpha$ is a maximum. Conversely, $\alpha$ will be a minimum when the eccenter is at the innermost position. It is often desirable for surface control and for effective slurry distribution to introduce a bias that shifts the sweep angle either to the right $\beta$ or to the left $\gamma$ of the spindle. Another possible adjustment is either to pull the pin forward so that the center of the tool will pass in front of the center of the workpiece spindle or to to push it back so that the tool will pass behind the spindle axis.The latter option is not often used.

### 5.5.3.  Radial Motion

Radial motion machines are used for grinding and polishing steeply curved convex and concave surfaces including hemispheric and hyperhemispheric convex blocks and single lenses. The larger radial machines can handle

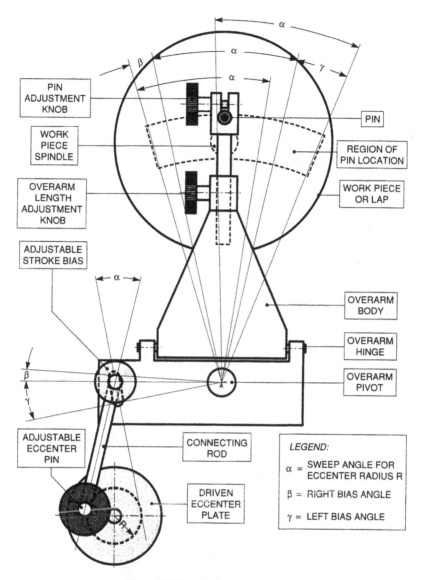

**Figure 5.109.** Sweep-type overarm.

blocks and parts up to 100 mm (4.0 in.) in diameter, but most of the radial grinding and polishing machines are designed for smaller lenses and lens blocks in the diameter range from 3 to 20 mm (0.120 to 0.800 in.).

High-volume, high-speed radial arm machines are most often ganged together with the radial sweep angle the same for all spindles. The work spindles are either ganged in pairs or in groups of three. Some of these machines have the radial arms ganged for only two or three spindles in order to directly correspond to the grouping of the work spindles. The radial sweep angle can be changed and biased to either the left or the right of the spindles. The pin position can be changed front to back so that the tool runs forward of center or in back of center. The speeds of the work spindles and the corresponding radial arms can be independently varied, at least for a ganged group of spindles. The tool spindle can also be driven in a counterrotating mode relative to the rotation of the work spindle.

Figure 5.110 shows the spindle and overarm of a radial grinding and polishing machine [90]. The overarm is spring loaded to provide uniform pressure to the pin without the detrimental effects of inertial forces. A variant of the swing lever is found on contemporary high speed polishing machines that drive several spindles in a radial motion by a common yoke [91]. These high-speed machines are particularly well suited for the mass production of high-quality lenses, especially when the polishing is done with synthetic pads or foils. The swing yoke moves the pins for three spindles simultaneously. The uniform pressure, which remains constant in every possible overarm position, is provided by a spring or a compressed air cylinder. The machines are also equipped with pumps for recirculating slurry. The polishing pressures on these high-speed machines is approximately 10 $N/cm^2$. The spindles revolve at speeds from 300 rpm to 2000 rpm.

Another variant of the radial arm grinding and polishing machine is the stick polisher. This type of machine is especially designed for steeply curved small lenses and lens blocks, both convex and concave. The grinding or

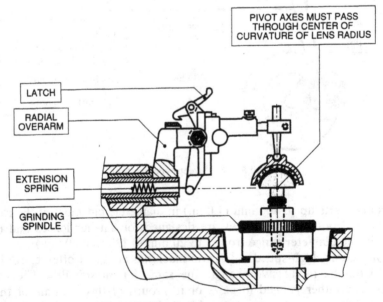

**Figure 5.110.** Radial-motion overarm.

polishing tool is mounted on the spindle, and the workpiece runs almost always on top. This arrangement is quite common for concave surfaces, although it runs counter to the way convex surfaces are normally processed. The pivot point, which must be the center of curvature of the lens radius lies on the axis that connects the workpiece with the overarm. To make this motion possible, the pin must engage the workpiece extension rod beyond the radius of curvature.

### 5.5.4. Cylinder and Toric G&P Machines

Cylindrical optics are used to produce a line image of a source to either illuminate a slit or focus light on a linear detector. A long radius, straight cylinder lens can be ground and polished on a machine where the workpiece is fixed, and the tool is moved by the overarm axially in the x-direction (front to back) and longitudinally in the y-direction (side to side). The cylindrical surfaces can be convex or concave. The convex component, either the workpiece or the tool, is fixed on the bottom. The matching concave component runs on top. Toric surfaces can be generated on a similar machine where the axial motion is not linear but permits the tool to follow a radius. A toric surface has two distinct radii at right angles to each other. It produces two orthogonal line images that are separated by a distance that is proportional to the difference between the two radii. A very precise convex cylinder can be ground and polished when the radius is small enough to permit the making of a full cylinder (rod) that can be rotated about its axis.

### 5.5.5. Draper Machine

The Draper machine is an old design that was specifically created for the polishing of large astronomical mirrors. Machine of this type with workpiece capacities of up to 3.125 m (125 in.) have been built for commercial use. Even larger machines have been built for special projects.

A sturdy table accepts the workpiece, which is usually a large mirror blank. The table slowly rotates at a speed that is proportional to the diameter of the table. The peripheral table speed ranges from 150 to 200 mm/sec (6 to 8-in./sec) at the low speed setting to up to ten times that speed at the high-speed range. The speed can be infinitely adjusted over that range while the machine is running. The large work table can be tilted from the horizontal to the vertical position for inspection and testing the large workpiece without having to remove it from the machine.

The driven tool spindle rides on a rail above the workpiece. It is typically driven back and forth by a reversing motor by means of a lead screw drive. In this configuration the tool will follow a straight line motion across the slowly rotating mirror blank. This creates a petal-like stroke pattern. A second motion component is added when the rail is pivoted on one end, and this causes the tool to eccentrically oscillate from side to side in addition to its axial motion. A more complete random motion of the lap is achieved by a third eccentric at the opposite pivot end. A sideways offset and a bias along the rail axis are two additional ways by which the tool motion can be affected. Drapers have up to six motors: table spindle, tool drive, eccentric drive, stroke drive, table tilt, and offset adjustments.

### 5.5.6.  Historical Perspective

Old G&P machines had only one large motor to provide the power to drive the main shaft that ran the entire length of the machine. Each spindle had a large friction wheel that was mounted on the main shaft. A small pickup wheel that could engage the friction wheel by means of a lever transferred some of the power to a work and eccentric spindle pair. The work spindle drove the eccentric over a simple pulley system. This meant that a fixed speed ratio had to be established that could only be changed in discrete steps by the pulleys. The rotational speed could be varied by raising or lowering the pickup wheel. This type was used for many years in optics manufacture in the United States, Germany, and elsewhere. In recent years, with the development of superior friction wheel materials, a few small capacity machines have been built for special projects that use the old friction wheel principle.

After World War II new G&P machines were introduced that had individual motor drives with independently variable speed transmissions for each spindle or for groups of spindles. The eccenter was still tied to the work spindle by a stepped pulley system. Split sheave belt drives or variable speed gear reducers were used to provide a limited range of stepless variable spindle speeds. With split sheave pulleys, stepless variable speed changes are possible by altering the pulley diameter through mechanically or hydraulically changing the separation between the pulley sheaves. Some of these machines are still in use today. These machines were typically equipped with simple mechanical speed indicators or tachometers.

Contemporary volume production grinding and polishing machines for commercial quality optics use ganged drives by which all spindles or groups of spindles are driven identically by one motor. Each spindle can be disengaged from the gang drive by a simple clutch mechanism. This permits the inspection of the work on one spindle while the other spindles of the ganged unit continue to run.

Other modern machines for high-volume polishing provide a limited range of individual spindle speed adjustment. These machines are typically equipped with constant speed motors that can be stepped down over a three- or four-step pulley arrangement. The overarm oscillations are taken from the same drive but stepped down in speed to 5% or 10% of the spindle speed by means of a gear reducer. These machines usually run quite fast, with spindle speeds ranging from 50 to 900 rpm.

### 5.5.7. Modern G&P Machines

Modern precision grinding and polishing machines are characterized by independently driven work and overarm eccentric spindles. Individual motors are used for each spindle. Their rotational speed is infinitely variable over a wide range of speeds. The drive motors are either ac motors, which have variable torque at varying speeds, or dc motors, which maintain a constant torque over the range of speed variations. Although dc motors are more expensive than ac motors, they offer speed control without loss of torque. This is the main reason why most opticians prefer dc motor drives. These machines have overhead control panels with switches, timers, and parameter displays at eye level. Most provide pneumatic overarm pressure and integral slurry recirculating systems. The various features makes these machines more expensive than their simpler predecessors, but they give the skilled operator a much wider latitude of variables for quick response to varying block conditions to make the added investment worthwhile.

The ac or dc motor speed is controlled by turning a knob on the overhead control panel. This changes the voltage applied to the motor and results in a speed change. Typical variable speed ranges are 10:1 or 20:1. The actual speed is electronically displayed on the panel either by an analog galvo tachometer (needle) or, as is more commonly preferred today, by an electronic digital tachometer. The display is often a dual display, one for the work spindle speed rpm the other for the overarm rpm. These speeds can be independently controlled over wide range, and this can result in speed settings that would set up synchronous motions between the spindle rotation and the swings of the overarm. Such synchronization can lead to serious figure control problems during polishing. A simple speed adjustment for one of the spindles will break up the synchronism and correct that condition. The spindles can be directly driven by the motors, but that arrangement is often not desirable. It is better that the power is transferred from the motor to the spindle over a fixed timing belt drive. This arrangement provides for softer starts and stops, and it also reduces the transmission of motor vibrations to the work.

A hydraulically driven overarm machine was marketed at one time [92]. Although this design provided considerable force to the overarm so that it was impossible to stall it, the sudden reversal of the overarm at the end of each stroke caused the lap to "slap." This invariably led to the formation of zones in the surface figure. The decided advantage of this scheme was that it provided not only variable speed oscillation control but also permitted changes in the stroke length while the machine was running. The problems with this type of machine focused attention to the importance of a smooth speed transition as the overarm reverses direction at the end of each stroke. Only an eccentric drive can provide that type of smooth motion.

Another variation of standard spindle drives comes from Japan [93]. Not only are the work spindles and overarm spindles independently driven, the pins are driven to counterrotate relative to the work spindle as the overarm oscillates back and forth. The rotating pin engages the tool which is then driven over the workpiece. This design produces very high relative speeds that are necessary for efficient stock removal during the prepolishing or burn-in phase of the polishing cycle. However, there is a limit beyond which higher tool speeds result in reduced removal rates, for the slurry or coolant is flung off before it can become effective.

### 5.5.8.    Basic Design Contraints for Modern G&P Machines

With the wide range of different types of G&P machines available, it is quite difficult to generalize about design constraints. There are, however, some facts that limit most parameters. The values listed in the following paragraphs are averages from a composite evaluation of a wide range of different machines. There are many exceptions to these general values, and each machine must be judged on its own merits with the intended application clearly in mind.

**Block Capacity and Number of Spindles.** The block capacity is one of the constraints that determines the size of a grinding and polishing machine. The number of spindles is another constraint. Together, they define the physical size of the machine.

The block or lens capacity ranges for most machines from a low of 50-mm (2.0-in.) diameter to a diameter of 900 mm (36 in.). There are larger machines for special applications with spindle capacities of up to 3.0 m (120 in.). For most common optical fabrication requirements, however, the block capacity of grinding and polishing machines lies somewhere between 100 and 400 mm (4 and 16 in.) diameter.

Grinding and polishing machines range from single-spindle machines to multispindle machines with 16 spindles or more. The most common range is from 1 to 6 spindles.

**Pan Sizes and Center-to-Center Spindle Separation.** Slurry pan sizes used to be the measure that defined the spindle capacity. The maximum block that could be safely polished could be no larger than 80% of the pan diameter when all spindles were engaged. It is, however, much more meaningful to specify the center-to-center distance as the measure for spindle capacity because that determines whether a certain block size can be appropriately processed on the machine. The use of slurry pan sizes is no longer meaningful with many modern machines because they use slurry pans that are common to all spindles of the machine.

The center-to-center distance is, of course, a function of the maximum block diameter. Figure 5.111 shows that relationship. Safe spindle-to-spindle distances prevent overarm and tool interference with those of an adjacent spindle.

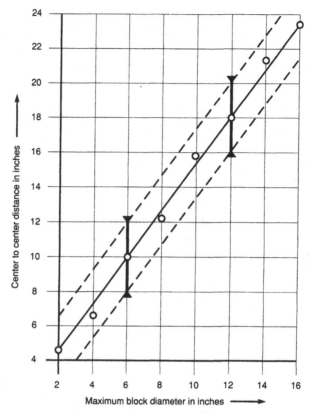

**Figure 5.111.** Typical block size limits as function of spindle spacing.

**Eccentric Stroke and Overarm Offset.** These are two additional settings that are of great interest for the optician. An adjustment in the eccentric position increases of decreases the stroke length of the overarm. A change in the offset introduces and additional bias to the stroke at either end. (see Fig. 5.109). For most machines, the eccentric stroke adjustment is about 50% of the maximum block diameter that can be run on the spindle, while the stroke offset is 25% of the maximum block diameter.

**Spindle Threads.** One of the areas where there is far too much individualization is in the size of spindle threads. The vertical spindles that accept the grinding or polishing tool, or the lens block when they run on the bottom, are threaded to fasten the tools to the spindles. The thread permits the firm mounting of the block, grinding, or polishing tool to the lower spindle. The G&P machine survey discovered an unusually wide range of spindle threads in use. They ranged from a low of 5/8-11 to a high of 2-4, with nearly every possible thread in the series represented in 1/8 in. increments. Although different size tools will require different threads, it is hard to imagine that such diversity is necessary. One would assume that the smaller threads are for small tools and that the larger threads are for larger tools. While this appears to be true in a general sense, there are numerous departures to this rule that one is forced to conclude that some of the thread choices are quite arbitrary. For instance, a 18-inch diameter maximum block capacity spindle can have a 3/4-10, 7/8-9, 1-8, 1 1/4-7, and a 1 1/2-12 thread.

The German optical industry has long recognized the problem that too much thread variety entails. In response DIN standards [94,95] specify a limited series of metric spindle threads and mating tool threads that are clearly related to block and tool diameters. These range from a M16 thread for small blocks to a M39 thread for large blocks. These standards also clearly define the required centering bosses on the spindles and the mating centering bores on the tools.

In addition to a variety of threads, many shops still prefer the use of tapered spindles. Besides the common AO taper that originated in the opthalmic optics industry, there are a number of other tapers in use. Some of these are so unique that they and the tools can be used in only one shop and only on one type of machine. The problem with this variety in threads and tapers can be found in the way that a shop or a company is locked into a specific tool design that would be useless on any other machine in another shop unless adapters are used. A strong case for a meaningful standardization can be made on this issue.

**Slurry Systems.** Many opticians prefer to brush-feed slurry to the lenses or blocks or to apply just the right amount of slurry to each spindle with a squirt bottle. While this is a good method for prototype and low-volume precision work, it requires too much attention from the optician for routine polishing in volume production. For high-volume requirements, many contemporary G&P machines come equipped with a common slurry basin and an integral recirculating slurry system. These systems are usually available as an option for all other machines.

There are several methods for recirculating slurry feed for polishing. The slurry dip application comes to mind for which the polisher or the workpiece is dipped into the slurry with every overarm swing. Another method utilizes impellers that raise the slurry from the pan by means of a blade mounted on the work spindle to funnel it to the polisher. The rotating pan method is still in use for which the slurry collects at the outer edge from where it is redirected so that is squirts onto the polishing workpiece (Fig. 5.112). With all these slurry application methods, the settling out of the slurry is a problem. Even when suspension aids are added, the slurry must be constantly stirred; otherwise the slurry would become too weak and the polishing effect would be diminished.

The modern machines are equipped with pumps that supply the slurry under pressure. The action of the pumps keep the returning slurry stirred up so that it remains in an approximately constant mixture. Such a slurry sys-

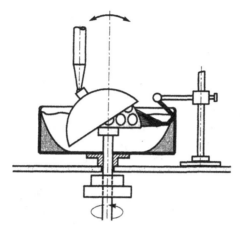

**Figure 5.112.** Continuous slurry application by the rotating pan method.

tem typically consists of a pump to recirculate the slurry, an agitator to keep the polishing compound in suspension, supply lines that bring the slurry to each spindle by means of nozzles, valves that permit the control of the flow and the shutoff of lines that are not needed, a large stainless steel slurry splash basin, slurry return lines, a settling tank, and a reservoir.

The slurry supply lines should be made from a clear flexible tubing that allows the operator to monitor the slurry flow and to line blockage before it becomes a problem.

The interior of the pump housing, the pump impeller, all piping, valves, and nozzles must be made from corrosion resistant materials. Slurry remains in constant agitation if it is allowed to pass through a bypass pump when the slurry is not needed at the spindles. This process is called *internal recirculation*.

For some radial overarm high-speed grinding and polishing machines, the polishing slurry is not only applied from the side to the lens block but also through the center of a hollow tool spindle that terminates in a perforated ball joint. This method ensures that the slurry is supplied to the center of the block for even and controlled polishing. To avoid undue contamination of the slurry, the steel ball of the pin rides in a plastic pin seat in the tool. The drawback is that this seat has only limited life and requires frequent replacement.

High-precision polishing machines can also benefit from a recirculating slurry system. Instead of flooding the block on each spindle with a steady stream of slurry, a slow feed–drip feed system is preferred. It relies on a peristaltic pump that forces carefully metered amounts of slurry from a reservoir through flexible plastic hoses by squeezing them and the slurry inside in a slow rotary motion in the direction of the spindle. The result is a slow but constant drip of slurry. The drip rate can be increased or slowed down by varying the speed of the pump motor. The slurry reservoir can be filled for a single-pass supply only, or it can serve as the settling tank for a closed-loop recirculating system. Single-pass flow is used for the final polishing of critical surfaces or when polishing with chemically active slurries. Each peristaltic slurry pump can serve up to six separate lines. Since each line exists independently of the others, it is possible to supply a different slurry through each line. This requires a separate slurry reservoir for each spindle. One way to do that is to draw the slurry out of the bottom of a specially modified stainless steel or plastic slurry pan that rotates with the block and to pump it back from there to the block.

**Down Pressure.** The only way to provide additional polishing pressure on the older grinding and polishing machines was by placing weights over the overarm pin. This method is still preferred by many opticians for the control of critical polishing operations on brush-fed sweep-motion machines. Modern machines often come equipped with down pressure provided either by a spring mechanism or by a pneumatic cylinder or air piston (Fig. 5.113). While the spring-loaded pressure option has the advantage of not requiring a supply of compressed air, it lacks the ease of adjusting the down pressure that is possible with the pneumatic option. Although the use of pneumatic down pressure is optional, there are many machines in use, mainly in the United States, that are equipped this way.

The air piston delivers well regulated and stepless variable pressure from 0 psi up to full line pressure of about 80 psi. The pressure is adjusted by means of a regulator and is monitored on an air pressure gauge. The air piston also responds to a lift switch that raises and lowers the overarm. The switch that activates a valve can be a foot switch but it is most often a toggle switch that is mounted on the easily accessible control panel of the machine.

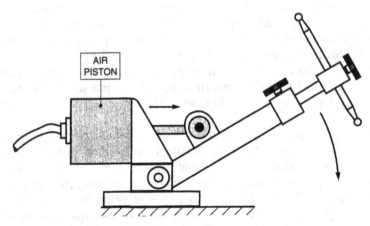

**Figure 5.113.**  Pneumatic down pressure.

The advantage of using pneumatic down pressure on sweep motion machines is the total elimination of inertial effects on either the tool or the block that are unavoidable when weights are used. These inertial effects can lead to zonal irregularities such as a down-turned edge on the polished surface. The real advantage, however, of using pneumatic down pressure lies in its application on high-speed radial overarm grinding and polishing machines. The pressure is uniformly applied over the entire overarm swing and is always normal to the lens or block surface. Externally applied weights cannot provide the same pressure control.

Pneumatic pressure can also be used for very high precision optics. A Japanese machine manufacturer [96] has an optional negative pressure control that reduces the down pressure to near zero during the critical final stages of polishing.

**Load Meters.** It is advisable to use load meters on each spindle of the larger machines to monitor frictional loads and to detect overload conditions as early as possible. This feature can alert the operator to take corrective action and avoid machine, tool, or workpiece failure. Load meters are typically ammeters that measure the amount of electrical current. An increase in current is required when the motor has to work harder to overcome increasing frictional forces that result from excessive down pressure or from insufficient slurry flow. The increase in current is measured and displayed by the load meter.

**Timers.** Many modern grinding and polishing machines are equipped with automatic cycle timers. While most timers control the cycle for each spindle independently, there are some machines where each timer controls several spindles. Once the timer is set, the cycle will be stopped automatically after the preset time interval. Timers are a particularly useful feature for multi-spindle, high-speed, high-volume polishing operations that use polishing pad or other synthetic polishers. The cycle can be interrupted at any time by using the manual override function.

**Overhead Controls.** One obvious feature of a modern grinding and polishing machine that differentiates it from the older models is the overhead control panel. On older machines, all control functions were low on the front of the machine where they were hard to see. All of the control functions on contemporary machines are at eye level and in easy reach of the operator.

Some machines are equipped with fluorescent work lights that are mounted on the underside of the overhead control where they provide very good lighting to all spindles. The fluorescent white light lamps can be exchanged with black light tubes, which are preferred by some opticians as sources of nearly monochromatic light for test plate testing right on the spindle. Since they are often used without filters, several color bands are usually seen.

One set of controls serves a group of spindles on some ganged production machines. For most versatile precision polishing machines, however, each spindle has its own set of controls. These controls are switches, indicator lights, and either analog or digital displays for down pressure, work spindle speed, stroke offset, eccentric motion, and overarm stroke speed.

**Spindle Bearings.** Since abrasive slurries are used on grinding and polishing machines and since most of the forces are acting on the work spindle, it is very important to pay close attention to the spindle bearings to make sure that they are properly designed for the intended application. Since the work spindles are often subjected to high cyclic lateral forces, all spindles should have several heavy duty bearings.

The more serious problem in spindle-bearing design is the encroachment of abrasive slurries into the races of the bearings. Once this occurs, rapid bearing failure is almost guaranteed. The first line of defense to keep slurry away are simple slingers and seals. But since it is nearly impossible to completely keep slurry away from bearings, only sealed bearings must be used that are specifically designed for use in aqueous environments.

One manufacturer [97] has attempted to reduce spindle failure due to slurry encroachment by using an inverted spindle design that reduces the risk of premature bearing failure by sealing out moisture and contaminants. This particular design, makes the spindles much higher than on standard machines; however, this can lead to difficulties in tooling and in operation. Another response to spindle wear is the use of cartridge type spindles that can be easily replaced in the shop whenever spindle failure occurs.

**Machine Base.** Older machines had frames that were constructed from angle stock that was bolted or welded together into a machine base. These frames were not particularly stiff and tended to transmit motor and bearing vibrations to the spindles. Modern machines use unitized, heavy-duty welded steel bases. This type of solid boxlike construction provides the strength and rigidity that is required for vibration-free operation, which, together with low-vibration power trains, make the fine-grinding and polishing of high precision optics easier to control.

**Special Features.** The many diverse applications for optics G&P machines and the endless demands for custom features have produced a number of unique machine options:

A small amount of random motion can be introduced by offsetting the overarm pivot so that it oscillates a little bit eccentrically [98]. A greater amount of randomness can be obtained by a double eccentric arrangement that is integrated either into the spindle drive [99] or by two separately driven eccenter spindles [100]. These arrangements yield a continuously changing stroke pattern that results from the randomly changing length of the overarm stroke. A slightly elliptical overarm stroke reduces the lifting of the tool by eliminating a complete stop and start at the end of each overarm stroke [101]. This feature also permits higher polishing speeds, which can lead to higher polishing rates.

Adjustable overarm risers [102] permit the adjustment of the proper overarm attitude for unusually thick workpieces. They can also be useful in raising the overarm to compensate for the higher inverted machine spindles. The overarm can be raised as much as 100 mm (4.0 in.) this way.

Tiltable work spindles have proved to be quite effective for the grinding and polishing of steep curves on sweep motion machines. The tilt angles range from the standard vertical position at 0° to 30° tilted front to back. The sometimes bothersome dead spot at the lens center of single surfaces and domes is eliminated when the lower workpiece spindle can be tilted. The radius range that can be polished on such a machine is also increased substantially as compared to a similar fixed spindle machine.

A driven lap or polisher adds to the relative speed between the tool and the workpiece. This feature can greatly increase the lapping or polishing efficiency, but it is typically only desirable on high-speed pellet laps or on polishers with synthetic polishing surfaces for high-speed polishing of commercial quality optics. On some machines [103,104], the lower workpiece spindle and the upper tool spindle that is held by the overarm counterrotate to produce high relative speeds. This can be done for the traditional sweep-type machine or for the radial-motion machine.

For some specialty machines such as stick polishers for small lenses with near-hemispheric surfaces, the lower workpiece spindle and the upper tool spindle can be independently raised and lowered along their axes. This is a necessary feature for setting up the machine in the proper manner for each condition.

### 5.5.9. Inspection of Polished Surfaces

Polished surfaces are inspected for surface quality and surface accuracy. A surface is judged clean when the grinding pits are polished out and there are no scratches, sleeks, or comparable flaws. The surfaces are inspected with a loupe under a strong light. The preferred light uses either a clear glass lamp or a projection lamp of 100 to 200 watts in conjunction with condensor lenses (Fig. 5.114). The loupe most often used is the 6× Steinheil loupe (Fig. 5.115a). In many cases a 4× or 5× loupe will suffice (Fig. 5.115b). They also do not need to be achromatized. The determining factors are the visual acuity and the experience of the inspector.

The polished surface is held under the lamp in such a way that one sees the border between light and dark. In this attitude a pseudo-darkfield effect is created which effectively enhances surface flaws. Experienced inspectors examine this way the center and edge of a surface and they know at once if it

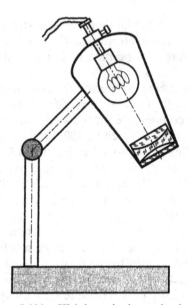

**Figure 5.114.**  High-intensity inspection lamp.

pays to inspect further. If one sleek is found on the surface, then other sleeks are to be expected. Surface flaws are marked with a grease pencil, and the workpiece is returned for repolishing.

At the surface inspection of polished parts, no flaws should be passed that can be removed by additional polishing. Even those flaws that are permitted by the surface specification should be removed because additional flaws can be induced during subsequent operations, so the error limit would be exceeded.

To inspect for surface accuracy, the polished single surfaces are cleaned with a sponge and cloth and carefully dried off. The surface is then brushed with a camel's hair brush, and the test plate is placed on the surface. Newton interference fringes can be seen. When testing strongly curved surfaces, the

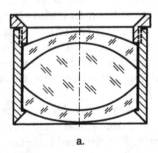
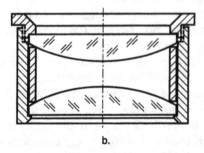

a.                                    b.

**Figure 5.115.**  Surface inspection eye loupes. (*a*) 6-Power Steinheil loupe. (*b*) 5-Power standard loupe.

test plate fit can change as the surfaces of test plate and lens are brought closer together. The test plate may show at first several fringes low, but if it is pressed down more on the surface, the true fit will become apparent as several fringes high. This effect can be easily explained by two circular arcs that approach one another.

Only one lens on the edge of a block needs to be tested with a test plate. The opticians typically wipe the lens clean with the ball of their hands and lay on the test plate. This method of testing is sufficient when the lens surface can deviate by 3 to 5 fringes from the curvature of the test plate. Also fringe irregularities for the edge row of lenses can be recognized in this way. The lenses at the center of the block are not covered by this technique. It is known, however, that surfaces that polish irregular typically exhibit sleeks. When the surface is free from sleeks, it is safe to assume that it has the same curvature in the center as on the edge.

It is quite common for the test plate fit to change with decreasing temperature. The lens is tested after it is deblocked from the lens block so that the magnitude and direction of the change in test plate fit can be determined. In general, one can assume that the fit as measured on the just polished and still warm lens will closely approximate the true fit.

Plano surfaces, especially prism surfaces, are preferably tested in an interferometer. Since these instruments use monochromatic light, interference bands can still be seen when there is an air gap between test plate and prism surface of 0.5 mm (0.020 in.) or more. Interferometers are discussed in Chapter 6.

### References for Section 5.5 (G&P Machines)

[88] Loh double eccenter.

[89] Leico, Lohnberg/Lahn, Germany.

[90] Kärger, Berlin.

[91] Loh, Models PM-2 and PM-3.

[92] Rogers & Clarke Mfg. Co., Rockford, IL.

[93] Mildex (Ugadawa).

[94] DIN 58 725.

[95] DIN 58 726.

[96] Mildex (Ugadawa).

[97] Strasbaugh.

[98] Strasbaugh.

[99] Rogers & Clarke.

[100] Loh double eccenter.

[101] Mildex (Udagawa).

[102] Strasbaugh.

[103] Mildex (Ugadawa).

[104] DAMA.

## 5.6.  LENS CENTERING

To ensure the proper performance of a lens system, all lenses must be accurately centered with respect to the optical axis. The level of accuracy depends on the intended application. Centering and edging to finished diameters is one of the last operations performed in the fabrication of spherical lenses. Nonspherical surfaces are centered by more complex means, and plano surfaces need not be centered at all.

During the centering stage, the edge of a lens is ground in such a way that it forms a cylinder that is concentric about the optical axis. The optical axis is an imaginary line that connects the two centers of curvature of the lens surfaces. Accurate bevels can be ground on the lens after it has been edged and while it is still mounted on the mandrel. This ensures that the bevels are concentric and even all around (Fig. 5.116).

Even a relatively small centering error influences the image quality of an objective in a serious way. Such errors can rapidly multiply in a lens system and lead to astigmatism and coma. Constant demands on improved image quality have resulted in a continuous improvement of centering processes and machines.

One of the most important requirements for the precision centering of lenses is the use of a true running centering spindle that cannot have more than 2 $\mu$m (<0.0001 in.) of runout. A centering chuck is mounted on the spindle, and the edge of the chuck is cut true to the spindle axis. Only after that has been done can the lens be mounted on the chuck and centered. The

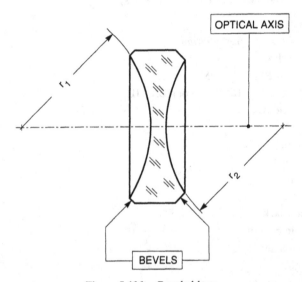

**Figure 5.116.**  Beveled lens.

spindle is then firmly mounted on the centering machine. While the mounted lens turns, its edge is ground round to the exact diameter.

Since the optical axis is an imaginary line, it is not possible to use it directly to center the lens. A physically measurable feature of a centered lens must be used to do this. This is found in the rotationally symmetrical body of the lens when it runs concentric to the optical axis. Lenses are centered by making use of this feature.

### 5.6.1. Dimensioning for Centering on Lens Drawings

There are some standard ways [105] and many nonstandard ways to define how the centering dimensions and the permissible tolerances are to be specified on lens drawings. The centering accuracy consists of two basic components. One concerns the physical dimensions such as the lens diameter and the bevel sizes. The other concerns the position of the optical axis relative to the cylinder axis of the lens edge. In this context the latter component is generally designated the *centering accuracy*.

**Diameter Tolerances.** Lens diameters are accurately ground to within ±0.01 mm (±0.0004 in.) on centering machines equipped with diamond wheels and precision spindles. In the past, when wider tolerances were typical, the seat in lens mounts had to be turned to fit each individual lens. But the tightly toleranced lenses for today's high-definition lens systems can be mounted directly in prefabricated lens cells. Lenses used for many other applications like infrared (IR) systems have diameter tolerances of 0.5 mm (±0.002 in.). Most other requirements fall between these extremes.

Lens diameters may soon be toleranced on the part drawing according to ISO standards [106[. The most often used tolerance series is designated "h" for which the nominal diameter is the maximum dimension (Fig. 5.117).

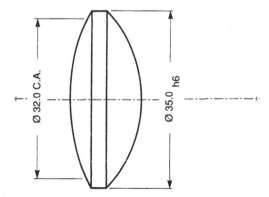

**Figure 5.117.** Tolerancing of lens diameter per International Standards Organization (ISO).

Lenses for high-precision objectives are centered to the diameter quality level h6 for which, depending on lens diameter, the tolerances are approximately ±0.01 mm (±0.0004 in.). The quality requirements can be relaxed to level h11 for lenses used for simpler applications. However, these rather broad tolerance ranges are generally not used because modern centering machines can hold closely toleranced diameters without difficulty.

In addition to the lens diameter, the clear lens aperture is frequently dimensioned. The areas of the lens that lie outside the clear aperture are not optically used. They may exhibit flaws without causing rejection of the lens. If at all possible, the centering chucks should contact the lens surface only in this zone. Any surface damage the chucks may cause will then appear outside the critical area.

**Permissible Centration Error.** The permissible centration error is shown on lens drawings in one of several ways. The magnitude of the wedge error can be shown in arc minutes, for instance, 4/2.5′ as in Fig. 5.118. The permissible error in arc minutes is designated with $\gamma$. It is formed by tangents that contact the lens surfaces where the cylindrical axis of the lens intersects them (Fig. 5.119). This definition assumes that the two lens surfaces form a prismatic wedge. The refractive effect is influenced by the refractive index n of the glass according to the relationship $\varepsilon = \gamma (n - 1)$. The centering error can also be defined as a tilt error to one or the other lens surface.

In the DIN standards version which was valid until 1972, the centering error z was measured as the distance in millimeters between the optical axis and the mechanical axis of the lens (Fig. 5.120). This version of specifying centering error can still be found on many of the older part drawings. A third method measures the variations of the edge thickness, which ideally should be zero. This method, which is often referred to as the *true indicator reading*

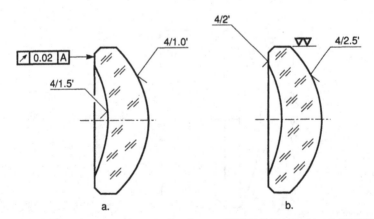

**Figure 5.118.** Centration and wedge specification per DIN 3140. (*a*) Preferred. (*b*) Still in use.

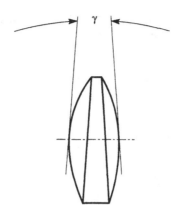

**Figure 5.119.**  Centering error as angle $\gamma$.

(TIR) method, is useful only for loosely toleranced centration requirements (Fig. 5.121).

The angle $\gamma$ is measured by optomechanical means. To do that, the lens is held in a true running mount as it is turned. The centering error $z$ is then read off in arc minutes in a vertically mounted autocollimating telescope (Fig. 5.122). According to the old definition, the centering error $z$ is measured with a dial indicator fixture [107]. The lens under test is rotated between two centering chucks, while the dial indicator contacts the lens edge. Twice the centering error, (i.e., $2z$) is measured with this device (Fig. 5.123). The runout of fine-ground bevels can also be tested with the same fixture. The center-

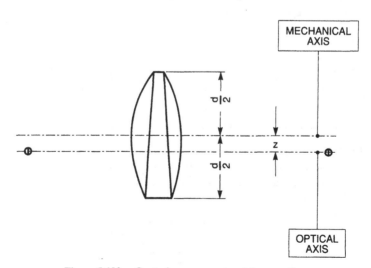

**Figure 5.120.**  Centering error $z$ as axial separation.

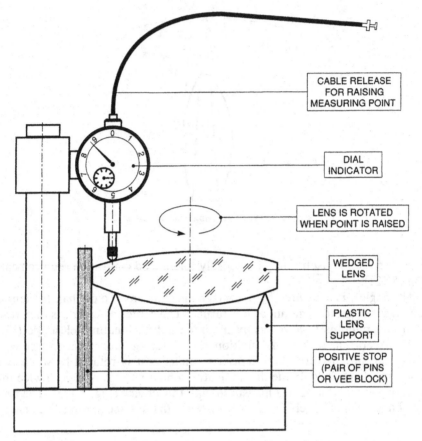

**Figure 5.121.** TIR setup for measuring wedge.

ing chucks for this device are made of plastic material to prevent damage to the lens surface.

### 5.6.2. Calculation of the Centering Error

The centering run-out $z$ in millimeters and the centering error $\gamma$ in arc minutes, where $\gamma$ (in arc min) = 3437 $\gamma$ (in arc min) = 3437 $\gamma$ (in radians), have the following relationships (see Figs. 5.119 and 5.120):

$$\gamma \text{ (arc min)} = 3437 \left( \frac{z}{r_1} + \frac{z}{r_2} \right),$$

$$\gamma \text{ (radians)} = \left( \frac{z}{r_1} + \frac{z}{r_2} \right).$$

(5.15)

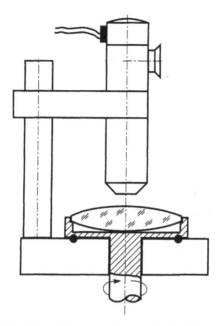

**Figure 5.122.** Use of an autocollimator for checking centration.

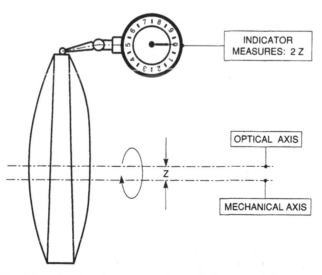

**Figure 5.123.** Measuring the edge run-out of an optically centered lens prior to edging.

When converted,

$$z = \frac{\gamma \text{ (arc min)}}{3437(1/r_1 + 1/r_2)},$$

$$z = \frac{\gamma \text{ (radians)}}{(1/r_1 + 1/r_2)},$$

(5.16)

where

$z$ = centering run-out in millimeters,
$\gamma$ (arc min) = centering error in arc minutes,
$\gamma$ (radians) = centering error in radians,
$r_1$ and $r_2$ = radius of curvature of the lens surfaces,
3437 = conversion constant.

The radii of convex surfaces are positive, and those of concave surfaces negative.

## Example

The allowable centering error on an old drawing is shown to be 0.02 mm. The biconvex lens has the radii $r_1$ = +20 mm and $r_2$ = +100 mm. The allowable centering error in terms of angle $\gamma$ must be calculated using Eq. (5.15):

$$\gamma = 3437 \left(\frac{z}{r_1} + \frac{z}{r_2}\right)$$

$$= 3437 \left(\frac{0.02}{20} + \frac{0.02}{100}\right)$$

$$= 4.1 \text{ arc minutes.}$$

**Overage Allowance for Centering.** The diameter overage of an uncentered lens must be at least as large as the anticipated centering run-out. This overage is almost always estimated. A lens with convex surfaces gets a diameter overage of 1 mm (0.040 in.), whereas lenses with concave surfaces get at least 1.5 mm (0.060 in.). In addition to the centering run-out, such lenses have protective bevels that must be removed.

It is best to calculate the required diameter overage to minimize the material that has to be ground off the lens edge, at the same time ensuring that the edge will clear out. This is especially important for meniscus lenses

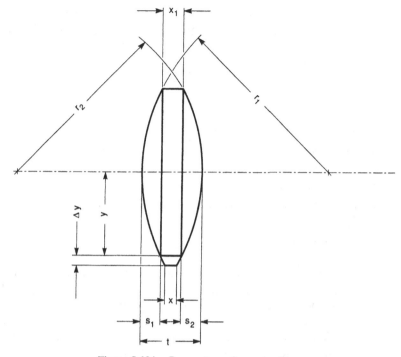

**Figure 5.124.** Parameters of a wedged lens.

made from expensive materials such as germanium (Ge) and for glass lenses with low refractive powers for which the run-out can easily exceed several millimeters. A proven approximation based on the condition represented in Fig. 5.124 will suffice for this.

The centering run-out $\Delta y$ that must be ground off depends on the edge thickness difference $\Delta x$, the lens diameter $2y$, and the lens radii $r_1$ and $r_2$. According to Fig. 5.124, the edge thickness $x$ of a lens can be calculated from the center thickness and the two sagittas of the lens surfaces:

$$x \approx t - \left(\frac{y^2}{2r_1} + \frac{y^2}{2r_2}\right).$$

The edge thickness of the opposite side, which differs by $\Delta y$, is then

$$x_1 \approx t - \left[\frac{(y + \Delta y)^2}{2r_1} + \frac{(y + \Delta y)^2}{2r_2}\right].$$

When both expressions for the edge thickness are subtracted from one another, we get $\Delta x = x_1 - x$, which is the edge thickness difference of a wedged lens. The approximate expression for the centering run-out can be calculated from this. Note that the relatively small value for $(\Delta y)^2$ was not considered.

$$\Delta y = \frac{(\Delta x)(r_1)(r_2)}{y(r_1 + r_2)}, \tag{5.17}$$

where

$t$ = center thickness,
$x$ = edge thickness,
$y$ = half of the lens diameter,
$r$ = radius of curvature, (concave radii are negative),
$\Delta y$ = centering run-out,
$\Delta x$ = edge thickness difference.

From a mathematical standpoint, this is only a first-order approximation. It is, however, sufficiently accurate in a practical sense. By using Eq. (5.17), one can calculate how much diameter overage is needed for a lens that has a known edge thickness difference $\Delta x$. Conversely, one can also determine how parallel a lens must be worked when only a limited overage can be provided. Depending on shape, size, and the fabrication method used, a lens can have an edge thickness difference ranging from about 0.1 to 0.5 mm (0.004 to 0.020 in.) (Fig. 5.125).

**Example**

The required diameter overage is to be calculated for a biconvex lens. The lens has radii of $r_1$ = +13.320 in. (338.3 mm) and $r_2$ = +20.000 in. (508.0 mm). The diameter $2y$ is 1.200 in. (30.5 mm). During fabrication, the lens has a maximum edge thickness difference $\Delta x$ = 0.009 in. (0.23 mm). Using Eq. (5.17), one can calculate the required overage $\Delta y$ as

$$\Delta y = \frac{(0.009)(13.32)(20)}{0.6(13.32 + 20)}$$

$$= \frac{2.3976}{19.992}$$

$$= 0.120 \text{ in.}$$

The diameter of the uncentered lens must be 0.120 in. (3.0 mm) larger than the finished diameter. To be sure, another 0.020 in. (0.5 mm) is added and the blank or pressing diameter is specified as 1.340 in. (33.5 mm). The same results are obtained when the metric equivalents are used.

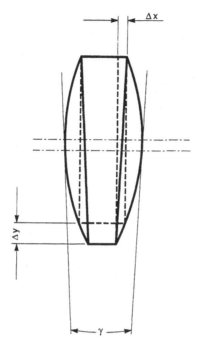

**Figure 5.125.**   Edge run-out of a wedged lens as function of edge thickness variation.


### 5.6.3.   Centering Methods

There are two basic ways to center lenses. The older transfer spindle method is still used for prototype and low-volume production and for centering lenses made from soft and sensitive materials. The other method is the *bell-chucking method*. Most lens centering of volume production is now done with this method.

**Transfer Spindle.**   For the transfer spindle method, the lens is mounted on a precise mandrel which is fastened by a thread to a precision spindle. This mandrel, also called a *centering bell*, is nearly always made from brass, although steel bells are sometimes used as well. The lens is either mechanically or optically aligned on the mandrel so that the optical and mechanical axes of the lens are coincident. The spindle with the centered lens is then mounted in a centering machine, which is used to grind the diameter (OD) of the lens concentric to the axis of rotation.

There are several ways in which a lens can be centered on the bell-shaped centering mandrel (Fig. 5.126a). In the first case, the lens and the mandrel are heated sufficiently so that a small amount of a specially formulated centering wax can be applied to the edge of the mandrel. The heated lens is then positioned on the mandrel by mechanical means or by an optical align-

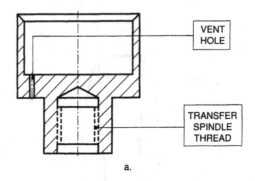

a.

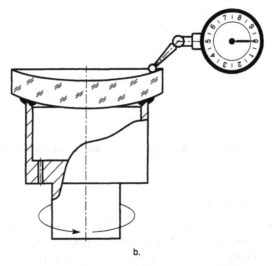

b.

**Figure 5.126.** Basic transfer spindle lens centering methods. (*a*) Typical centering bell. (*b*) Runout measurement (TIR).

ment method. For the mechanical method the edge thickness variations (ETV) must be nulled out or at least reduced to a minimum while the lens rotates. The run-out is monitored by means of a dial indicator, as shown in Fig. 5.126*b*.

The optical methods rely either on light reflected off the outer surface of the lens (Fig. 5.126*c*) or on the rotation of a target image that is projected through the lens (Fig. 5.126*d*). A laser beam alignment method is used for the most critical centration requirements. It is quite similar to the target image projection method. Both of these optical centration methods are only useful for lenses made from visually transparent materials. Projection methods are not possible for IR materials and metal optics because they are not transparent in the visible region of the spectrum.

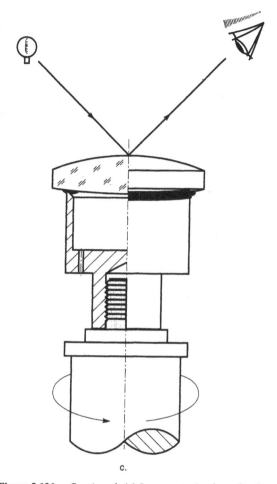

c.

**Figure 5.126.** *Continued.* (c) Lens centering by reflection.

**Bell Chucking.** The second lens centering method is bell chucking by which a suitably curved lens is self-centering when clamped between identical and precision aligned brass mandrels or bell chucks. This is shown in Fig. 5.127 in which the lens is shown in the preferred horizontal arrangement. Bell chucking is a purely mechanical method that relies on the fact that the lens will slide along the edge of the bell chuck until it seats itself at the zone of equal edge thickness. In this alignment the optical axis of the lens and the mechanical axis of the spindle become colinear. When the edges of the centering chucks are well rounded and polished, the lens surface will not be damaged, even at high chuck pressures.

Bell clamping can be used for lenses with diameters as small as 3 mm (0.120 in.) and as large as 150 mm (6.0 in.) with horizontal spindles. Larger

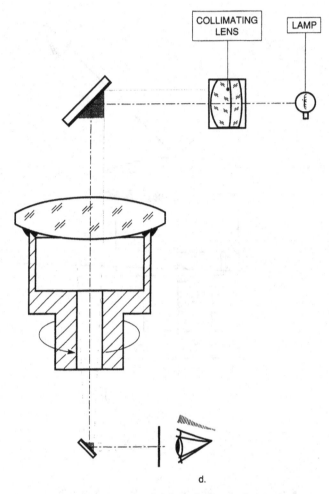

**Figure 5.126.** *Continued.* (*d*) Optical centering method.

lenses up to 250 mm (10.0 in.) in diameter can be centered with the bell-chucking method but vertically aligned spindles are required, as shown in Fig. 5.128. For larger lenses a vacuum assist method is necessary to reduce the otherwise required high clamping loads on the lens surfaces that can lead to damage.

**Centerability.** For the transfer spindle method there is no theoretical limit on centerability, although some practical limits do exist. Even lenses with very weak optical power can be centered with this method as long as the centration error can be optically or mechanically detected. There is a well-defined centerability limit, however, for the bell-chucking method. When the lens radii become too long, even strong clamping forces **F** will not prevent

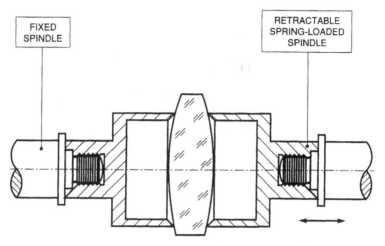

**Figure 5.127.** Bell chuck centering for small lenses.

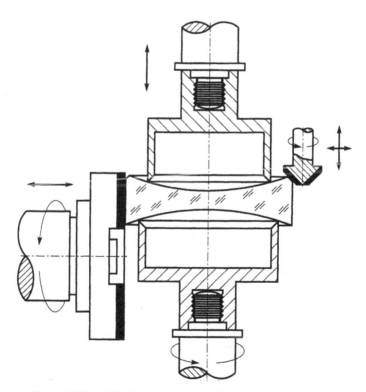

**Figure 5.128.** Bell chuck centering and beveling for large lenses.

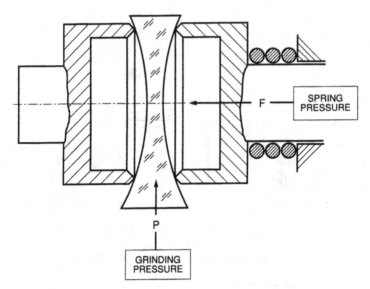

**Figure 5.129.** External forces acting on bell chuck centered lens.

the lens from being displaced by the appreciable side pressures **P** exerted by the diamond wheel (Fig. 5.129). Before a lens production run is committed to a bell-chucking machine, it must first be determined if the lens can be safely centered this way.

The limit of centerability with the bell-chucking method is a function of the slide angle of the lens relative to the chuck edge and the coefficient of friction between lens and chuck. The motion component **B** which can be derived from the chuck pressure **F** causes the lens to move between the bell chuck edges (Fig. 5.130). This motion component must be large enough so that it can overcome the friction between the bell chuck edges and the lens surfaces. Since the value of this component depends on the shape of the lens, it is possible for it to become smaller than the frictional component. The lens will then no longer slide between the bell chuck edges, and it can no longer be centered this way. The limit of centerability can be calculated from the lens diameter, the lens radii, and the coefficient of friction.

**Theory of Bell-Clamping Method.** The bell-clamping angle $\delta$ determines if a lens can be successfully bell chuck centered or if it must be centered and edged with the more traditional transfer spindle method. This angle $\delta$ is a function of the part geometry (see Fig. 5.131$a$). The angle $\delta$ can also be expressed mathematically as the sum of the included angles $\varphi$ for side 1 and side 2 of the lens. The included angle per surface is $\varphi = \arcsin (d/2r)$, where $d$ is the diameter of the bell chuck and $r$ is the radius of curvature of the lens (Fig 5.131$b$).

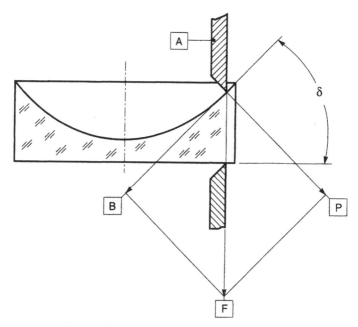

**Figure 5.130.**   Force diagram for a clamped lens.

These considerations lead to an equation that permits the calculation of the slide angle $\delta$:

$$\delta = \arcsin\left(\frac{d_1}{2r_1}\right) + \arcsin\left(\frac{d_2}{2r_2}\right) \tag{5.18}$$

The edge of the bell chuck is forced against the lens by the compression spring force $\mathbf{F}$ (Fig. 5.130). Since the force does not act normally on the surface, the pressure component $\mathbf{P} = \mathbf{F}\cos\delta$. The frictional force $\mathbf{R}$ of the bell chuck edge along the lens surface is then

$$\mathbf{R} = \mu\mathbf{F}\cos\delta, \tag{5.19}$$

where

$\mu$ = coefficient of friction (for polished glass on polished steel, $\mu \approx 0.14$),
$\mathbf{R}$ = frictional force,
$\mathbf{F}$ = pressure component parallel to axis of rotation,
$\delta$ = slide angle of the bell chuck edge.

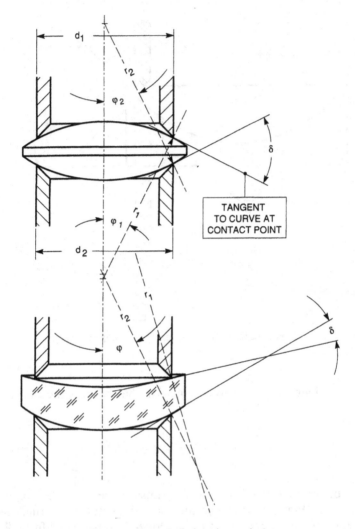

**Figure 5.131a.**   Bell clamping angle δ.

The motion component **B**, which permits the sliding of the lens relative to the bell chuck edge, is:

$$\mathbf{B} = \mathbf{F} \sin \delta. \tag{5.20}$$

The motion component **B** becomes effective only after the frictional force **R** has been overcome. The limit lies where **B** and **R** are of equal value. We therefore set both expressions to be equal and obtain from this relationship

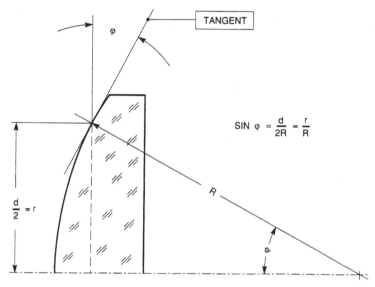

**Figure 5.131b.** Definition of the included angle $2\varphi$.

that the tangent of the angle $\delta$ equals the coefficient of friction:

$$\mathbf{F} \sin \delta = \mathbf{F} \cos \delta(\mu)$$

$$\frac{\sin \delta}{\cos \delta} = \tan \delta = \mu. \tag{5.21}$$

The tangent of $\varphi$ can be derived from the edge thickness $y$ of a lens as derived for Eq. (5.17).

$$x \approx t - \left(\frac{y^2}{2r_1}\right) - \left(\frac{y^2}{2r_2}\right), \tag{5.22}$$

where

$x$ = edge thickness of a lens,
$t$ = center thickness of a lens,
$y$ = semidiameter of the bell chuck,
$r_1, r_2$ = lens radii.

When Eq. (5.22) is differentiated to $dy$, we get

$$\frac{-dx}{dy} = \frac{y}{r_1} + \frac{x}{r_2}.$$

The differential quotient $dx/dy$ is the tangent of $\varphi$ at the point where the bell chuck edge contacts the lens surface. Therefore, according to Eqs. (5.21) and (5.22), the limit of centerability for a lens surface is

$$Z = \frac{y}{r_1} + \frac{y}{r_2} \geq 0.14. \qquad (5.23)$$

When this relationship is expanded so that the centering bell diameters $d_1$ and $d_2$ are used in place of the lens semidiameter, the equation must be multiplied by a factor of 2. If the equation is further expanded for a whole lens, which has two surfaces, then it must be multiplied by 2 again. In this case the criterion of centerability of a lens reduces to

$$Z = \frac{d_1}{r_1} + \frac{d_2}{r_2} \geq 0.56, \qquad (5.24)$$

where

$d_1$, $d_2$ = diameter of the bell chucks in millimeters,

$r_1$, $r_2$ = lens radii in millimeters (convex radii positive, concave radii negative).

The value $Z = 0.56$ is the limit below which the lens will no longer slide between the bell chuck edges. It is an approximate value that is affected in a limited way by the lubricating qualities of the cooling oils used.

## Example

Determine if a particular lens can be centered by the bell chuck method. The negative meniscus lens has the following dimensions:

$r_1$ = +63.0 mm,
$r_2$ = −27.8 mm,
$d$ = 36.0 mm.

For the convex surface a centering bell chuck with a diameter $d_1$ = 34.0 mm is chosen, and for the concave side $d_2$ = 32.0 mm is chosen. Using Eq. (5.24), we calculate

$$Z = \frac{34 \text{ mm}}{63 \text{ mm}} - \frac{32 \text{ mm}}{27.8 \text{ mm}}$$

$$= +0.54 - 1.15 = -0.61$$

Since the absolute value of $Z$ is greater than 0.56, the lens can be centered with the bell-chucking method.

When the two bell chucks have the same diameter, $d_1 = d_2$, then Eq. (5.24) can be rewritten to

$$Z = d \left( \frac{1}{r_1} + \frac{1}{r_2} \right) \geq 0.56. \tag{5.25}$$

This equation can be solved for the bell chuck diameter limit at which the lens will begin to slide:

$$d = \frac{r_1(r_2)(0.56)}{r_1 + r_2}. \tag{5.26}$$

Therefore, the limit value for the example is

$$d = \frac{(63)(-27.8)(0.56)}{(63 - 27.8)} = 27.9.$$

This result means that the bell chuck diameter must be at least 28 mm to be able to center the lens.

**Manually Assisted Bell Centering.** Lenses that have a value of centerability close to 0.56 can be centered only with difficulty when using the bell-chuck-centering approach. But when such a lens is mounted using only light spring pressure at first, it can be slightly turned during the mounting stage to overcome the initial friction. A skilled operator can feel if such a lens is seated properly. The limit for centerability can be reduced to $z \approx 0.5$ this way on a trial and error basis. To do this, it is necessary that the machine be equipped with centering spindles that provide for both light and heavy spring pressures. The heavy pressure is used only after the lens has been properly seated.

For machines that have horizontal centering spindles, large and consequently heavy lenses should be loaded into the machine in such a way that they slide into the centering position from the top. The motion component, which is transversely aligned relative to the spindle, might not be able to forcibly lift such a heavy lens when it is mounted below the spindle axis so that it will seat exactly between the bell chucks. As a result heavy lenses should be centered, if possible, on machines with vertical spindles [108, 109]. The lenses lie in a horizontal position in these machines, and their weight has little or no effect.

The run-out accuracy of the centering bells and the spindles determine how accurately the mounted lens can be centered. In the literature [110, 111], the attainable trueness for bell chuck centering is presented as a func-

tion of the lens shape. This effect, however, is not founded on mathematical
or physical proof.

### 5.6.4.  Optical Centering

Optical centering is generally understood to mean centering with the aid of
reflections from the lens surfaces. The term can also include centering by the
use of transmitted light. With either method, a projected or reflected target is
made to be stationary as the lens rotates. The position of the lens is adjusted
in this manner until the spindle axis and the optical lens axis coincide.

**Mounting of Lenses for Optical Centering.** Lenses that have reduced refractive powers are mounted with centering pitch to true running centering
chucks (Fig. 5.132). The lenses are aligned on the chucks by hand according
to the way the reflections from the polished surfaces move. The center of the
lens surface that contacts the chuck lies always on the axis of rotation of the
centering spindle. As long as the centering pitch is still warm and pliable, the
lens can be moved laterally until the center of the exterior surface lies also
on the spindle axis (Fig. 5.133). The lens is properly centered when the
reflections from both surfaces remain stationary as the centering spindle is
turned.

Particularly small lenses that can no longer be aligned by hand are centered with the use of a hardwood peg (Fig. 5.134). The peg rests on a small
platform and touches the outer lens surface near the edge. To be able to
observe the movement of the reflections, it is necessary in most cases to
magnify them with an eye loupe (Fig. 5.135).

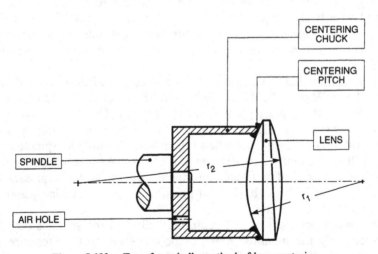

**Figure 5.132.**  Transfer spindle method of lens centering.

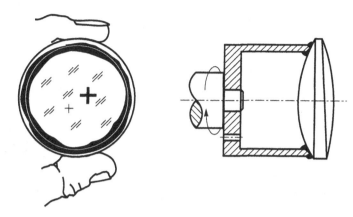

**Figure 5.133.** Lens centering by hand, using reflection off both lens surfaces.

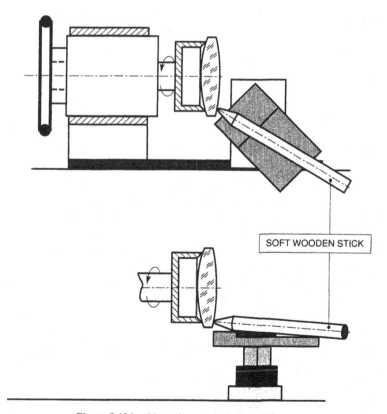

SOFT WOODEN STICK

**Figure 5.134.** Manual centering of small lenses.

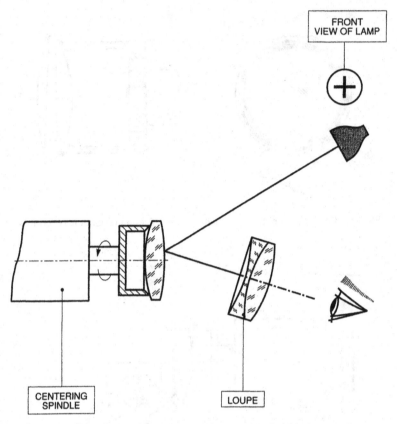

**Figure 5.135.** Optical centering of small lenses with the aid of an eye loupe.

When the lens has surfaces with different radii, the side with the stronger curve is mounted to the centering chuck. The reflection, as well as the image run-out of the longer radius, appears magnified. The run-out of the reflection is then more easily seen, and this increases the centering accuracy. The centering pitch is typically a blend of approximately equal parts of soft yellow or black pitch and shellac that is triple filtered to be free of any contamination that can damage polished surfaces.

**Centering Accuracy with the Unaided Eye.**  A lens cannot be very accurately centered with the unaided eye (Fig. 5.136). The limit of the centering accuracy that is recognizable with certainty is

$$\gamma \text{ (in radians)} \leq 0.0005 + \frac{0.025 \text{ mm}}{r_2 \text{ (mm)}}, \tag{5.27}$$

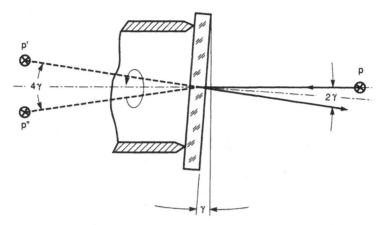

**Figure 5.136.** Reflection off a plane parallel glass disk mounted on an incorrectly cut centering chuck.

where

$\gamma$ = largest expected centering error in radians when centering with the unaided eye,

0.0005 = ⅛ the recognizable run-out of the centering chuck,

0.025 mm = lateral run-out of the reflected image off the exterior lens surface, which is just discernable at a standard viewing distance of 250 mm,

$r_2$ = radius of the exterior lens surface in millimeters.

The maximum exterior radius is 500 mm. For longer radii the reflection no longer lies within the standard viewing distance of 250 mm (Fig. 5.137). It should be noted that the lens diameter has no influence on optical centering methods.

**Example**

A lens with an outer radius of 50 mm is to be centered by the unaided eye. What is the expected centering accuracy? According to Eq. (5.27),

$$\gamma = 0.0005 + \frac{0.025}{50} = 0.001 \text{ radian.}$$

This converts to less than 3.4 arc minutes. A smaller centering error for this type of lens cannot be resolved with the unaided eye.

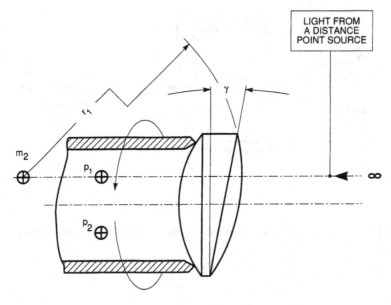

$p_1$, $p_2$: Images of the distant point source

**Figure 5.137.** Limits to visual centering.

**Optical Centering Instruments.** Since many lenses cannot be centered with the unaided eye to the required accuracy and since it also requires skilled opticians to center lenses reliably with this method, a variety of instruments were developed with which lenses can be centered to better than 1 arc minute without much difficulty.

The simple loupe has already been mentioned as an effective centering aid for strongly curved lens surfaces for which the reflected image lies close to the surface. Corresponding to the magnification of the loupe, the motion of the image is seen enlarged, and the centration can be achieved with greater confidence (see Fig. 5.135). But centering with the aid of a loupe is a special case which cannot be applied in every situation. Furthermore this method does not provide any measurement of the centering error.

The magnitude of a centering error in arc minutes can be determined with the autocollimator (Fig. 5.138). The light that emerges from the instrument is reflected back from the test surfaces and reenters the autocollimator where an image of the illuminated reticle is formed in the reticle plane of the eyepiece. When the lens is properly centered, the image remains stationary even when the lens is rotated, but when it is not fully centered, the surface will describe a wobbling motion. This causes the reflected image to wander along a circular path. The diameter of this circular motion is proportional to the centering error. With a suitable reticle scale of the autocollimator, the error can be directly read off in arc minutes. The surface curvature of the

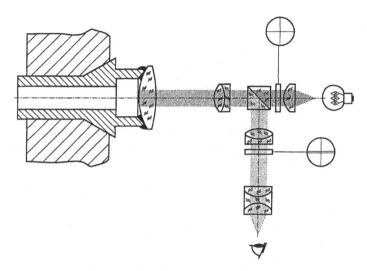

**Figure 5.138.**   Centering autocollimator.

lens under test must be compensated for by correction lenses which are mounted in front of the autocollimator objective to recollimate the beam path.

When the centering bell is bored through and the centering spindle is hollow, it is possible to additionally test the run-out of the pitched-down lens surface. Instruments that serve this purpose use two collimators [112]. Through the use of a flip mirror, either the upper or the lower lens surface can be tested for centering run-out.

To avoid having the expensive centering machine sit idle during the lens-centering stage, the lenses are pitched to the bell chucks on special alignment benches and are centered there. Interchangeable spindles or exchange bell chucks are used in this mode. Alignment benches are described in Section 5.6.5.

The alignment fixture is built into a table. It has a seat for the interchangeable spindle or it has a spindle that accepts the exchange bell chuck. The optical alignment device is mounted above the spindle axis so that its eyepiece is conveniently accessible. The lenses and the bell chucks are prewarmed on a hot plate. Only the contact edge of the chucks should be warmed up, while the mounting thread should remain at ambient. The centering pitch is heated at the same time in a suitable container so that the contact edge of the chuck can be dipped into it, thus wetting the edge with melted pitch.

A unique alignment device [113] operates on the basis of light interference. The light of a spectral lamp is monochromatically filtered and directed onto the surface under test. The reflected light interferes with the reflection from an accurately centered reference surface and interference bands are seen in the eyepiece. The number of bands increases when the surfaces are mutually decentered. When the surfaces are perfectly centered relative to each other, no interference bands are seen at all. The centering accuracy can be deduced from the number of bands. This centering process can be very accurate because of the inherent sensitivity of this method.

**Centering Chucks and Spindles.** The lenses are supported during the centering process by the true running edges of the centering chucks. The requirement that the center of curvature of the contacting lens surface must lie on the rotational axis of the spindle can be achieved only when the centering chuck runs true.

The centering bell chucks used on many centering machines are made from brass (see Fig. 5.126a). They are threaded to screw to the centering transfer spindle. An air hole in the shank prevents the entrapped air from exerting an pneumatic effect on the lens. The supporting edge of the brass chuck is usually turned with a hand-held bit on a simple bell trueing lathe. These operations are described in Section 5.6.5. The sharpened bit must remove only very little material with this method, so there are no chips generated but only fine brass dust. Any edge irregularities must be carefully removed in this manner.

The centering chucks for the newer production machines are made from steel. The supporting edge has sometimes been plated with a hard metal coating. Centering chucks become resistant to damage this way, and they do not tend to trap contaminants that could otherwise be pressed into the soft brass. A type of centering chuck with a tapered bell [114] has found broad acceptance in some European shops (Fig. 5.139).

The centering chucks should have the largest possible diameter, which ideally should be only 0.5 mm (0.020 in.) smaller than the finished lens diameter. When bevels are simultaneously ground, the height of the bevel segment on the grinding wheel must be considered. Lenses with centered bevels are mounted on centering chucks with different diameters (Fig. 5.140).

The supporting edge of the steel chuck is accurately ground in while it rotates on the centering spindle. The spindles can be removed from the centering machine for this purpose. This is why these spindles are generally referred to as transfer spindles. Only when the chuck diameter is appreciably larger than the diameter of the spindle is it necessary to remove the chuck from the spindle. The seats on the spindle and the chuck must then be marked and the chuck remounted according to this marking. This method sacrifices accuracy for practicality, but large lenses rarely need to be centered to a high degree of accuracy. The centering chucks are ground in on the spindle on a running bell trueing lathe that has been converted to an ID/OD grinder. Damaged chucks are also reground on the lathe.

Some of the spindles for optics machines have been standardized [115]. The spindle diameters and spindle lengths are usually stepped. As a result

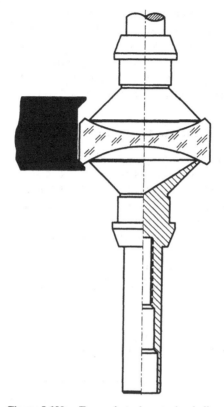

**Figure 5.139.** Tapered steel centering bells.

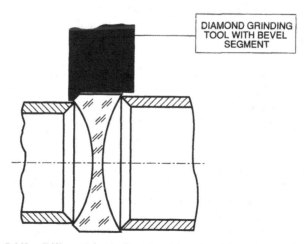

**Figure 5.140.** Different chuck diameters for different beveling requirements.

there is a long and a short spindle for every spindle diameter. The short spindles are used for grinding, while the long spindles are used for centering.

Slide bearings are preferred for slowly rotating spindles. Fast running spindles are equipped with roller bearings such as the popular "Müller-Spindle" [116]. Diamond wheels that have been specifically designed for use on precision centering machines are described in detail in Chapter 4.

### 5.6.5.   Centering and Edging Machines

Centering and edging machines are used to work the edges of the lenses so that the lens edge and the bevels run concentric to the optical axis of the lens. These machines have horizontal spindles for lenses to about 75 mm (3.0 in.) diameter: machines for larger lenses usually have a vertical spindle arrangement. Many centering and edging machines are dual function machines that are convertible from bell chucking (high volume) to the transfer spindle operation (low to moderate volume). Most center and edging (C&E) machines are single units, but dual spindle production machines have been built. Production machines operate on an automatic process cycle once set up and initiated.

**Evolution of Centering Machines.** Early centering machines were constructed on the same principle as a lathe. The spindle, which had a conical slide bearing, rotated at the end of a center, and it could be adjusted to have no play. A long round belt drive enabled the spindle to run vibration free. The spindle had a threaded end to which the centering chuck was fastened. The lens was cemented to the chuck and centered by reflected image. An adjustable mount which supported a grinding plate made from brass was fastened to the machine frame. This plate could be easily positioned with a thumb screw to bring it into contact with the edge of the lens. The rotating lens was then ground on the plate with loose abrasive. To prevent the lens edge from assuming a barrel shape, the spindle had to be moved longitudinally over the grinding plate by turning a handwheel that moved the spindle on a sliding bearing.

In the next generation of machines this simple concept was advanced to incorporate an abrasive grinding wheel to replace the messy grinding plate. The lens on the centering chuck rotated at about 300 rpm. The grinding wheel, which rotated at a higher speed, was moved into close proximity of the lens. The wheel was supported on a swing arm behind the centering spindle and was cooled with water. A linkage with a ratchet moved a threaded adjustable stop so that the swing arm slowly approached the lens. Only light grinding pressure was generated because of the slow infeed of the grinding wheel. A second positive stop limited the infeed distance of the wheel.

These old machines were replaced in the 1930s by machines using diamond as an abrasive medium. At the same time the bell compression chucking was also developed.

**Automatic Centering Machines.** The horizontal construction of modern bell-chuck centering machines are well represented by a few of the more popular automatic centering machines [117, 118]. These machines are equipped with a coolant reservoir with a recirculating pump in the machine base which also houses the motor that drives the centering spindle. The centering spindles are located above that in the cast machine housing. The right spindle slides in a linear bearing and is retracted with a lever when a lens is inserted. An extension spring, which supplies the compressive force, acts on this lever by which the lens is held between the centering bells. The left spindle rotates in a fixed position. For some machine models the left spindle can be removed from the housing when the machine should run with transfer spindles.

The diamond centering wheel is mounted to a pivoting support behind the centering spindle. The support is drawn toward the mounted lens by means of an extension spring. A cam-operated stop determines the feed distance. A second stop limits its forward motion when the lens diameter has been ground to size. The entire support, together with the centering wheel, also moves laterally. The diamond wheel is utilized in this way across its entire width, and the bevel rings on the wheel can be moved against the lens edge. To ensure that the bevel is sharply delineated from the lens edge, the centering wheel is moved back by about 0.1 mm. This is accomplished automatically by the stop pin which runs against a slight step in the stop anvil mounted to the pivoting diamond wheel support.

The centering spindle drive method varies from machine to machine. Although it was quite common until fairly recent times that the centering spindle revolved at a constant speed of about ten revolutions during the centering cycle, it is more desirable with modern machines to let the spindle rotate about 50 times. Some of the older machines have multistep pulleys that permit speed settings between 30 to over 200 rpm. The modern machines are typically equipped with infinitely variable speed controls ranging from 3 to 100 rpm. The higher speeds reduce the grinding pressure on the lens, since only a small amount of material is removed for each revolution of the lens. This change has become necessary because centering wheels are now made with finer diamonds that produce better edge finishes, though they do this at the expense of abrasive efficiency.

The diamond wheels used are peripheral wheels, typically 150 mm (6.0 in.) in diameter and between 12.5 and 25 mm (0.5 to 1 in.) wide. Tool spindle speeds range from 2500 to 3500 rpm, which yields peripheral speeds from about 20 to 27.5 m/sec (4000 to 5500 SFPM). Smaller diameter wheels must rotate faster, and larger ones slower to maintain a similar range of peripheral speeds.

The cycle times are a function of lens diameter, lens edge thickness, and the amount of diameter overage. This is so because like any other abrasion method, the edging time is limited by the diamond wheel capacity to abrade a certain volume of material. Therefore a longer time is required to edge a large diameter lens with a wide edge and excessive diameter overage than is needed for a smaller lens with only minimal overage. Figure 5.141 represents

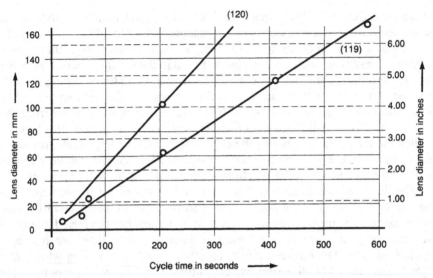

**Figure 5.141.** Relationship of cycle time to lens diameter for a typical centering machine.

this relationship graphically for two different center and edging machines [119, 120]. In general typical cycle times on centering and edging machines range from 0.5 to 3 minutes for small-, 1 to 7 minutes for medium-, and 3 to 10 minutes for large-diameter lenses.

Production centering machines are often built with vertical centering spindles that facilitate the loading and unloading of parts. These machines can range from a small unit that can center lenses up to 25 mm (1.0 in.) in diameter [121] to large, hydraulically operated and controlled machines that can handle lenses of up to 250 mm (10.0 in.) in diameter [122]. Intermediate versions of these machines have also found widespread use [123].

The vertical arrangement of the centering spindle has permitted the development of fully automatic centering machines that use a caroussel loading method with which new lenses are continuously fed to the centering spindle. There are two systems for the unloading from the carroussel. One, shown in Fig. 5.142, uses a vacuum-activated gripper that takes the lens, loads it into the bell chuck, and deposits it back again after the lens has been edged [124]. The second method [125] lifts the lens with the lower bell chuck out of the carroussel, centers it, and replaces it back into the carroussel afterward (Fig. 5.143). The carroussel advances to the next lens after each centering cycle has been completed.

When properly set up, the machine will center, edge, and bevel a magazine or carroussel load of lenses automatically. One operator can readily operate several production machines. Not surprisingly, the number of lenses that a carroussel can hold is a function of the lens diameter. This relationship is shown in Fig. 5.144.

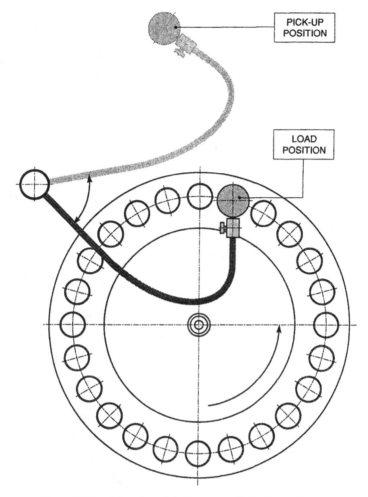

**Figure 5.142.** Magazine plate for automatic loading of lenses.

An automatic lens centering cycle using a circular magazine part feed method [126], proceeds as follows:

- Lens magazine indexes into position.
- Lower spindle lifts lens out of magazine and presses it against upper spindle, and the lens automatically centers itself between the two bell chucks.
- Once properly clamped, the spindles begin to slowly rotate as the rapidly spinning diamond wheel feeds in until it bottoms out against a preset stop.

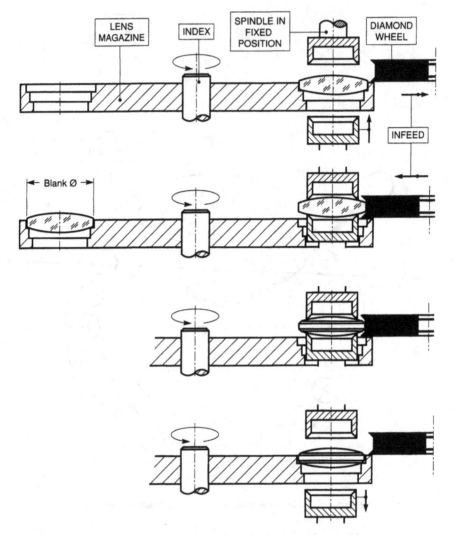

**Figure 5.143.** Fully automated production centering machine.

- Bevels are ground on the same operation when suitable compound wheels are used. The edging segment of the wheel moves away from the lens edge by a small distance. The lens (or wheel) is then moved laterally by a predetermined distance first to one side then the other to grind on the bevels to precise tolerances.
- A very important feature is the automatic compensation for part thickness variations. This feature, which maintains the proper bevel width, is available with the bell-clamping method only. The machine senses thickness differences from lens to lens and adjusts the lateral infeed and

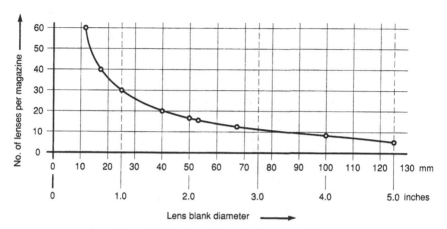

**Figure 5.144.** Load sizes for lens magazines.

the longitudinal feed accordingly to keep the face width of the bevels constant.

· The lower spindle will retract and redeposit the centered and beveled lens back into the magazine of the part caroussel.

**Laser Centering Method.** Modern centering and edging machines utilize thru-the-lens laser alignment, CNC controls, and automatic compensation for lens thickness variations. Optical through-the-lens centering can detect very small centering errors that would be very difficult to see with the unaided eye. Spot image projection systems (see Fig. 5.126d) have been used for many years. Some contemporary machines [127] incorporate this method. It functions on the basis of aligning a fixed and a reflected cross hair image that are displayed on a video screen. This method permits accurate centering of very flat surfaces with the bell-chucking method. For very weak radii a vacuum-assisted hold down must be used for one side of the bell chuck pair.

For the most precise centering, however, a laser centering method was developed for lenses with low refractive powers that cannot be centered with the bell-chucking method [128, 129]. Previously such lenses had to be cemented to the centering chuck and centered on a transfer spindle. These steps are entirely eliminated with this new method.

A typical laser centering system [128] is shown in Fig. 5.145. The laser is mounted above the machine. Mounted in line with the laser is a focusing system that compensates for the refractive powers of the lens that is being centered. A plano mirror redirects the laser beam through the hollow spindle of the centering machine. The beam passes through the bore of the upper spindle and then through the lens, which is captured under very light pressure between the centering bells. The beam passes from there through the

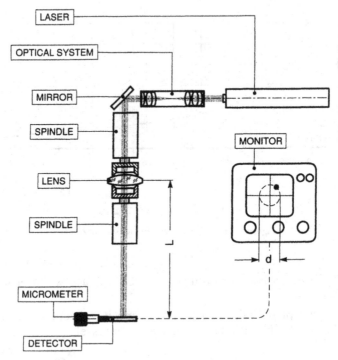

**Figure 5.145.** Laser centering system—diagram.

lower spindle and impinges as a light spot on an underlying detector disk. The location in $x$ and $y$ of this spot relative to the fixed reference center is determined to a high degree of accuracy. The spot is sensed by the vidicon detector, and it is displayed as a bright dot on an video monitor screen.

As the decentered lens rotates, the spot describes a circle on the video monitor. The diameter of the circular path that the spot traces is directly proportional to the magnitude of decentration. The diameter of this circle is a function of the refractive index of the lens material and the wedge angle or tilt. The centration accuracy of the laser unit is 10 arc seconds. The laser method can work both by reflection for nontransparent materials and by transmission for transmissive elements. Laser centering is useful for clamping angles of less than 17°. The lens is adjusted by hand until the bright spot remains stationary as the spindle is turned. It is then firmly clamped between the chucks ready to be edged. The electronic controls of the display screen are adjusted so that the bright dot for a centered lens lies at the center of the cross target on the screen. Subsequent lenses are then easily centered with reference to that target.

The laser light used for this application radiates intensively. It emerges as a thin beam that can be easily directed through small apertures. The light of any other source such as a projection lamp could not be collimated that finely so that it could be passed through the long bore of a spindle axis. Since the laser light passes through the lens, the centering condition of both lens surfaces is encompassed and the error is shown functionally, as it would affect the performance during actual use.

Since the measurement is made in transmission through the lens, it represents a centering error which is affected by the refractive index of the lens material. The accepted standard for centering errors [130] considers only the geometric shape of the lens. Therefore, for an identical centering condition, the magnitude of the centering errors for the two measurement techniques will be different. One of the manufacturers [128] has provided diagrams for conversion of the measurement results. These are based on an equation that deduces the magnitude of the centering error expressed in arc minutes from the deviation of the laser light, the refractive index of the glass from which the lens is made, as well as magnification factors which are set on the instrument.

$$\alpha = \frac{1720d}{VL(n-1)}, \tag{5.28}$$

where

$\alpha$ = centering runout in arc minutes per DIN 3140,
$n$ = refractive index of the lens material,
$d$ = diameter of the circle traced by the spot on the display screen,
$V$ = magnification set on the screen,
$L$ = distance between lens and detector disk.

**CNC Centering Machine.** There is now one CNC centering and edging machine built in the United States [131] and another in Germany [132]. As this technology finds more applications in optical machinery, other such machines are bound to follow. The U.S. machine has a two-axis CNC control which permits about 150 programmable steps for as many as nine different operations. The most commonly used functions are

- OD grinding with plunge feed
- OD grinding with incremental feed
- OD grinding with constant feed
- Beveling steps
- Face flat grinding on concave surfaces
- ID grinding with a special attachment

The German CNC centering machine is a six-axis machine that uses bell clamping and laser centering. It has an automatic loading feature and uses an integral lens diameter and center thickness measurement system that automatically corrects for tool wear and adjusts for different lens thicknesses. It is also programmable to grind complicated bevels to precise dimensions.

**Beveling with Compound Wheels.** Centering and edging machines can be used effectively as beveling machines for circular parts such as lenses. In fact there is no better way to mechanically bevel round parts than when they are centered on a mandrel. The bevels are uniform and completely concentric to the lens axis. Face bevels can also be ground on on a suitably modified center and edging machine. The bevels are perpendicular to the lens axis, concentric to the diameter, run true to the opposite radius, and are square to the edge.

With the bell-chucking method of lens centering, appropriate compound diamond wheels, consisting of a central peripheral edging wheel flanked on either side by beveling segments, can be used to simultaneously edge and bevel. Figures 5.146 and 5.147 show how the wheels are used to grind the OD and grind on bevels on both sides of the lens in one sequenced operation. The bevel segments can be angled at 45° for standard bevels or at 90° for grinding on face flats on concave surfaces. Other unusual bevels can also be ground on with specially made bevel segments.

Beveling can also be accomplished with transfer spindle machines. After edging the lens to the required diameter, the transfer spindle is placed into a cradle that can be adjusted to any angle between 0° and 90° relative to the edge of the diamond wheel. Standard bevels can be put on when the cradle is aligned at 45°, and face flat bevels are easily ground on the concave side of the lenses when the cradle alignment is 90°.

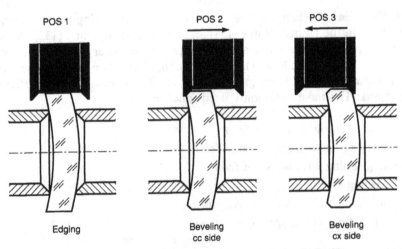

**Figure 5.146.** Edging and beveling with compound wheels.

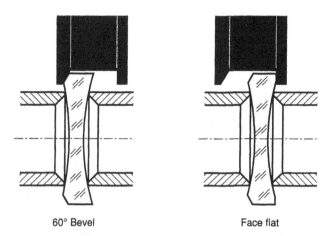

60° Bevel                           Face flat

**Figure 5.147.** Special bevels and face flats ground with compound wheels.

Noncircular parts can also be shaped on some center and edging machines [133, 134]. They can be set up to shape truncated lenses or square or rectangular lens shapes. This is done through the use of templates and special cams. The machines can be adapted for using a similar cam-follower method to shape rounded ends on porro prisms and similar components.

**Centering Support Machines.** Centering benches and bell chuck lathes are auxiliary machines that support a centering and edging operation. Even though auxiliary machines may not be essential in a lab, they are very important for the efficient operation of a production line.

One such machine [135] shown in Fig. 5.148 can be set up to serve as a vertical bell chuck lathe or as a lens centering bench for transfer spindles. This machine can be used to turn brass centering bells on the transfer spindle with tool steel cutters or to grind hardened steel bell chucks with a high-speed grinder. It can also be used for precision centering of lenses by the reflected image or the transmitted target method. The latter approach requires the use of a projection telescope. Another useful application is for precision cementing of lens doublets to a centering accuracy of 15 arc seconds. This method requires an integral collimator.

Another quite similar machine [136] is a horizontal bench for bell chuck finishing and for manual lens centering (Fig. 5.149). Bell chuck finishing is required each time a brass bell chuck is mounted on a transfer spindle. It consists of the trueing of the bell chuck by shaving a minimal amount off the edge that will support the lens. These shavings should never be in the form of discernable chips or strands but rather in the form of very fine brass dust. The shaving is done with the sharp edge of a tool bit that is mounted on the tail stock. Sometimes hardened steel chucks are preferred for volume cen-

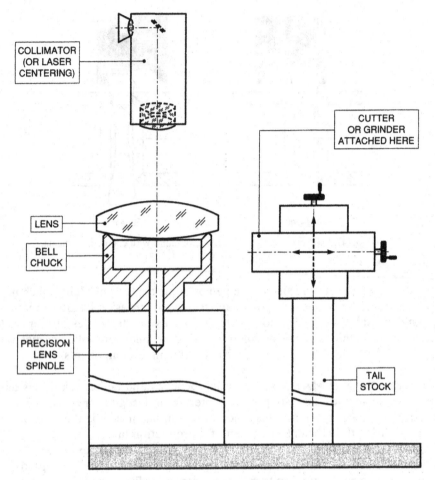

**Figure 5.148.** Vertical centering bench.

tering and edging. These are finished by light grinding with a small high-speed tool grinder mounted on the tail stock. The setup can also be used for manual wax centering and testing. This may be required for lenses with centering angles $\gamma$ of less than 17° because they cannot be easily centered and edged with the bell-chucking method.

### 5.6.6. Beveling

The edges of glass parts are protected with bevels against chipping. Bevels, also called *chamfers*, are most often specified to be at approximately 45° to the lens or prism surfaces, but other angles can also be specified.The size of the bevel can either be called out by the leg dimension $L$ or the face width $F$. This is shown in Fig. 5.150. The face width specification is the preferred one

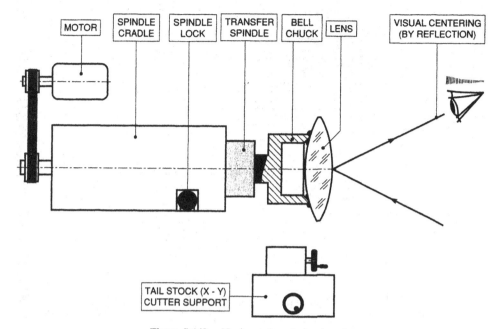

**Figure 5.149.** Horizontal centering bench.

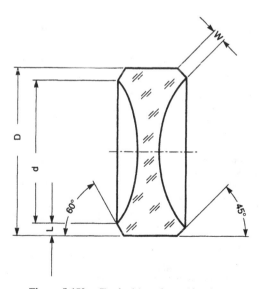

**Figure 5.150.** Typical bevel specifications.

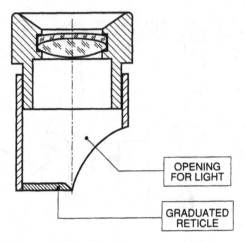

OPENING
FOR LIGHT

GRADUATED
RETICLE

**Figure 5.151.**   Comparator, measuring loupe.

because this dimension can be measured directly with a comparator. The bevel width can be held to a 0.1 mm (0.004 in.) accuracy, which is verified with a 0.1 mm accuracy graduated eye loupe (Fig. 5.151).

A DIN specification [137] recommends bevel face widths as a function of lens diameters. These recommendations are listed in Table 5.10. Although not binding, these values can serve as a useful guide when the bevel specification is not clear. The recommendation in the table suggests that the bevel width should be about 1% of the diameter. It would be more correct from a technical standpoint, however, to make the bevel width a function of the weight of the optical part. The bevel can also serve as aperture, but such bevels are generally wider than the protective bevels and they are afterwards blackened.

The bevel angle is the angle formed by the bevel and the edge of the lens [138]. The bevel width is measured as represented in Fig. 5.150. When the bevel has to perform a specific function such as serve as an aperture stop, the remaining clear aperture of the lens surface must also be specified. Most

**Table 5.10.   Bevel Face Width per
DIN 58 160**

| Workpiece diameter | | Face width of bevel | |
| mm | inch | mm | inch |
| --- | --- | --- | --- |
| 0   to 30 | 0.00 to 1.20 | 0.3 | 0.012 |
| 30 to 63 | 1.20 to 2.50 | 0.5 | 0.020 |
| 63 to150 | 2.50 to 6.00 | 1.0 | 0.040 |

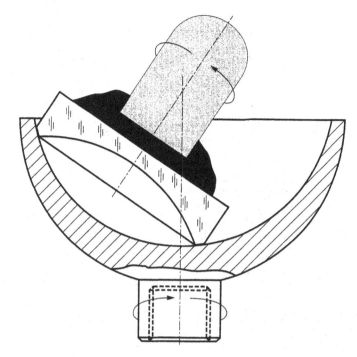

**Figure 5.152.**   Beveling tool for manual beveling.

bevels are generated during the centering operation as already described. Occasionally, it is not possible to bevel both sides of a lens during centering. In that case the missing bevel must be ground in a concave grinding tool (Fig. 5.152). Depending on the bevel width, loose abrasives with grain sizes ranging from 12 to 45 $\mu$m are used.

Electroplated diamond tools are preferred by many opticians, especially on hard materials such as fused silica and silicon. These tools are usually too aggressive on softer materials which they can damage with unacceptable microchips. The lenses can be blocked to wooden handles. In order for the lenses not to shift after blocking but to remain relatively well centered on the handles, they are held in a vertical position in a peg board. With some skill, it is also possible to bevel the blocked lens on a plano tool.

The bevels generated in concave grinding tools are annular zones of a spherical surface. They are typically quite narrow, so they can be seen as the surface of a cone segment (Fig. 5.153). The radius of the grinding tool with which a predetermined bevel angle is generated can be calculated per Eq. (5.29).

$$r = \frac{d}{2 \cos \alpha},$$  \hfill (5.29)

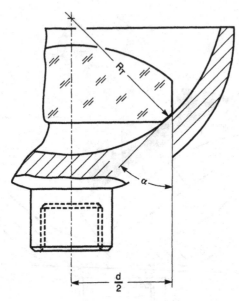

**Figure 5.153.** The beveling tool radius $R_T$ is a function of bevel angle $\alpha$ and lens diameter.

where

    $r$ = radius of the grinding tool,
    $d$ = lens diameter,
    $\alpha$ = the bevel angle.

Since the cosine of 45° is 0.707, the radius of the concave grinding tool for the most common 45° bevel is about 0.7 times the lens diameter. It is a rule of thumb that $r = 0.7d$ is adequate for all but the most tightly specified 45° bevels.

    The edges of concave rectangular lenses, such as for camera viewfinder lenses, can be beveled on convex grinding tools (Fig. 5.154). The radius of the convex grinding tool can be calculated according to Eq. (5.30). It is assumed that the lens is held horizontally in the 45° region of the grinding tool.

$$r_s = \frac{r_1}{2 \cos \alpha}\left[ \sqrt{\left(1 - \frac{a^2}{r_1}\right)} \right] \qquad (5.30)$$

where

    $r_s$ = radius of the convex grinding tool,
    $r_1$ = lens radius,

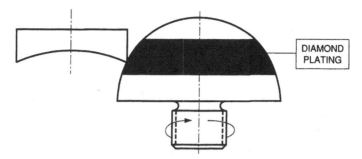

**Figure 5.154.** Beveling concave rectangular lenses on a convex tool.

$a=$ half the edge length of the lens,
$\alpha =$ bevel angle, typically 45°.

Mirrors or prisms that receive straight bevels are advantageously beveled on a plano tool which is electroplated with 20-$\mu$m diamond. The tool has a diameter ranging from 100 to 150 mm (4.0 to 6.0 in.), depending on the size of

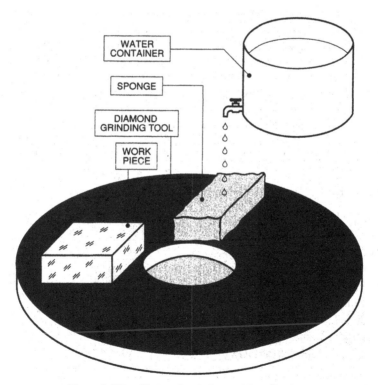

**Figure 5.155.** Electroplated diamond beveling lap.

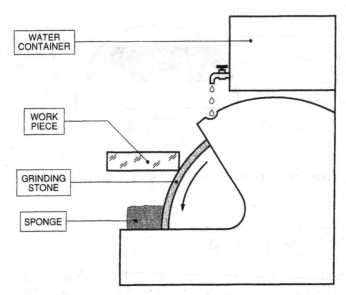

**Figure 5.156.** Beveling wheel.

the parts to be beveled. The diamond plating is applied only to an annular region, and this leaves the central 40 to 50 mm (1.6 to 2.0 in.) free. The cooling medium is water, which is dispensed from a drip pot. A sponge resting on the bevel tool distributes the water evenly (Fig. 5.155).

Simple parts can also be beveled on a grinding stone. It is a simple machine that was popular with eyeglass opticians in the past (Fig. 5.156). Such a machine has a grinding stone of about 250 mm (10.0 in.) diameter. The bound abrasive is typically silicon carbide with a grain size of 280. The stone is wetted from a drip pot, and the water is distributed by a sponge. The stone runs in the opposite direction from that of most other stones.

### References for Section 5.6 (Centering)

[105] DIN 3140, sheets 1 and 6.

[106] DIN 7160.

[107] W. Bothner, Germany.

[108] Bothner, Model B38.

[109] Loh, Model LZ-80A.

[110] G. Schulze, *Feinwerktechnik,* vol. 68, h8.

[111] Feinoptik Glassbearbeitung, W. Zschommer, Verlag C. Hanser.

[112] Carl Zeiss, double reflection tester.

[113] Carl Zeiss, Oberkochen, Germany.

[114] Carl Zeiss.

[115] DIN 58 740.

[116] George N. Müller Kugellagerfabrik KG, Germany, GNM spindles.

[117] Loh, Model WG.

[118] Bothner, Model B10.

[119] Wilhelm Bothner GmbH, Germany.

[120] Mildex, Saida, Japan.

[121] Loh, Model LZ-25.

[122] Bothner, Model B40HY.

[123] Loh, Model LZ-80.

[124] Loh, LZ-25A.

[125] Bothner.

[126] Bothner.

[127] Saida, Japan.

[128] Loh, laser centering.

[129] Bothner, laser centering.

[130] DIN 3140.

[131] Strasbaugh, Model 7AC.

[132] Loh Centromatic 240 CNC.

[133] Bothner.

[134] Wilhelm Loh, Wetzlar, Germany.

[135] Bothner.

[136] Loh.

[137] DIN 58 160.

[138] Definition per DIN 3140.

[139] Loh, Models M-1 and M-2.

## 5.7.  CEMENTING

Lens or prism surfaces that are designed to function in contact are joined with a clear, transparent optical cement. A lens doublet example is shown in Fig. 5.157. Since the cement acts as an index matching fluid, there is little or no light reflected from the cemented surfaces. Surfaces that are to be cemented must not be coated.

### 5.7.1.  Optical Cements

Until a few decades ago canada balsam was often used for cementing. Canada balsam is the natural resin of a Canadian evergreen. The raw resin is prepared in chemical factories where it is purified with tolulene. By controlling the degree of evaporation of the solvent during the purification process, harder or softer varieties are obtained.

Almost all cementing is done today with synthetic resins. Polymers are preferred, and they are available in a variety of types. They differ in their

**Figure 5.157.**   Cemented doublet.

refractive properties, in their elasticity, and in the manner by which they
cure. Two component cements such as epoxy resins or polyester resins are
now only used on rare occasions. These somewhat viscous cements cure
through the addition of an accurately measured catalyst or hardener. They
tend to entrap air during this process, and they must be outgassed in a
vacuum jar. UV-curing optical cements are now the preferred choice because they are easy to use, and the cementing process can be greatly accelerated. Natural and synthetic optical cements are described in greater detail in
Chapter 3.

**Working with Optical Cements.**   Prior to cementing achromatic doublets, the
lens sets are, first, paired up according to the actual center thickness so that
the thickness of the cemented doublet will be near nominal. This means that
a thick crown lens is paired with a thin flint lens, and vice versa. The doublet
will then have a uniform thickness, which is especially advantageous for the
mounting of lenses.

The lens surfaces are cleaned, brushed free from dust, and paired together
on a soft (desirably black) pad. As the lenses are laid together, the crown
lens must float in the concave surface of the flint lens. If the lens gets hung
up, then the surfaces are not clean enough and the process must be repeated.

The assembly of the lenses should be done in a dust-free workplace, such
as provided by a laminar flow bench (Fig. 5.158). These workbenches have a
hood into which filtered dust-free air is blown. Because of the positive air
pressure in the hood, the dusty workshop air cannot migrate into the laminar
flow box. The thoroughly mixed and fully outgassed cement is filled into a
hypodermic-type dispenser from which it is applied to the lenses. For small
lenses, the cement is removed from a container and then applied with a glass
rod.

The upper convex lens of the doublet is lifted up, and a drop of cement is
applied to the concave surface of the flint lens. For large lenses, one drop

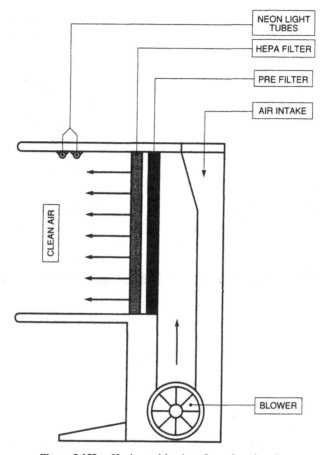

**Figure 5.158.** Horizontal laminar flow clean bench.

will be insufficient. In that case the cement is applied in the shape of the letter Z or like a plus sign so that no air pockets can form. The upper lens is lowered again, and the excess cement is gently pressed out. With a cork or a soft wide blade tool, the upper lens is moved relative to the bottom (Fig. 5.159).The cement is uniformly distributed in this way and any air bubbles are forced to the edge. The lenses still rest on the soft pad during the cementing procedure. Too much pressure on a hard surface can cause stress deformation. The properly pressed out cement layer is only 0.01 to 0.015 mm (0.0004 to 0.0006 in.) thick.

The excess cement that runs over the edge is wiped off before it can cure. A solvent is used for this purpose. The solvent must be used sparingly or the soft cement will dissolve in the solvent, and that can lead to cementing defects. The solvent used for lenses cemented with UV curing cement is acetone. As little of it should be used as possible.

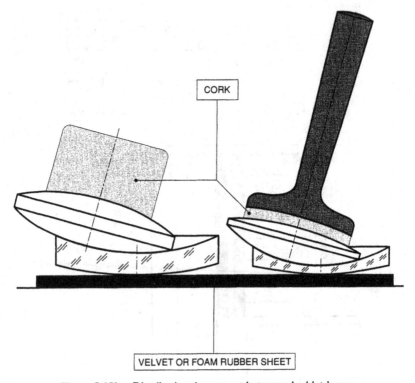

CORK

VELVET OR FOAM RUBBER SHEET

**Figure 5.159.**   Distributing the cement between doublet lenses.

When optical components are cemented with epoxy resin or polyester resin, the cemented doublets must be cured in an oven at about 40°C (104°F). The doublets cemented with a two-component cement are cured at about 70°C (158°F) in an oven. Curing is achieved after about 2 hours at that temperature. At higher temperatures the cement frequently takes on a crystalline structure. The curing cycle can be interrupted for the adjustment of the lenses.

The UV-curing polymer cements are especially easy to work with. Therefore they are used extensively today. Time-consuming mixing is not required for these cements, and they are less viscous than other cements. In this case the lenses are also prepared by laying them to color in matching pairs. The cement is applied and pressed out. The cement is precured before the doublet is aligned. This is done by illuminating the lens pairs for a few seconds with the light of a mercury lamp or by irradiating them with an electronic flash unit. A short exposure to a long wave black light (UV) source also works well. The bonded parts can be moved still with some difficulty after this precuring step. While still in the precured state, the doublets are cleaned with acetone and inspected in transmission. Flawed bonds are easily separated when they are only precured. After the lenses have received their final alignment with respect to the optical axis, the bond is fully cured in a UV light box for an exposure of about 1 hour. Additional curing takes place over time.

### 5.7.2. Centering of Cemented Doublets

The two lenses of a cemented doublet should have a common optical axis. This is accomplished by moving them slightly with respect to one another by 0.1 to 0.2 mm (0.004 to 0.008 in.) before the final curing of the cement. To ensure that during this centering alignment, the edge of the convex lens never protrudes beyond the edge of the flint lens, the convex lens should always receive a smaller diameter. When the lenses are aligned optically, the concave lens is edged during the centering process to a diameter that is sufficiently smaller than that of the convex lens to allow the lenses to be adjusted.

For noncritical doublets it is sufficient to align the lenses by their common diameter. The doublet is put into a vee-jig for that purpose (Fig. 5.160). The cemented doublet is then centered as accurately as the individual lenses. The centering tolerances can be additive, however.

A somewhat more accurate method uses a fixture in which the doublet is held in a mount. Mounted above the doublet is a centering bell on the same axis as the lens holder (Fig. 5.161). This bell is lowered until it contacts the doublet, which will then automatically align to equal edge thickness. Align-

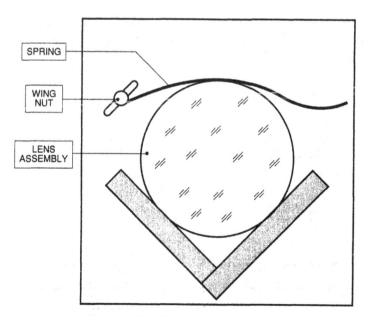

**Figure 5.160.** Simple cementing fixture.

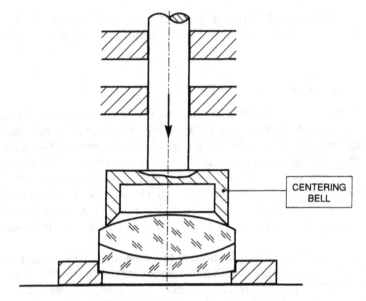

**Figure 5.161.** Self-aligning lens cementing fixture.

ment to equal edge thickness is also the principle on which an older instrument was based (Fig. 5.162).

Doublets for high quality objectives are optically aligned. A transmitted light method is well known for which the setup is equipped with a collimator and a telescope (Fig. 5.163). Systems are also used in which the light from an autocollimator passes through a doublet and is reflected back by a plano mirror (Fig. 5.164).

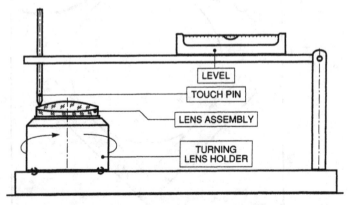

**Figure 5.162.** TIR-type lens-centering fixture.

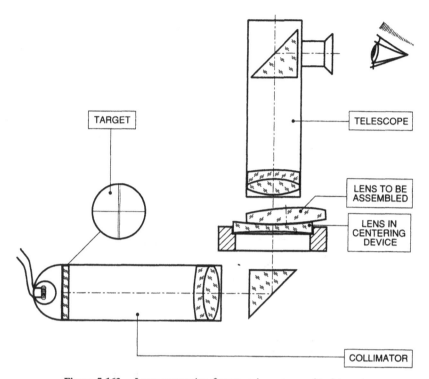

**Figure 5.163.** Lens-cementing fixture using a transmitted target.

Doublets can also be cemented and aligned to their common optical axis on the laser centering test unit [139]. The lenses are laid to color, and the cement is applied and evenly distributed. This method works best with UV cement. The lens pair is then placed on the centering chuck of the instrument where it is held by light vacuum pressure. The lower, typically concave, lens is pushed against a laterally attached vee stop which has been previously adjusted with a centered lens and locked into position.

The laser passes through a focusing system that compensates for the refractive powers of the doublet. A plane mirror redirects the beam from there through the doublet, which then passes through the hollow spindle and impinges as a light spot on a detector plate. The detector is electronically coupled to an image screen that displays the light spot as a bright dot. A change in position of the light spot on the detector plane is seen magnified many times on the image screen.

The upper lens is then shifted in such a way that the dot on the screen remains stationary as the spindle is rotated. The centered doublet is now irradiated with UV light. The cement sufficiently precures in a few seconds, so then the cemented doublet can be removed from the centering chuck. If

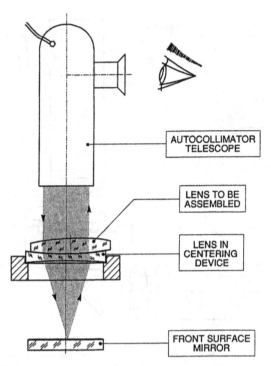

**Figure 5.164.** Lens-cementing fixture using autocollimation.

flaws such as bubbles, lint, or drawn-in air pockets are detected at this stage, the precured bond can be easily separated. On occasion, electronic flash units for cameras have been used as UV light sources. The doublets must be fully cured in an UV light box.

When a cemented triplet has to be made, first two lenses are cemented, centered and fully cured. After that, the third lens is cemented in the same fashion as a doublet lens.

**Decementing.** Every effort expended on care and cleanliness pays off when cementing. Improperly cemented elements with synthetic resins are often not salvagable. To decement doublets and to clean them requires the expenditure of a great deal of time with uncertain results.

Typical cementing errors are air bubbles in the cement, entrapped dust particles, or lenses that have shifted out of alignment. The bond can only be broken at temperatures above 200°C (428°F). To accomplish this, the doublets can be heated in boiling castor oil or in any other high boiling oil and separated with the parts held between tweezers or clamps. They can also be held over a gas flame with tweezers. On occasion they can be placed on an electric hot plate where a mount aids in good temperature transfer. The glass components will separate with an audible crack, and they must be separated at once. With all of these methods, the flint lens of the doublet is the most vulnerable.

The residual cement on the glass surfaces dissolves when the lenses are immersed for a prolonged time in dichloromethane or in concentrated nitric acid. However, the polished surfaces also suffer from this treatment. The optical cement manufacturer does provide information on the best way to remove cured cement. Whenever aggressive chemicals are used, every safety precaution must be observed. Decementing should not be attempted until all safety issues have been properly addressed.

### 5.7.3. Other Cementing Options

In addition to the cementing of lens and prism pairs, there are a number of other options where a cementing approach can achieve the desired optical effects.

**Cementing of Ground Surfaces.** The refractive index of polyester resin is close to that of plate glass. The resin acts like an index matching fluid when ground plate glass is cemented and the grey surface can no longer be seen. However, the surfaces must not be ground "black." Fresh grinding compound should be applied once more near the end of the grinding cycle so that the grinding pits will be free of abraded material from the grinding tool.

**Cementing with Canada Balsam.** Although canada balsam is rarely used today, its application will be described here for the sake of completeness, and because it offers some advantages over commercial cements for prototype assembly of sensitive materials.

The glass elements are prepared as already described. The matched lens pairs and a wide mouth jar containing canada balsam (Fig. 5.165) are placed on a large surface hot plate, which is set at approximately 90°C (194°F). The upper lens is lifted up, and a drop of softened balsam is put between the lenses with a glass rod. The cement is evenly distributed through the motion of the upper lens, and air bubbles are forced to the edge. The lens edges can then be cleaned with alcohol. A skilled person knows how to apply just enough balsam so that very little excess cement runs over the edge and unneccessary cleaning is avoided.

The cemented parts are aligned in a vee-jig; the alignment is maintained until the parts have cooled. When precise optical alignment is required, the cooled and cleaned doublets can be reheated. The doublets that require optical alignment are cemented with hard balsam. A sufficient amount of the cement is melted on a highly heated lens. One way to do that safely is to fuse a piece of canada to a metal pin that is fastened to a large cork (Fig. 5.165). Stressed doublet lenses can be annealed when they are cemented with canada balsam by reheating them and letting them cool off very slowly.

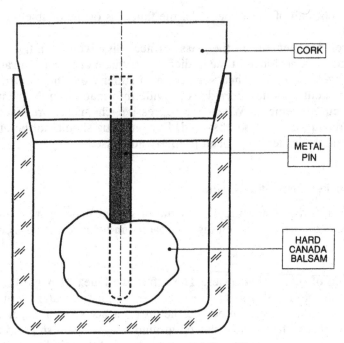

CORK

METAL PIN

HARD CANADA BALSAM

**Figure 5.165.** Canada balsam jar.

**Edge Blackening.** Lens edges are blackened to prevent stray light in lens systems. An alcohol based lacquer is typically used. Graphite has also been used with success. The lens or doublet is placed on a rotary spindle where it is held with a sticky wax, or better yet with vacuum pressure. The edge of the rotating lens is then painted black with a soft brush (Fig. 5.166).

### 5.7.4. Contacting

Contacting is a special blocking method that is used when very accurate angles have to be held or when plano parts have to be polished parallel to a few arc seconds or less. This method requires very skillful handling, a very clean environment, and nearly infinite patience. It could have been described as a blocking method, but since it is different from all other blocking methods and requires the same environment and care that is needed for cementing, contacting is described in this section.

The well-known interference colors will be seen at first when two exactly fitting and perfectly clean glass surfaces are placed in contact. The colors will disappear when the entrapped air between the surfaces is forced out by light pressure on the glasses, and the surfaces will tend to be drawn together by mutual attraction. This process is known as *optical contacting*. When these conditions are met, adhesion forces that hold matter together, or that

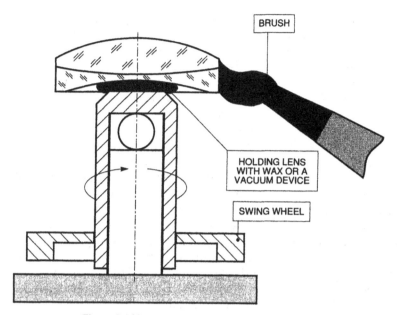

**Figure 5.166.** Rotary chuck for edge blackening.

cause a liquid to creep up the walls of a capillary, come into play. The two contacted glasses act as if they were one piece. They adhere tightly to one another, and the reflection from the contacted interface is eliminated. The location at which contacting is initiated appears dark because no light is reflected there, and all of the light is transmitted.

With the contacting method two glasses can be joined together without the benefit of any adhesive medium. Contacted glasses can be ground to a sharp edge and polished free of edge roll or edge chips. That is especially desirable for roof prisms for which the roof edge must remain absolutely sharp.

Optical parts should be contacted for which angular accuracies in the range of arc seconds are required or for which it is essential that the edges be free of edge roll and edge chips. This applies primarily to roof prisms or plano parallel plates. Spherical surfaces, however, can also be contacted, although with much difficulty.

Contacting is generally an indirect method, for workpieces are contacted to contact blocks that have exact surfaces and angles. The surfaces, in preparation to contacting, must be perfectly clean and free of organic contaminants. They must be thoroughly cleaned with reagent grade acetone followed by a cleaning with anhydrous ethanol.

A soft, lint-free cloth and a clean chamois carry the cleaning solvents. The contact block is cleaned first, and then the contact surfaces of the workpiece. Following that, the workpiece surface is brushed dust free with a camel's hair brush, and the cleaned surface is turned over at once and held in that position so that no new dust can settle on it. Next the contact block surface is brushed clean and the workpiece, such as a round glass part, is laid to color. The workpiece must be able to move freely back and forth on the contact block without getting hung up. It should freely float on the entrapped cushion of air. If even the slightest resistance is noticed, the contact surfaces are not clean and should be recleaned before proceeding.

While the workpiece can still be moved, it is pushed into the desired position. As that is done, the workpiece is lightly pressed down so that the entrapped air is forced out. This causes the interference fringes on plano surfaces to move toward one side as straight bands, and a white color appears uniformly over the surface. The actual contacting takes place when further constant pressure is applied. It can be seen then at the spot where the greatest pressure is applied that the white coloration of the surfaces suddenly changes to dark and the workpiece can no longer be moved. Once the contacting process has begun and especially when the contact surfaces are truly clean, it can propagate through the effect of the adhesion forces without further external pressure. Frequently, however, it is necessary to assist the process with light but constant pressure.

Roof prisms are often contacted to the sides of a contact bar which rests on a fine-ground glass plate (Fig. 5.167). Several bars with an even number of roof prisms contacted to their sides are then contacted to a plano contact plate to form a prism block. The edges of the contacted parts are protected with a lacquer so that water cannot touch the joint during processing. A thick lacquer made from shellac flakes dissolved in ethanol has proved to be superior to commercial formulations in sealing out moisture.

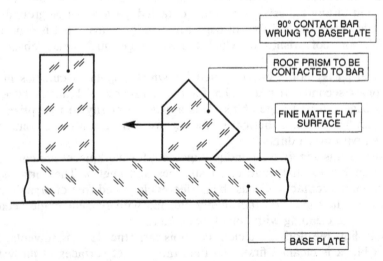

**Figure 5.167.** Contacting of roof prisms.

The pressures excerted during grinding and polishing are easily endured by the contacted surfaces. They are only sensitive to sudden mechanical or thermal shock. The larger the contacted surface, the more tenaciously the contacted parts cling to each other. The tensile strength of the contact bond, which is the force that pulls vertically to the surfaces, is 2 to 3 kp/cm$^2$ (29 to 43 psi). The shear strength, which is the force that acts parallel to the surface and would tend to push the part off the block, is about 8 kp/cm$^2$ (114 psi).

After the contacted glasses are ground and polished, they must be separated from the contact block. Prisms can be removed by a light blow with a small wooden mallet. However, it is better when the contact edge is touched by the finely adjusted point of a bunsen burner flame. The stream from a hot air gun is also useful. The surfaces separate at once with an audible crack as if the glass has fractured. The prism must be removed right away before it has a chance to recontact. Circular plano parts can also be decontacted in this manner.

Thin glass parts decontact easily this way, but they tend to recontact immediately. The lacquer must be removed from the edges first. A little bit of water is applied to the edge. As the flame of the bunsen burner touches the part, it is lifted slightly with a small wooden stick. This permits water to be drawn in between the contact surfaces to separate the parts.

### References for Section 5.7 (Cementing)

[139] Loh, Models M-1 and M-2.

## 5.8. CLEANING OF OPTICS

Optical glasses are cleaned repeatedly during fabrication to remove adhering blocking pitch and other contaminants. They are cleaned the first time when one side of the lens has been polished. They are cleaned again after the second side has been polished. The third cleaning is done on the centered lenses prior to their final inspection. From there, the lenses proceed to the vacuum coating area where they are cleaned again. These repeated cleaning operations claim a considerable portion of the fabrication cost, and they deserve particular consideration for that reason alone.

### 5.8.1. Manual Cleaning

Manual cleaning is a simple operation that is typically performed by support personnel. However, since appreciable costs have already been invested in these parts and since the parts are often very sensitive, the cleaning of polished optics is an activity that requires care and responsibility. All the work that has already been invested up to that point can be easily wasted.

In the past optical components were almost exclusively cleaned by hand. Manual cleaning is today performed only for limited quantities. The lenses that must be cleaned manually are placed into a shallow tray that is lined with lens tissue. The tray is filled with just enough solvent to cover the lenses. When several layers of lenses are placed in the same tray, a sheet of firm lens paper must separate the layers. The tray is covered with a lid, and it is put aside for several hours until the adhering pitch has dissolved. Following that, each lens is individually cleaned with a soft cloth. The polishing compound deposits that are frequently found on the lens edges must not be wiped across the polished surfaces. It is recommended that the edge be cleaned with a small brush or a piece of felt before the lenses are placed into a tray with clean solvent.

For larger quantities the parts are placed into racks made of steel or plastic. These racks are submerged in an approved room-temperature, chlorinated solvent such as chloroethane to soften the pitch. This method is especially effective when several such baths are used in succession. The parts will emerge from the last bath already quite clean. All soluble contaminants will have been flushed away and only a small amount of the polishing compound deposits will still adhere to the parts. In this condition the lenses still in the rack are immersed in a tray filled with either acetone or methanol, washed clean and then wiped dry with a clean cloth. Depending on size, it requires from 30 to 60 minutes to clean 100 lenses in this way. A more thorough cleaning is achieved with sequential baths. The parts are washed in clean acetone first, followed by a submerged wash in methanol. Some cleaners mix acetone and methanol to achieve better results but this practice is not founded on any chemical principle.

### 5.8.2. Ultrasonic Cleaning

Ultrasonic cleaning has been used for many years as an effective method for cleaning lenses. The ultrasonic energy sources are either magneto-strictive nickel or piezoelectric PZT (lead zirconte titanate) transducers that oscillate at frequencies from 20 KHz to 40 KHz. The ultrasonic energy becomes useful through the process called *cavitation*. Small gas bubbles are generated in the liquid. The bubbles collapse implosively, and by doing so, they knock off contaminants adhering to the parts. However, it is not only the ultrasonic action that cleans the parts but also the passage through a succession of suitably chosen baths of which only a few need to be ultrasonically agitated. Aqueous detergent solutions are now commonly used in ultrasonically agitated cleaning baths, but cleaning solvents are still in use because of their effectiveness on many types of soils found in optics. The cleaning solvents now in use are mostly fluorinated hydrocarbons (CFC). They have all but replaced the chlorinated solvents that were once in widespread use in every optical shop. The use of chlorinated solvents has been restricted for some time because they are suspected carcinogens and only chloroethane 1,1,1 can be used with relatively few restrictions. The fluorinated solvents, known as Freon [140] or Genesolv [141], have in recent years become a source of concern because they have been identified as the primary chemical agents responsible for the alarming depletion of the protective ozone layer in the polar region of our earth.

The Clean Air Act in the United States and the Montreal Protocol require a timed phase-out of the use of CFC products that can escape into the atmosphere. The current schedule, calls for CFC's use as cleaning solvents to be completely phased out by the year 2000. Chlorinated solvents such as 1,1,1 will no longer be available after the year 2002. There are signs that the phase-out could even be accelerated. Regional air quality regulations might also limit the evaporation of volatile solvents, so that even alcohols, acetone and other similar solvents might be restricted as well.

The emphasis is now on the development of substitute solvents, semi-aqueous and all-aqueous cleaning systems. Aliphatic hydrocarbons, of which kerosene is the best known, have shown promise to become useful substitute solvents. But these solvents are flammable, and this creates a new set of problems. Environmental restrictions are going to present a serious challenge to the optical manufacturing community because many of the soils that are generated during the manufacture of an optic may not clean well enough with new solvents, and they do not respond well to an all-aqueous cleaning process. If no effective substitute solvents can be developed to deal with these organic soils, a thorough rethinking of the entire manufacturing process will be required to ensure that only minimal levels of soils are present on the parts at the end of the process. Some of the standard operations such as wax or pitch blocking may have to be altered or even entirely eliminated to reduce the introduction of large amounts of organic and particulate soils. What will replace these vital steps is not yet clear.

One way to deal with some of the foreging problems, however, does not require a major process change. The application of good manufacturing practice coupled with some common sense and attention to detail can go a long way to keep parts nearly free of soils throughout the process: good housekeeping of work areas, regular cleaning of the interior of machines, frequent washing of lenses and blocks during the grinding and polishing stages, laquering of ground edges where soils can accumulate, and periodic change of coolants and slurries. There are still other steps that can be taken to reduce the buildup of soils on optical parts as they pass through the manufacturing process. The problem is that shop management in the highly competitive world of today does not encourage such cleanup practices because they are seen as interfering with productivity. This is a rather myopic perspective that the new realities of environmental concerns may force into sharper focus.

With the emphasis on all-aqueous cleaning in optics, the manner in which the cleaned parts are dried will become a very important issue. A highly effective part drying method that does not recontaminate the parts is required at the end of the clean line. There are currently two basic methods of water displacement drying. One of these methods [142] relies on a 3% to 5%

isopropyl alcohol content in a fluorinated hydrocarbon (CFC) solvent to displace, absorb, and carry away residual water. The other method [143] uses a surfactant blended in a CFC solvent to displace the water, which then rises to the surface because it is lighter than the blend and is skimmed off by a weir separator. Both of these methods require an immersion in distilled solvent followed by a slow passage through pure solvent vapors to emerge dry and spot free. But these solvents fall under the restrictions for CFC's and face total elimination in the near future. There are not too many good options for the spot-free removal of water droplets. High-speed, centrifugal force spin drying that is assisted by hot nitrogen blowoff is one consideration. Whether hot air or nitrogen gas dryers fill the need remains to be seen. A great deal of work needs to be done to improve the spot-free drying efficiency of these units.

For the ultrasonic energy to act unimpeded on the lenses, the cleaning racks in which the lenses are submerged in the solvent baths should be made of stainless steel wire (Fig. 5.168). The lens seats must be shaped such that the wire contacts only the lens edge and not the polished surface. Any clean rack design for ultrasonic cleaning should not use plastics because most of them are very good attenuators of ultrasonic energy. It is best to use an open design stainless steel rack.

The lenses are put into these cleaning racks and are first precleaned in the customary manner by static immersion in a solvent to remove gross contaminants before they go through the clean line. This precleaning step should be done thoroughly so that the tanks of the clean line are not prematurely contaminated. It has been clearly demonstrated [144] that the cleaning effi-

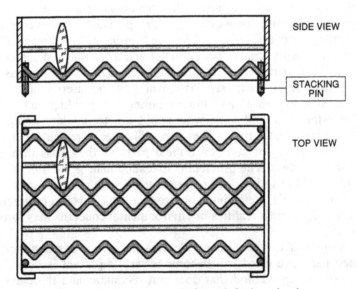

**Figure 5.168.**   Stainless steel wire rack for lens cleaning.

ciency of an aqueous line is greatly enhanced when the presoaked racked parts are first immersed in a static detergent bath at room temperature. An empty rack is immersed in a static room-temperature deionized (DI) water bath. The lenses are then taken one by one from the rack in the detergent bath, gently wiped with a cotton ball on both polished surfaces, and loaded into the empty rack in the DI bath. A small piece of felt can also be kept in the detergent bath to clean off compacted slurry from the ground edges. This step should be done prior to wiping the polished sides. In this manner all lenses are wiped on both sides and transferred to the rack in the DI bath. During this stage and any subsequent stages in the aqueous line, the lenses must not be allowed to dry, even temporarily. They must be kept wet or submerged throughout the cleaning cycle until they can be properly dried in a water displacement dryer or any other qualified drying system.

The manual wiping with the cotton balls accomplishes what even a prolonged and potentially destructive immersion in ultrasonic baths cannot do. It strips very small, tightly adhering particulates from the surfaces, and the detergent solution sequesters them so that they are unable to reattach themselves to the surfaces. Many of the residual particles are then removed during subsequent ultrasonic baths and rinses as they pass through the clean line. Very large numbers of these tiny particulates are the source of haze, stains, and spots that can be greatly enhanced after a thin film coating has been applied to the surfaces. The sooner and the more effectively these particulates are removed from the surface, the easier it will be to get the parts clean. Unfortunately, the introduction of this manual operation is often seen by production workers and shop management as an unnecessary and time-consuming step, and the wisdom of adding it to the cleaning process is often questioned.

The questions that should be asked are: Which approach is more cost-effective in the long run? Are one or two additional labor hours added to a lot of parts more expensive than repeated recleans and reinspection and possibly damaged parts, or do they cost more than a coating lot that is stained to the point of rejection? These questions, however, are almost never considered.

In the passage of the racks through the tanks of the aqueous clean line, a carefully timed cleaning cycle is essential. The composition of the baths in the individual tanks are chosen on the basis of the type of contaminants present and on the chemical resistance of the glass types of the lenses. For instance, eyeglass lenses can be cleaned in strongly alkaline solutions, whereas photographic lenses are made mostly from chemically sensitive glass types that can only be cleaned safely in neutral detergent solutions. For that reason alone it is not possible to develop a cleaning system that is useful for all purposes. The cleaning experts provide guidelines, but their guidelines have to be modified for specific requirements.

An example of such applied guidelines provides that after the precleaning and prewash cycle, the lenses in the wire racks are immersed for one minute in each of two tanks filled with an aqueous detergent solution. The tanks are heated to 30°C (86°F), and one of the tanks is ultrasonically agitated. After that, the residual detergent is flushed away under a spray rinse. The wet lenses are passed through a water displacement dryer charged with a dewatering agent that separates the water from the lenses. The residues of this agent are flushed away in two pure solvent baths, one of which has ultrasonic agitation. The parts are actually clean at this stage. They only need to be dried spot free. This is facilitated by a bath that contains only pure solvent vapors. As was already mentioned, the use of this drying step will soon be no longer a viable option. The solvent process is described here only because it is still being used by many manufacturers and because reliable alternatives have not yet been developed.

A deep tank is used for the final stage, in which pure CFC solvent is vaporized at about 47°C (117°F). The upper third of this tank is wrapped with a cooling coil that contains the vapors in the tank and that recondenses the solvent. The wire rack with the lenses is submerged into the vapor zone and then assumes the temperature of the vapors. It is slowly withdrawn upward through the cooling zone, and the lenses emerge clean and dry from the drying unit. The cooling zone must be traversed with an experimentally determined speed. This requires that the wire rack be raised by mechanical means. If the rack is raised too quickly, the drops of solvent evaporate on the lens surface, and this can lead to spotting.

It is recommended that the wire racks be raised and lowered with the lenses in the ultrasonic baths to bridge zones of weak ultrasonic energy caused by shadowing in the tank. Low-boiling solvents such as Freon are preferred for the cleaning of thick lenses. Such glasses tend to fracture if they are subjected to the thermal shock of higher-boiling chlorinated solvents.

The surfaces of ultrasonically cleaned glass parts have a sticky feel, but they can be smoothed out with a chamois. It has been shown that this roughing of the surfaces is not detrimental. Such surfaces can be coated with an evaporated coating, and they behave then like a smooth surface. But these sticky surfaces should not be used as exterior surfaces in instruments because streaks will develop on them where they are touched. Experience has shown that the surface quality of glass components has been degraded when the glass has been subjected to prolonged ultrasonics. This is especially true if the cleaning was done in an alkaline solution. Residual scratches and subsurface damage not seen prior to cleaning become visible. This occurs most likely because the silica gel layer on the polished surfaces has been attacked or destroyed. Optical glasses must not be exposed to ultrasonic agitation for more than one minute per bath. Any arbitrary extension of the immersion time can lead to surface degradation without additional cleaning advantage.

The operation of a modern cleaning system (Fig. 5.169), which can involve twelve different baths or more, is an important assignment. The success of the cleaning process can be compromised when the temperature

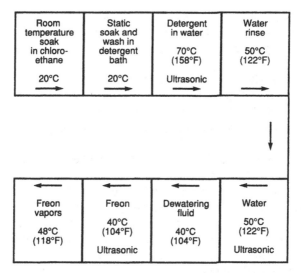

**Figure 5.169.** Example of a simple optics clean line.

of one bath is allowed to fluctuate unreliably or when the pH value has changed in a bath. Likewise it is imperative that the baths be always well maintained, very clean and replenished or renewed when necessary. The ultrasonic transducers, however, do not generally present a problem; since they can run for long periods of time without special attention.

Optical glass components can be cleaned well enough in clean lines with twelve or more tanks, so they can be coated without further cleaning. To ensure a particularly well-regulated passage through the line, high-quality cleaning systems utilize transport mechanisms that automatically move the racks from bath to bath.

### 5.8.3. Polishing Stains

Caked on polishing compound is especially difficult to remove from ground or polished surfaces. Such stains are the result of prolonged polishing cycles during which the blocks have not been thoroughly cleaned on a regular basis. Compacted polishing compound tenaciously adheres to the edges of lenses or prisms.

Ground surfaces can be easily cleaned on a piece of felt with water, detergent and a little bit of grinding compound. Polished surfaces require repolishing. A somewhat daring process uses a 10% solution of hydrochloric acid to wipe off the compacted polishing compound. The compound dissolves at once, but the lens must be rinsed in water very quickly to avoid damage to the polished surfaces. When using this or any other acids, rubber gloves should be worn and all relevant industrial safety directives should be consulted. An obviously better solution is to protect the edges of the lenses with a suitable lacquer and to keep the blocks throughout the grinding and polishing process clean so that compound compaction and stain formation never become a problem.

**Stains on Optical Glasses.** Stains can form on glass surfaces that cannot be cleaned off with a soft cloth, water, and alcohol. The glass surface is in most cases affected only locally where the stain has formed a thin layer. This layer has a different refractive index than that of the underlying glass. In severe cases the stains are visible as colorful interference phenomena. However, in most cases this layer is so thin that only a trained eye can discern the subtle differences in the reflection off the glass surface. During the inspection the glass component is held above a dark surface that itself does not emit light, and light from a fluorescent lamp is reflected for evidence of discoloration. The difficult-to-see stains are dreaded because they are enhanced after the surfaces are coated. Although the stain does not have an effect in transmission and, as a result, is not detrimental from a performance point of view, components with such stains are typically rejected in final quality assurance for cosmetic reasons. Therefore such stains must be either removed or their formation must be prevented.

For a long time there was no good explanation for this type of surface flaw, especially since these stains can take on many different appearances. This type of flaw is often referred to as *oxide* because there are occasional similarities to the corrosion of metals, but this actually is not the case. Recent investigations [145] have resulted in a satisfactory explanation of the chemical processes during the changes on the glass surfaces. The formation of the film results from aqueous solutions accompanied by chemical transitions during which alkali and metallic components are leached out of the surface. The silicate structure of the glass remains intact, but it absorbs hydrogen and converts to silica gel, which has a different refractive index than the underlying glass. Therefore the thin layer exhibits interference colors. Furthermore the gel layer is softer than the glass, and it is scratched easily when the surface is wiped.

Very dilute acids and bases can also attack the surfaces of optical glasses. Even pure water exchanges a portion of its free hydrogen ions with ions of alkaline constituents of the glass and thus becomes a base that acts destructively on the glass surface. This explains why a thin layer of water is more readily enriched and attacks glass more aggressively than an immersion in a water bath. The base is very much diluted in the bath and is no longer active. The chemical behavior of optical glasses is described in greater detail in Chapter 1.

Since glass is in constant contact with water throughout the fabrication stages, it is always in danger of being attacked. It is easy for a drop of water to evaporate from the surface only to leave a water stain behind. During certain temperature conditions, water can also condense from the air. It is especially dangerous when fruit is eaten in the vicinity of polished surfaces since droplets of fruit juice can result in severe stains.

Sensitive glasses are protected with a spray-on lacquer immediately after they are polished. The lacquer should not form cracks after evaporation of the volatile solvents because water can then penetrate to the surface. Polished prism surfaces, which are especially at risk during plaster blocking, receive a layer of paraffin on top of the protective layer. This layer is applied by dipping the prism into melted paraffin.

Sensitive glasses are also endangered during blocking for working the second lens surface. The water condensate that adheres to the cold lens surface is pressed into the hot blocking pitch together with the lens. The elevated temperature promotes dissolution processes and chemical reactions and in this manner encourages the formation of stains. Stain formation can be avoided when the glasses are heated prior to blocking. The surfaces should be free of water vapors at about 70°C to 80°C (158°F to 186°F). It has been recommended on occasion that particularly sensitive glasses be optically coated prior to centering. However, optical coatings provide only a weak protection against stain formation.

Sensitive glasses tend to form stains more readily when they were polished with coarse and aggressive polishing compounds. Presumably the optical surfaces are roughened up in this manner by a process that can no longer be proved by optical means to make them more susceptible to chemical attack.

Stains are typically removed by repolishing the affected surfaces. To maintain the proper surface shape, the polishing tool must contact the surface uniformly right away. Therefore polishing is done with a soft polisher, and the polishing compound is broken down first with a well-fitting dummy piece. Accurate plano surfaces are repolished on a stationary polisher by a process that is called *hand polishing*. For less critical surfaces with a surface figure accuracy of about 5 fringes, well-broken-in cloth or pad polishers have proved adequate for repolishing.

### References for Section 5.8 (Cleaning)

[140] Du Pont.

[141] Allied Chemical Corp.

[142] Crest Ultrasonic Corp., Trenton, NJ.

[143] Baron-Blakeslee, and Branson.

[144] Karow, H. H., "Cleaning of optical components," Short course notes, 1982 Spring Conference on Applied Optics.

[145] Prof. Dr. Kaller, Jena, Germany.

### 5.9. THIN FILM COATINGS

The surfaces of optically polished components are almost always coated with thin films to either raise or lower their surface reflection at specific wavelengths or wavelength ranges.

### 5.9.1. Reflective Coatings

Light is reflected from polished surfaces, especially those of a metal substrate. While only about 5% of the incident light is reflected by glass surfaces, silver has the ability to reflect about 95% and aluminum about 88%. Mirror substrates therefore are only carriers of the reflecting metal films.

Silver films react with the sulfur dioxide constituent of the air, which causes it to tarnish. An effective protective layer that does not restrict the optical properties has not yet been found. As a result only the second side of mirrors are silver coated. Likewise prism faces, which reflect light back into the prism, are frequently coated with silver films. Aluminum has a lower reflection than silver, but it can be protected quite well with an overcoat of silicon dioxide ($SiO_2$) against corrosion and mechanical damage. It therefore is well suited for first surface reflectors. For partially transparent mirrors, a metal film can be deposited so thin that it remains transparent, though a portion of the incident light is absorbed.

Reflective coatings made with materials that have a high index of refraction are practically free of absorption. Suitable substances are titanium oxide ($TiO_2$) with a refractive index of $n = 2.66$ or zinc sulfide (ZnS) with $n = 2.40$. The reflected portion of the normally incident light is calculated according to the Fresnell reflection equation, Eq. (5.31):

$$R = \left(\frac{n - 1}{n + 1}\right)^2, \tag{5.31}$$

where

$R$ = reflected portion of the incident light,

$n$ = refractive index of the reflecting thin film in air.

The reflection off the coating substrate interface must be added to the reflection off the coating. To prevent the latter reflection from being canceled by destructive interference, the coating thickness must be at least half the wavelength of the reflected light. The reflection can be further enhanced when an intermediate layer of magnesium fluoride ($MgF_2$) with $n = 1.38$ is evaporated. The light is then reflected by three planes that are close to one another. Through further addition of layers, the reflection can be increased still further. The coating materials are evaporated onto the substrate in a high vacuum.

**Chemical Silvering.** Several methods are described in the literature for chemical or wet silvering. All of these methods depend on the dissolution of silver nitrate. After the parts that must be silvered have been submerged in the solution, a reduction mixture is added that separates out metallic silver from the solution. The reduced silver deposits on the surfaces of the parts. One particular chemical silvering process is represented in Table 5.11.

Table 5.11. Chemical Silvering Solution (Miethe)

| Solution | Ingredient | Status |
|----------|-----------|--------|
| First solution | 900 cubic cm distilled water<br>30 grams silver nitrate | Mixed solution |
| Second solution | 900 cubic cm distilled water<br>20 grams potassium hydroxide | |
| Third solution | 1000 cubic cm distilled water<br>50 grams dextrose | Reduction solution |

Before the parts can be silvered, they are cleaned with a cotton ball and nitric acid, followed by a rinse with distilled water. Occasionally, the use of tin oxalate or tin oxide is recommended for cleaning. A mixture of equal parts of a premixed solution and alcohol serves as final cleaning. The parts are again rinsed with distilled water. As can be seen by this thumbnail description, the cleaning of the parts plays a decisive role in successful silvering.

The cleaned parts are placed in a shallow dish. Prisms can also be suspended in the dish (Fig. 5.170). A mixed solution has previously been prepared. Ammonia is slowly added to five parts of "solution 1" until the solution just becomes clear. Six parts of "solution 2" are added to this and then again ammonia, which is added drop by drop until the solution is clear. Finally, one part of "solution 1" is added, and this gives the finished mixture a brown color. Only as much of this mixture is prepared as can be utilized on

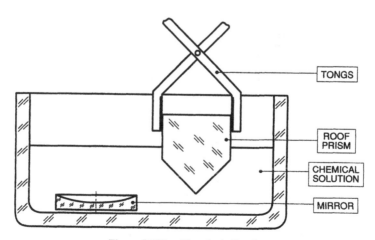

Figure 5.170. Chemical silvering.

the same day. The mixture is filtered into a brown bottle and is protected from direct exposure to sunlight. The exact composition of these solutions is often considered proprietary, so it is difficult to get a clear definition.

The mixed solution is poured into a dish until it is about 1 cm (0.400 in.) deep. The parts should be just covered or submerged. As the dish is rocked back and forth, the reduction solution is added. One part of reduction solution is added to two or three parts of mixed solution. Because of the sudden fallout of the silver, the final liquid mixture appears black. The chemical silvering process is completed when the liquid mixture has become clear again. The silvered parts are then removed and rinsed with a fine stream of distilled water.

For the parts to dry quicker, they may be placed on a slightly warm hot plate that has been covered with an absorbent paper.The chemically deposited silver is initially still quite sensitive against physical contact. After the parts have dried, the silver coating will become tougher. The silvering process can be made more efficient when large quantities of parts must be chemically silvered. In such a case the solutions are combined in a mixing chamber [146] from which the resulting blend is dripped or sprayed onto the parts, as shown in Fig. 5.171. Excess silver deposits such as on bevels or mirror edges are wiped off with a felt strip or with cotton balls that have been dipped in diluted nitric acid.

Second surface silver layers are protected by an overlaying copper layer that is deposited galvanically (Fig. 5.172). A current of about 0.3 amps flows through a copper salt solution. The resulting electrolysis causes a copper layer to be deposited onto the silvered parts which are held in a wire basket

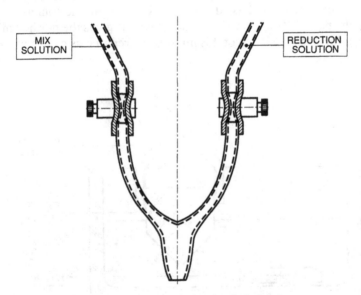

**Figure 5.171.**   Mix nozzle for chemical silvering.

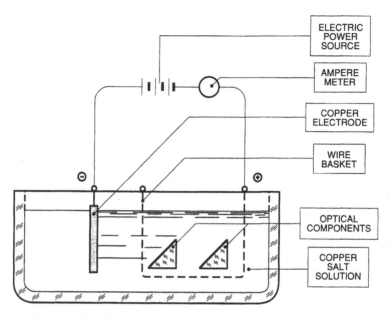

**Figure 5.172.** Galvanic bath for depositing copper on silvered optics.

that serves as the anode. The previously deposited silver should extend a little beyond the edge so that it makes electrical contact with the wire basket. A copper plate serves as the cathode. When the plating is complete and the surfaces have dried, a black mirror lacquer is applied to protect the metal layers. The lacquer must not have any pores after drying; otherwise, moisture can penetrate to attack the silver.

### 5.9.2. Physical Fundamentals of Vacuum Technology

In the evaporative coating technology the vacuum pressure is measured in millibars (mbar). One mbar = 100 N/m². The millibar unit is relatively new in vacuum technology, and the dials of many gauges are still calibrated in the old unit "torr" (one torr is the gas pressure that raises a column of mercury by 1 mm). The following relationship exists between these two units: 1 mbar = 0.75 torr. Torr and millibar can be considered as being equal in value because only the negative powers of ten are of interest for evaporative coating technology.

According to Avogorado, $2.7 \times 10^{19}$ gas molecules are contained in 1 cm³ of volume at 0°C and normal pressure. At $10^{-5}$ mbar there are still $2.5 \times 10^{11}$ gas molecules that move in the reduced pressure at a speed of about 500 m/ sec. The higher the gas pressure, the more frequently the gas molecules will collide. Statistical calculations provided the result that at normal pressure such collision must be expected after about 0.001 mm of travel. In a vacuum of $10^{-5}$ mbar, in contrast, the free path can be as long as 0.5 m before a gas molecule collision occurs. As a result a stream of evaporants is only slightly impeded in a high vacuum, and a considerable percentage of the evaporant will reach the substrate. At $10^{-5}$ mbar and a 0.5-m path length, about 40% of the evaporated material reaches the substrate. The remaining 60% collided on the way with residual gas molecules and were deflected by those collisions.

Carefully controlled amounts of certain gases can be added by a needle valve to the vacuum during the pump-down cycle. The residual gas that is still in the vacuum chamber consists then predominantly of molecules of the bled in gas. For instance, the oxidation of the evaporant can be prevented by bleeding in a noble gas, or the oxidation process can be promoted when pure oxygen is added.

**High Vacuum Systems.** Reflective and antireflective coatings are evaporated in the same high-vacuum chambers. Therefore the systems are described first before their application is examined.

A modern box coater is basically a steel vacuum chamber that is supported on a sturdy steel base. The chamber can be lifted pneumatically or hydraulically. Large chambers are constructed in separate parts that can be opened to the front. The chamber can be heated so that its walls give up entrapped water vapors more readily during the evaporation cycle. A sight glass permits the observation of the processes in the interior of the chamber. A shaft that is located on top penetrates into the chamber to provide rotation to the tooling planets. A needle valve can be connected to another feedthough the vehicle through which carefully controlled amounts of gasses are bled into the vacuum. Finally, a flood valve is connected to the chamber, which is opened after completion of the work.

The planetary drive, which holds the coating tooling, is located at the top of the vacuum chamber. It is nearly hemispherical in shape. It enables the vapor stream to strike the substrate surfaces nearly vertically, and this causes the coating to adhere well. When flat racks are used, additional sheet metal shutters must be built in to intercept and block the vapor streams that are incident at angles that are too oblique.

The base has several feedthroughs for feeding the heating and glow current into the chamber, as well as monitors and illumination. The evaporant material is heated through the heating power source by means of resistive heating. It has a potential of only 10 to 20 volts but a correspondingly high current. The cables for the heating current therefore are quite thick, and they end in hefty copper terminals. The evaporant source boats are mounted between these terminals. The evaporant material is put into the boats made from either tantalum (melting point of 3000°C) or tungsten (melting point of 3410°C) sheet metal; these boats can also be made from an electrically conductive ceramic material. For reflective coatings, the evaporant material is suspended over a coiled tungsten wire.

Multilayer coatings are quite common today. Therefore the chamber has several connections for the heat source power, which can be alternately activated. Since different substances evaporate at different temperatures, the voltage of the source can also be regulated. Some materials cannot be evaporated through resistive heating. Such materials are evaporated by a direct electron beam.

A potential exists across a glow bar inside the chamber. This glow bar which acts as the cathode, is in most cases a piece of aluminum sheet metal. With the chamber walls acting as anode, a glow discharge originates from the cathode at $10^{-1}$ mbar and 3000 to 10,000 volts of electrical potential. A second piece of sheet metal that is isolation mounted under the cathode protects the lower part of the chamber from the discharge. The discharge causes an ion stream which frees the glass surfaces from the last vestiges of adhering gas and water vapor residuals just prior to the evaporation process. When glowing exceeds 15 minutes, traces of the cathode material are transferred to the glass surfaces. These traces form a glue layer to which the subsequently evaporated optically active layers adhere well.

**Vacuum Pumps.** The high vacuum systems are run with two pumps that are matched to one another. The first is a rotary pump which is driven by an electric motor. It has two or three movable vanes with which it captures air, compresses it, and forces it through a valve. To prevent water from condensing at the exhaust valve, a "gas ballast" is added to the pumped-out air. Such a rotary pump can pump down a vacuum chamber to $10^{-3}$ mbar in a few minutes. However, in actual use the rotary pump is used only to pump the chamber down to $10^{-2}$ mbar, after which the pump-down process is continued with an oil diffusion pump. For this reason the rotary pump is called a *roughing pump* in the coating shops.

The oil diffusion pump can operate only in the prevacuum conditions provided by the roughing pump. In the diffusion pump an oil is vaporized by electric heat. The oil vapors emerge at high velocity from a nozzle located in the upper portion of the pump. The nozzle directs the vapor stream downward, and the oil vapors carry along any air molecules that it encounters. When it reaches the bottom, the vapor stream strikes the water-cooled jacket of the pump and causes the vapors to condense back to oil. The oil then returns to the heated vaporization chamber. The air that was absorbed by the oil is freed by the renewed heating and is extracted from the pump. Vacuum pressures of $10^{-5}$ to $10^{-7}$ mbar can be achieved with a diffusion pump.

For the vacuum above the diffusion pump to remain oil free, a small cooled vapor trap is located there. It is designed to suppress any escaping oil vapors. This ring shaped device is also called a *baffle*.

**Vacuum Measurements.** The air pressure of the prevacuum of $10^{-3}$ to $10^{-1}$ mbar is measured by monitoring the thermal conductivity of the rarefied air.

The lower the pressure, the less heat is transmitted. The gauge, named after Pirani, has an electrically heated wire that is a part of a bridge circuit. Depending on the density of the vacuum, the wire transmits more or less of its heat to its surroundings, and it changes its electrical resistance accordingly. The supplied voltage is adjusted in such a way that the current in the bridge circuit remains in equilibrium. Since the required voltage changes are influenced by the vacuum pressure, the scale of the gauge can be directly calibrated in millibars.

A somewhat coarser measurement is done with an instrument in which a heated wire stimulates a thermal element. Depending on the condition of the vacuum, the thermal detector receives more or less heat. This generates a current that depends on the vacuum pressure. The milliammeter that measures this current can therefore also be calibrated directly in millibar.

The high vacuum of $10^{-7}$ to $10^{-3}$ mbar is measured by means of an electron flow between two poles. More electrons flow from the cathode to the anode at higher pressures than at lower pressures. The gauge is powered by several kilovolts of direct current. It can only be used for pressures below $10^{-2}$ mbar because a glow discharge occurs at $10^{-1}$ mbar. The electron flux is measured on a scale calibrated in millibars.

### 5.9.3.  Coating Technology

The surfaces must be absolutely clean prior to coating. As was mentioned earlier, ultrasonically cleaned parts can be coated without further cleaning. The surfaces of automatically cleaned parts feel dull, but that does not impede the coating process. Nevertheless, most parts are still manually cleaned with solvent soaked wipes. The cleaning personnel should not have sweaty hands and should not treat their hands with fatty creams. That would cause the wipes to become useless and the coated surfaces to exhibit stains.

Surgical gloves or at least talc-free finger cots should be worn when the cleaned parts are placed into the coating insert rings (Fig. 5.173). The parts remain in these rings when they are later loaded into the coating racks. This eliminates the need to touch the parts again. The insert rings are made of either aluminum or steel. They have an inner diameter (ID) that is about 0.25

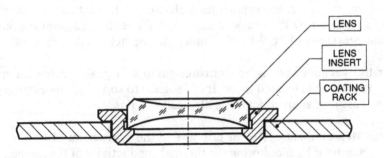

LENS

LENS INSERT

COATING RACK

**Figure 5.173.**  Typical coating rack with lens inserts.

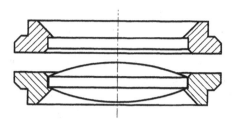

**Figure 5.174.** Flip tooling for coating of lenses.

mm (0.010 in.) larger than the part diameter and a support lip of about 0.5 mm (0.020 in.). When the first side is coated, the lenses are transferred by flipping them into a second ring. This way the second lens surface can be coated without the need for a reclean (Fig. 5.174).

To avoid breaking the vacuum in the chamber when turning over the parts, flip tooling was developed. It is essentially a coating rack that has sectors that can be mechanically flipped. But a percentage of the load capacity is lost with this design, and as a consequence flip racks have been used mostly for small diameter parts. However, the sectors can be removed from the racks, to permit advance loading of sectors. The loaded sectors are kept in a dust-free cabinet or on a dust-free laminar flow bench from which they are drawn as needed. It is also possible to load sectors with different parts into the same rack, and this is rarely possible for a simple rack design.

Mirrors are held in the racks by leaf springs. Because of their irregular shape prisms must be mounted in the racks with the help of special clamps. The back surfaces of prisms should be covered during coating. The vapor stream can produce shadowlike deposits on the back surfaces through improperly shielded gaps.

The loaded rack is mounted in the coating chamber. To facilitate the loading process, the rotary motion in the chamber has two fixed and one spring-loaded position for the racks. The spring-loaded position is swung away during rack loading and is later fastened in place. Just prior to closing the vacuum chamber, the glass surfaces are dusted off with a broad dust brush. The brush is moved slowly to prevent the creation of air turbulence, which can redeposit the dust. The dust can also be gently blown off with a filtered air gun.

In the meantime the coating material has been prepared. This material is typically magnesium fluoride ($MgF_2$) for antireflective coatings. A carefully weighed amount that corresponds to the desired coating thickness is put into an evaporation boat. Only pure materials are used, such as are commercially available from reliable vendors. Great care is exercised in handling these materials to prevent contamination.

For first-surface mirrors pure aluminum is evaporated. The aluminum is used in strips or rods. Pieces of equal length, or rather equal weight, are cut off from these. The aluminum is generally evaporated above a heated tungsten spiral wire. The silicon monoxide (SiO) which is used as a protective overcoat on aluminum is a brown material. It is weighed in empirically determined quantities and placed into a tantalum boat for evaporation.

The internally reflecting surfaces of prisms are frequently coated with silver. The reflectivity of aluminum is too low for this application. Aluminum coatings appear darker when they are coated in a low vacuum. The protective copper layer over the silver is better when it is galvanically deposited, as described under chemical silvering. It will be thicker and thus protect better. But copper can also be evaporated.

After the chamber is loaded with workpieces and coating material, the door is closed. Since compressed air is typically found in coating shops, the heavy chamber can be pneumatically lowered. The rotary pump is turned on to produce the prevacuum. When the chamber is pumped down to $10^{-1}$ mbar, the glow current is turned on. The process causes the cathode to glow. The cathode should be kept very clean because dust particles on its surface can cause a lightninglike phenomenon that can lead to coating flaws. The glow process repels residual contaminants and water vapors from the glass surfaces.

After a prescribed time, which can be as long as 15 minutes, the glow process is ended by shutting off the glow current. The prevacuum has become by now deep enough so that the diffusion pump can become effective. The shutter that leads to the diffusion pump is opened. The oil reservoir had already been preheated so that the diffusion pump can continue to withdraw air from the chamber.

The activated pressure gauge for high vacuum indicates the reducing air pressure. The evaporation process can begin when, depending on the coating instructions, $10^{-5}$ to $10^{-7}$ mbar has been reached. The heating current is turned on and the evaporant in the crucible begins to heat up. The voltage is increased slowly so that the coating material remains quiet and does not spatter. Coating material spatter are flaws much feared among coaters. The spatter does not only adhere to the surface, it can even cause local fractures in the glass. A part with such a flaw is almost always a reject. Depending on the nature of the coating material, it becomes liquid either before it evaporates or it sublimes which means that it evaporates before it melts.

As already indicated, the thickness of the coated layer is influenced by the original quantity of the coating material in the crucible. The layer thickness can be deduced from its interference color. The color can be observed through a sight glass and can be compared to a color standard. When the correctly colored hue is obtained, the heating current is shut off, and a shutter is placed above the source. After that, the valves to the pumps are closed, and air is slowly bled into the chamber. The chamber is then opened, and the coated parts are removed. The chamber is again reloaded with new parts and coating material to repeat the process on the next batch.

Next the finished coated parts, which are for the most part coated on both sides, are tempered in an oven at 300°C for several hours. This final bake makes the layers resistant to damages from wiping and handling. The chemical changes that take place in the layers at that high temperature are required for the optical and mechanical quality of the coating. For instance, silicon monoxide, which is coated over aluminum to serve as protective layer, if not already oxidized during coating, will be converted during the bake into the very stable silicon dioxide which is also known as quartz.

The reflection-coated or antireflection-coated surfaces have a sticky or dull feel. They can be burnished, however, with a chamois. Mirror coatings must be burnished until bright. In principle this dullness is not detrimental for antireflection-coated surfaces. However, stripes and pressure points can develop where these dull surfaces are contacted. Surfaces of lenses intended for the outside of instruments must be burnished for this reason.

**Coating Thickness Measurement.** The process steps for automatic coating systems are controlled by computers. Many different coating processes can be controlled when previously established programs are stored on computer disks.

The thickness of the deposited layer must be measured accurately for all of these automatic cycles. The automatic controller switches over to the next operation when the required layer thickness has been reached. Two methods are used to determine the coating thickness.

A crystal quartz monitor connected to an electronic circuit is positioned in the vacuum chamber. During the evaporation cycle the quartz monitor is coated, and this alters its inherent frequency. The change in frequency is sensed by the electronic circuit, which then alerts the system controller. For the second method the interference color of the layer is monitored. When the layer reaches the color of an interference filter, a strong light flux passes through the filter. This flux is detected photoelectrically which then triggers the controller to advance to the next process step (Fig. 5.175).

For manually controlled coating an empirically determined amount of coating material is evaporated. The layer thickness, for example, on mirrors, can be accurately enough controlled with this approach. Also the interference colors of the layers can be observed through a viewing port. When the color coincides with that of a reference standard, the desired layer thickness has been obtained. The heating current is shut off, and a shutter is closed over the source for added assurance.

**Dip Coatings.** Large area parts such as glass sheets can be antireflection coated with a liquid process called a *sol-gel process*. The layer material is dissolved, and the parts are dipped into the solution or are flooded by it. After the solvent has evaporated, the remaining material adheres to the surfaces of the parts to form an antireflection layer.

The wet process produces layers that are optically effective as well as chemically and mechanically durable. The layer thickness is determined by the viscosity of the solution, and the velocity with which the parts are withdrawn from the solution or the solution is drained away from it. The

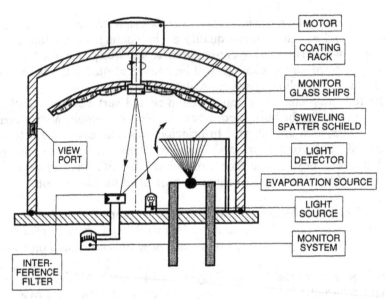

MOTOR

COATING RACK

MONITOR GLASS SHIPS

SWIVELING SPATTER SCHIELD

LIGHT DETECTOR

EVAPORATION SOURCE

LIGHT SOURCE

MONITOR SYSTEM

VIEW PORT

INTER- FERENCE FILTER

**Figure 5.175.**  Optical monitoring system for an automatic coating machine.

parts must also be very clean for this process and the workplace dust free; otherwise, layer flaws will develop.

With this dip process irregularities in the layer must be expected over at least the last 2 mm (0.080 in.) of the bottom edge from which the solution drains off. A buildup near the edge is not detrimental for large parts such as glass disks. For smaller parts, however, such an irregularity could cover a large part of the surface and deposit an asymmetric layer on round parts. To overcome these problems, a special method [147] was developed in which lenses are spun on a spindle about their axis at several thousand rpm as the solution is dispensed at the lens center. As a result the liquid distributes itself uniformly, while the excess is flung off. At the present time these wet processes are used only for special cases.

**Inspection of Coated Parts.**  Designers of instruments demand that coatings have certain properties, and they carry these demands into the drawings and specification sheets. While most coating houses use their own coating standards, there have been some efforts made to standardize the more important aspects of optical coatings [148]. Also required are clear specifications [149] of optical properties such as transparency or reflectance at specific wavelengths.

The mechanical strength of coatings must also be determined. There are no units of measurement for mechanical strength but that it should be greater for exterior surfaces than for surfaces in the interior of the instruments. The chemical solubility of the layers is covered by still another standard [150].

The chemical stability of coatings is usually tested only in special cases where materials were evaporated for which the coating shop had no prior experience.

In general the coated surfaces are inspected in transmission and by reflection. Pinholes caused by dust particles on the surfaces and by coating material spatter are considered coating flaws, and they are treated like other surface flaws. Stains in the coating, which are discernable by reflection, are subjectively evaluated. They are generally not detrimental when the optics parts are in use. It looks bad, however, when the first lens surface of an objective lens exhibits discernible stains.

The durability of the coatings is tested on a sample basis on witness parts. The coated surfaces on such test parts are rubbed with a chamois or cheese-cloth, and any changes in the coatings are observed. To make the test reproducible, an instrument can be used with which the test can be performed at constant pressure and equal time intervals.

Reference samples, and when necessary also limit samples, are fabricated to meet the physical and optical requirements. Their transparency and reflection properties are measured photometrically. They are also visually inspected for changes in interference colors. The coatings on production parts are visually compared against these color references.

**Decoating.** Coatings that are flawed or that do not meet transmission or reflection specifications must be removed to prepare the optical components for recoating. Antireflection coatings adhere so tenaciously that they can be polished off only with great difficulty. The surfaces are also irregularly engaged during polishing, so lenses are easily decentered. Therefore chemical etch processes have been developed with which the coatings can be dissolved. The etchant also attacks the glass surfaces. But the surfaces can be easily repolished.

A proven etchant solution consists of 1 liter (1.06 qt) of sulphuric acid (85% to 88%) which is heated to 140°C (284°F). About 50 gm (1.6 oz) of crystalline boric acid is carefully mixed into the sulphuric acid. The lenses are dipped into this hot solution. The dip time must be empirically determined. Particularly resistant coatings dissolve only when the etchant solution is heated to 160°C (320°F).

### References for section 5.9 (Coating)

[146] W. Zschommler.
[147] Schott & Gen., ALS Alflexil method.
[148] DIN 58 195.
[149] DIN 58 197.
[150] DIN 58 196.

# 6

# Optical Shop Testing—Methods and Instruments

The testing of optical surfaces is a critical part of the fabrication process because the optician must verify that the specified precision and level of quality has been achieved. Testplate testing is the oldest and most commonly used method to verify the flatness or sphericity of optical surfaces during the fabrication phase. Highly versatile shop interferometers are used now in most shops for final acceptance of surface figure and for measuring the distortion of the transmitted wavefront. Spherometers have been in use for a long time to accurately determine the radius of curvature of spherical surfaces. Modern spherometers utilize the most recent advances in precision metrology in conjunction with computer technology. The autocollimator has also been used by opticians for many years to measure small angular deviations and the parallelism of plane-parallel surfaces. The qualitative determination of surface roughness is now possible with a number of sophisticated surface analysis instruments. This is a fairly recent technology that has made the development of polishing methods to produce super-smooth surfaces possible. These methods and instruments are described in this chapter.

## 6.1. TESTPLATE TESTING

### 6.1.1. Introduction

Testplating is the simplest and therefore the most common method of testing the flatness or the sphericity of optically polished surfaces. It utilizes the interference effects known as *Fizeau fringes* or *Newton's rings*. It is the oldest technical use of the interference phenomenon. In the early years of the nineteenth century, Fraunhofer [1] described this phenomenon, and he is credited with being the first person to test polished optical surfaces in this manner. Testplating is also sometimes called *contact interferometry* because the surfaces of the testplate and the workpiece are in intimate contact. This description differentiates it from all other commonly used interference methods, all of which do not involve contact with the workpiece.

Despite the widespread use of shop interferometers, testplates continue to be an indispensible tool for the optician, who uses them to monitor the flatness or the sphericity of the work and makes the necessary corrections during fabrication to achieve the goal of producing a close and regular surface fit. This section examines the underlying concepts, design, fabrication, use, fringe interpretation, and special applications of testplates.

### 6.1.2. Basics of Interference Fringes

The circular interference fringes that are observed when using testplates are commonly called *newton rings* in honor of of the eighteenth-century scientist Issac Newton [2] who was the first to observe and document the existence of this phenomenon. Straight fringes when formed in an air wedge between two flat and transparent surfaces are called *fizeau fringes* after the nineteenth-century French scientist [3] Armand fizeau who, among other things, developed the first interferometer. Newton's rings or fizeau fringes represent special cases of interference in that they seem to be localized within an air film trapped between two closely fitting transparent surfaces. This condition exists when testplates are used on matching optically polished surfaces.

Newton's rings were originally described as "colors of thin layers" which can be explained by the wave nature of light. For testplates this thin layer consists of the air that is trapped between the matching surfaces. This thin layer of air has a refractive index of $n = 1.0$. When light falls on such a thin transparent layer, the light is reflected from the top and the bottom surfaces (Fig. 6.1). The light that is reflected off the top surface precedes the light reflected off the bottom surface by about twice the thickness of the layer. In other words, the wave trains of both rays are offset to one another. These wave trains interfere because the light is coherent, which means that they

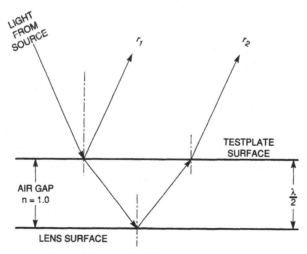

**Figure 6.1.** The interference phenomenon.

originate from the same incident ray and because the distance between the two surfaces is so small.

When the light is also monochromatic or "of one color" then, depending on the wavelength of the light and the thickness of the layer, it is either extinguished or reinforced. This results in the formation of the characteristic alternating light and dark bands when the thin layer is wedge shaped. When the two wave trains are offset in such a way that their amplitudes swing to the same side, they are reinforced and a light fringe is formed. When, on the other hand, they swing in opposite directions, they will cancel each other out and this creates a dark fringe.

Another way of describing the same phenomenon [4] is to say that one of the rays travels a slightly shorter path than the other, and this causes a phase difference between the two. If the phase difference resulting from this optical path difference (OPD) is exactly one wavelength of the monochromatic light source used, then constructive interference occurs and a light fringe is seen. If the OPD is one-half wave, then destructive interference occurs, and a dark fringe is seen.

More specifically, in the case when the layer thickness is exactly half the wavelength of the light, and the light traverses this distance twice, the two wave trains will be offset by one whole wavelength. This causes their amplitude to swing to the same side. Therefore the first bright fringe appears at that point. Another bright fringe is formed each time the thickness of the thin layer increases by another half wavelength. The phase shift that occurs at each air-glass reflection is not considered in this description. In the case where white light illumination is used, the interference in the wedged thin layer is displayed as alternating multicolored bands.

The thin layer discussed so far is the air film that forms between two testplates. The layer of equal thickness is circular between curved test glasses. Therefore the interference bands are circular fringes or rings. Straight fringes will show up for a testplate match of very plane surfaces or when suitably designed radius testplates are used that allow the light rays from the source to strike the surfaces at the interface at normal incidence. More will be said about this later.

Much of the following has been discussed by the author in a previously published paper [5]. When two nearly plano and reasonably parallel polished pieces of a transparent optical material are placed in intimate contact and are illuminated with a monochromatic light of wavelength $\lambda$, nearly straight or slightly curved fringes are formed. The formation of these fringes can best be explained by the diagram in Fig. 6.2.

Dark fringes are formed when the optical path difference OPD = $2t_N$, as shown in Fig. 6.2, is

$$t_N = \frac{N(\lambda)}{2},$$ (6.1)

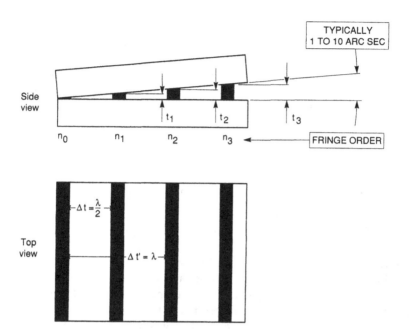

**Figure 6.2.** Fizeau fringes.

where $N = 0, 1, 2, 3, 4, \ldots$, and $N$ is the order of the fringe, which is always a positive integer. It is assumed here that the refractive index $n = 1$. From Eq. (6.1) and Fig. 6.2, the following conclusions can be drawn:

$$\Delta t = t_1 - t_0 = (1) \left(\frac{\lambda}{2}\right) - (0) \left(\frac{\lambda}{2}\right) = \frac{\lambda}{2},$$

$$\Delta t = t_2 - t_0 = (2) \left(\frac{\lambda}{2}\right) - (0) \left(\frac{\lambda}{2}\right) = \lambda.$$

In other words, the thickness difference of the air gap from one dark fringe to an adjacent dark fringe is $\lambda/2$, whereas the difference between any three adjacent fringes is $\lambda$. These are important points to remember because they are fundamental for the fringe analysis of most commonly used tests.

Light fringes are formed when the optical path difference $2t_N$, as shown in Fig. 6.2, is

$$t_N = N \left(\frac{\lambda}{4}\right), \tag{6.2}$$

where $N = 1, 3, 5, 7, \ldots$, and $N$ is the order of light fringes. This order is always an odd integer. Since these fringes are rarely, if ever, used for test-plate measurements, there is no further need to discuss them.

The preceding equations are strictly valid only for normal or near-normal incidence. If the angle of incidence is appreciably inclined to normal, the following relationship [6] applies

$$OPD = 2nt \cos \theta, \tag{6.3}$$

where

$n$ = refractive index of air film ($n = 1$),

$t$ = physical thickness of air film,

$\theta$ = angle of incidence to normal

**Spherical Surface on Plano Testplate.** When a slightly convex surface is placed on a flat surface, circular fringes are formed by interference of light from a source with wavelength $\lambda$. The development of these fringes is graph-ically represented in Fig. 6.3. This geometric relationship can be mathemati-cally expressed by using the approximate sag equation $s = d^2/8R$ as follows:

$$t_N = \frac{(D_N)^2}{8R} \tag{6.4}$$

where $D$ is the diameter of the fringe and $R$ is the radius of curvature of the spherical surface. This only holds true for slightly spherical surfaces that are tested against a plano testplate. There are separate equations for the condi-tion where a spherical testplate is placed against a matching spherical sur-face that has a slightly different radius. This is the typical condition when using radius testplates. Those equations will be discussed separately.

By equating Eq. (6.1) (dark fringe fizeau) and (6.4) (dark fringe newton), a relationship for dark fringes results that can be used to calculate all parame-ters that are of interest to the optician:

$$N\left(\frac{\lambda}{2}\right) = \frac{(D_N)^2}{8R}. \tag{6.5}$$

By solving for the various components in this equation, three very useful equations result:

| | | |
|---|---|---|
| Fringe diameter of order $n$ | $D_N = 2/N\lambda R$ | Eq. (6.6) |
| Radius of curvature | $R = (D_N)^2/4N\lambda$ | Eq. (6.7) |
| Number of fringes (order) | $N = (D_N)^2/4R\lambda$ | Eq. (6.8) |

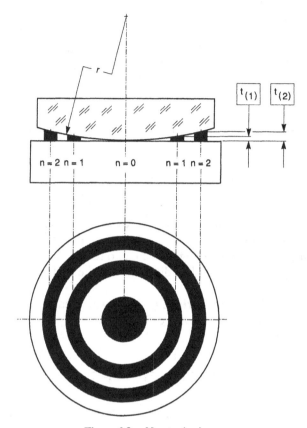

**Figure 6.3.** Newton's rings.

The use of these equations is best explained by suitable examples. The light source is a mercury green light with $\lambda = 5.461 \times 10^{-4}$ mm ($21.5 \times 10^{-6}$ in.). A more detailed description of commonly used monochromatic light sources can be found in Section 6.1.5.

The dimensions used for these and all other examples in this book are either in metric or in English measure. It does not matter which method is used as long as all dimensions in the equations are of the same type.

**Example 1**

What is the diameter $D_{12}$ of the twelfth fringe as counted from the center when the radius $R$ of the spherical surface is 3718 in.? Note that the central fringe is the zero order fringe which is not counted.

Using Eq.(6.6),

$$D_{12} = 2\sqrt{[12(21.50)(10^{-6})3718]}$$
$$= 2\sqrt{959.244 \; (10^{-3})}$$
$$= 1.9588 \text{ in.}$$

With this arrangement, the twelfth fringe has a fringe center to fringe center diameter of nearly 2 in.

**Example 2**

A more likely or practical application exists when the radius $R$ must be calculated on the basis of a testplate fit. Note that for this example the spherical surface is referenced against a plano testplate. The same approach can be used to calculate the radius difference of the surface relative a known testplate radius.

The diameter of the fourth fringe ($N = 4$) was measured to be $D = 22.5$ mm. What is the radius of curvature $R$?

Using Eq. (6.7),

$$R = \frac{(22.5)^2}{4(4)(5.461)(10^{-4})}$$
$$= \left(\frac{506.25}{87.376}\right)(10^4)$$
$$= 57,939.25 \text{ mm.}$$

The radius of the spherical surface is almost 58 meters!

Another example for this equation is the calculation of an extremely long radius: A 6-in. diameter flat window shows 1/10 wave of convex spherical power when tested against a perfectly plano testplate. Therefore, $D = 6$ in. and $N = 0.1$. What is the radius of the window?

$$R = \frac{6^2}{4(0.1)21.5(10^{-6})}$$
$$= \left(\frac{36}{8.6}\right)10^6$$
$$= 4,186,047 \text{ in.}$$

When this number is divided by 12 in./ft and 5280 ft/mi, then the radius is found to be 66.068 mi! This example demonstrates the extreme flatness that can be achieved and measured with testplates on very high precision optics.

**Example 3**

It may sometimes be required to calculate a value for the fringe order $N$, which represents the number of fringes over a given diamter $D_N$ when the radius $R$ of the spherical surface is known. This of course assumes that the testplate surface is perfectly flat, and the test part surface is perfectly spherical.

Using Eq. (6.8) and the values from example 1, (i.e., $D_N = 1.9588$ in. and $R = 3718$ in.), the fringe order N is

$$N = \frac{(1.9588)^2}{4(3718)\ 21.5\ (10^{-6})}$$

$$= \left(\frac{3.8369}{319,478}\right)10^6$$

$$= 11.9997.$$

Therefore N = 12 as in example 1.

It should be understood that the same equations hold true for slightly concave surfaces as well. The order of the fringes, however, is reversed. This means that the zero order fringe is the one around the edge of the part where it contacts the plano testplate.

When a concave spherical testplate with radius $R$ is placed on a convex lens surface with a slightly different radius $R + e$, concentric fringes will be observed. The value $e$ is a small radius error in the part relative to the spherical testplate. For this condition the following equations apply:

$$N = \frac{D^2 e}{4\lambda R^2}, \tag{6.9}$$

$$D_N = \frac{2R\sqrt{(N\lambda)}}{e}, \tag{6.10}$$

$$e = \frac{4N\lambda R^2}{(D_N)^2}. \tag{6.11}$$

These are approximate equations that will yield results with an error not exceeding 1%. This makes them accurate enough for most common applications encountered in the shop [7]. If higher accuracy in the results is required, there are three corresponding equations that are more precise but also much more complex [8].

Note that for the preceding equations, $D_N$ refers to the diameter of the fringe of order $N$, while $D$ refers to any chosen diameter over which the number of fringes $N$ must be calculated for a given radius $R$ and wavelength $\lambda$. Also note that the radius difference $e$ can have a positive and a negative value. The sign is determined by the conditions given in Table 6.1. The use of these equations is also best explained by an example.

**Example 4**

What is the radius difference $e$ of a concave spherical surface with respect to a matching convex spherical testplate with radius $R = +16.000$ in. when the diameter of the fourth fringe is 2.75 in. It has also been established that the tested surface fits convex to the testplate. A mercury green monochromatic light source is used with $\lambda = 21.5 \times 10^{-6}$ in.

Using Eq. (6.11) and the values in Table 6.1, the radius difference $e$ between the two surfaces can be calculated.

$$
\begin{aligned}
e &= \frac{4(4)(21.5)(10^{-6})16^2}{(2.75)^2} \\
&= \frac{88064(10^{-6})}{7.5625} \\
&= 11{,}645(10^{-6}) \\
&= 0.011645 \ in.
\end{aligned}
$$

Since the testplate is convex ($+16.000$ in.) and the testplate fit is also convex, $e$ is negative, as shown in Table 6.1. The radius of the tested surface is $R_p = R + e$. But since $e$ is negative,

$$
\begin{aligned}
R_p &= 16.000 - 0.011645 = 15.988355 \\
&= 15.9884 \ in.
\end{aligned}
$$

Similar equations can be used for white light fringes. The thickness of the air film changes by a half wavelength of the light from one interference band to the next. This means that for a white light source, four blue rings correspond to a thickness change in the air film of about 1 $\mu$m. The same holds true for three red rings. This is so because the wavelength (or two rings) of blue light is about 0.5 $\mu$m and that of red light is about 0.65 $\mu$m. This

**Table 6.1.  Sign of Radius Difference $e$**

| Testplate radius | Testplate fit to lens surface convex | concave |
|---|---|---|
| Convex | e is negative | e is positive |
| Concave | e is positive | e is negative |

information can then be used to calculate how far the radius of a tested surface deviates from that of the testplate when the match shows a certain number of rings. Taking an approximation results in (6.12):

$$\delta r = \frac{4R^2 N\lambda}{D^2},$$
(6.12)

where

$\delta r$ = radius deviation in millimeters,
$R$ = testplate radius of curvature in millimeters,
$N$ = number of rings over $D$,
$D$ = diameter in millimeters,
$\lambda$ = wavelength of light in millimeters.

**Example 5**

A surface with a diameter of $D = 25$ mm and a radius of $R = 50$ mm is tested with a testplate in sodium light of $\lambda = 5.89 \times 10^{-4}$ mm. Five rings are seen over the diameter (i.e., $N = 5$). By how much does the lens radius deviate from that of the testplate?

$$\begin{aligned}
\delta r &= \frac{4(50^2)(5.89)(10^{-4})}{25^2} \\
&= \frac{4(2500)(5.89)(10^{-4})}{625} \\
&= 0.047 \text{ mm.}
\end{aligned}$$

If the rings were high, indicating convex testplate match, then the radius difference is 47 $\mu$m less than nominal for a convex lens surface and 47 $\mu$m greater than nominal for a concave lens surface. If the match had been low (i.e.,concave), then the radius difference would still be 47 $\mu$m, but the convex lens surface would have a longer and the concave surface a shorter radius than that of the testplate.

Another equation can be used to calculate the radius of curvature of the spherical surface under test when fringes are counted over a specific diameter and the radius of the testplate is known [9].

$$\frac{1}{R} = \frac{1}{R_0} + \frac{(N_n - N_0)\lambda}{(D/2)^2},$$
(6.13)

where

$R$ = radius of test surface,

$R_0$ = testplate radius,

$N_n - N_0$ = number of fringes between center and edge,

$\lambda$ = wavelength of light source,

$D/2$ = semidiameter over which fringes are counted.

Testplating with white light illumination results in color fringes. The thickness of the air film between the surfaces increases from violet to blue to yellow to red corresponding to the increasing wavelengths [9]. This provides the optician with an unambiguous measure whether a surface match is high (convex) or low (concave). The optician selects the blue fringes from the multicolored bands or rings as reference. The OPD or $\lambda/2$ of blue fringes is about 0.25 $\mu$m. If the red fringes lie inside the blue fringes, then the testplate match is convex. For a concave match the red fringe will lie outside the blue fringe [10]. Convex or concave conditions are also indicated by the way the fringes will move when the testplate is wedged slightly relative to the test surface. This will be discussed in greater detail later in this chapter.

Still another way to calculate $\Delta R$ is expressed by the following equation [11]:

$$N(\lambda) = \left(\frac{1}{R} - \frac{1}{R_0}\right) r^2 = r^2 \left(\frac{\Delta R}{(R_0)^2}\right), \qquad (6.14)$$

where:

$N$ = order number of fringes,

$\lambda$ = wavelength of light source,

$R$ = radius of the test surface,

$R_0$ = testplate radius,

$r$ = half the diameter over which the fringes were counted,

$\Delta R = R - R_0$.

The zero-order fringe is not counted because $N_0 = 0$. It is the central dark fringe where the surfaces contact when the testplate fit is convex. The zero order fringe for a concave fit is the outermost dark ring where the surfaces are in contact. When the testplate is plano, then $R = r^22/N(\lambda)$. This equation is equivalent to Eq. (6.7).

For the sake of clarity, it was assumed for the preceding examples that the fringe order or fringe count is always a whole number or integer. This is almost never true in the real world, since there are always partial fringes in every testplate match. These partial fringes cannot be counted with any degree of accuracy and are usually ignored. But for very critical testplate matches, these partial fringes can introduce significant errors.

Modern phase interferometers that are equipped with sophisticated fringe analysis systems can account for these partial fringes and such measurements are much more precise than even the best results obtained with testplates. Therefore the use of testplates should be restricted to the shop floor as a convenient in-process check, while interferometers must be used for the final buy-off. Interferometers are discussed in Section 6.2.

### 6.1.3. Testplate Design Considerations

Testplates are fabricated from transparent, low expansion optical materials that must be also sufficiently wear resistant since they are used in contact with the work. There are two basic types of testplates, namely, plano and spherical.

For testing the flatness of plano surfaces, plano testplates must be used. They are usually fabricated singly if they can be tested against a calibrated master testplate that is often referred to as being traceable to the National Bureau of Standards or to the equivalent standards agencies of other countries. What is not always clear is how many copies of plano master plates lie between the one used and the one that was actually calibrated originally by the NBS. For all but the most stringent flatness requirements any independently verified plano testplate by other interferometric means can serve as a good master flat. There are several detailed descriptions on how to make master flats with the three-flat method with which three plano plates are checked against one another in a carefully prescribed sequence until the possible combinations of plate matches gives the same flatness. Although this may be a very challenging exercise, it is doubtful in light of other available options that the extraordinary amount of time for this approach makes much economic sense.

While the reference surface of test flats must be polished to an accuracy of $\lambda/20$ for master flats and $\lambda/10$ for standard flats, the flatness of the back surface is not as critical and can be several fringes high or low. The reference surface must be identified on the edge of the test flat by an arrow. The actual, measured flatness should also be permanently marked.

Radius testplates are used for testing spherical surfaces. They must be made in pairs to the exact radius specified in the lens design. One set of radius testplates is required for each radius of curvature. This is true even though two radii may only differ by a small amount. The actual, measured radius, not the design radius, should be permanently inscribed on the back surface.

It is recommended [12] that master testplate sets, which are made to be very accurate, never be used for in-process testing. Professionally run optical shops first fabricate standard testplates from these masters and then make working plates from the standard plates. Only the working plates are used in the shops at the machine. The standard plates are used in the quality assurance (QA) lab, where this is still done. Smaller shops, typically under pressure of time and lack of enough skilled workers, make working plates directly from master plates. The working plates have quite often been made to a diameter that is only slightly larger than the lenses they are intended to test. It has become customary in recent years for only one set of testplates to be made for each radius. These plates are then used as master, standard, and working plates all in one. An independent radius reference really never exists, so it is obvious that a great deal of the precision is lost in this rather questionable approach.

Master plates are made as accurately as can be measured, both in terms of radius accuracy and surface figure. The actual radius should be as close to the nominal design radius as precision spherometry will allow. Once the master testplate radius has been established, it then becomes the nominal radius of the lens surface. It is this actual value that should be logged into the testplate list. If a small radius difference exist between the master testplate radius and the original design radius and further radius correction cannot guarantee an improvement, the actual master radius will become the nominal radius, and the lens designer will have to recalculate the design on that basis.

Because of the appreciable cost of making master, standard, and working testplates, lens designers are always encouraged to first use available radii from the testplate master list before they commit their design to new radii of curvature. This is why these lists are so important. They form the basis of very expensive lens designs. The inventory of lens manufacturers master radius plates is one of their most important assets. It should be clear to everyone that to try and save money by circumventing this process can lead to some serious and very costly consequences.

The match between the two master plates for a specific radius of curvature must be as nearly perfect as can be achieved. With white light illumination, the testplate fit must be "white" or essentially colorless. In the monochromatic light of an interferometer, the sphericity must be better than $\lambda/10$. The problem with this approach is that the reference surfaces of the transmission spheres used in the interferometer are usually limited to $\lambda/10$. This leaves an uncertainty that may have to be resolved by an independent test method such as white light interferometry.

The standard radius testplates that are used by quality assurance for final acceptance are referenced to their corresponding master plates to a surface accuracy of about $\lambda/5$. The working testplates that are used by the optician at the polishing machine are made against the standard set to an accuracy that is about 1/3 of the maximum permissible fringe count. In other words, if three fringes are the permissible limit for the lens surface, the working plate figure must be held to less than one fringe or $\lambda/2$ relative to the standard plate.

Optical test flats, unlike radius testplates, are commercially available in sizes ranging from 50 mm (2 in.) up to 300 mm (12 in.) in diameter. They are tested in helium light (0.588 $\mu$m or 23.1 $\mu$in. per wave). Table 6.2 lists standard flatness specifications [13].

Testplates are used in two basic attitudes. Either the testplate is placed on top of the workpiece or the workpiece is placed on top of the testplate. In the case where the fringes are viewed through the testplate, the testplate material must be sufficiently transparent and homogeneous so that the fringes are not distorted. This is usually the case with radius testplates. The selection of testplate material is not nearly as critical for the case where the testplate supports the part and the fringes are viewed through the workpiece (which is in most cases made from optical quality material so that fringe distortion from this source is not a problem). Testplates that are used in this mode are almost always plano, and since weight is not a prime consideration, they can be quite large. Plano testplates 300-mm (12 in.) diameter are not uncommon. Figure 6.4 shows a few testplate arrangements.

Testplates must be designed in such a way to allow for testing of the entire surface of the optical component, namely, a lens or a prism. As a result they must be slightly larger than the surface to be tested. This of course is practical only as long as the lens diameters do not exceed 4 in.; otherwise, the testplates will become too large for proper handling as their weight increases quite rapidly as they become larger. Technically, when the testplate diameter doubles, the testplate weight increases by a factor of $(2)^3 = 8$. But this is true only if the diameter to thickness aspect ratio is the same and the plates are made from the same material.

The aspect ratio is an important consideration because it defines the testplate thickness for a specific testplate diameter. Therefore the thickness is not an arbitrary choice. The edge thickness on convex and the center thickness on concave radius testplates range from $\frac{1}{5}$ to $\frac{1}{3}$ of the diameter. If the thickness is less than $D/5$, then there is danger that the testplate surface will be distorted by outside forces, namely, weight and pressure. If the thickness is greater than $D/3$, the testplate could become too heavy or unwieldy. The choice of thickness depends on the steepness of the testplate radius, the material used, the weight, the required accuracy, and the intended application. The typical aspect ratio for plano testplates is $t = D/5$.

Table 6.2.   Grades of Optical Test Flats

| Test flat grade | Max. departure from true flat | Flatness at 23.1 micro-inch |
|---|---|---|
| Commercial | 8 micro-inch | $\lambda/3$ |
| Working flat | 4 micro-inch | $\lambda/6$ |
| Reference flat | 2 micro-inch | $\lambda/12$ |
| Master flat | 1 micro-inch | $\lambda/25$ |

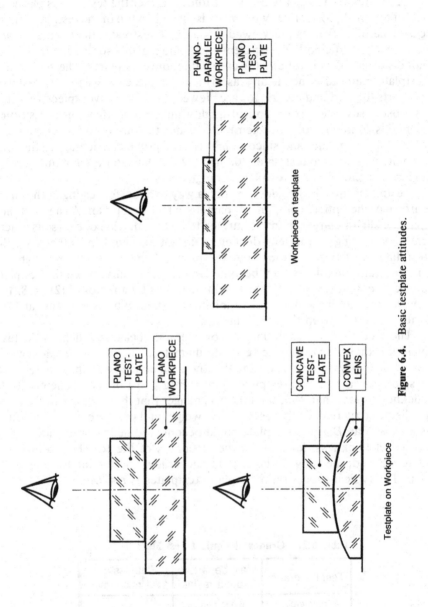

**Figure 6.4.** Basic testplate attitudes.

The edge thickness is directly related to the center thickness and the curvature of the radius side. Within the constraints of the $D/t$ aspect ratio and the sag of the radius, the edge thickness must be chosen such that the testplate can be easily held and used. This requires that the edge thickness be not less than 12.5 mm (0.5 in.) for testplates up to 75 mm (3 in.) in diameter and no more than 38 mm (1.50 in.) for the largest testplates.

Liberal protective bevels must be provided to prevent chipping and breakage during use of the testplates. The bevels should be about 1 mm (0.040 in.) for testplates up to 75 mm (3 in.) in diameter and up to 1.5 mm (0.060 in.) for larger testplates.

Testplate materials should be optically transparent at visible wavelengths, although optical transparency is not a requirement for some applications. They must, however, be wear resistant, and they must have a low coefficient of expansion. The final choice of materials is determined by the optical properties, wear resistance, coefficient of expansion, required accuracy, availabilty, and cost. All of these factors must be considered, and the best possible choice within these constraints made, for each requirement. Typically used testplate materials are listed in Table 6.3.

- BK 7 is rarely used anymore for making testplates because it is a relatively soft glass with a high coefficient of expansion. The advantages are its low relative cost, excellent optical properties, and easy availability. Although it readily takes on a good polish, controlling a precise figure may prove difficult. It can be a good choice for working plates used for noncritical applications.
- ZK 7 was once used for standard plates. It is, however, difficult to work because it is quite brittle. It also has excellent optical properties, good wear resistance, and an acceptable coefficient of expansion. Since it is not readily available and also difficult to polish, it is hardly in use anymore.
- Pyrex has become the material of choice for making testplates, and most testplates made today for use as standard or working plates are made

**Table 6.3.   Testplate Materials**

| Material | Ref. | Optical transparency | Wear-resistance | Coeff. of expansion ($\times 10^{-7}/°C$) | Specific weight (gm/cm$^3$) |
|---|---|---|---|---|---|
| BK-7 | (14) | excellent | marginal | 71 | 2.51 |
| ZK-7 | (15) | excellent | good | 45 | 2.49 |
| Pyrex | (16) | marginal | good | 35 | 2.23 |
| Fused silica | (17) | excellent | best | 5 | 2.20 |
| Cervit | (18) | poor | good | 0 | 2.50 |

from this commonly available low expansion material. Pyrex may not be the best choice if fringes have to be viewed through the testplate because its optical transparency and homogeneity is not very good. It is, however, an excellent choice for large plano testplates. Cast circular disks ranging from 75 to 300 mm (3 to 12 in.) in diameter with commonly used aspect ratios are commercially available at relatively low cost. These are typically called *mirror blanks*.

• Fused silica is a very hard, wear-resistant optical material that has a very low coefficient of expansion. The optical properties range from good to excellent depending on the grade selected. The grades selected are "commercial quality" for standard testplates and the significantly more expensive "optical quality" fused silica is reserved for master plates where the added cost can be justified. Fused silica polishes readily and takes as well as holds a precise figure over a wide range of temperatures. Its primary drawback is cost.

• Cervit is a low-expansion glass ceramic. It is quite wear resistant but it has undesirable optical properties. It has yellow or yellow-green coloration and material inhomogeneities that makes it not useful for transmissive testplates. It is an excellent choice, however, for large reference quality flats.

• Other commercially available glass-ceramics are Zerodur [19] and ULE [20]. The relative cost is high and large diameters may not be readily available.

In addition to the previously listed material characteristics, materials for testplates must above all be strain free. This requires that they be fine-annealed; otherwise, partial strain relief during testplate manufacture could distort the resulting surface figure.

**Reflectivity Matching.** An optically polished glass surface reflects about 4% of the incident light. When a testplate is brought in contact with such a surface for the purpose of testplate testing, fringes of good contrast are formed in the air wedge between the surfaces as shown in Figs. 6.2 and 6.3.

If, however, the surface of the workpiece is highly reflective, such as polished silicon, germanium, or other metallic materials, or it is a reflection-coated glass substrate, the fringes will be difficult to see because the fringe contrast is greatly reduced. The best way to overcome the considerable mismatch of the reflectivities of the contact surfaces is to apply a semitransparent coating to the reference surface of the testplate, (i.e., the one in contact with the workpiece) to decrease the magnitude of the mismatch. This poses another problem in that the coating must be very thin and of uniform thickness so that it has little or no effect on the sphericity and radius of the reference surface, and it must be hard enough to withstand the frequent cleaning required during testing.

The measure of reflectivity is expressed in terms of percent of the total light energy that falls onto the surface. If there is no reflection off the surface, nor absorption and surface scatter, all the light will pass through the surface. Transmission then is said to be 100%. This is only a theoretical value because there will always be a fraction of a percentage reflected off each surface, even on highly efficient antireflection coatings. There is also a small amount of the light absorbed by the surface and an additional small amount is scattered by the surface. If we choose to ignore the absorption and scatter at the surface, then it can be assumed that 70% of the light is transmitted when a 30% reflective coating is applied to the surface of the testplate.

The required reflectivity ($R_t$) of the testplate reference surface is a function of the testpart reflectivity ($R_p$) and the transmission ($T_t$) through the testplate. This relationship can be expressed by

$$R_t = (T_t)^2 \ (R_p) \tag{6.15}$$

**Example**

If the reflectivity of the back surface is 4% (i.e., 0.04) in visible light, then the transmitted light is 96% (0.96). The reflectivity of polished germanium is 56% (0.56). What should be the reflectivity of the coating applied to the reference surface in order to maximize fringe contrast?

$$R_t = (0.96)^2 \ (0.56) = 0.516.$$

Therefore a 52% reflective coating applied to the reference surface will optimize the fringe contrast.

To further improve the visibility of fringes, an antireflection coating (AR) can be applied to the back surface if the fringes are viewed through it. The coating design must be optimized at the wavelength of the monochromatic light source used. The AR coating will reduce bothersome reflections, and this tends to improve the clarity of the fringe pattern.

The foregoing is mentioned here only for the purpose of providing a complete description of the available options. The information is not intended to describe a widely accepted practice. In reality the cost of coating the testplates and the time involved usually prevents this from being done on a routine basis. Antireflection coating has been done with success for very critical applications.

**Aplanatic Backs.** Under certain conditions such as viewing angle, viewing distance, and fast $f/\#$'s for radius plates, the testplate design may have to be modified to overcome certain limitations that can lead to significant distortions of the fringe patterns. This is particularly true for strongly curved surfaces that are steeper than $f/4$ when $\lambda/10$ accuracy is required for the test. This means that the $R/D$ ratio is $4:1$ or less. For these conditions the testplates must be provided with aplanatic backs. Instead of a plano back surface, the testplates will have a spherical back side that is concave for convex testplates and convex for concave testplates. The radius of the curved or aplanatic backs and the distance at which they must be viewed are calculated by Eqs. (6.16) through (6.19). These relationships are shown in the corresponding Fig. 6.5a and b [5]:

For convex testplates,

$$R = \frac{n(R_0 + t)}{n + 1}, \tag{6.16}$$

$$D = R(n + 1) - t, \tag{6.17}$$

and for concave testplates,

$$R = \frac{n(R_0 - t)}{n + 1}, \tag{6.18}$$

$$D = R(n + 1) + t, \tag{6.19}$$

where:

$R$ = back radius of the testplate,
$R_0$ = nominal testplate radius,
$t$ = testplate center thickness,
$n$ = refractive index of the testplate material,
$D$ = position distance for convex surfaces or viewing distance for concave surfaces.

**Internal Interference.** For plano testplates that will be used in laser-illuminated test setups, a small wedge angle of a few arc minutes is purposely introduced to eliminate the bothersome interference that occurs between the flat and parallel surfaces of the testplate in addition to the interference between the reference surface of the testplate and the test part. This internal interference sets up stationary fringes that distort the fringes from the surface to make it impossible to interpret them. This is called the *herringbone effect*.

Because of the short coherence length of all commonly used monochromatic incoherent light sources, internal interference is not a problem when the testing is done in this way. But it becomes a real problem when coherent laser light is used as light source because the long coherence length of the laser light causes often severe internal interference that makes it impossible to interpret the fringe patterns. However, a wedge of at least 30 arc minutes

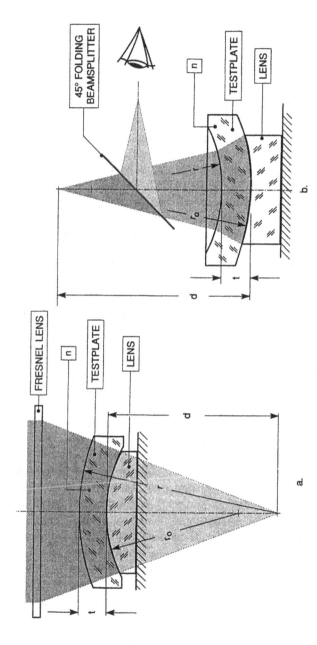

**Figure 6.5.** The use of aplanatic testplates. (*a*) For testing convex surfaces. (*b*) For testing concave surfaces.

between the plano faces of the testplate will reduce this effect almost entirely
[21].

It should be kept in mind that the various design options discussed in this
section should be considered guidelines for generally accepted practice.
Other approaches are also possible. Much depends on the intended application.

### 6.1.4. Testplate Manufacture

Even in the current age of computerization and automation, the ultimate
precision of testplates still originates from the skillful hands of the master
optician. Machine-polishing methods are commonly used only during the
initial stages to speed up the polishing phase. Machine polishing of larger
testplates can be successfully used if proper controls are carefully maintained. In most cases, however, the prepolishing phase must be followed by
a skillful application of the optician's art to figure the surfaces to the desired
level of perfection.

Testplate fabrication methods vary from optician to optician. In fact there
may be as many individual methods as there are opticians who know how to
do this. This makes it difficult to describe a particular way without inviting
criticism from other fellow opticians who are convinced that their methods
are superior.

With the exception of plano testplates, which are almost always tested
against plano reference plates that have been calibrated by independent
means, all other testplates for spherical surfaces are made in pairs to ensure
not only that the testplate has the correct radius of curvature but that the
spherical surface is also highly regular.

The precise parameters of testplates such as diameter, thickness, and
choice of material have to be defined at the design stage. This could be as
formal as making a testplate drawing or as casual as selecting them on the
basis of available testplate blanks. The radius of curvature for spherical
surfaces is a design specification that must be held well within the radius
tolerances, which must always be given by the lens designer. In the ideal
case the testplate radius is the same as the nominal design radius. This
almost never happens because of manufacturing and measurement constraints. The deviation from the nominal value, however, should not be
greater than 10% of the total radius tolerance range.

The radius tolerance for a finished lens surface is usually less than 0.3%
(0.003) of the radius. If no radius tolerance is given, it should be assumed
that it is 0.1% (0.001). This tolerance is still relatively easy to hold but it
provides a sufficiently close range to meet almost any required radius accuracy. It requires more effort, however, to hold the radius this close during
the manufacturing and testing steps, and this practice may be more expensive than it needs to be. Therefore it is always advisable to clearly specify
the radius tolerance. For very precise lens surfaces, a radius tolerance of

0.05% (0.0005) or often less is called for. In that case the fringe power call-out must not exceed the number of fringes that the radius tolerance permits.

One exeption to these general tolerance limits exists for any long radius, and especially for those that exceed 1 m (40 in.). A 0.1% radius tolerance for a 50-m (2000-in.) radius is about 50 mm (2 in.). This tolerance range represents only fractions of fringes over the typical range of diameters. For instance, it can be shown that a ±0.1% radius tolerance for a 50-mm (2-in.) diameter surface with a 50-m (2000-in.) radius means that the total available radius tolerance range is less than one fringe. Assuming that the testplate radius is exactly nominal, a high or low testplate fit of more than half fringe would bring the part radius outside the tolerance range. Obviously, different radius tolerances are required. Since there are no standards or accepted rules, every situation must be evaluated by considering the specific application and the limitations imposed by making and measuring such long radii. This requires that the designer consults the fabricator before the design is finalized.

The actual preparation of the testplate blanks proceeds usually according to established optical fabrication methods, as described in Chapter 5. These preparation steps include rounding, milling, curve generating, rough and fine-grinding, polishing of the back side, and beveling. Once the desired shape has been formed from the selected material, the final fine-grinding, polishing, and figuring can be done in one of several ways:

- By manual fine-grinding, hand polishing, and figuring which is limited to 4-in. diameters or less for practical reasons.
- By fine-grinding, polishing, and figuring mechanically using standard optical machines or annular machines for flatwork.
- By combining machine and manual methods for reducing time and optimizing control during the figuring stage.

**Plano Testplate Manufacture.** A surface with a radius of curvature that approaches infinity is called a *plane*. The word plano has been derived from this which is commonly used in practical optics to describe a flat or non-curved surface.

There are many optical components that have plano surfaces. Most notable are mirrors, windows, and prisms. Their surfaces must be checked during manufacture and for final acceptance for departure from flatness. Although most plano surfaces today are tested in a plane-wave interferometer, plano testplates are still very useful for testing surfaces right on the polishing machine.

A calibrated reference flat of sufficient diameter must be on hand for plano testplates to be measured against it. They are commercially available from a number of sources, and they range in size from 25 to 300 mm (1 to 12 in.) in diameter. Larger plano reference plates can be made on request. The aspect ratio ranges from 5 : 1 to 6 : 1 for most flats. They are available in all of the previously discussed materials. The flatness is graded as listed in Table 6.2.

Most commercially available test flats are made from fine-annealed Pyrex or from optical quality fused silica or fused quartz. They come in wooden storage boxes supplied with proper certification and unique identification. If a master flat is available, reference and working flats can be made in-house. They can be hand polished if they are less than 4 in. in diameter, but they will be much too difficult to handle if they are larger than that and must be machine polished.

The secret of successfully polishing a plano testplate, whether done on a machine or by hand, lies in a very thorough fine-grind and careful maintenance of the polishing laps. The flatness of a properly fine-ground plano testplate blank should be well within 0.5 $\mu$m (0.0005 mm) over any 100-mm diameter when measured with a Millimess tripod spherometer (see Fig. 6.37) which has been calibrated against a reference flat. The final fine-grinding cycle should be done with the finest abrasive size that is still effective to control the flatness. A 9-$\mu$m fine-grind usually does a good job. Finer abrasive are usually not very effective on typical testplate materials.

**Manual Polishing.** Hand polishing must be done with care on carefully pressed out pitch laps using only small amounts of slurries made from closely sized polishing compounds. There is no specific reason to prefer one type of compound over another as long as the compound is effective on the material and is closely sized. The hand-polishing cycles must be frequently interrupted to monitor the progress toward the goal of the required flatness. This makes hand polishing a relatively slow but potentially very accurate method. The reasons for the frequent interruptions are, besides giving the optician's hands an occasional rest, the prevention of premature overcorrection of surface flatness and the control of heat buildup from the optician's hands.

The flatness test must be preceded by a thermal stabilization period that can range from 5 to 15 minutes depending on the thermal properties of the testplate materials. Low-expansion materials require a shorter time than optical glasses such as BK 7. If no time is allowed for thermal stabilization, and flatness readings are taken while the testplate is still thermally distorted from the combined effects of polishing and thermal transfer from the hands, corrections in the polishing stroke and direction will be made for a nonexisting surface condition. It is not difficult to see that such an approach will lead the optician on a time-consuming wild goose chase. The process of short hand-polishing cycles followed by longer testing cycles continues until the required flatness has been achieved and the surface is sufficiently polished out.

The experienced optician will usually hand polish two or even three different flats in rotation to make good use of the available time and to keep the pitch laps flat and at optimum temperature. When only one flat is worked on, it is advisable to press out the polishing tool while the flat is being tested. The pitch polisher is placed with the polishing surface down on a plano press-out tool and weighted. Either a light dusting with dry compound or a thin film of slurry that is allowed to dry on the press-out tool first will act as release agent. Placing the pitch lap on a wet press-out tool should be avoided since considerable suction can be created that makes it very difficult to remove the lap without damage.

Larger plano testplates, ranging in diameter from 100 to 300 mm (4 to 12 in.), must be machine polished by using a carefully controlled process that is based on generally accepted fine-grinding and polishing methods. As is true for hand polishing, any brute force approach must be avoided. A true flat surface can only be established through a series of small incremental changes that follow a carefully controlled test sequence. One of the key requirements in any high precision surfacing is the preparation of the surfaces. Milling (Blanchard grinding) and rough grinding with loose abrasives should be avoided if possible. If these operations must be performed, they must be followed by an extensive multistage fine-grinding cycle to completely neutralize the often significant surface stresses that are introduced by these roughing operations.

When using a Millimess tripod or ring spherometer, as described in Section 6.3, that has been calibrated first against a reference flat, the final fine-ground surface should measure no more than 0.5 $\mu$m over a 100-mm diameter. The surface should ideally be concave so that a plano polisher will contact the flat first around the edge. This condition corresponds to an optical flatness of less than 2 fringes over any 100-mm diameter when the parts are tested in mercury green light (0.5461 $\mu$m).

The polishing tools must be carefully pressed out prior to use. This is necessary to avoid an improper match between the polisher and the flat during the initial stages of the polishing cycle and thus the formation of surface irregularities that are difficult to correct. It is desirable to have two or even three 300-mm (12-in.) diameter pressing tools made of a low-expansion material. The surface need not be polished but an inspection polish is usually applied so that the surface flatness can be tested. One of these should be convex by about 2 fringes over the 300-mm (12-in.) diameter, one must be flat within 1/8 wave, and the optional third pressing tool should be 2 fringes concave. By using these tools, the polishers can be conditioned to closely fit the particular testplate surface as was revealed by the last thermally stabilized test.

For flats up to 150 mm (6 in.) in diameter, the testplate usually runs on top of the polisher. For those that exceed that size, it is usually better to run them on the spindle with the polisher running on top. These are common-sense guidelines only. Other arrangements are possible, and there is quite often a great deal of variability from shop to shop on how to best run this type of work. The basic considerations are the size and weight of the flat and the required size of the polisher. These considerations determine not only the ease of handling but the control of the figuring process as well. The height of the pin above the polisher could become too great if a large flat runs on top because larger diameter flats are also thicker. The pin height should be kept to a practical minimum, since if it is too great, undesirable instabilities are set up that make figure contol during machine polishing very difficult.

As was already described for hand polishing, the machine-polishing cycles must be kept relatively short. This is especially important during the final figuring stages. Thermal buildup in the flats can distort the surface figure during polishing and testing. This type of distortion, coupled with accelerated testing resulting from impatience, can easily result in an uncontrolled process.

Commercially available plano test flats are usually not fabricated with the methods discussed so far. The most commonly used process utilizes the annular pitch polishing technique which, when properly controlled, can yield plano surfaces of very high precision. Figure 6.6 shows such an arrangement. Since the final figuring is essentially done by the machine, even operators with limited skills can produce very flat surfaces on large as well as on difficult-to-polish shapes.

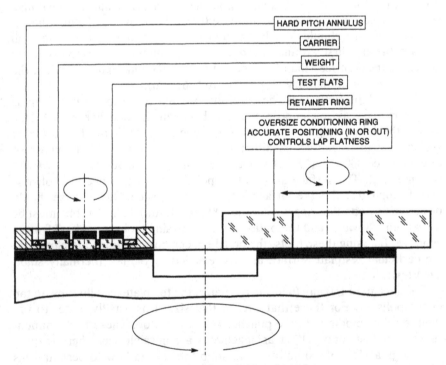

**Figure 6.6.** Annular polishing of test flats.

Despite a considerable investment in equipment cost, the annular pitch polishing method has been proved to produce flats of reference quality more economically than can be done by any other method. To amortize the high cost of the equipment and to control the flatness of the lap, a continuous supply of flat parts must be polished on such a machine. The key to success in operating this type of machine lies in the close thermal and positional control of the lap. Rotational synchronism between lap and parts also play a major role. Annular polishers are discussed in greater detail in Chapter 5. In brief, the plano testplates are individually weighted down, and they float on the pitch annulus which is kept flat by the accurate positioning of the conditioning ring. The testplates are laterally constrained by a carrier, called a *septum,* which is allowed to rotate freely with the retaining ring. The entire annulus rotates slowly so that a planetary type motion is created for the flats. The polishing action is provided by commonly used slurries that are supplied by a closed-loop recirculating slurry system.

When estimating the cost of fabricating large flats, the actual costs often exceed the estimated values. This is especially true for the larger flats.The cost is a function of the diameter of the flats which varies linearly with diameter for parts up to 200 mm (8 in.) and with the square of the diameter for parts larger than that. Therefore the fabrication costs for a 150-mm (6-in.) flat is twice that of a 75-mm (3-in.) flat. But a 300-mm (12-in.) diameter flat costs four times more to make than a 150-mm (6-in.) flat.

With the often-mentioned three-flat method of making plano testplates, three plano plates are polished and tested in such a manner that they alternately fit one another, no matter which two plates are put in contact. It is an interesting option that can have a practical value when a calibrated master flat is not available. But from a practical optics manufacturing standpoint the economic advantages of this method are by no means obvious.

**Radius Testplate Manufacture.** The radius requirements for lenses can assume any value between near zero and infinity. They can be either concave or convex. In addition the diameter can vary from less than 12.5 mm (0.5 in.) to 400 mm (8 in.) or more. Given this variability, it becomes quite clear why radius testplates are not commercially available. Therefore each shop must fabricate its own radius testplates to suit each specific need. As a result independent radius standards do not exist, and the quality of radius testplates ranges from superior for a few professionally run shops, to marginal for most others, and to decidedly poor for a number of commercial quality shops. The detailed description in this section is designed to convey long-established methods that have proved to result in master-quality radius testplates.

An optical shop that values quality will always attempt to produce radius testplates of master or at least reference quality. It is always good practice to fabricate a set of highly precise reference plates first for each radius against which the working testplates are then calibrated.Because of the significant cost involved, however, this desirable approach is usually limited to high-volume applications, and then only when the lead time permits this. If a shop fabricates a lens system that is sold as a product based on a proprietary lens design, it is foolish for that shop not to invest the time to produce a complete set of master, reference, and working testplates for each radius. The cost of doing this at the onset of the preproduction phase is insignificant in comparison to the cost of a product failure due to inconsistent quality and performance. For prototype or low-volume work there is usually time to make only one set of testplates for each required radius. This set should be of reference quality because it will be used for final buy-off and serve as working plate as well.

The manufacture of radius testplates follows in a general sense the pattern already discussed for plano testplates. There are, however, unique differences. Since a spherical testplate must not only be faultlessly spherical but must also have the correct radius of curvature, it is necessary to always fabricate a set of radius testplates that is made up of a matching convex and a concave plate to permit the testing of sphericity interferometrically.

Spherometer readings are used to establish the exact radius values, unlike for plano plates where the spherometer was used only as a comparative reference. Spherometers are described in greater detail in Section 6.3. Since the surfaces are spherical, different polishing techniques are used, and there are no press-out tools to help keep the polishers properly conditioned. Hand polishing is often preferred for small testplates up to 75 mm (3 in.) in diameter. It is done on a pedal, or motor bench, or on a regular polishing machine spindle. Testplates larger than that are generally machine polished following accepted methods.

Round disks of the selected optical material, usually fused silica or Pyrex, are ground flat on one side and polished to a commercial finish so that they become transparent on that side. The polished surface must then be protected by tape, paint, or wax during the grinding of the radius. The curve can be rough ground by hand for long radii and relatively small diameter testplate blanks. For these, the radius to diameter ratio exceeds 1.5. For short radii or for ratios less than 1.5, a curve-generating step is advisable unless the diameter is fairly small. The chosen method will not appreciably affect the quality of the finished product, though it will have an impact on the time involved and ultimately on the cost of the testplate. Therefore it is up to the optician to judge each requirement on its own merits.

The following is a description of the grinding and polishing process for making radius testplates that was translated from a German book on optics fabrication [22]. "The fine-grinding laps must be checked and selected with a radius template. The set of laps must be prepared in such a manner that when the glasses are rubbed together after grinding, they should contact lightly at the edge." This condition is shown in Fig. 6.7. "Small plano blockers are blocked against the plano side of the radiused testplate blanks so that they can be fastened to the grinding spindle. One of the blanks,

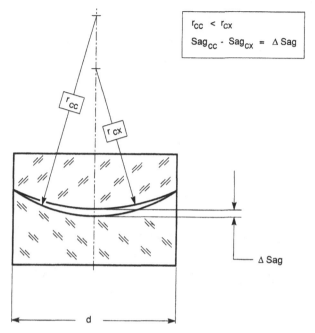

**Figure 6.7.**   Preferred match of a fine-ground testplate pair.

usually the convex one, is mounted on the spindle, and the other is ground against it. One grinds this way glass against glass until the two surfaces have conformed to each other. The abrasive used for this must not be too coarse because it would otherwise attack the outer zone of the glasses, making the radius too short. After the surfaces have been finely ground, one has to measure the previously calculated sag with a spherometer.'' The construction and use of spherometers have been described in detail in this chapter.

**Measuring Testplate Pairs.** Spherometers are used to measure the sag on testplate pairs. The testplate sets are worked in such a manner that they show on the spherometer a previously calculated sag sum that verifies whether the plates have the desired radius of curvature (see Fig. 6.8). As compared to other techniques, the spherometer can also be used to measure the curvature of nonreflecting, dull surfaces. Therefore accurate spherometer readings can be taken during the grinding stages.

The testplate pair consists of a convex and a concave glass. When the plates are completely polished, the two surfaces fit to a white color, which means that there are no interference fringes. To accomplish this fit, the glasses are worked so that they yield the precalculated sum of the two sags when measured on the spherometer. Equation (6.20) must first be solved for

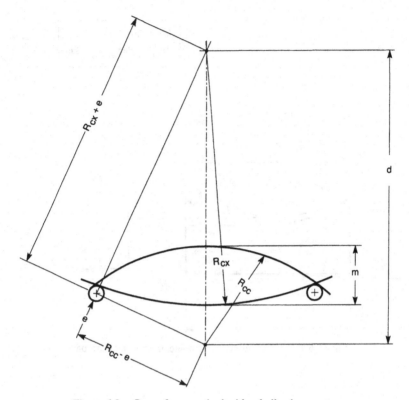

**Figure 6.8.** Sum of sag method with a ball spherometer.

the sag $s$ to make this calculation:

$$S = R - \sqrt{R^2 - \frac{D^2}{4}}. \tag{6.20}$$

Following the conventions described in the section on ring spherometers, $D_0$ is the outer ring diameter which is used for concave surfaces, while for convex surfaces $D_i$ is the inner diameter. This distinction need not be made for ball spherometers.

The two calculated sag heights are added, and the testplates are worked in such a way that they measure this sag height sum when measured on the spherometer. The concave and convex radius plates are alternately placed on the spherometer during these measurements. A calibration measurement against a plano testplate is not required with this technique. The resulting measurement is more accurate than the previously described determination of the radius of curvature of a single spherical surface.

The sag height sum method for making radius testplates can also be used with a ball spherometer. For that purpose Eqs. (6.21) and (6.23) must be solved first for the sag height $s$, which results in Eqs. (6.22) and (6.24), respectively:

$$R_{cx} = \frac{a^2}{2s} + \frac{s}{2} - e, \tag{6.21}$$

$$s_{cx} = (R + e) - \sqrt{[(R + e)^2 - a^2]}; \tag{6.22}$$

$$R_{cv} = \frac{a^2}{2s} + \frac{s}{2} + e, \tag{6.23}$$

$$s_{cv} = (R - e) - \sqrt{[(R - e)^2 - a^2]}. \tag{6.24}$$

Both sag heights are added (Fig. 6.8) to yield the combined sag height sum $m$:

$$m = s_{cx} + s_{cv} \tag{6.25}$$

$$m = 2R - \sqrt{(R + e)^2 - a^2} - \sqrt{(R - e)^2 - a^2}, \tag{6.26}$$

where

$m$ = sag height sum,
$R$ = radius of curvature of the spherical testplate pair,
$e$ = ball radius,
$a$ = radius of the circle on which the three balls are located.

The accuracy of radius testplates increases with their diameter. The highest accuracy is obtained with the largest spherometer ball circle or testplate diameter. The relative error in the measurement [11] is

$$\frac{R}{R_0} = \frac{N\lambda R_0}{d^2/4}, \tag{6.27}$$

where

$R$ = measured radius,
$R_0$ = nominal radius,
$N$ = order of fringes,
$\lambda$ = wavelength of source,
$d$ = testplate diameter.

All dimensions are either in millimeters or in inches. The test is possible with residual errors as low as 1 in 50,000, and this requires a temperature stabilization of both parts of 1°C or less.

**Measurement of Very Small and Very Large Testplate Radii.** For testplate sets with radii of less than 25 mm, the convex plate is typically ground and polished into a hyperhemisphere. The diameter of such a testplate can be easily measured to an accuracy of 1 μm (0.00004 in.). This means that the testplate radius can be measured to an accuracy of 0.5 μm (0.00002 in.). Even a very small testplate with a 1.0-mm (0.040-in.) radius can be measured with a suitable dial indicator and gauge blocks to an accuracy of 0.5 pro mille (0/000) or 0.0005.

Radii larger than one meter are measured on an optical bench using the autostigmatic method described in section on spherometry (Fig. 6.41). Since in this case, the accuracy of 0.5 (0/000) translates into an error of 0.5 mm, it is sufficient when the scale on the optical bench can be read with a vernier to an accuracy of 0.1 mm.

The raw glass disks are ground plano on one side and polished to make them transparent. These polished sides are then protected by a lacquer or shellac. The required curvatures are ground in with coarse abrasives on the opposite sides. Specially prepared radius grinding tools are used for this purpose; they have been verified with appropriate radius gauges. During the final stages of the rough grinding, the testplate is ground over the edge of the tool. This causes the tool to contact the testplate surface around the edge for the next finer grind. It is then easier to grind with the same tool using finer abrasives if the testplate is later guided more over the center of the tool. After this pre-fine grind, the testplate curvature is measured for the first time on the spherometer. It will not yet show the previously calculated sum of the sags. However, if the difference between the sag sum and the readings is less than 0.005 mm, then the final correction can be made by polishing. If the differences are greater than that, then one or both of the grinding tools must be corrected and the fine-grinding step must be repeated.

It was customary in the past to grind testplate pairs against each other. When the radius of the pair was too long, the convex plate was mounted on the machine spindle and fine-ground with the concave plate running on top. Both surfaces became more strongly curved, however slowly. When the curvature had to be flatter (i.e., less curved), the position of the plates was reversed. When the testplates, which are fine-ground in this manner, are rubbed together, they contact in the center. This condition requires that the surfaces be polished slightly from the center upon completion of the fine-grinding. However, a new testplate set must typically be made to test a new type of lens. It is advantageous, in that case, to have a well-fitting and a well-ground-in fine-grinding tool available for the prototype lenses. It is primarily for this reason that glass to glass fine-grinding is not done very often anymore.

The properly fine-ground testplates are then polished on pitch polishers to make them fit to one another. Since surface regularity is more important for testplates than surface quality, a harder pitch is used for polishing testplates than is customary for the same type of surface on lenses.

When the glasses are polished and fit well to each other, the testplate set is again measured on the spherometer. The sag reading that is necessary for the exact radius of curvature will have changed by a few microns and a correction may have to be made. When the sag is too large, the convex surface is polished from the center and the concave surface from the edge. The order must be reversed when the sag difference is too small. To avoid too many unnecessary spherometer measurements during this correction phase, the white light interference fringes of the testplate match must be carefully monitored. We know from the previous discussion on white light testplating, that four blue or three red interference rings correspond to a sag difference of 0.001 mm (1 $\mu$m). For instance, when the testplate pair during the radius correction phase measures 0.002 mm too much sag, then the convex plate is polished four blue rings low and the concave plate is polished four blue rings high to fit.

The working testplates used in the shops are frequently made from hard fused quartz or fused silica. They typically have the diameter of the lenses that are to be tested with the testplates. Their curvature is polished relative to the standard set or, occasionally, to the master testplate set. The quartz plates must be fine-ground accurately enough during their manufacture so that they fit "white" after the initial prepolishing because it is quite difficult to polish the hard quartz a few rings high or low.

The finished radius testplates are inscribed with the value of the measured radius on the polished backside using a diamond stylus. It is customary to incribe the radius value to two and sometimes three decimals. However, according to some error analyses, it would be more meaningful to incribe a row of five significant digits, such as $r = 33.333$ mm or $r = 333.33$ mm.

### 6.1.5. Using Testplates

The optician prefers to measure the shape of the surfaces with testplates in daylight. Bothersome reflections are eliminated when the light passes through a ground glass diffuser. For very accurate testplate matches, conclusions can still be drawn from the transition from one color to another. The testing in monochromatic light is less sensitive but quite sufficient for testplate matches down to two rings.

As already mentioned earlier, testplates can be used in one of several modes, depending on the size of the test part or testplate, the nature of the test, and the required accuracy. The most rudimentary testplating is performed when lenses are tested on the block during the polishing stage to test for sphericity and power match using a radius testplate. The same of course applies to checking the flatness of plano components on the block, although this test is usually performed in most shops today with a plano shop interferometer (Fig. 6.17).

The customary procedure for lens blocks that can be easily handled is to remove one of the blocks from the machine after completion of a polishing run. The block is held in the left hand, and the slurry is wiped off a selected lens, usually from the outside row of the block, with the palm of the right hand. The slurry is transferred by a reflex motion from the hand to a towel which is slung over the right shoulder of the optician. The experienced optician is no longer conscious of this motion. The matching testplate is then picked up with the right hand, and with one smooth rotary motion the spherical contact surface is wiped clean with the palm of the left hand. In doing this, the optician has to be careful not to drop the lens block or chip the testplate by allowing it to strike the block. The cleaned testplate is then placed carefully in contact with the cleaned lens, and the optician observes the interference fringes that form.

The testing can be done either in white light or with a portable monochromatic source, as shown in Fig. 6.9. This description of the process is obviously from the perspective of a right-handed person. Lefties may have to reverse the order of their hands in performing this task, but there is no reason to believe that testplating done by a left-handed person is any different than if it were done by a right-handed person.

By the very nature of this type of contact interferometry, the cleanliness of the work environment is of utmost importance to prevent unnecessary

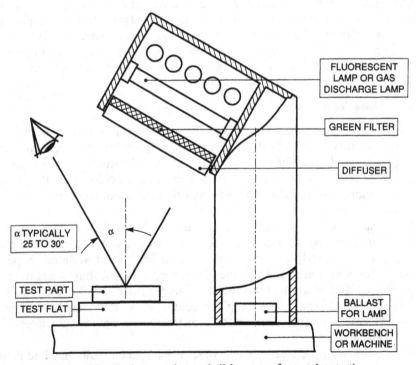

**Figure 6.9.** Basic monochromatic light source for testplate testing.

downgrading of the surfaces of the test part or the testplates. A good optician will not only keep the machine clean but also the hands by frequent washing and by changing the work coat when it has becomed soiled. By taking these basic precautions (which should always be observed whether there is testplating or not) in conjunction with good testplating techniques, downgrading of polished surfaces can be entirely prevented.

With some practice the average optician will have reduced the entire procedure to a series of automatic moves and a meaningful test is obtained in less than one minute every time. Experienced opticians can do it in less than half the time. Despite the availability of sophisticated shop interferometers which can test spherical surfaces to a high degree of accuracy without the use of testplates, most optical shops that produce lens optics still depend on testplate testing as the only practical in-process testing method. There are several reasons for this. On-the-block testing in the interferometer is very difficult and requires complicated manipulating stages. But even if that were not a problem, the easy availability of the testplate where it is needed (at the machine) and the speed with which the test can be performed make it very hard for the interferometer to compete, even though the manufacture of testplates represents a significant investment.

For larger spherical and plano blocks the procedure is changed in that the blocks that usually run on the bottom remain on the machine. Again, one surface is cleaned, the matching testplate is also cleaned and placed in contact with the test part. Since the surfaces cannot be tilted to yield the best reflection, it is advisable to take advantage of a portable monochromatic test light that can be easily adjusted to provide optimum illumination.

This type of on-the-block testing is in most cases adequate to yield fringe readings of sufficient accuracy since the fringe call-out seldom requires a power match of less than a couple of fringes. If a better accuracy is required, however, the previously described test processes cannot guarantee reliable results because distortions caused by thermal effects and by microscopic particles trapped between the surfaces often make a precise reading impossible. Under these conditions more refined testplating methods must be used.

Since very tight fringe call-out for spherical surfaces, such as 1/4 wave or less, usually precludes block polishing, they are polished as single surfaces and they must be tested accordingly. The casual wiping away of the slurry by hand is not acceptable for testing such lenses. The lens must be removed from the machine and thoroughly cleaned and dried. The lens, preferrably still mounted, is then placed on the platform of a test station. The matching testplate, which must be somewhat larger than the lens, must then be carefully cleaned with alcohol and a suitable wipe. The lens surface and the testplate are then gently brushed with a clean camel's hair brush to remove loose dust particles and lint, and the surfaces are carefully brought into contact. This operation can never be forced, and it may have to be repeated a number of times until the fringes move freely. This indicates that there are no particles trapped between the surfaces. The testplate should "swim" on the lens surface on an airfilm without getting hung up.

When testing convex surfaces, the testplate must be constrained to keep it from sliding off. No such precautions are required for concave surfaces. Because of the more complex and time-consuming nature of this type of test, shop interferometers can be used here quite competively. A special fixture may have to be made to support the blocked lens. This could be difficult for large and heavy optics, especially if the lens has to be held in a vertical position. A down-looking interferometer arrangement will make the testing of large lenses much easier.

Particular care must be exercised when placing reference or master-quality testplate sets in contact for the purpose of verifying the match. Both surfaces must be thoroughly and carefully cleaned with a fast evaporating solvent such as alcohol or acetone. Acetone is usually better because alcohol tends to promote moisture condensation on the surfaces and thus can cause the surfaces to stick together tenaciously when pressed together. The concave surface is placed on the bottom, while the convex surface is brushed with a clean camel's hair brush. The convex surface is held with the left hand facing down, while the concave surface is gently brushed in a similar fashion. When both surfaces are clean, they are brought into contact. The convex part should be allowed to float into the equilibrium position by letting go while it is still somewhat off center. If the convex plate settles down in a smooth but rapidly decaying oscillatory motion without getting hung up, the testplate set can be allowed to thermally stabilize for an accurate test. If the convex plate does not oscillate freely, the testplates must be separated, and the entire sequence must be repeated.

The techniques for plano surfaces are essentially the same, except that parts of a high degree of flatness can be polished on a multiple part block and also tested while still on the machine. There may be a variety of reasons why that is not desirable, but under the right conditions it can yield accurate results. The interferometers can also play an important role, and they often outperform plano testplate testing.

As is true with every production process, there are many different solutions possible, and much depends on the requirements, the available equipment, and to a large degree the preferences of the opticians. The testplate matches are specified according to the number of interference rings allowed. When the testplate rests on the edge and three rings show across the surface, the optician calls this "three rings low." When the testplate contacts at the center, it is called "three rings high."

For a low testplate match, the rings will run toward the center when light pressure is applied. When this occurs and only a blue spot remains in the center of the lens, this is called "blue dot". Colored fringes can only be detected when a white light source is used. When the blue dot also vanishes and a nearly colorless surface with a pale yellow spot in the center is seen, the match is called "white." The perfect match between the two surfaces is achieved when the yellow spot also disappears and only a uniform shiny grey surface remains.

It is also determined with the testplate if a surface is truly spherical or plano or if it deviates from the ideal shape. As was already discussed, surface irregularities are more detrimental for the image formation than simple radius deviations. When lenses are worked as single surfaces, such deviations are usually rotationally symmetrical. The surface errors that occur on lens surfaces that were polished on blocks are much more damaging. This is especially true on spot blocks. In this case a rolled edge on the block can deform the surfaces asymmetrically on all lenses in the outer row.

Several types of irregularities are recognized. They are easily seen when the testplate is lifted slightly and gently pressed down again. This causes the interference fringes to move, and one can see if they advance more at one spot or lag behind at another. The commonly encountered irregularities in testplate matches for plano surfaces are shown in Figs. 6.10*a* through 6.10*d*. Under these conditions the interference fringes should run in the direction of the arrows.

### 6.1.6. Light Sources

The most readily available light sources are the sun, ambient daylight, and fluorescent lamps. The spectral makeup of the latter closely resembles that of the sun. This type of light is composed of all the wavelength of the visible spectrum, so it is called *white light*, since the mix of all colors produces white. As a result interference fringes observed in white light are composed of the cardinal colors of the visible spectrum. Although the fringe definition in white light illumination appears poor at first glance, the sequence of certain color fringes can provide the experienced master optician with surface data that are much more precise than those obtainable with the standard dark and light fringes produced by monochromatic illumination. If, however, fringe definition is more important, then the light of a single wavelength (monochromatic light) must be used. This results in dark fringes against a lighter background, which appears in the color of the monochromatic source used.

A variety of light sources and viewers, which are typically called *test stations*, have been devised. They not only emit the desired monochromatic light but also permit the fringes to be viewed at nearly perpendicular incidence. This arrangement reduces any fringe distortion to a minimum. Most of these light sources are called *extended sources* because the light is emitted over an extended area like that emitted by a fluorescent lamp. By contrast, light that originates from a single point is called a *point source*. These are not commonly used in test stations, although some test arrangements may incorporate this type of source.

Gas discharge lamps are sources of monochromatic light. They are charged with one or two basic elements that are vaporized if they are not

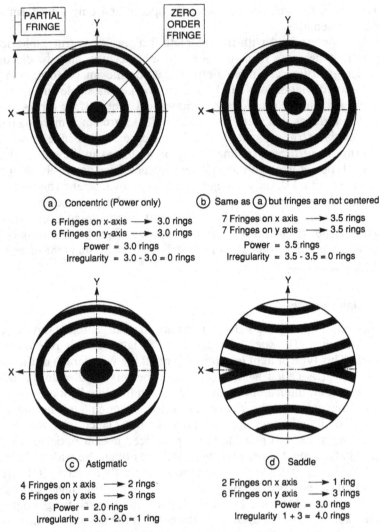

**Figure 6.10.** Basic fringe pattern with closed fringes or rings.

already gaseous so that they can be excited electrically. The electrical discharge excites the individual atoms of these elements. The elements then radiate light at certain discrete wavelengths determined by their atomic characteristics. The commonly used monochromatic light sources for testplate testing are listed in Table 6.4.

Sodium D light is a lively orange color. It was the most commonly used source of monochromatic light a few decades ago. But because the lamp was rather small, it did not serve well as an extended source for testing larger surfaces. It did, however, do quite well as a source for on-the-block testing of relatively small lenses which are less than 50 mm (2 in.) in diameter right at the machine because of its compactness and portability.

Mercury light became more common as a monochromatic light source. The characteristic wavelength of 0.5461 $\mu$m became the standard wavelength of measurement prior to the widespread application of helium-neon (HeNe) based interferometers. It is a subdued green color that can be extracted from long wave black light (UV) and from standard fluorescent white light lamps with the use of special green filters. Specially designed extended mercury green sources have also been developed. This wavelength is preferred when hardcopy interferograms need to be taken with Polaroid film since the emulsion is most responsive in the green region of the spectrum. Helium light sources have also found widespread acceptance because it is the least expensive extended source that can be found in most if not all commercially available test stations.

Most of these monochromatic sources play only a limited role today. So much of the testing is now done with the highly versatile and sophisticated interferometers that use a helium-neon laser as a light source. The light of the He-Ne laser has a wavelength of 0.6328 $\mu$m (24.9 $\mu$in.). But the long coherence length of the laser light and its high intensity can cause disturbing reflections and spurious fringes that make testplate testing nearly impossible, to say nothing about the uncontrolled reflections of laser light in the shop. Consequently lasers make poor monochromatic light sources for testplate testing, and the old sources still play a useful role in the modern optical shop. Only when modern interferometers can check lenses on the block right at the polishing machine as quickly and easily as is done with testplates, can we discontinue using testplate testing as an in-process control. Only then can we discard the old test stations.

There is usually considerable confusion when talking about the various wavelengths of light. This is especially true among those in the business who do not usually deal with these values on a daily basis. The wavelength can be expressed in terms of Ångstroms (Å), millimicrons (m$\mu$), nanometers (nm), microns ($\mu$m), and finally micro-inch ($\mu$in.). For clarification of the various measures and their relationship to each other, refer to Table 6.5:

Monochromatic light sources are commercially available. Many of these are small portable units that use helium sources [23, 24]. One other manufac-

**Table 6.4.  Primary Monochromatic Extended Light Sources**

| Source | Color | Wavelengths in µm | in µin |
|--------|-------|-------------------|--------|
| Sodium D | Bright orange | 0.5893 | 23.2 |
| Mercury | Dull green | 0.5461 | 21.5 |
| Helium | Pale orange | 0.5876 | 23.1 |

**Table 6.5.   Conversion Factors for Wavelengths.**

| | | | |
|---|---|---|---|
| 1 mm | = $10^{-1}$ cm | = $10^{-3}$ m | = $3.937 \times 10^{-2}$ in |
| 1 µm | = $10^{-4}$ cm | = $10^{-3}$ mm | = $3.937 \times 10^{-5}$ in |
| 1 nm | = $10^{-6}$ mm | = $10^{-3}$ µm | = $3.937 \times 10^{-8}$ in |
| 1 mµ | = $10^{-6}$ mm | = $10^{-3}$ µm | = $3.937 \times 10^{-8}$ in |
| 1 Å | = $10^{-7}$ mm | = $10^{-1}$ nm | = $3.937 \times 10^{-9}$ in |
| Note:  1 mµ  =  1 nm !! | | | |

turer [25] produces a monochromatic source that uses two clear (uncoated) fluorescent lamps. The light from these lamps is filtered by a 546.1-nm green filter, and it is diffused by an opal glass diffuser. The light is directed downward at an angle of about 25° to provide near perpendicular viewing of the fringes.

**Test Stations for Testplate Testing.** In its simplest form a test station is merely a suitable sunlit spot where testplates can be used to observe white light fringes. When the testplate is held at the proper angle, where the light can be viewed against a neutral background such as the reflection of the sky or of a white wall, strong fringes of vivid colors can be seen. Such a test can yield valuable and accurate information about the condition of the test surface. Many master opticians prefer this type of interferometric test, especially when making testplates. They attribute a greater sensitivity to white light testing as compared to standard monochromatic testing whether it be with testplates or with an interferometer. Figure 6.11 shows a typical test arrangement [6].

A similar arrangement can be used for viewing white light fringes in the light of a normal fluorescent lamp. Unless the light is well diffused by an opal glass or a similar type of diffuser, a narrow strip of light that is the reflected image of the fluorescent tube will produce dull colored fringes, and no fringes can be seen outside the strip. The fringe definition can be somewhat improved by using a so-called blacklight tube which is technically defined as a long wave UV lamp. But since the UV tube emits light over a fairly broad range, fringes of several colors are seen that are not easily interpreted. These facts make standard fluorescent lamps not very suitable for meaningful tests of medium to large surfaces, although it may be an adequate way for in-process testing of small lenses or similar parts on the block. In addition fluorescent lamps do not have the necessary intensity, especially when dif-

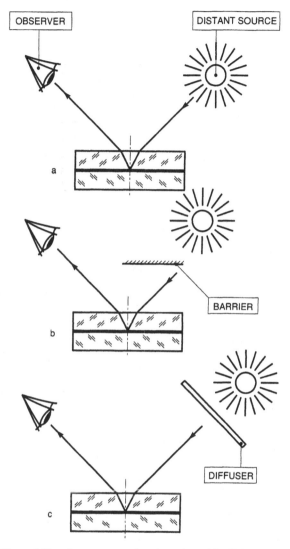

**Figure 6.11.** Arrangements for observing white light fringes.

fused, to produce fringes of sufficient definition and strength to use them for anything other than a quick in-process reference.

The test arrangements discussed so far force the observer to view the fringes at oblique incidence. As will be discussed later, oblique incidence viewing of fringes can under certain conditions cause considerable distortion of the actual fringes. This can lead to errors when figuring surfaces to very close tolerances. For most production type work where three to four fringes

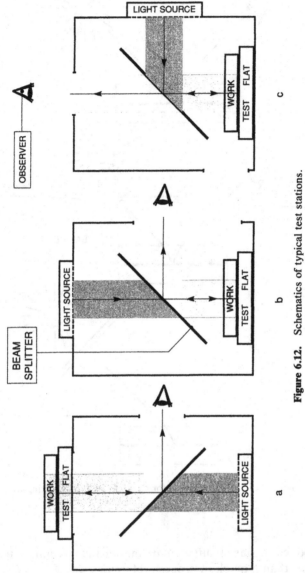

**Figure 6.12.** Schematics of typical test stations.

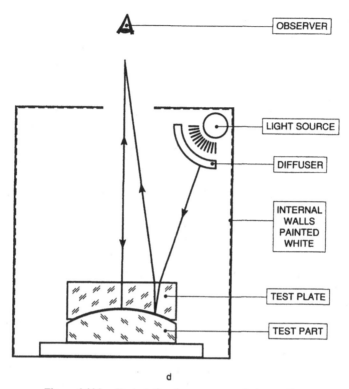

**Figure 6.12d.** Test station for near normal observation.

are acceptable, this type of distortion is negligible. Fringes must be viewed at or very close to perpendicular incidence for any close tolerance work. This condition can be created by designing test stations that allow for viewing fringes in this preferred attitude.

There are a number of commercially available monochromatic test lights that still require viewing of fringes at several degrees. One such light is shown in Fig. 6.9 [23]. It consists of a helium gas discharge tube, a power supply for the tube, and a ground glass diffuser. It is a very compact unit, which makes it easily portable. The portability makes it particularly attractive as an in-process test light that can be used right at the machine where it is needed. These lights are designed for the measurement of flatness of small optical or metallographic surfaces where fringe distortion caused by oblique incidence is not significant.

Other, more commonly used arrangements for testing larger surfaces at perpendicular incidence are shown in Figs. 6.12a, b, c, and d. In Fig. 6.12a the light source is at the bottom of the test station, and the testplate and workpiece directly above it. Between the source and the testplate is a 45° beam splitter that permits the viewing of the fringes horizontally. In this fashion the reflected image of the fringe pattern can be viewed at perpendicular incidence. While this arrangement offers some conveniences in manipulating the test and in viewing the fringes, it also has a serious drawback in that the heat emitted from the lamp can cause thermal distortion of the fringes. This, however, can be easily remedied by providing for adequate ventilation to carry the heat away.

Figure 6.12b shows a more common arrangement, whereby the source is on top and the testplate is on the bottom. The beam splitter remains as in Fig. 6.12a, and the fringes are viewed perpendicularly, although the image is viewed from the side. Because the light is on top of the test station, proper ventilation can be provided by merely having holes drilled in the housing that holds the lamp. The ballast transformer for the gas discharge lamp can be mounted on top totally external to the test station so that any heat generated can be carried away by room air flow.

Another variant of this basic scheme is shown in Fig. 6.12c. Here the position of the observer and the light source are interchanged as compared to the arrangement in b. The thermal effects of this arrangement can also be reduced or even eliminated by adaquate ventilation.

These three preceding test station designs have one common drawback. Since the fringes are viewed indirectly by observing their image as reflected by the beam splitter, the viewing distance generally exceeds the normal viewing distance of 250 mm (10 in.). This can give rise to some uncertainties when very small fringe variations must be judged. The accuracy of the fringe interpretation is further affected by the optical quality of the beam splitter when it acts as a reflector. Although this effect is usually small for most routine work, it can cause some added uncertainties when very close fringe tolerances are required.

While designs shown in Fig. 6.12b and Fig. 6.12c can be used for testing the sphericity of lenses, the test station shown in Fig. 6.12a is used for plano testing only. There is usually no great difficulty in observing fringes of curved surfaces as long as they are not too strongly curved. The test station shown in Fig. 6.12d is a variation of design c, with the light source relocated and the beam splitter removed. It has proved to be useful for testing lenses with strongly curved surfaces. The white reflecting walls of the test station create a more evenly distributed light, which then makes it easier to view the fringes right to the edge of the parts. However, strongly curved surfaces can add considerable fringe distortion that this test station cannot correct for. Aplanatic backs on the testplates, as already discussed in this section, can eliminate most if not all of the fringe distortion.

A special test station design [26] is shown in Fig. 6.13 . It is designed for photographing interference fringes when testplating large convex surfaces. This arrangement has been referred to as the *Fizeau-Fresnel tester*, since it is based on the Fizeau interferometer and utilizes Fresnel lenses. The light source, usually a UV penlight point source, is located at the focal length of a long focal point Fresnel lens. This collimates the light over a diameter that is nearly as large as the diameter of the Fresnel lens. Therefore the Fresnel

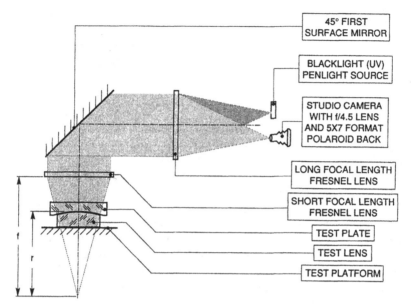

**Figure 6.13.** Fizeau-Fresnel tester for convex surfaces.

lens diameter must be chosen so that the lens surface under test is fully illuminated. The collimated beam is then directed downward by an aluminized first-surface mirror that is inclined at 45°. The length of the mirror must be 1.4 times the diameter of the Fresnel lens and must be at least as wide as the lens diameter to prevent any obscuration of the collimated beam. Beneath the folding mirror but above the lens under test is another Fresnel lens with the same diameter but with a shorter focal length. The function of this lens is to focus the collimated beam.

The similarity of this test arrangement to the modern Fizeau-type shop interferometers should be pointed out here. The penlight source has been replaced in the interferometer by the HeNe laser. The HeNe beam is expanded and collimated, just as the light from the penlight source is collimated by the Fresnel lens. While the interferometer does not normally need a folding mirror, such arrangements are sometimes used. The Fresnel lens that focuses the collimated light plays the same role as the transmission sphere of a modern interferometer.

When the lens under test is introduced at the proper position such that the converging rays strike the convex lens surface normally, the light is reflected back into the system along its original path. This can occur only when the focal point of the lower Fresnel lens and the center of curvature of the test surface coincide. The reflected rays carry the image of the test surface, and since interference fringes appear with a testplate in contact, they too are projected back into the camera lens which is located in the plane of the penlight. Since the light emitted by the UV penlight is not sufficiently mono-chromatic, a green isolation filter is required in front of the camera lens to isolate the strong mercury line at 0.5461 $\mu$m (546.1 nm). High-contrast black-and-white fringe interferograms can be recorded on Polaroid film.

### 6.1.7.  Fringe Distortion

The theory and practice of testplate testing discussed so far assumed ideal conditions, but the real world is not quite that simple. The ideal fringes are usually distorted to some greater or lesser degree by a variety of external influences. Such influences can be actual surface irregularities, but they can also be foreign particles trapped between the surfaces, weight effects, thermal effects, material inhomogeneities in the testplate through which the fringes are viewed, and optical effects caused by viewing the fringes at oblique angles. When the test specification permits up to two or three fringes, these effects play a negligible role, and they can be ignored. When, however, sphericity or flatness in fractional waves are required, the combined effects of these distortions can easily overwhelm the information in the fringes on the actual condition of the surface shape.

Surface irregularities are not really considered distortions since the determination of their nature and magnitude is the reason for the testplate measurement. Distortions caused by foreign particles trapped between the surfaces can be eliminated by following accepted testplating practice, as described in preceding paragraphs. The effects of heavy testplates can be minimized by making sure that the surfaces are clean and that the testpart is properly supported. Distortions caused by testplate material inhomogeneity can be greatly reduced by selecting a good-quality material when the test-plates are used in transmission. Oblique incidence viewing of fringes is all but eliminated when using test stations designed to ensure perpendicular line of sight. This leaves only thermal effects in need of further explanation.

**Fringe Distortion Caused by Thermal Effects.** When a lens is freshly polished, or when a testplate has been extensively handled, or when either or both have been subjected to an unevenly distributed external heat source, changes can take place in their shape that then show up as an often significant distortion of the fringe pattern when they are tested. A thermal stabilization period is usually required before a meaningful test can be performed. This period varies as a function of the mass and type of the material of the lens and testplate and by how much the component's temperature varies with respect to the ambient temperature of the testing environment. Thermal stabilization periods may vary from a few minutes for small lenses that have to meet a fringe call-out of two to three rings to several hours for large lenses of very high precision.

The effects of thermal conditions on fringe distortion can be calculated, at least in theory, but there is usually insufficient information to obtain accurate results. The simplest way to look at thermal effects is to assume that one component, such as the testplate, is at one uniform temperature, while the other, in this case the lens, is at another uniform temperature. For this condition the following equation holds true [5]:

$$\Delta s = (\alpha_1 - \alpha_t)(T_1 - T_t)s, \tag{6.28}$$

where

$\alpha_1$ = coefficient of thermal expansion of lens,
$\alpha_t$ = coefficient of thermal expansion of testplate,
$T_1$ = uniform temperature of lens,
$T_t$ = uniform temperature of testplate,
$s$ = nominal sagitta of spherical surface in contact,
$\Delta s$ = sagittal difference (fringe distortion).

Using standard optical materials for the lens and Pyrex for the testplate, it can be shown that minor thermal differences of 5°F to 10°F between the lens and the testplate can introduce fringe errors of several fringes. This effect is particularly noticeable on large diameter precision surfaces of 1/4 wave or better.

If a surface requires a flatness (or sphericity) to within 1/4 wave at 546.1 nm, then we must be certain of the fringe pattern to within 1/8 wave or better. This means that fringe distortions must not be greater than 1/8 wave. If we assume that the only fringe distortion is from thermal effects, then how much thermal difference is acceptable between lens and surface? Solving Eq. (6.28) for $\Delta T = T_1 - T_t$, we get:

$$\Delta T = \frac{\Delta s}{s(\alpha_1 - \alpha_t).} \tag{6.29}$$

Since $\Delta s = 546.1/8 \times 10^{-6}$ mm, then for the above example, when $s = 12.7$ mm, $\alpha_1 = 84 \times 10^{-7}$, and $\alpha_t = 35 \times 10^{-7}$, we have

$$\Delta T = \frac{68.26 \times 10^{-6}}{12.7(4.9 \times 10^{-6})} = 1.1°C.$$

This means that thermal differences will have to be of the order of 1°C or less to ensure that fringe distortion from thermal effects will not exceed 1/8 wave. This fact clearly demonstrates that good thermal control is essential when fabricating and testing lenses or plano surfaces of high precision.

The preceding examples assume that the testplate and the test part are at slightly different but otherwise uniform temperatures. This condition rarely exists in real life. For instance, the temperature of a surface point is almost always different from that of a point in the interior of the element because thermal energy typically flows from the interior to the surface where it dissipates in the cooler surrounding air. When this condition exists, it is said that a thermal gradient or a continuous thermal difference exists between those points.

Suppose that a thermal gradient exists in a testplate such that the test surface is warmer at temperature $T_1$ (in °C) and the back surface is cooler at temperature $T_0$. Then some change in the testplate geometry will take place. This change will manifest itself as a spherical error (i.e., power error) in the testplate match. The equation for this condition is [27]

$$\Delta T = (T_1 - T_0) = \frac{8t\Delta s}{\alpha d^2},\qquad(6.30)$$

where

$\Delta T$ = thermal difference in °C,

$t$ = testplate thickness in millimeters,

$\Delta s$ = sagittal difference in millimeters,

$\alpha$ = coefficient of expansion in 1/°C,

$d$ = testplate diameter in millimeters.

**Example**

What is the maximum allowable thermal difference between the front and back of a 150-mm diameter Pyrex testplate that has an aspect ratio of 6:1 to ensure that thermal distortion from this source does not exceed 1/10 wave at mercury green light?

· The aspect ratio is 6:1, and the diamter is 150 mm. This means that the testplate thickness $t = 150/6 = 25$ mm.

· The wavelength of mercury green is $546.1 \times 10^{-6}$ mm. The sagittal difference cannot exceed 1/10th of this. Therefore $\Delta s = 54.61 \times 10^{-6}$ mm or ($55 \times 10^{-6}$ mm).

· The coefficient of thermal expansion for Pyrex is $35 \times 10^{-7}$ per °C (see Table 6.3).

· Then, using Eq. (6.30),

$$\Delta T = [8(25)(55)\,10^{-6}]/[3.5(150^2)\,10^{-6}] = 0.14°C$$

This result shows that the thermal difference for this set of conditions cannot exceed 1/7°C between the front and the back of the testplate before the thermal distortion exceeds 1/10 wave.

It can be deduced from Eq. (6.30) that a greater thermal difference $\Delta T$ would be required to produce the same amount of thermal distortion if the testplate thickness $t$ were increased without a corresponding change in testplate diameter. A change to a material with a lower coefficient of thermal expansion will also produce a similar effect.

Normally, thermal gradients do not behave as nicely to be treated mathematically in this simple fashion. Irregular thermal gradients can and often do exist within the material and on the surfaces of the element in question to create totally unpredictable distortions. Instead of trying to figure out what the effect of such conditions may be, it is much more practical to allow for proper thermal stabilization away from external drafts and heat sources prior to testplate testing, as discussed in this section.

In general, studies have shown that for reliable and accurate testplate testing for close fits (1/4 wave or less) on large surfaces (100 mm or 4 in. in diameter and up), the following thermal conditions must be met [27]:

- Long-term air temperature variations of the test environment must not exceed ±2°C.
- Short-term (cyclic) air temperature variations as from air-conditioning system must not exceed 1°C.
- The lens element and the testplate should be thermally stabilized at ambient temperature to within 1°C before testplate testing.

The thermal stabilization time is typically in the order of 15 to 20 minutes for large elements when the thermal difference between the lens and the testplate is only a few degrees. If the temperature difference is much greater than that, for instance if lens was polished hot, then the time required for proper thermal stabilization to ensure that there is no fringe distortion can be more than one hour. This fact is quite often overlooked when tests are performed on lens elements that are 100 mm (4 in.) in diameter or larger and for which high precision of 1/4 wave or better is required. These thermal effects are not critical for smaller lens surfaces and for those applications where high precision is not essential.

The thermal stabilization period can be cut in half if both the testplate and the lens are soaked in a constant velocity and thermally controlled airflow. Figure 6.14a shows two curves that represent fringe distortion due to thermal differences between the matching parts as a function of time. The set used for this test were large 200-mm (8-in.) diameter testplates, one of which was kept at room temperature while the other was soaked in warm water that was about 5°C (9°F) higher to simulate the thermal condition of a recently polished surface.

The upper curve in Fig. 6.14a shows the condition when parts are stabilized at room temperature, and the lower curve represents the condition when the parts are kept in a constant airflow. In both cases the maximum

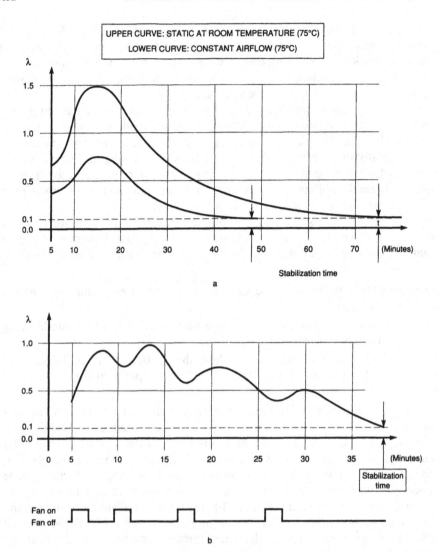

**Figure 6.14.** *(a)* Thermal stabilization. *(b)* Stabilization in intermittent airflow.

fringe distortion occurs after about 15 minutes and then slowly decays. The less than 1/10 wave fringe distortion target was reached with the constant airflow after less than 50 minutes, while the set stabilizing at room temperature took nearly 80 minutes to reach the same level of residual thermal distortion. The stabilization time can be further reduced when the thermally controlled airflow is not constant but cyclic according to a carefully devised plan. Figure 6.14*b* shows such a cyclic stabilization curve for the same example as shown in Fig. 6.14*a*.

Apparent fringe distortions can also result from changing angles of incidence. This is why it is essential that fringe patterns from large surfaces are viewed as close to the normal as possible. For the same reason strongly curved surfaces must be tested with testplates that have aplanatic backs. Such testplates were described in Section 6.1.3.

**Haidinger Fringes.** When testing optically transparent plano parts that are no more than 10 mm (0.400 in.) thick and are of good flatness and parallelism in a monochromatic test station, one can observe circular fringes that move when the observer moves or when the part moves. Although this effect is usually bothersome when attempting to measure the flatness of the surfaces, the movement of fringes can be used as a sensitive measurement technique of parallelism. This test can also be performed on blocked plane-parallel parts as long as fringes of sufficient contrast are generated.

Haidinger fringes are fringes of equal inclination that are seen as concentric fringes when looking into the part while the eye is focused at infinity [9]. They can be seen directly when testing the part in a test station because the beam splitter affords normal viewing of the fringes. A lateral movement of the part past a fixed aperture, or a movement of the observer's head relative to the part, will cause fringes either to emerge from the center and move outward toward the thicker end of the wedge or collapse into the center toward thinner part of wedge [28].

From one dark Haidinger fringe to the next represents a change of the OPD by two fringes or one wave. This corresponds to a change in physical thickness of $\lambda/2n$, where n is the refractive index of the material. Therefore one counts the fringes that are seen across the part diameter $d$, and from this information the wedge error in radians can be calculated as in Eq. (6.31):

$$\varepsilon = dN\left(\frac{\lambda}{2n}\right), \tag{6.31}$$

where

- $\varepsilon$ = wedge error in radians,
- $d$ = diameter over which fringes are counted,
- $N$ = number of dark fringes,
- $\lambda$ = wavelength of light source,
- $n$ = refractive index.

### 6.1.8. Testplate Fringe Analysis

Although the fringes seen when testplating can be analyzed by one of the many interferometer fringe analysis systems now commercially available as long as a hard copy interferogram can be produced, most testplate fringes needed are not of the accuracy that computerized fringe analysis is neces-sary. Therefore a similar but sufficiently different approach is used, and this approach warrants a more detailed description.

When testplates are properly used for testing lens surfaces, the resulting fringe pattern is undistorted by external influences such as mechanical or thermal deformation. The fringe pattern then represents the true deviation of the test part surface relative to the reference surface of the testplate. This is the information the optician needs to know to determine if the surface has been polished to the correct shape or if further correction is necessary. The interpretation of the resulting fringe pattern is not always easy and often leads to divergent results. A clarification of the basics of fringe analysis might be helpful.

There are basically two types of fringes. Fringes can be open, or they can be closed. Within these two types there are regular and irregular fringes. This basic classification of fringes applies for all interference patterns whether they result from testplate testing or from an interferometer test. Fringe analysis methods associated with modern shop interferometers will be discussed in Section 6.2. The description in this section concerns itself with the interpretation of testplate fringes.

Open fringes occur when the test part surface differs from the testplate reference surface by less than two fringes (1 wave). For this reason they are found only on high-quality surfaces. Figure 6.15 shows a typical example.

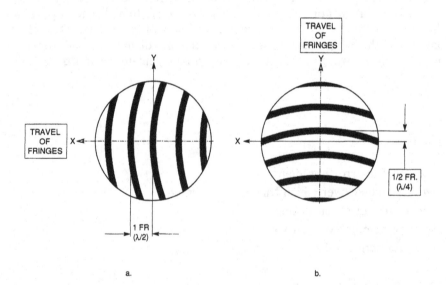

Irregularity: 1 - 1/2 = 1/2 rings

**Figure 6.15.** Evaluation of open testplate fringes. *(a)* 1 Fringe convex power in *X* axis. *(b)* ½ Fringe power in *Y* axis.

The testplate match shows one fringe power in Fig. 6.15*a* when the fringes are oriented as shown. When the fringes are rotated 90° to run parallel to the *x*-axis, they show one half fringe power in Fig. 6.15*b*. The fringe rotation is accomplished by applying slight pressure to one side of the testplate. The testplate or the part are not rotated. The difference in the curvature of the fringes indicates that the surface is not spherical but is astigmatic by 1/2 fringe (or 1/4 wave). The power is the larger of the two readings, so the analysis of this fringe pattern concludes that the surface exhibits 1 fringe power and 1/2 fringe irregularity. This is an acceptable test for many applications. A difference of 1/4 fringe (1/8 wave) can still be judged with some degree of accuracy, but it is generally accepted that this represents the accuracy limit of testplate measurements. If no curvature in the fringes is detected, which means that they are straight, then we say that the test is at least 1/10th wave. Surfaces of such degree of flatness or sphericity must be tested in an interferometer to get a more accurate analysis of the fringe pattern.

The direction in which curved fringes move when pressure is applied to the testplate shows whether the test is concave or convex. When the fringes are curved in the direction of fringe motion, the testplate match is convex, which means that a convex testpart has a slightly shorter radius or that a concave testpart has a slightly longer radius than the testplate. When they curve in the opposite direction, the test is concave, and the radius differences are reversed.

An accurate surface figure specification must not only state the maximum allowable power and irregularity of the testplate match in terms of fringes or waves, but the wavelength of the monochromatic source and the diameter over which the test applies must also be clearly defined. Even though there is no clear consensus how to specify surface figure, the widespread use of laser-based interferometers has encouraged designers to provide more details. It is generally true that the maximum irregularity is either 1/2 or 1/4 of the allowable power call-out. The wavelength is usually 0.633 $\mu$m (HeNe) for laser-based interferometers or 0.546 $\mu$m (Hg) for testplate stations. The diameter over which the fringe call-out has to be met is most often the clear aperture (CA) of the test part surface that should be called out on the part drawing. For larger surfaces the specification may require that the fringe call-out applies over a specific diameter anywhere on the surface within the clear aperture, for instance, over any radial inch (another way of saying over any 2-in. diameter area) or over several 3-in. diameter areas whose locations are clearly defined. The same definition applies to metric values as well.

**Wavelength and Diameter.** The need to specify both the wavelength and the diameter of a test is a direct consequence of Eq. (6.5) [5]. Fringes can be scaled for one wavelength as a function of diameter over which the fringe pattern applies, or they can be scaled over a specific clear aperture as a function of different wavelengths.

**Diameter Scaling.** When a large plano part must be tested over a smaller aperture or with a smaller testplate, it is possible to calculate the flatness over the entire part by scaling the flatness measured over the testplate diameter. The assumption is that there is only spherical departure from planeness.

Using Eq. (6.5) and solving for $R$, we can set up an equality for two different diameters $D_0$ and $D_1$. This assumes that there is only a purely spherical deviation with a very long radius $R$. Since $R = D_0^2/4N_0\lambda = D_1^2/4N_1\lambda$, then

$$\frac{D_0^2}{N_0} = \frac{D_1^2}{N_1}$$

and this leads to

$$N_1 = N_0 \left(\frac{D_1^2}{D_0^2}\right), \tag{6.32}$$

where

$N_1$ = number of fringes over $D_1$,
$N_0$ = number of fringes over $D_0$.

### Example 1

Two fringes ($N_0 = 2$) are seen over a 3-in. diameter testplate ($D_0 = 3$). How many fringes ($N_1$) can we expect to see if we tested over the full diameter of 12 in. ($D_1 = 12$)?

$$N_1 = 2 \left(\frac{12^2}{3^2}\right) = 2 \left(\frac{144}{9}\right) = 32.$$

While two fringes may not seem too bad, a supposedly plano surface with 32 fringes is nothing to be proud of. This example illustrates the danger of evaluating the figure of an optical surface with a testplate or interferometer aperture that is too small.

### Example 2

How many fringes would be allowed over the 75-mm diameter testplate if the specification called for no more than two fringes over any 100-mm diameter aperture within the clear aperture (CA)?

First, Eq. (6.32) has to be solved for $N_0$:

$$N_0 = N_1 \left(\frac{D_0^2}{D_1^2}\right)$$

$$= 2\left(\frac{75^2}{100^2}\right) = 2\left(\frac{5625}{10,000}\right) = 1.125 \text{ fringe.}$$

To meet the specification, the 75-mm diameter testplate placed anywhere within the CA should not show more than 1 and 1/8 fringe.

**Example 3**

A 4-in. diameter part is tested in an interferometer, and two fringes of spherical power (circular rings) are seen over the full aperture of the part. What would the flatness of 1-in. diameter parts be that could be cut from this part, assuming no effect on the surface figure from the operation?

$$N_0 = N_1 \left(\frac{D_0^2}{D_1^2}\right) = 2\left(\frac{1^2}{4^2}\right) = 2\left(\frac{1}{16}\right) = 1/8.$$

The resulting 1-in. diameter parts would be quite flat at 1/8 fringe.

Equation (6.5) also determines that the diameter over which the fringes are to be evaluated must be defined. Since the diameter is a squared quantity in the equation, it is even more important than the wavelength. Assuming that a surface with diameter $D_0$ must be tested with a testplate with diameter $D_1$ and that $N_1$ fringes are counted, what is the fringe count over the part diameter $D_0$? The curvature $R$ is the same in both cases, since the test part surface is assumed to be spherically out of flat. Solving for $R$, two separate equations result:

$$R = \frac{D_0^2}{4N_0\lambda} = \frac{D_1^2}{4N_1\lambda}.$$

This reduces to

$$\frac{D_0^2}{N_0} = \frac{D_1^2}{N_1}.$$

Solving for $N_0$,

$$N_0 = N_1 \left(\frac{D_0^2}{D_1^2}\right).$$

The equation can be represented graphically as in Fig. 6.16.

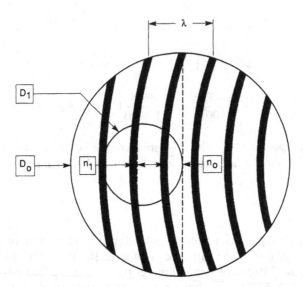

**Figure 6.16.** Fringe count as a function of diameter.

The fringe pattern can be analyzed using basic manual fringe reduction methods where the fringe to fringe spacing and and the fractional fringe deviation is measured over the diameter of interest. For instance, in Fig. 6.16, the fringe-to-fringe spacing ($\lambda/2$) is 12.7 mm. This means that one wavelength is equivalent to twice that distance, or 25.4 mm. The fringe deviation $N_1$ was measured to be 1.5 mm over testplate diameter $D_1$ and the deviation $N_0$ of the adjacent fringe was found to be 6.5 mm over part diameter $D_0$. This means that the flatness over the smaller ($D_1 = 45.5$ mm) diameter testplate is $1.5/25.4 = 0.059$ or about 1/17 wave, while the actual flatness over the larger ($D_0 = 95$ mm) diameter part is $6.5/25.4 = 0.256$ or about 1/4 wave.

Equation (6.32) must first be solved for $N_1$ before it can be used on the same example:

$$N_1 = N_0\left(\frac{D_1^2}{D_0^2}\right)$$

$$= 0.25\left(\frac{2070}{9025}\right)$$

$$= 0.057 \text{ or } 1/17.5 \text{ wave.}$$

This result is in rather good agreement with the graphic solution.

**Wavelength Scaling.** Wavelength scaling is often required when a surface is tested at one wavelength but the specification applies at another wavelength. To scale the fringes from one wavelength to another, we proceed as follows:

Using Eq. (6.5) we set up two equations with the same radius of curvature $R$ measured over the same diameter $D$, one for fringe number $N_0$ at wavelength $\lambda_0$, and one for fringe number $N_1$ at wavelength $\lambda_1$:

$$\frac{D^2}{8R} = N_0\left(\frac{\lambda_0}{2}\right) = N_1\left(\frac{\lambda_1}{2}\right)$$

and this resolves to:

$$N_0\,(\lambda_0) = N_1\,(\lambda_1). \tag{6.33}$$

**Example 1**

A print calls out a 1/60 wave flatness at 10.6 $\mu$m for an IR reflector. Can this level of flatness be achieved and measured?

Since we cannot measure at 10.6 $\mu$m unless we have the use of an IR interferometer, we must test the surface at visible wavelengths. Most modern interferometers use a HeNe laser ($\lambda = 0.633$ $\mu$m).

$$N_0 = \left(\frac{1}{60}\right)\left(\frac{10.6}{0.633}\right) = 0.279 \text{ wave or } 0.56 \text{ fringe.}$$

This is a rather good surface, and one-half fringe is not impossible to polish and measure.

**Example 2**

Sometimes, an older print calls out a flatness at 541.6 nm, which is the mercury green line that was widely used in optical test instruments and the earlier interferometers. If two fringes are permitted over the CA of the part, how much of an adjustment do we have to make for flatness if we test in a HeNe laser interferometer?

$$N_1 = 2\left(\frac{541.6}{632.8}\right) = 1.71 \text{ fringes.}$$

The surface tested at the longer wavelength must meet a more stringent test by a factor of the ratio of the shorter wavelength over the longer wavelength. In this case the factor was 0.856, or about 14% fewer fringes are permitted.

If a surface is tested for figure over a specified diameter at one wavelength but the specification calls out that the figure applies at another wavelength, the above equality can be further simplified by canceling out the diameter $D$.

Then

$$N_1 = N_0 \left(\frac{\lambda_0}{\lambda_1}\right).$$

This simple equation permits the scaling of fringes from one wavelength to another. For instance, IR optics are usually specified to be tested at visible wavelengths.

## Example 3

A surface must meet 1/10 figure at 1.06 $\mu$m. The only available test station operates at 0.546 $\mu$m. How good does the test have to be?

$$N_1 = 0.1 \left(\frac{1.06}{0.546}\right) = 0.194 \text{ or about } 1/5 \text{ wave.}$$

It is also possible to use these equations to combine the effects of both diameter and wavelength scaling. For instance, the test procedure could call out for a one-fringe test using a 100-mm (4-in.) diameter testplate with mercury green light (0.546 $\mu$m), but the optician might only find a 75-mm (3-in.) diameter testplate and he or she must test with HeNe illumination. Assuming again the residual curvature is purely spherical with radius $R$, from Eq. (6.5) we have

$$4R = \frac{D_0^2}{N_0\lambda_0} = \frac{D_1^2}{N_1\lambda_1},$$

$$N_0 = N_1 \left[\frac{\lambda_1(D_0^2)}{\lambda_0(D_1^2)}\right]$$

$$= 0.5 \left[\frac{0.546 \times 10^{-3}(5625)}{0.633 \times 10^{-3}(10,000)}\right]$$

$$= 0.24 \text{ wave at } \lambda = 0.633 \ \mu\text{m.} \tag{6.34}$$

The optician must polish the surface to an accuracy of 1/4 wave to over the 3-in. diameter testplate to meet the test specification.

If the flatness specification for a specific component calls for the power not to exceed 1/8 wave but neither the wavelength nor the diameter over which the call-out applies is specified, what kind of errors can occur when it is left up to the optician to decide how best meet the specification? The designer assumed the "customary" mercury light source at 0.546 $\mu$m and that the test applied over the full diameter of the part. The optician, however, decided to use a HeNe laser based test (0.633 $\mu$m) and test over the clear aperture, which is typically 90% of the full diameter. Using Eq. (6.7) and solving for $4R$ yields

$$4R = \frac{D_0^2}{N_0 \lambda_0} = \frac{D_1^2}{N_1 \lambda_1};$$

also for this example $D_1 = 0.9 \, D_0$. Then

$$N_1 = 0.81 N_0 \left( \frac{\lambda_0}{\lambda_1} \right)$$

$$= 0.81(0.125) \left( \frac{0.546}{0.633} \right)$$

$$= 0.087 \text{ (or } 1/12 \text{ wave).}$$

This result says that the optician should have polished the surface to 1/12 wave at 0.633 $\mu$m over the clear aperture in order to meet the 1/8 wave requirement at 0.546 $\mu$m over the full aperture. Even if the optician had polished it to 1/10 wave, it would not have been good enough.

**Fringe Analysis of Closed Fringes (Rings).** Closed fringes occur when the surface of the test part differs from the testplate reference surface by one fringe (1/2 wave) or more. For most good-quality photographic optics, the testplate match rarely exceeds three rings or 1.5 waves. A spherical mismatch of several waves is permissible for medium-quality optics, such as eyepiece lenses.

The central spot often seen in fringe patterns is the zero-order fringe which is not counted. This fringe is the location where a for convex test the part contacts the matching testplate and the air gap between the surfaces is zero. The zero-oder fringe for a convex test is located around the periphery of the clear aperture.

The ideal of perfectly circular and concentric rings is almost never really met. There is usually some residual irregularity that slightly distorts the symmetry of the rings. This is recognized by the fact that every surface figure specification has two components, namely, power and irregularity. For most conditions the allowable fringe irregularity is 1/4 of the maximum power. Therefore a maximum four ring power call-out limits the irregularity to one ring or less. Occasionally other ratios are also used.

Figure 6.10*a*, *b*, *c*, and *d* shows three typical closed ring interference patterns and the manner by which they are analyzed. Regardless of the complexity of the pattern, the process follows the same steps:

- Count the number of fringes across the clear aperture along the axis that contains the greatest number of fringes. Do not count the zero fringe.
- Determine the power by dividing the count by two. The resultant num-

ber, which may be a fraction, is the average number of rings that cross the axis. Record the number.

- Count the number of dark rings that cross an orthogonal axis , which is the one that lies at right angles to the first axis.
- Divide the result by 2 to obtain the power in rings. Record the result.

There are two things to remember when counting fringes over a specified diameter. For perfectly centered and circular rings, there are always partial rings just inside of the clear aperture. How are they accounted for and how are they measured? Since three to four closed fringes or rings are typically permitted for many surfaces, all that needs to be done is to make sure that there is one less ring than allowed. This will certainly cover the effects of any partial rings, though it will reduce the fringe tolerance, making it a little more difficult on the optician. The magnitude of these partial rings is difficult to judge because there is nothing to reference it to. One way to measure the magnitude is to set the rings slightly off center. Then the missing ring would be seen, and the ring spacing better judged. This procedure of offsetting the rings creates a problem in that some rings will be intercepted by the edge of the aperture such as the edge of the testplate. This condition does not change the fact that the fringes are closed. As long as a fringe portion on one side of the center has a matching counterpart at the opposite side, which is part of the same fringe, then that fringe is considered closed. If this condition is not considered, erroneous measurements will be made.

The maximum power value obtained in this way is representative for the surface and must be equal or less than the permissible power value. The magnitude of the irregularity is determined by subtracting the smaller ring count from the larger. The difference is the irregularity, and it must not exceed the specification.

It must be clearly understood that the power and irregularity values obtained by interpreting closed interference fringes or rings are strictly valid only over the tested area. In cases where the testplate diameter is smaller than the diameter of the test part, no firm conclusion can be drawn about the surface figure outside the covered area unless there is certainty that the surface figure is uniform over the entire test part surface. Such ideal conditions rarely exist. The example for Eq. (6.32) suggested that one can calculate the power over a large surface from the power measured over a smaller one. This was based on the assumption that the curvature $R$ of the surface (i.e., the power) was uniformly spherical over the entire part. This example was primarily intended to demonstrate that the diameter over which the test applies must be clearly specified. If the testplate is smaller than the test part, then several overlapping areas on the surface of the test part must be evaluated in order to fully understand the figure of the entire surface. This method is called *mapping*, and it is often part of a test specification.

Circular fringes or rings are a desirable surface figure for spherical lens surfaces as long as the number of rings do not exceed the maximum allowable count. Circular fringes result from spherical departure from the test-plate radius. There is little if any surface irregularity. Oblong fringes or rings represent the presence of a surface figure irregularity that must not exceed a maximum value. This type of irregularity is called astigmatism. Astigmatism is defined as the difference of fringe counts along two orthogonal axes. A third and most unfortunate surface figure irregularity is commonly known as *saddle*. It produces an optical distortion of an image or wavefront that is known as *coma*. It is in almost all cases an unacceptable surface shape that has a convex curvature in one axis and a concave curvature along the other axis. The irregularity is not the difference of the fringe counts along the two orthogonal axes but rather the sum, for in one axis the curvature is positive while it is negative in the other. With this constraint, saddle is almost never an acceptable condition unless it is so mild that the sum of the fringe counts does not exceed the limit value.

There have been many descriptions and illustrations provided in test specifications and other printed matter that attempt to show how to interpret a wild variety of possible and even impossible surface shapes. When a surface figure departs much from the examples given in this chapter, it is usually unfit for use in any optical imaging system. Elements with such surfaces are most often used only in illuminating systems where flatness or sphericity is of little consequence for the performance of the system. The best general case is usually a combination of the various basic surface figure shapes that have been discussed in this chapter.

## References for Section 6.1 (Testplate Testing)

[1] Josef von Fraunhofer, 1787 to 1826, German physicist.

[2] Sir Isaac Newton, 1643 to 1727, English physicist.

[3] Armand H. L. Fizeau, 1819 to 1896, French physicist.

[4] *Handbook of Industrial Metrology*, ASTME, 1967.

[5] Karow, H. H., "Interferometric testing in a precision optical shop: a review of testplate testing," SPIE, *Interferometry*, vol. 192, 1979.

[6] Smith, W., "How flat is flat?" *Optical Spectra*, May 1974.

[7] Private communication (notes on spherical test plates) S. I. Robinson, Itek Corp., 1968.

[8] Robinson, S. I., *Notes on Testplating*, Itek ODR-66-18, 1966 upublished manuscript.

[9] Francon, M., *Optical Interferometry*, Academic Press, N.Y., 1966.

[10] Pforte, H., *Feinoptiker, Teil II*, VEB Verlag Technik, Berlin, 1975.

[11] Naumann, H., *Optik für Konstrukteure*, Knapp Verlag, Düsseldorf, 1960.

[12] DIN 58 161.

[13] *Handbook of Industrial Metrology*, ASTME, 1967.

[14] BK 7 (borosilicate crown glass) is also identified as 517 642.

[15] ZK 7 (zinc crown) is also identified as 508 612.

[16] Pyrex is made by Corning (USA).

[17] Fused silica is made by Heraeus Amersil (Germany), Dynasil (USA), and Nippon Silica Glass (Japan).

[18] CerVit is made by OI Kimble (USA).

[19] Zerodur is made by Schott (Germany).

[20] ULE is made by Corning (USA).

[21] Melles Griot Optics Guide 2.

[22] Schade, H., *Arbeitsverfahren der Feinoptik*, VDI Verlag, Düsseldorf, 1955.

[23] Lapmaster International.

[24] Van Keuren Co.

[25] Rogers & Clarke, Model "Unilamp."

[26] Zimmerman, J., "Aid to viewing test plate interference fringes," *Applied Optics*, vol. 10, no. 9, September 1971.

[27] Barnes, W. P., Jr., "Thermal stabilization of testplate testing," Itek ATR 68-32, 1968.

[28] Tew, E. J., Jr., "Measurement techniques used in the optics workshop," *Applied Optics*, vol. 5, no. 5, May 1966.

## 6.2. INTERFEROMETERS

### 6.2.1. Shop Interferometers

Some form of interferometry has been used for the in-process testing of optically polished surfaces for a long time. Testplate techniques have been an integral part of optical fabrication for well over 100 years, and some form of interferometers such as the Fizeauscope have been in use for many years. It was not until the development of the laser that a light source was found that made shop interferometry practical. More recent developments in high-resolution photodetectors and powerful minicomputers have converted the interferometer into an indispensible analytical instrument.

Since the early 1970s a number of laser interferometer systems have been perfected for use in the optical shop and the quality lab. These laser interferometer systems are not only easy to set up and use, but their modular construction makes them very adaptable to accommodate almost any test that has to be performed in the fab shop or in the lab. Because of the ease of use and the powerful fringe analysis capabilities of these systems, it is essential for any modern shop to use a laser interferometer for in-process testing and for qualifying the product.

There are a number of different interferometers specifically designed for use in the optical shop or the optics lab. The primary types are laser interferometers, phase interferometers, IR interferometers, and special purpose interferometers. These will be discussed in this section.

### 6.2.2. Fizeauscopes

A simple interferometer for testing plano surfaces has been in use for a long time. This so-called Fizeauscope, shown in Fig. 6.17, was usually built by the optical shops where they were used.

Eventually a commercial instrument with a 125-mm (5-in.) aperture [29] became available, and it found widespread acceptance in the shops. Variations of this instrument were also built, some with apertures up to 300 mm (12 in.) for testing large plano parts, and others with apertures as small as 37 mm (1.5 in.) for small surfaces.

The original Fizeauscopes used a nearly monochromatic incoherent light source at 546.1 nm (mercury green). In the more recent past these old interferometers could be upgraded for use with a HeNe laser source (632.8 nm). A modern version of this type of HeNe plano interferometer [30] comes equipped with a video display option that uses a high-resolution CCD camera, a digital filar micrometer for measuring fringe spacings accurately, and a thermal printer to generate hard copy interferograms for later manual fringe reduction.

Fizeauscopes utilize a two-spot alignment scheme that ensures that the test surface is properly aligned to the output beam so that fringes can form.

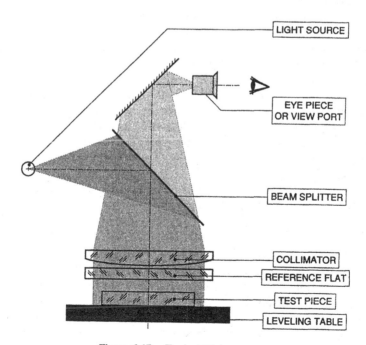

**Figure 6.17.** Typical Fizeauscope.

This scheme has also been incorporated, with some modifications, in today's interferometer mainframes. A stationary spot is the projected image of the pinhole that serves as the point from which the light seems to originate. The second, movable, spot is an image of the same pinhole that is reflected off the test surface. To achieve proper alignment, these two spots must be superimposed by tilting the workpiece. This is accomplished by a kinematic leveling table on which the testpiece rests that can be tilted in two orthogonal axes independently.

Another way to describe this [29] is as follows:

> The part is placed on the table, the eyepiece is flipped into viewing position—two or more images of an illuminated pinhole are seen—one remains fixed in the center of the viewing screen while the other is reflected off the test piece surface—the reflected pin hole image can be made to superimpose on the fixed image by adjusting the leveling screws of the table—when the alignment is complete, the eyepiece is flipped out of the way and fringes are seen—fringe spacing and orientation can be changed by small adjustments of the leveling screws.

**Plano Interferometer.** A variation of the Fizeauscope has found popularity in the optics industry of Germany and other European countries. It is a vertically arranged, uplooking instrument designed for accurately measuring the flatness of plano surfaces and the parallelism of nearly plane-parallel optically transparent parts. This type of interferometer is shown in Fig. 6.18.

The advantage of this type of instrument [31, 32] is that the testpart, such as a window or a prism, is placed face down on the three-ball support of the precision tilt table. This arrangement makes this a test with minimal contact between the test surface and the instrument, and this results in virtually no danger of surface damage to the part. The test piece is also not constrained, except by its own weight, and this reduces the distortion of the fringe pattern. The light sources used in this instrument are 546.1 nm or 632.8 nm, and the instrument can be converted to operate with either source.

The claimed system accuracy is 1/50 wave for plane surface measurements and 0.2 arc seconds for parallelism measurements. The reference plate must be removed for parallelism measurements. The clear aperture of between 100 to 150 mm (4 to 6 in.) diameter is comparable to that of similar interferometers.

The instrument is convertible by adding or rearranging some of the key components to operate in the Fizeau, Michelson, or Twyman-Green configuration. An additional 1:6 zoom feature simplifies the evaluation of small surfaces. An integral alignment and fringe acquisition system makes the setup and use of this instrument a relatively easy matter. The fringes can be analyzed by a computer-assisted fringe analysis system that evaluates peak-to-valley (P–V), optical path differences (OPD), and root-mean-square (rms) values plus several other parameters.

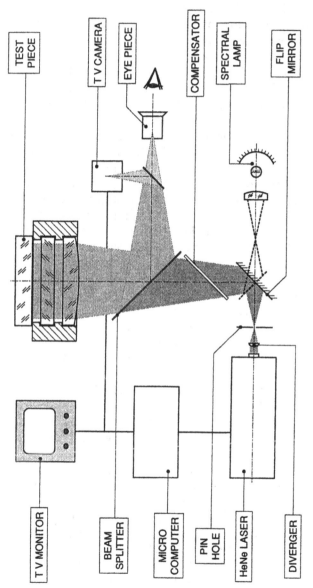

**Figure 6.18.** Fizeau interferometer for plano surfaces.

TEST PIECE

T V CAMERA

EYE PIECE

COMPENSATOR

SPECTRAL LAMP

FLIP MIRROR

T V MONITOR

BEAM SPLITTER

MICRO COMPUTER

PIN HOLE

HeNe LASER

DIVERGER

### 6.2.3.  Laser Interferometers

The first laser interferometer specifically designed for shop use was introduced in 1970 [33]. This instrument ushered in the era of modern interferometry. Prior to this, Fizeau type interferometers were in use for many years, and Twyman-Green unequal path laser interferometers (LUPI) were used in various shops for measuring larger optics. But unlike the modern shop interferometers, these earlier devices were limited to evaluate plano surfaces in visible (nonlaser) light as in the case of the Fizeauscope, or they were large, expensive, and difficult to line up as was the case for the earlier LUPI which were used for testing large astronomical optics.

The new shop interferometer was a departure from these constraints. It was an easy to use and very versatile instrument that did not require a great deal of expertise in optical alignment. The component and system alignment method used a fixed image of an illuminated pinhole as target. An image of a second pinhole that was reflected off the test surface had to be superimposed on the target. This brought all components in the system on axis and when a movable beam splitter redirected the beam to the viewing mode, interference fringes were seen at once. Once the basic alignment method had been mastered, it was easy to set up the next part for testing.

The mainframe is the heart of the laser interferometer. An expanded and collimated laser beam which is typically 100 mm (4 in.) in diameter emerges from the output port. The snap-in components are mounted on this port to produce plane or spherical reference wavefronts for either Fizeau fringes or for high-finesse fringes. In this arrangement the interferometer can be set up to test plano or spherical specular (reflective) surfaces in reflection to evaluate the flatness or sphericity of the surfaces, or to test components and systems with transmitted wavefronts. The internal layout of the mainframes is shown in Fig. 6.19. The limits of measurement are determined by the accuracy of the reference surfaces of the snap-in components. These are typically $\frac{1}{20}$ wave at 633 nm for plano and $\frac{1}{10}$ wave for convex or concave surfaces.

Other features were soon added that made this type of interferometer much more versatile than all others and helped to make it the standard of the industry. Some of these are discussed here. For instance, a 1:6 zoom feature makes it possible to measure small diameter parts or to enlarge the interference patterns of relatively small, long radius parts so that they can be analyzed. An associated feature permits limited focusing of the fringe pattern, called *aperture focusing*, which is especially helpful in ensuring that the fringes are sharp right out to the edge. A hand-held remote operator control permits switching from the alignment to the view mode without having to be near the interferometer. It incorporates the zoom and focus controls as well.

The conversion from projecting the fringe pattern on a screen to the use of an integral video system was another important step in the evolution of the

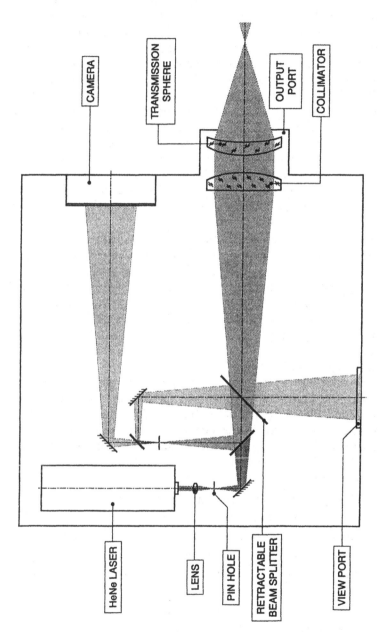

**Figure 6.19.** Schematic of an early shop interferometer (Zygo GH).

CAMERA

TRANSMISSION SPHERE

OUTPUT PORT

COLLIMATOR

HeNe LASER

LENS

PIN HOLE

RETRACTABLE BEAM SPLITTER

VIEW PORT

shop interferometer. The projected fringes in the original system were of low contrast, which made it difficult to see them. A photographic hardcopy interferogram (Polaroid) had to be first generated before the fringes could be manually reduced. The video system produces high-contrast fringes with very good definition. This major improvement opened the door for the automatic fringe analysis that is now quite common and highly refined.

Mainframes with laser sources other than HeNe are available for special applications. For instance, argon laser sources operating at either 488 nm or at 514 nm, near-IR laser interferometers operating at the Nd : YAG wavelength of 1.06 $\mu$m, and IR laser interferometers at the $CO_2$ wavelength of 10.6 $\mu$m are available. The IR interferometers are discussed separately

Although several generations of laser interferometers for use in the optical shop have been introduced since the early 1970s, the basic mainframe has remained essentially unchanged. What has changed greatly is the way the fringes are observed and measured and how they are analyzed and displayed.

One considerable advantage that interferometers have over the more traditional testplates is that spherical power can be nulled from the fringe pattern so that only the actual irregularities are exposed. This permits the accurate independent measurement of critical fringe parameters such as power and irregularity, and it makes it possible to display them as separate surface errors. It is the sum of all irregularities in the optical systems that determines to a large degree its resolution and image-forming capabilities. The effect of spherical power can in most cases be entirely eliminated by axial positioning of the elements.

Even strongly curved surfaces can be viewed at normal incidence with properly matched divergers or convergers. It was already discussed in the previous section that testplates must be provided with carefully designed compensating curves to avoid the considerable fringe distortions that are unavoidable when such steep radii are tested with plano-backed testplates.

Among the more important interferometer capabilities are accurate measurements of the following:

- Parallelism of optically transparent plane-parallel parts
- Surface flatness of plano surfaces
- Transmitted wavefront through optically transparent windows
- Index homogeneity of optically transparent materials
- Surface figure (sphericity) of spherical surfaces
- Surface figures of parabolas
- Radius of curvature of concave and convex surfaces
- Single lens or lens system performance tests
- Determination of precise prism angles
- Optical system collimation and alignment
- As a totally noncontact method, there is no damage to the most sensitive surfaces and no distortion of thin parts

Despite numerous advantages interferometers have some limitations:

- Finite aperture, typically 100 mm (4 in.) diameter
- Larger diameters available at greatly increased cost
- Physical constraints imposed by f/# of spheres and R/D of lenses
- Maximum radius of curvature, length of rail for cc, BFL of sphere for cx
- Reflectivity mismatch reduces fringe contrast
- Measurement accuracy system limited to λ/10 sphere, λ/20 plano

In addition there are a number of special testing capabilities to be considered:

- Transmitted wavefront fringes for the calculation of parallelism
- Testing of large diameter plano surfaces as long as the diameter is less than the clear aperture of the interferometer reference flat
- Functional transmitted wavefront through assembled prisms
- Transmitted wavefront through ground glass samples using oil-on plates for testing the homogeneity of transparent materials
- Checking prism angles by retroreflection (right angle prism, corner cubes)—bends or curvature in fringes reveal angle errors
- Transmitted wavefront through lenses

**Interferometers versus Testplates.** One of the questions that is often asked is why interferometers have not yet eliminated the need for radius testplates as has already happened for plano surfaces. This was one of the primary selling points when the mainframe interferometers were first introduced. Now, however, 20 years later, spherical testplates are still in widespread use for the manufacture of lens optics.

In theory, at least, it is possible to grind and polish a spherical surface and measure the exact radius of curvature and the departure from true sphericity with a shop interferometer without the benefit of radius testplates. This approach has been demonstrated without the use of testplates [34] when single-surface prototype meniscus lenses were successfully produced using a shop interferometer (Zygo Mark II) which was equipped with a 1-m (39.4-in.) radius rail. The experiment failed the economic test, however, because the interferometer could not be set up in the shop in close proximity to the polishing machine, and this caused the optician to spend an inordinate amount of time transporting the lenses between the machine and the interferometer.

The same interferometer setup was used to reduce the cost of making radius testplates by just making the required testplate rather than the pair of plates with the more traditional approach. The manufacture and use of testplates is described in great detail in Section 6.1. For instance, only the concave half of a testplate pair was made for testing a convex lens surface.

With the modified testplate making method, the matching half of the test-plate pair is provided by the spherical wave front that emerges from the output lens of the interferometer. A spherometer for measuring the radius of curvature is also not needed because the radius rail will provide a highly accurate measurement as described later in this section.

This approach was also successful from a technical standpoint. It also provided insufficient economic incentives for a change, even though the cost of making testplates was potentially only half of the testplate pair method. In the end the optician preferred the traditional testplate method because it turned out to be the more practical and cost-effective approach. How much this decision was based on traditional bias is hard to assess.

The primary limitations lie in the difficulty in getting a shop interferometer set up right at the machine and in the relative complexity of making the measurement as compared to the traditional use of testplates. The objections of setting up an interferometer in the shop are not based on technical merit but rather on the mistaken assumption that such a delicate (and expensive) instrument does not belong in the "dirty" environment of an optical shop. These decisions are made in most cases by people who are not opticians. To them an interferometer is a sensitive instrument that must be set up in the quality lab. These objections, however, can be overcome by an enlightened management. It is harder to justify the use of interferometers in lieu of testplates on economic grounds. The reason is the greater complexity of setting up the testpiece and making the measurements compared to using testplates.

The polishing of optical surfaces is an iterative process during which several measurements of the surface are made and corrective polishing parameter adjustments such as relative speed, swing angle, swing bias, and pin position follow on the basis of these measurements. The measurement and the necessary corrective steps should be made within the span of one minute or less to maintain the critical thermal equilibrium between the lap and the lens surface. If this step were allowed to take much longer, the thermal equilibrium would be lost, and the resulting minute changes in the shape of both the lens and the tool surfaces would lead to mismatch between the surfaces. Such mismatch would introduce surface errors and could also lead to the formation of polishing sleeks.

Even under the best of circumstances an interferometer test will take much longer than 1 minute. During a series of tests with production parts [35], it was determined that the actual time to mount the lens, align the interferometer, make the radius measurement, take the hard copy interfero-gram and analyze it, and demount the lens was closer to 6 minutes for the 100 mm (4-in.) diameter lenses. If this test were to be performed as part of the normal polishing process, the lens shape would have changed sufficiently by the time the test is completed that the result of the measurement would differ in a measurable way from the shape of the lens surface when it was still on the machine. During the same time the tool would have changed in a similar manner. In effect the optician would end up making corrective adjustments on the machine settings for a nonexisting surface condition.

The use of an interferometer undoubtedly offers some important advantages. An additional benefit of using the interferometer for measuring spherical surfaces is the more advantageous utilization of the radius tolerance which lies between 0.1% and 0.3% of the nominal radius value for most lens surfaces. Once a spherical testplate has been made with the measured radius at or near the nominal value, all lens surfaces measured against it must be held to within the maximum allowable fringe power, which is usually only a small fraction of the radius tolerance. With the interferometer approach, however, the test must simply show that the radius lies within the radius tolerance range and that the surface irregularity does not exceed the specified limit. Spherical power is usually nulled out, and only residual power becomes part of the measurement.

The preceding discussion was based on the measurement of single-surface lenses for which the interferometers are well suited. However, it is much more difficult to effectively use them for testing lenses on the block. This is the most common method by which lenses are ground and polished. It was already pointed out that the only meaningful way to utilize the interferometer for in-process metrology is to set it up in the shop next to the polishing machines. This will expose it to contamination, mechanical vibration, and environmental perturbations, which can affect both the instrument and the test results. Also, unless the blocks are small, testing lenses on the block requires sturdy but sufficiently sensitive mounts so that the necessary adjustments can be made. For steep convex blocks a complicated goniometer-type mount may be required that makes it possible to swing the block around its center of curvature and to rotate the block around its axis so that all lenses in the outer rows can be tested. These mounts are quite expensive.

Another way to make the interferometer adaptable for testing lenses on the block involves the modification of a Bridgeport-type milling machine. This approach has been suggested by a German interferometer manufacturer [36]. Figure 6.20 shows this schematically. A rotary stage is mounted on the x–y cross slide of the mill. The lens block is mounted on the rotary stage. The laser is mounted vertically downlooking on the mill head, which can be adjusted either up or down along the z-axis. By tilting the head and adjusting its vertical position, the focus of the transmission sphere can be made to coincide with the center of curvature of the block. If the block position is adjusted with the x–y cross slide, it should be possible to get an interference pattern at any point on the block. When the block is then rotated on the rotary stage, any zone on the block, including the edge zone, can be evaluated.

Concave blocks are limited to a radius that is equal to the length of the radius rail which is typically 1 m (39.4 in.). The limitations are more severe

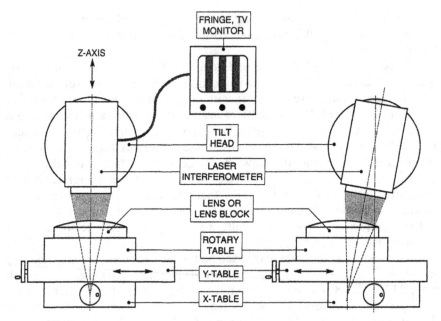

**Figure 6.20.** Vertical interferometer setup for block inspection.

for convex blocks because they are restricted by the working distance of the transmission spheres. This distance is a function of the focal length and the clear aperture of the transmission spheres. For a standard lens with a 100-mm (4-in.) diameter aperture, the working distance is as short as 112 mm (4.5 in.) for an f/1.5 lens and about 1050 mm (41.3 in.) for an f/11 lens [37]. In addition the converging beam precludes the full surface measurement of convex lens blocks that are larger than 4 in. in diameter. Only the central region will be represented in the interferogram, since the block surface will be underfilled.

In the mid-1980s a compact version of a shop interferometer was introduced that was specifically designed for use right at the polishing machine [38]. This small table unit accepts all commonly used standard accessories such as reference flats or transmission spheres. This small interferometer, which comes equipped with a simple video system to display the fringes, can be set up in a vertical position, either up- or downlooking, or it can be set up horizontally. Despite its compactness, ease of use, and attractive price, this interferometer has not become the standard in optical shops.

Because of these practical, not strictly technical limitations, spherical testplates continue to be used in most optical shops. In recognition of this, a sizable part of this chapter (Section 6.1) has been devoted to the making and the use of test plates.

### 6.2.4. Phase Interferometers

Commercial phase interferometers became available in the early 1980s [39]. This was the next major improvement of the laser interferometer. Since then these new interferometers have been vastly improved by using high-resolution photodiode or CCD arrays and by introducing complex and powerful phase fringe analysis systems and color graphic displays [40, 41, 42, 43, 44].

The mainframe of this type of interferometer is basically the same as that for the original laser interferometers. One major difference, however, is the phase modulator unit that is attached to the output port of the mainframe to move the reference surface axially by a minute distance. A special control unit that includes a powerful microprocessor for fringe analysis completes the interferometer configuration. Other interferometer manufacturers [42, 43] opted for the Twyman-Green configuration, as shown in Fig. 6.21. The internally mounted reference mirror is modulated by the PZT transducer.

There are several important advantages with a phase interferometer:

- They permit real time evaluation of interference fringes. This eliminates some of the problems that are typical for the simpler laser interferometers such as fringe drift and the negative effects of mechanical vibrations and thermal distortion.

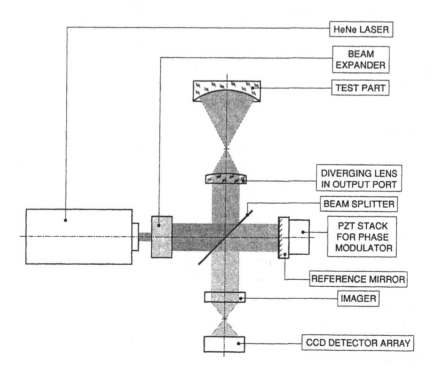

**Figure 6.21.** Typical Twyman-Green phase interferometer.

- Unlike their simpler predecessors, phase interferometers can evaluate closed fringes and highly irregular fringe patterns with fringe spacings as small as 2 mm (0.080 in.).
- They will also unambiguously determine the sign of the error; that is, they can tell convex from concave fringes.
- Phase interferometers can evaluate the fringe patterns from discontinuous or interrupted surfaces and from oddly shaped surfaces as well.
- Phase interferometers are inherently more accurate than the earlier laser interferometers because of their high-resolution CCD cameras and their powerful fringe analysis capabilities.

In phase interferometers the reference flat or lens is installed into the output port of the mainframe. However, the phase modulation unit is firmly mounted to the mainframe on one side, and it supports the output port on the other side. This is the only obvious difference between the phase and laser interferometers. All other features such as attachments and accessories are identical to those used on the laser interferometers. The setup and test methods are also the same. The way phase interferometers work is as follows:

Piezoelectric transducers mounted in the phase modulation unit are energized. This moves the entire unit, including the reference flat or lens a small but measurable distance. When the energy that is applied to the transducers is cycled, the process is called *modulation,* and the reference surface is moved repeatedly back and forth. This results in a controlled motion of the reference surface that modulates the interference fringes, causing them to vary sinusoidally in intensity. The fringe intensity at any point and time is the phase difference between the two interfering wavefronts (reference and test).

The variations in light intensity are sensed by a detector array. The analog or continuously changing signal is electronically converted into digital form so that the microprocessor can calculate the optical path differences for each pixel in the array. Phase maps are then created from these individual calculations which accurately represent the shape of the tested surface. From these data all other parameters are calculated and displayed on the video display terminal.

The resolution of fine fringe detail is a function of the number of pixels in the array. Initially $100 \times 100$ arrays were used with about 10,000 pixels [39]. More recent phase interferometers have either a $244 \times 388$ CID array [40] or a $256 \times 256$ array [41]. The advantages of high resolution are offset by the need for larger, faster, and more expensive computers and a slower response time because the OPD must be calculated for a much larger number of pixels. For most precision quality optics, however, a $100 \times 100$ array will provide acceptable resolution. Higher resolution may be called for when more complex interference patterns must be analyzed. These are rarely found in precision optics.

### 6.2.5 IR Interferometers

Most infrared (IR) components are made from materials that are opaque in the visible range of the spectrum. While the polished surfaces of IR lenses and windows are usually tested by reflection with visible light interferometry, typically HeNe laser light at 633 nm, the opaque nature of these materials precludes the evaluation of material homogeneity and transmitted wave front tests.

These are often very important tests since the evaluation of the material prior to fabricating an expensive lens can save a great deal of otherwise wasted effort and ensure the proper performance of the finished lens when it is tested in a functional transmitted wave front test. Both of these tests must be done at the design wavelength of the lens. For the standard germanium (Ge) lenses that are commonly used in IR systems, the optimal operating wavelength range lies between 8 and 12 $\mu$m. The light emitted by a $CO_2$ laser at 10.6 $\mu$m falls right into this range. Therefore most IR interferometers have a $CO_2$ laser source. A HeNe laser is sometimes used in these systems for visually aligning the internal interferometer optics and also for positioning the testpart relative to the interferometer. There are several IR interferometers commercially available [45, 46, 47].

For testing near-IR optics in the 1 $\mu$m region, a Nd:YAG laser that operates at 1.06 $\mu$m is used as an IR interferometer source. It can be used for evaluating the internal quality of materials that are transmissive in this region. These are usually opaque in the visible region, so HeNe interferometry cannot be used. It is also useful to test components made from these materials at the design wavelength in transmission.

The claimed system accuracy is $\frac{1}{10}$ wave at 10.6 $\mu$m over a 290-mm aperture and $\frac{1}{20}$ wave at 1.06 $\mu$m over a 100-mm (4-in.) aperture [45]. With expensive beam expanders, the useful aperture can be as large as 406 mm (16 in.) [47]. In addition to performing transmitted wave front tests of material and finished components, IR interferometers can be also useful in aligning and functionally testing IR lens systems, performing surface metrology of difficult to test surfaces such as diamond-turned mirrors and aspheric lenses, and permitting the evaluation of fine-ground lens and mirror surfaces prior to polishing.

IR phase measuring interferometers are also commercially available on special order [48, 49]. The standard Twyman-Green configuration can be converted by driving the reference mirror with a piezoelectric transducer (PZT). The fringes are detected by the 120 × 120 pyroelectric array of an IR video camera. This arrangement is shown in Fig. 6.22. as set up for a reflected wave front test. For testing in a transmitted wave front the test part is placed between the output port and a suitable reference quality reflector, which can be either plano or concave. The nature of the test determines the exact configuration.

The standard IR interferometer has no phase-measuring capabilities. A pyroelectric video camera detects the fringes, and the resulting image is

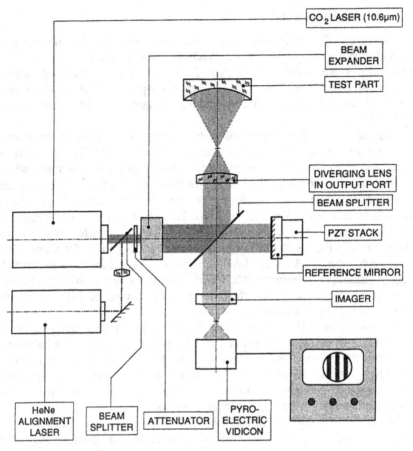

CO₂ LASER (10.6μm)

BEAM EXPANDER

TEST PART

DIVERGING LENS IN OUTPUT PORT

BEAM SPLITTER

PZT STACK

REFERENCE MIRROR

IMAGER

HeNe ALIGNMENT LASER

BEAM SPLITTER

ATTENUATOR

PYRO-ELECTRIC VIDICON

**Figure 6.22.**  IR interferometer (Twyman–Green phase shifting).

displayed on a TV monitor. The resulting fringes are well defined and of high contrast. A hard copy (Polaroid or thermal printer) is needed for the computer-assisted fringe evaluation. Standard fringe analysis software are used for the analysis of fringe patterns. The computer data can be displayed as contour maps or as three-dimensional (3-D) surface plots. All other parameters can also be calculated and displayed, same as done for the HeNe laser interferometers.

### 6.2.6.  Special Interferometers

Laser interferometers can also be equipped with an argon laser source for evaluating surfaces by reflection or components and systems in transmission at the primary argon wavelengths of 488 and 514 nm [50]. Large-aperture interferometers are available; they use a standard HeNe laser mainframe with suitable beam directors [51] and beam expanders that can expand the

100-mm (4-in.) standard output beam up to 800 mm (32 in.) or even more. These large test systems, which are often set up for dual aperture testing (standard and enlarged), require vibration-isolated optical tables. These systems are used to perform all common interferometer tests on large systems and surfaces.

Small aperture interferometers [50] are created by reducing the standard 100-mm (4-in.) aperture with a beam reducer to 12-mm (0.5-in.) diameter. These special systems are designed for testing small surfaces and parts. They are particularly useful for testing the ends of laser rods, especially when the 1:6 zoom feature is employed. With this arrangement, a 2 mm (0.080-in.) diameter rod end can be evaluted for flatness or sphericity at the full aperture of 100 mm (4 in.). In other words, the resulting interference pattern will fill the screen.

Other laser interferometers have been built for special applications. For instance, a couple of laser interferometers are designed for evaluating the output of laser diodes at a wavelength of 840 nm [52, 53]. Another special purpose instrument is a grating interferometer for measuring cylindrical surfaces to an accuracy of $\frac{1}{8}$ wave at HeNe wavelengths [54].

### 6.2.7. Interferometer Accessories

The original shop interferometer [33] introduced a series of external accessories that snapped into the output port of the mainframe. These accessories, such as reference or transmission flats or a series of transmission spheres, convert the mainframe into a multifunctional interferometer.

The accessories transform the collimated output beam of the mainframe into a plane wave front for testing plano surfaces and plane-parallel parts or into a spherical wave front for testing lenses and lens surfaces. A sufficiently reflective test piece surface introduced into these wave fronts will reflect some of the light back into the interferometer, where it will interfere with the reference wave front. The resulting interference pattern can be observed directly when it is projected onto a ground glass screen, or it can be indirectly seen when it is detected by a photodetector array and displayed on a video monitor. The earlier fringe projection method is no longer used because the fringe contrast was poor and the fringes had to be indirectly reduced from a hardcopy which was usually a Polaroid picture. All modern interferometers use advanced photodetector arrays to acquire the interference pattern and display the high contrast fringes on a video monitor.

Two types of fringes are typically seen. Either they are the commonly seen broad ($\cos^2$) fringes when the surface reflection of both the reference and test surfaces is low, or they are high-finesse fringes that are the result of multiple beam interference when both surfaces are highly reflecting (>40%). A mismatch in the surface reflection between the reference surface of the accessory and the test surface results in poor fringe definition which becomes increasingly poorer as the mismatch becomes greater.

The original easy-to-use fringe acquisition method [33] has never been radically changed. It is still used in essentially the same manner today. The focused spot image of the reference surface is centered on the screen by adjusting two screws on the output port until the spot image is positioned at the center of a fixed cross target. The focused image of the test surface is then superimposed on the center of that target and the position of the test surface adjusted. When this alignment is complete, a beam splitter is swung into position to redirect the beams, and this causes them to interfere. The original accessories have also remained unchanged over the years and have become the standard for all similar interferometers. The only significant change has been the expansion in size and in the versatility of applications.

**Interferometer Flats.** Plane wave accessories are stress-free mounted precision flats that are designed to permit testing in either transmission or by reflection. The reference surface of a transmission flat is uncoated, while the opposite exterior surface is antireflection coated. The reference surface of a reflection flat is coated with a 90% reflecting coating. The reflection flats are used only for testing highly reflecting plano surfaces in multiple beam interferometry. There are no equivalent high-reflection elements for spherical surfaces.

The standard 100-mm (4-in.) diameter clear aperture flats are made from low-expansion materials such as fused silica, ULE [55], or CerVit [56]. The reference surface is polished to an accuracy of $\frac{1}{20}$ wave or better at 633 nm. The output beam can be reduced to 33-mm (1.3-in.) diameter which provides for a $3\times$ magnification of the fringe pattern. This makes it easier to view fringe patterns of small diameter surfaces. Flats with 150-mm (6-in.) clear aperture are also available [39]. Transmission and reflection flats have been made up to 813 mm (32 in.) in diameter [50].

**Transmission Spheres.** Transmission spheres are highly corrected, multielement lens systems that convert the collimated output beam emerging from the mainframe into a spherical wave front. The typically concave surface that faces the output beam is the highly accurate reference surface of the transmission sphere. The rated precision of this surface is $\frac{1}{10}$ wave over the 100-mm (4-in.) diameter aperture.

The radius over diameter ($R/D$) ratio of the test surface must match the f/# of the transmission sphere in order to obtain the best interferogram. When the $R/D$ ratio is less than the f/#, then the test surface is overfilled. In the reverse case where $R/D$ is greater than f/#, the test surface is underfilled. The conditions discussed in the following paragraphs are graphically represented in Fig 6.27.

A graph shown in Fig. 6.23 [57] relates the lens or block radius to the lens or block diameter and the commonly used transmission spheres. This graph is helpful in selecting the best transmission sphere for each test condition. For instance, if the $R/D$ ratio falls below a transmission sphere line, then the

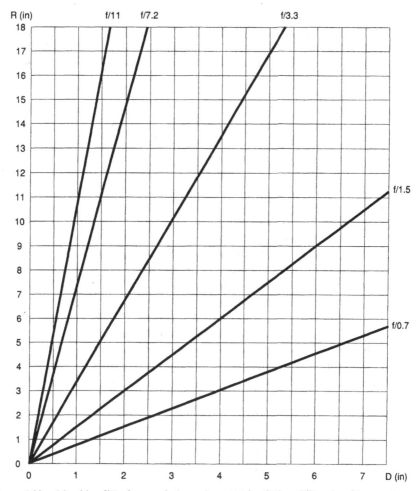

**Figure 6.23.** Matching f/# of transmission sphere (4.0 in. CA) to R/D ratio of lens under test.

surface will be underfilled. If it lies above it, then the surface is overfilled. The worst case occurs when the R/D ratio falls below the lines. In that case it is best to use the next faster sphere and test the part in the overfill condition. The ideal case of course is when the R/D ratio falls right on a transmission sphere line.

In the overfill condition the cone angle of the light that emerges from the transmission sphere is greater than the acceptance angle of the test surface as defined by its R/D ratio. This means that only a portion of the light is intercepted by the test surface, and this results in a reduced fringe contrast and a fringe pattern that is smaller than the full aperture on the video screen.

When the surface is underfilled, the cone angle of the spherical wave front is not large enough to cover the entire test surface. As a result the fringe pattern that will fill the clear aperture of the video screen represents only the central region of the test surface that is illuminated. This condition often leads to the erroneous conclusion that the entire surface is represented in the fringe pattern when in fact the important edge condition is not represented at all.

**Measurement of Long Radii.** For the measurement of very long radii of 1 m (39.4 in.) or greater, a series of 100 mm (4-in.) clear aperture transmission spheres are available ranging from f/15 to f/45. They are configured as either convergers form the measurement of convex surfaces or as divergers for concave surfaces. Figure 6.24 shows the general relationship between the f/# of the lens and the longest radius that can be measured this way. Divergers can be used to extend the range of a radius rail for measuring concave surfaces. Because divergers have a virtual focal point, the measuring range of the rail can be extended by this virtual focal length.

For convex or concave surfaces with a radius that exceeds 5 m (197 in.), a zoom lens system can be useful that can be adjusted from +f/50 through zero power to −f/50. The minimum concave radius that can be measured this way is 5.2 m (205 in.) at a clear aperture of 112 mm (4.4 in.) and the minimum convex radius is 6.3 m (248 in.) at a clear aperture of 127 mm (5.0 in.) [39]

Transmission cylinders for the interferometric evaluation of concave and convex cylindrical surfaces are also available. These 150 mm (6.0 in.) clear aperture diameter transmission lenses are multielement cylinder lens systems with a claimed reference accuracy of 1/4 wave at 633 nm [58]. These lenses must be used with specially designed mounts that ensure that the lens axes are lined up in the correct attitude.

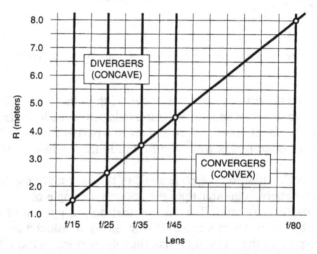

**Figure 6.24.** Radius limits as function of f/#.

**Radius Rail.** The typical interferometer when equipped with a radius rail can be used as a very accurate spherometer that measures the radius directly and has the added advantage that the surface figure can be determined at the same time when the power is nulled out.

The radius of curvature measurement with an interferometer is based on the autostigmatic method that is discussed in Section 6.3. The accuracy of the measurement depends on the R/D ratio of the test surface. The smaller the ratio, the faster the surface, and the more accurate the radius measurement. Errors are introduced into the measurement because it is difficult to judge the straightness of the fringes at the null positions and because the resolution of the digital linear encoder is 0.0125 mm (0.0005 in.), this adds an additional limitation to the measurement accuracy.

The straightness of fringes can be judged to an accuracy of 1/8 wave. With experience this limit can be reduced to within 1/10 wave. If the fringe straightness can be reliably judged to within 1/10 wave, then the measurement accuracy can be as good as 2 $\mu$m (0.0001 in.) for a f/2 surface or 32 $\mu$m (0.0013 in.) for a f/8 surface [58]. This means that if the f/# decreases by a factor of 4, the radius accuracy declines by a factor of 16.

The typical radius rail is 1 m (39.4 in.) long with a 0.013-mm (0.0005-in.) resolution of the digital linear encoder. This gives the rail a radius accuracy of 0.01%, which is potentially more accurate than a precision spherometer.

**Lens Mounts.** Mounts are an essential part of any interferometer. They are needed to firmly hold or support the test parts or interferometer components. The mount system originally developed for the first interferometer of this kind [33] has become with some additional refinements the standard of the industry.

Two-axis tilt mounts are used for holding plano parts. Either the test surface or the reference surface or both are tilted in one or both axes to align them so that they are square to the interferometer axis. Only when both surfaces are squared on and nearly parallel to each other can fringes be seen and measurements be performed in transmission or by reflection.

Two- or three-axis translation stages must be used for spherical surfaces to position them for interferometric measurements. The $x$–$y$ stage is used to center the surface to acquire the fringe pattern. The third axis, called the $z$-axis, is used to position the spherical surface along the interferometer axis. This positioning permits the removal of spherical power component of the fringe pattern, which results in straight fringes if there are no other fringe irregularities. By finding this null position, one can clearly expose residual fringe errors and separately evaluate them. This positioning feature is also essential for making radius measurements with the radius rail.

By combining the two-axis tilt mount and the three-axis translation mount, a five-axis mount can be constructed that can be used for almost all types of test setups. These mounts accept either the standard 100 mm (4-in.)

or 150-mm (6-in.) accessories. The test parts are most often held in a self-centering three-prong lens holder that is attached to one of the mounts.

For testing cylinder surfaces or the performance of cylinder lenses, the standard tilt and translation mounts alone are not sufficient. A rotation feature must be added to allow rotation of the component under test around the system axis for proper alignment and orientation.

**Beam Directors.** Flip mirror assemblies, called *MUX cubes* [51], are used to redirect the beam path by 90°. The beam can be redirected in either a horizontal or in a vertical direction. With the mirror folded out of the clear aperture, the beam will pass straight through the cube unimpeded. The MUX cubes are used to alternate the beam between two parallel setups that share the same mainframe. This is often done for testing at two different apertures. For instance, one beam operates at the standard 100-mm (4-in.) diameter for normal testing, while the other is expanded to 300-mm (12-in.) diameter for testing large surfaces. This way either beampath can be used for testing without having to disturb the setup.

**Attenuators.** Attenuators are needed to balance uneven reflectivities between the test surface and the reference surface. A significant mismatch in surface reflectivities will result in poor fringe definition. A pellicle attenuator is used with most modern interferometer systems for this purpose.

Pellicle attenuators are very thin, optically coated Mylar [59] membranes that have been tautly stretched across a round mounting ring very similar to cloth stretched across an embroidery hoop. These attenuator screens that range in size from 100-mm (4-in.) diameter up to more than 800 mm (32 in.) are very fragile because they are so thin. The larger they get, the more fragile they become. To protect them from inadvertant damage, the pellicles are mounted on a firm stand, and movable covers protect them when they are not in use.

Attenuators are used to reduce the light intensity in one arm of the interferometer by absorbing some of the light to match it to the light level in the other. Attenuators are important for maintaining optimum fringe contrast in that they balance the reflectivities of the test and reference surfaces to make them nearly equal. This balance is especially necessary when one surface is much more reflective than the other. For instance, a polished silicon surface reflects more than 40% of the incident light, while the reflectivity of the standard reference surface is about 4%. This rather significant mismatch will result in very weak fringes that cannot be easily evaluated. If the mismatch is greater, the contrast can be nearly entirely destroyed to the point where fringes cannot be detected.

Because the pellicle is so thin, it has little or no effect on the transmitted wavefront. There is also no reduction of the focal distance for even very fast converging or diverging beams.

### 6.2.8. Testing Arrangements

The mainframe and the lens mounts can be arranged in one of three configurations as shown in Fig. 6.25. The horizontal arrangement shown in Fig. 6.25a is the most common way to use the interferometer. Vertical arrangements are possible as shown in 6.25b and 6.25c. Both of these are useful for in-process testing of optical components. The vertical uplooking arrangement is useful only for testing plano parts, while the downlooking mode is

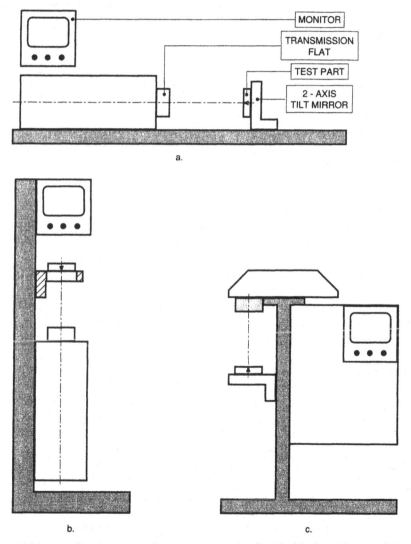

**Figure 6.25.** Interferometer mounting arrangements. *(a)* Standard horizontal setup, *(b)* Vertical up-looking. *(c)* Vertical down-looking.

preferred when testing lenses or small lens blocks. The test part mount must be adjustable along the interferometer axis to compensate for changing radii of curvature. Typical test setups are shown in Figs. 6.26 and 6.27 for the horizontal test arrangement. Figure 6.26 shows the most common plane wave test setups, and Fig. 6.27 shows those for spherical wave front testing.

The basic procedure for plane wave testing starts with the installation of the chosen transmission flat in the output port or mounting the reflection flat in a two-axis tilt mount. These components must first be squared on to the interferometer axis in the alignment mode by adjusting the two screws. Once this alignment has been done, the adjustment must not be touched unless the flat has gone out of alignment. All other adjustments should be made on the mount that holds the test part.

The same procedure holds for spherical wave testing. Once the selected transmission sphere is installed and aligned, it must not be adjusted again. Since, unlike plane surfaces, spherical surfaces must also be located on axis, it is best to use a radius rail. A simple optical rail against which the mounts can be registered will keep them on axis. It is even better, however, to invest in a 1-m (39.4-in.) radius rail that is designed for use with these mounts. A radius rail with a digital linear readout is an inexpensive way to acquire a high-precision spherometer capability.

## 6.2.9.  Fringe Reduction

Once a fringe pattern is obtained, it must be analyzed to determine the type and magnitude of surface errors. This can be as simple as looking at the fringes as is done when testplating and counting fringes and determining their shape. This is a qualitative method that is useful for in-process tests and for final acceptance when the acceptance criteria are several fringes. However, the method lacks the precision needed for final acceptance for more stringent requirements. The fringe analysis must be more quantitative for these more common requirements.

**Manual Reduction.** The simplest way to get quantitative data is to take a photograph of the interference pattern. Such photographs are called *interferograms*. It was quite customary until recent times to use an instant film such as Polaroid® to obtain an interferogram for subsequent analysis. The analysis was a manual process with which the fringe spacing and the fringe deviation from straightness were graphically determined. A line was drawn through the end points of a center fringe. Another line was drawn through the endpoints of an adjacent fringe. By definition, the distance between these lines is 1/2 wave. Any deviation from straightness was then measured from the center of the fringe to the line corresponding to that fringe. This distance could then be expressed in terms of the line separation that determined a 1/2 wave or 1 fringe. With some skill slight astigmatic conditions could also be defined this way. This method, which is described in greater detail in Section

| INTERFEROMETER SET-UP | ALIGNMENT | DISTANCE D | SPECIAL CONSIDERATION | TYPE OF TEST |
|---|---|---|---|---|
| 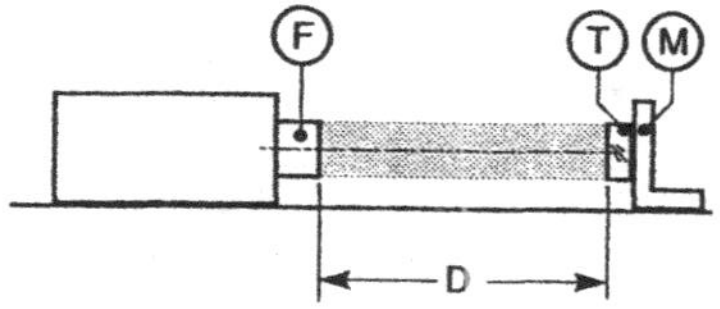 | (F) TRANSMISSION FLAT<br>(T) TEST PART NORMAL TO BEAM<br>(M) 2 AXIS TILT MOUNT | NOT CRITICAL | • IF (T) IS NOT COATED, USE 4%R FLAT (F)<br>• IF (T) IS HIGHLY REFLECTING, USE 90% FLAT (F)<br>• OR USE 4% (F) WITH ATTENUATOR | FLATNESS (FIRST SURFACE) |
| 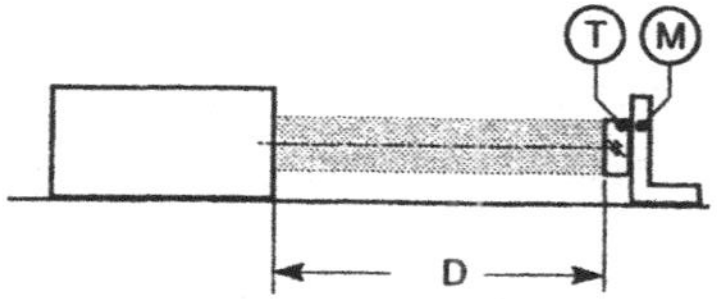 | (F) NOT REQUIRED<br>(T) NEARLY NORMAL TO BEAM<br>(M) NO ADJUSTMENT NEEDED | NOT CRITICAL | USE GRAPH FOR INTERPRETATION | PARALLELISM OR WEDGE |
| 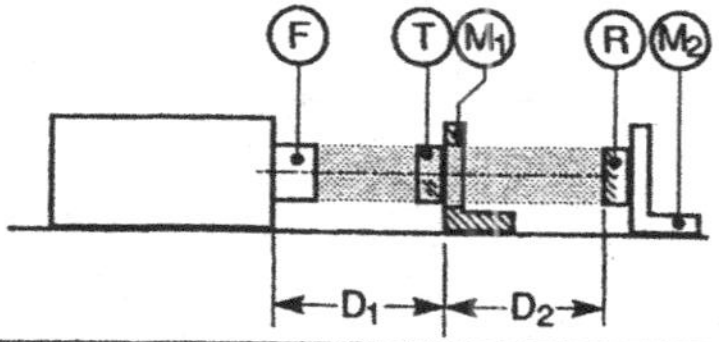 | (F) TRANSMISSION FLAT (4% OR 90%)<br>(T) NEARLY NORMAL TO BEAM<br>(M1) NO ADJUSTMENT NEEDED<br>(R) REFLECTION FLAT (4% OR 90%)<br>(M2) Z-AXIS TILT MOUNT | NOT CRITICAL FOR BOTH $D_1$ AND $D_2$ | (R) MUST BE SQUARED ON FIRST, THEN INTRODUCE (T) | TRANSMITTED PLANO WAVEFRONT |
| 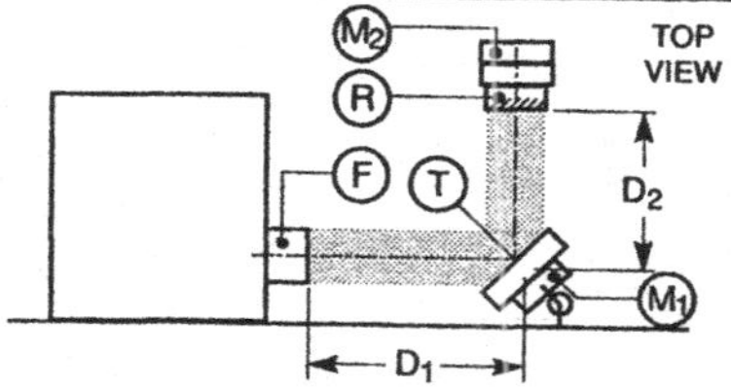 | (F) TRANSMISSION FLAT (4% OR 90%)<br>(T) AT ANGLE θ AND NORMAL TO BEAM<br>(M1) 2-AXIS TILT MOUNT<br>(R) REFLECTION FLAT (4% OR 90%)<br>(M2) 2-AXIS TILT MOUNT | NOT CRITICAL FOR BOTH $D_1$ AND $D_2$ | (R) MUST BE SQUARED ON FIRST, WITH 90° MIRROR AT θ (T) IS INTRODUCED AND ADJUSTED TO BEAM AND (R) | REFLECTED WAVEFRONT AT ANGLE θ |

Figure 6.26.  Basic plane wave interferometer arrangements.

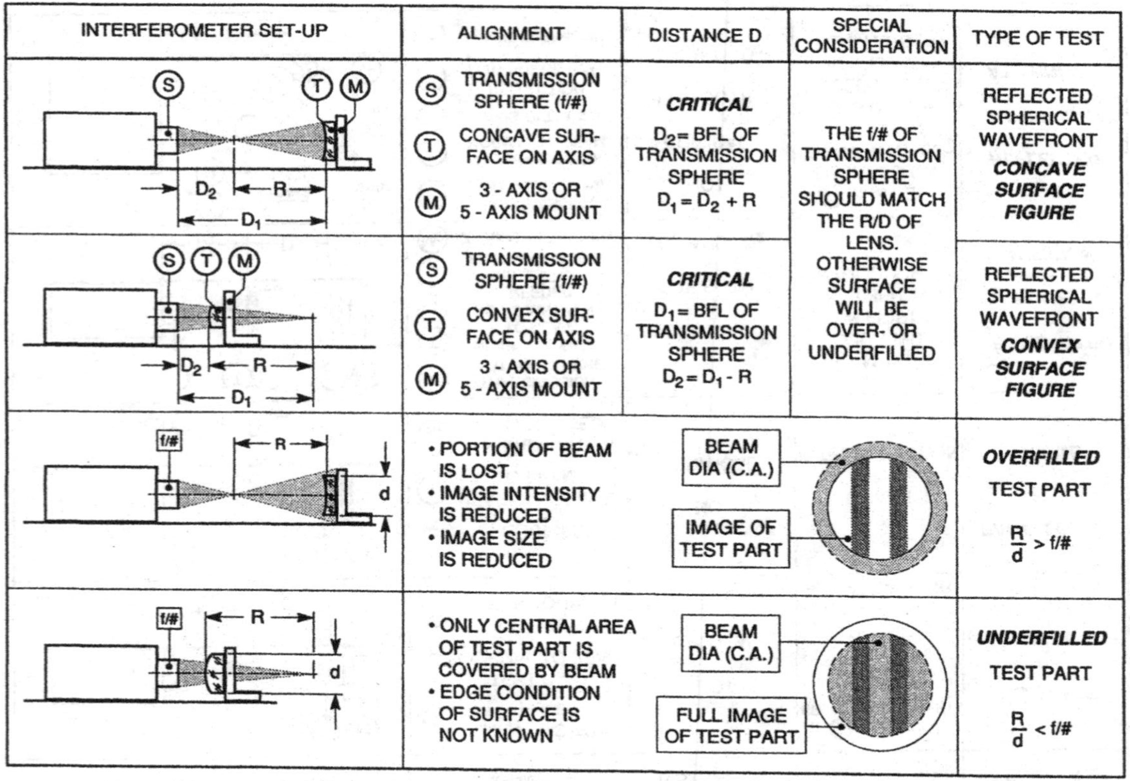

| INTERFEROMETER SET-UP | ALIGNMENT | DISTANCE D | SPECIAL CONSIDERATION | TYPE OF TEST |
|---|---|---|---|---|
| | (S) TRANSMISSION SPHERE (f/#)<br>(T) CONCAVE SURFACE ON AXIS<br>(M) 3-AXIS OR 5-AXIS MOUNT | *CRITICAL*<br>$D_2$ = BFL OF TRANSMISSION SPHERE<br>$D_1 = D_2 + R$ | THE f/# OF TRANSMISSION SPHERE SHOULD MATCH THE R/D OF LENS. OTHERWISE SURFACE WILL BE OVER- OR UNDERFILLED | REFLECTED SPHERICAL WAVEFRONT<br>*CONCAVE SURFACE FIGURE* |
| | (S) TRANSMISSION SPHERE (f/#)<br>(T) CONVEX SURFACE ON AXIS<br>(M) 3-AXIS OR 5-AXIS MOUNT | *CRITICAL*<br>$D_1$ = BFL OF TRANSMISSION SPHERE<br>$D_2 = D_1 - R$ | | REFLECTED SPHERICAL WAVEFRONT<br>*CONVEX SURFACE FIGURE* |
| | • PORTION OF BEAM IS LOST<br>• IMAGE INTENSITY IS REDUCED<br>• IMAGE SIZE IS REDUCED | | BEAM DIA (C.A.)<br>IMAGE OF TEST PART | *OVERFILLED* TEST PART<br>$\dfrac{R}{d} > f/\#$ |
| | • ONLY CENTRAL AREA OF TEST PART IS COVERED BY BEAM<br>• EDGE CONDITION OF SURFACE IS NOT KNOWN | | BEAM DIA (C.A.)<br>FULL IMAGE OF TEST PART | *UNDERFILLED* TEST PART<br>$\dfrac{R}{d} < f/\#$ |

**Figure 6.27.** Basic spherical interferometer arrangements.

6.1, is very time-consuming and not sufficiently accurate when residual fringe irregularities are present.

In the early days of shop interferometry, a mechanical pantograph was marketed [50] for simplifying manual fringe reduction. It was constructed like a hinged parallelogram. Equal-spaced wires were tautly stretched between two opposite sides of the parallelogram. When the angle of the parallelogram was changed, the spacing between the wires would change proportionately. By overlaying the wires over the fringe pattern, the 1/2 wave spacing could be quickly determined.

**Computer-Assisted Fringe Reduction.** One of the earliest computer-assisted fringe analysis systems [60] was only capable of evaluating open fringes that were nearly parallel and equally spaced with maximum fringe irregularity of one wave. The fringe pattern generated by the interferometer was either photographed with an integral Polaroid camera or it was scanned by a built-in video camera and displayed on a video monitor. The video image could be analyzed directly, while the hard copy interferogram had to be placed into an interferogram reader. The video image suffered from fringe drift that made it difficult to get a good analysis. This of course was not a problem with the hard copy interferogram.

The next generation of interferogram analysis systems [61] provided an objective fringe analysis automatically. With such a system the fringe centers are automatically located and the coordinates of the points that define the centers are referenced to a fixed coordinate system. The computer uses this information to calculate the data points, computes from these the power and irregularity components of the fringes, and displays the results superimposed on the interferogram on the monitor. Figure 6.28 shows such an arrangement schematically.

To evaluate the fringes only over the clear aperture of the test part, four electronically adjustable cursors are positioned over the interference pattern

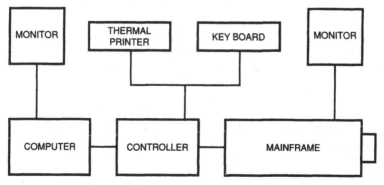

**Figure 6.28.** Basic interferometer system.

to define the area over which the fringes should be analyzed. This feature makes it possible to ignore edge conditions that lie outside the clear aperture but can adversely affect the results. The four orthogonally aligned but independently adjustable cursors form square or rectangular areas over which the fringes are analyzed. The typically round interference patterns are best analyzed by a superimposed square area that is defined by the cursors. The internal video camera displays the fringe pattern on a video monitor for observation by the operator. The pattern is also projected onto a camera for hard copy interferograms. In the past few years these cameras have been replaced with the more cost-effective thermal printers.

The data from the fringe pattern analysis are processed by a computer module and the results are displayed as peak-to-valley (P–V), power and root-mean-square (rms) values. Three-dimensional plots of the surface can also be displayed with a vertically exaggerated isometric view to highlight surface errors. By subtracting the power from the P–V value, the fringe irregularity can be isolated and displayed as P–V-PWR. This is a powerful feature since the P–V value includes the effect from all departures from the ideal surface shape, while P–V-PWR is a quantitative value for the more detrimental surface irregularity.

The latest big step forward in fringe analysis occurred with the introduction of the phase interferometer during the first half of the 1980s [62, 63, 41]. The refinement of these analysis capabilities is an ongoing effort. Phase interferometry offers some decided advantages over the older Fizeau interferometry:

- Multiple sampling of the fringes at real time rates has eliminated fringe drift and mechanical vibrations as sources of errors.
- Fringe analysis of closed fringes and of highly irregular fringe patterns is now routinely done.
- Fringe patterns can now be analyzed over noncircular apertures such as ellipses, squares and rectangular surfaces.
- Discontinuous surfaces such as those found on laser gyro mirrors can also be analyzed.
- The program can detect the maximum rate of slope change, and it can identify the location with a marker on the interferogram.
- Most important for the optician, the sign of the fringe curvature whether high or low can be unambiguously defined.

**Fringe Reduction Programs.** There are a number of fringe reduction programs available. Some of these are specifically designed to function with a specific interferometer system [64, 65, 66, 67], while others are adaptable to function with some of these systems [68, 69]. These programs provide some or all of the following features:

### For Geometric Analysis

- Peak-to-valley (P–V)
- Power (PWR)
- Root mean square (rms)
- Strehl ratio
- Best-fit Zernicke polynomials
- Seidel aberrations
- Spot diagrams
- Encircled energy distribution
- Slope error calculations

### For Diffraction Analysis

- Point spread function (PSF)
- Modulation transfer function (MTF)
- Radial energy distribution

While all of these features are important for testing components and optical systems, only some of these are of interest to the optician. This still incomplete list points to the versatility that modern interferometers possess.

The current most advanced phase interferometer [70] combines all of the preceding features with powerful menu-driven, multilevel software programs and couples that with high resolution CCD cameras and color video displays. The high data density of this new breed of interferometer makes it possible to accurately analyze all but the most complex interference patterns.

Despite this inherent complexity even these advanced systems are easily set up for routine shop testing, since the mainframe, the accessories, the mounts, and rails are for the most part identical to the earlier shop interferometers with which most opticians are familiar. The higher resolution of the cameras, the new powerful computer programs, and the detailed color video displays gives the opticians analytic tools that were not available before.

## REFERENCES FOR SECTION 6.2 (INTERFEROMETERS)

[29] Davidson Optronics D 305 plano interferometer.
[30] Graham Optical Systems, 1988.
[31] Carl Zeiss, model Direct 100.
[32] Möller-Wedel, model V-100, 1988.
[33] Zygo Corp., model GH, 1972.
[34] Test conducted by author (OCLI 1978).
[35] Test conducted by author (Coherent 1987).

[36] Möller-Wedel "Spheroscope."

[37] Zygo Corp., model Mark II, 1976.

[38] Zygo Corp., model PTI, Production Test Interferometer, 1986.

[39] Zygo Corp., model Mark III, 1978.

[40] Zygo Corp., model Mark IV, 1986.

[41] Wyko Corp., model SIRIS 630, 1986.

[42] Breault Research Organization, 1986.

[43] Wyko Corp., model 6000, 1988.

[44] Zygo Corp., model Mark IV xp, 1990.

[45] Zygo Corp., 10.6 $\mu$m IR interferometer.

[46] Zygo Corp., 1.06 $\mu$m IR interferometer.

[47] Wyko Corp., model IR-3 (10.6 $\mu$m IR interferometer).

[48] Breault Research Org., IR phase interferometer.

[49] Wyko Corp., IR phase interferometer.

[50] Zygo Corp., 1982.

[51] Zygo Corp., MUX cube.

[52] Zygo Corp., model Wavefront Analyzer, 1986.

[53] Wyko Corp., model SIRIS 800, 1986.

[54] Gomez Development Corp., 1988.

[55] ULE (Corning USA).

[56] CerVit (OI Kimble, USA).

[57] Adapted from Zygo Corp. operating instructions for model GH.

[58] Zygo Corp.

[59] Mylar.

[60] Zygo Corp., ZIPP.

[61] Zygo Corp., ZAPP, 1984.

[62] Zygo Mark III, 1982.

[63] Möller-Wedel, model III-70 (plano only), 1984.

[64] Zygo Corp., ZAPP (modified).

[65] Zygo Corp., FRINGE, 1982.

[66] Wyko Corp., WISP, 1986.

[67] Möller-Wedel, INTOMATIC, 1984.

[68] CVI Laser Corp., AIRS, 1986.

[69] Phase Shift Technology, FAST, 1988.

[70] Zygo Corp., model Mark IV xp.

## 6.3.  SPHEROMETRY

Spherometry is the collective term for a number of measurement techniques through which the radius of curvature of a spherical surface can be determined. There are three basic techniques:

First, there is an indirect technique by which the value of another parameter, the sagitta, is measured from which the radius of curvature can then be calculated. Second, by using certain optical techniques, direct measurement of the radius of curvature is possible. Finally, radius gauges and reference surfaces are often used for a quick radius determination.

The most commonly used instrument is the spherometer, which utilizes the indirect technique. The testplate is the most common form of radius gauge. The more direct measurement techniques make use of optical benches, height gauges, and interferometers. Hand spherometers, diopter gauges, and radius templates are often used for quick referencing.

### 6.3.1. Theory

The spherometer utilizes the indirect measurement technique. It is designed specifically to measure the sagitta ($s$) of a spherical surface which is defined by the radius of curvature ($R$) and the diameter ($d$) of the circle over which the measurement is made. This relationship is shown in Fig. 6.29. In plane geometry, the sagitta $s$ is a function of the radius $R$ of the circle and the length of the chord $d$. For a spherical surface, $s$ remains the sagitta (sag for short), $R$ is the radius of curvature, and $d$ is the diameter of the circle over which the measurement is made. The relationship is mathematically expressed by the "sag equation":

$$s = R - \sqrt{R^2 - \left(\frac{d}{2}\right)^2}.$$  (6.35)

Every optician should be familiar with Eq. (6.35) and how to use it because it helps solve a number of commonly encountered situations in the optical shop. Note that there is a term $(d/2)^2$ in the equation, where d is the

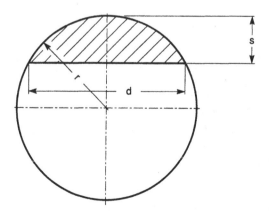

**Figure 6.29.** Definition of sagitta.

diameter of the spherometer ring used. Since $d = 2r$, where $r$ is the radius of the measuring circle, then $(d/2)^2$ is just another way of saying $r^2$. Equation (6.35) is typically expressed in terms of r which is the semi-diamter of the spherometer ring $d$.

$$s = R - \sqrt{(R^2 - r^2)}. \tag{6.36}$$

Since the measured value is the sag s and the value that must be calculated is the radius $R$, Eq. (6.36) must first be solved for $R$ to yield:

$$R = \frac{r^2 + s^2}{2s} \tag{6.37}$$

The value $r$ determined by the spherometer ring used, and the sag value $s$ is the result of the measurement. The spherometer is first nulled against a plano reference surface after which the sag height of the spherical test surface is measured. The sag value is the difference between these two measurements. From these data radius $R$ can then be calculated using Eq. (6.37). Examples will demonstrate the use of these equations later in this section.

For very long radii, where the sagitta becomes quite small, an approximate form of Eq. (6.37) will be sufficiently accurate. Since the value of $s^2$ becomes so small to be practically zero, it can be safely dropped from the equation.

$$R = \frac{r^2}{2s}. \tag{6.38}$$

Another equation for calculating the radius of curvature that is sometimes seen in the literature is

$$R = \frac{d^2}{8s} + \frac{s}{2}. \tag{6.39}$$

This equation is just another form of Eq. (6.37) which results from the substitution of $r$ by $d/2$. It leads to the term $r^2/2s = d^2/8s$, to which we must add the term $s^2/2s = s/2$.

Assuming a precision spherometer, the accuracy of the calculated value of $R$ depends on the accuracy with which the radius $r$ of the measuring circle is known and the care with which the sagitta s has been measured.

### 6.3.2. Basic Spherometer Types

There only two ways to define the diameter of the measuring circle that yields the value $r$. One is the ring spherometer for which the measuring circle is a perfectly round and sharp edged precision ground ring-shaped bell called a *spherometer ring*. The other is the ball spherometer for which the measuring circle is defined by the centers of three perfectly matched balls that form the corners of an equilateral triangle. There are a number of variations for either type, each with its own advantages, limitations and special applications. Most good spherometers are convertible from one type to the other.

The measuring point lies exactly at the center of the measuring circle for either the ring or the ball spherometer. The circle need not be closed, as is the case for the bar spherometer which will be discussed later in this section. Figure 6.30 shows three typical ways how the measuring cicrle can be defined.

The accuracy of the instrument depends not only on how small a value can be resolved by the micrometer but also to a large degree on the precision with which the measuring circle has been defined and how accurately the measuring point has been centered with respect to the circle. The manner by which the measurement is made varies from instrument to instrument. Some spherometers use a counterweighted measuring rod. Others use differential micrometers and precision transducers with electronic readout.

**Ring Spherometer.** One type of ring spherometer that has proved itself over many years for its utility and reliability is the Abbé type [71] which is still being produced by several European manufacturers. The basic construction of this old instrument is described because it is very similar to that of all other more modern spherometers. The body of this type of spherometer is a sturdy casting that firmly supports the interchangeable spherometer rings. The centrally located plunger with the measuring point at its end rises vertically under the influence of a counterweight until it gently contacts the surface that is being measured. The vertical motion of the plunger is nearly frictionless as long as the instrument is properly leveled. The total range of travel is 30 mm (1.2 in.)

The distance traveled by the plunger is precisely measured by an integral 50× microscope through which the exact position of the point can be read off

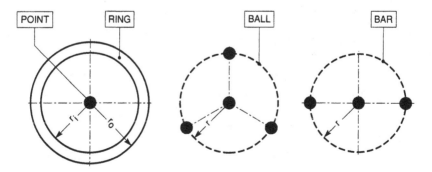

**Figure 6.30.** Definition of measuring circle.

on a precision vernier glass scale that is attached to the plunger. The potential reading accuracy is 0.5 $\mu$m (20 $\mu$in.), but this level of accuracy is usually not achieved under normal measurement conditions. A minimum value of 1 $\mu$m (40 $\mu$in.) is more realistic. For light lenses that might be lifted off the ring by the force of the point, a weight can be used to provide the necessary counterforce to assure proper contact with the ring. A schematic representation of this type of spherometer is shown in Figure 6.31. Digital readout spherometers are preferred today. It should not be assumed that these instruments are inherently more accurate than the older spherometers, but it is likely that reading errors can be almost totally eliminated.

The measuring circles of the ring spherometer are defined by an accurately made hardened steel bell. The rotational center of this bell is the location of the axially movable plunger. The steel bell can be shaped with a flat rim and straight or tapered sides, as shown in Fig. 6.32. Although the straight sided bell can be made very accurately because the inner and outer diameters can be precisely measured, the sharp edges of this design can be easily damaged due to misuse or ordinary wear. When this occurs, the ring can nevertheless be easily reground without changing the values for the inner and outer diameter. The sharp edges can also damage the polished surfaces where they contact. This danger can be reduced when the edges are

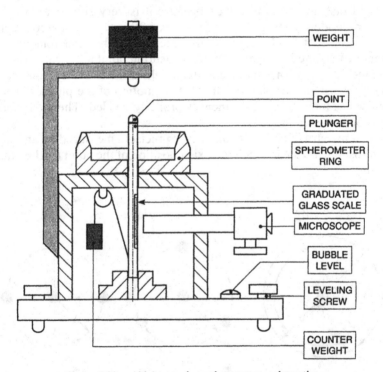

**Figure 6.31.** Abbé-type ring spherometer, schematic.

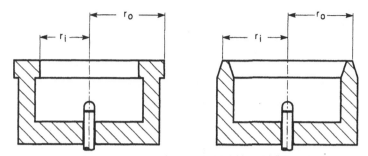

**Figure 6.32.** Typical ring configuration.

lightly polished by holding a piece of felt that has been soaked with polishing slurry against the rotating bell. This operation is best performed immediately after recutting.

The tapered spherometer ring is preferred because it is much less vulnerable to damage. However, it is much more difficult to accurately measure the diameter of the circles to a similar degree of precision. If resurfacing of a tapered ring should become necessary due to damage or wear, it will necessitate a redefinition of the values of the inner radius ($r_i$) and the outer radius ($r_o$) because $r_i$ will become smaller and $r_o$ will become larger.

As shown in Fig. 6.32, the ring spherometer has two distinct measuring circles. The inner circle with radius $r_i$ is used for measuring the sag of convex surfaces, and the sag of concave surfaces is measured with the outer circle with radius $r_o$. This leads to two variations of Eq. (6.37).

For convex

$$R = \frac{r_i^2 + s^2}{2s},$$

(6.40)

and for concave,

$$R = \frac{r_o^2 + s^2}{2s}.$$

(6.41)

Without going into the details of calibration and actual measurement at this point, the following two examples in Fig. 6.33 show how Eq. (6.40) and (6.41) are used to determine the radius of curvature based on the sag measured with a ring spherometer. Results are given in both metric and inches. The importance of using the correct measuring circle for the calculation cannot be overemphasized. If, for instance, in Fig. 6.33b an error had been made by using the radius value of the inner circle instead of the correct value for the outer circle, the calculated radius of curvature $R$ would have been 7.239 mm (0.285 in.) too short.

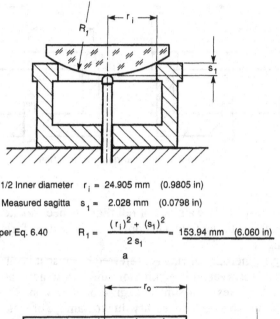

1/2 Inner diameter   $r_i$ = 24.905 mm   (0.9805 in)

Measured sagitta   $s_1$ =  2.028 mm   (0.0798 in)

per Eq. 6.40        $R_1 = \dfrac{(r_i)^2 + (s_1)^2}{2 s_1} = 153.94$ mm   (6.060 in)

a

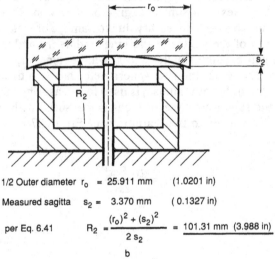

1/2 Outer diameter  $r_o$  = 25.911 mm     (1.0201 in)

Measured sagitta   $s_2$ =  3.370 mm   ( 0.1327 in)

per Eq. 6.41        $R_2 = \dfrac{(r_o)^2 + (s_2)^2}{2 s_2} = 101.31$ mm  (3.988 in)

b

**Figure 6.33.**  Examples for ring spherometer. *(a)* Convex. *(b)* Concave.

Reliable results can be expected when the radii of the circles are accurately known, the resolution of measurement of the instrument is sufficiently high, and good measuring techniques are employed. The circles can be defined to an accuracy of 2.5 $\mu$m (0.0001 in.), and the resolution of the sag measurement can be as high as 1 $\mu$m (0.00004 in.) with a quality instrument. The only limiting factors are the condition of the rings at the time the measurement is made and the care with which this was done.

One commercially available instrument is a high resolution ring spherometer with LVDT (linear voltage differential transducer) probe [72]. The
measured sag value is shown on a LED (light-emitting diode) display. The
precision ground hardened steel rings have a range of diameters as listed in
Table 6.6. The wall thickness of these rings ranges from 0.75 mm (0.030 in.)
for small diameters up to 5 mm (0.200 in.) for the largest rings. The range
selector has 12 sensitivity ranges with six calibrated in mm and six calibrated
in inches. The measuring resolution ranges from 10 microns (0.0005 in.) to
0.2 $\mu$m (0.00001 in.). Sets of standard spherometer rings are commercially
available from several other sources [73, 74, 75]. These are shown in Table
6.6.

**Table 6.6.  Spherometer Ring Dimensions.**

| Ring dia. in mm ID | OD | R&C [72] | TaLiang [73] | Bothner [74] | Loh [75] |
|---|---|---|---|---|---|
| 5 | 8 | X | X | X | X |
| 6.3 | 10 | X | | X | |
| 8 | 11 | | X | | |
| 8 | 12.5 | X | | X | X |
| 10 | 15 | | X | | |
| 10 | 16 | X | | X | |
| 12.5 | 16 | X | | X | X |
| 15 | 20 | | X | | |
| 16 | 20 | X | | X | X |
| 20 | 25 | X | X | X | X |
| 25 | 30 | | X | | |
| 25 | 32 | X | | X | X |
| 28 | 34 | | X | | |
| 30 | 40 | | X | | |
| 32 | 40 | X | | X | X |
| 34 | 44 | | X | | |
| 40 | 50 | X | X | X | X |
| 50 | 60 | | X | | |
| 50 | 63 | X | | X | X |
| 60 | 70 | | X | | |
| 63 | 71 | X | | X | X |
| 70 | 80 | | X | | |
| 71 | 80 | X | | X | X |
| 80 | 90 | X | X | X | X |
| 90 | 100 | X | X | X | X |
| 100 | 112 | X | | X | X |
| 112 | 125 | X | | X | X |
| 125 | 140 | X | | X | X |
| 140 | 160 | X | | X | X |
| 160 | 180 | X | | X | X |
| 180 | 200 | X | | X | X |

**Ball Spherometer.** The ball spherometer differs from the ring spherometer in that the measuring circle is defined by precision balls. The centers of the equidistant balls are located on the circle. Unlike for the ring spherometer, there is only one measuring circle for both convex and concave surfaces. The difference lies in the way the effect of the balls is accounted for. The ball radius must be subtracted from the standard equation, Eq. (6.37), for convex surfaces and added to it for concave surfaces. Two new equations result:

$$R = \frac{r^2 + s^2}{2s} - b, \tag{6.42}$$

$$R = \frac{r^2 + s^2}{2s} + b, \tag{6.43}$$

In this form the radius of the measuring circle is $r$ and the radius of the balls is $b$. Sag $s$ is the measured value. Two typical examples for using these equations are shown in Fig. 6.34 .

Most modern spherometers are ball spherometers. This may be due more to the popularity of these instruments than the ease of manufacturing them. It is more difficult to locate the balls on the correct ball circle to micron accuracies than to precision grind a steel bell to exact diameters. In addition the accuracy with which the radius of the balls can be defined affects the precision of the sag measurement. All three balls in a set must have the identical ball radius; otherwise the measuring circle will be incorrect. Fortunately the methods of producing perfectly spherical balls to a very high degree of precision from very hard materials such as sapphire, ruby, or ceramics are well established. Most modern spherometers use matched sets of sapphire balls. Since most precision spherometers are convertible from one type to another, the Abbé-type spherometer shown in Fig. 6.31 could be converted to a ball spherometer by replacing the spherometer ring with a suitably designed ball circle.

As an example of a ball spherometer, let us examine a typical representative of a modern instrument. The three-ball circle is screwed into the drum housing of a large diameter precision micrometer. The housing is supported by a bracket that also serves as a stand for the instrument. This arrangement is shown in Fig. 6.35. The vertical plunger is at the exact center of the ball circle and is movable by turning the micrometer spindle. The plunger has an added feature that gives it considerable sensitivity in that it incorporates a transducing element, called a *pressure transducer,* which changes its electric characteristics when a very small amount of pressure is applied. This electrical change is sensed, amplified, and read out on a galvo type gauge. The resolution for the sag measurement is as small as 0.25 $\mu$m (0.000010 in.) with this type of instrument.

The simplest ball spherometer [76] uses a set of aluminum ball rings of various diameters and a 0.0001-in. resolution standard unidirectional mi-

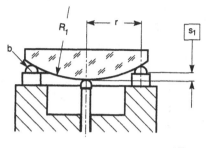

1/2 Ball circle diameter:   r = 25.408 mm        (1.0003 in)

Measured sagitta:        $s_1$ = 2.023 mm        (0.07964 in)

Ball radius:             b = 3.183 mm            (0.1253 in)

per Eq. 6.42   $R_1 = \dfrac{(s_1)^2 + (r)^2}{2(s_1)} - b = $ 157.39 mm   (6.196 in)

a

1/2 Ball circle diameter r and ball radius b same as example a.

Measured sagitta:        $s_2$ = 3.370 mm        (0.13267 in)

per Eq. 6.43        $R_2 = \dfrac{(s_2)^2 + (r)^2}{2(s_2)} + b = $ 100.649 mm   (3.963 in)

b

**Figure 6.34.**   Examples for ball spherometer. *(a)* Convex. *(b)* Concave.


crometer. This arrangement limits the sag measurement accuracy to about 0.0002 in. The balls are a matched set of three highly spherical polished ceramic spheres that can be easily replaced when worn or damaged. A lengthy table must be consulted to determine the radius of curvature based on the ball ring diameter used and the measured sag value. A suitable hand calculator can greatly facilitate and speed up this process, as well as do it more accurately. The unidirectional micrometer is easy to use for measuring the sag of convex surfaces. For concave sag measurements, however, the value must first be subtracted from 1 (for a 1-in. range micrometer) before a calculation to determine the radius $R$ can be made.

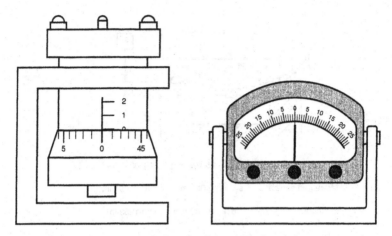

**Figure 6.35.** Modern ball spherometer with electronic readout.

A high-resolution drum micrometer that is bidirectionally calibrated is much better suited to measure either convex or concave surfaces. It reads from the center in both directions with one scale being positive from the null position, and the other is negative. The nominal resolution is 0.0001 in. which gives it a measurement accuracy of 0.0002 in. for the sag.

A further refinement is the electronic spherometer [77]. A drum micrometer is used to measure the sag to an accuracy of 0.0002 in. An analog gauge permits the refinement of that initial measurement through a linear differential transducer (LDT) element by nulling the indicator needle. A total of four consecutive sensitivity levels increases the measurement accuracy to 25 $\mu$in. ($25 \times 10^{-6}$ in.). With several repeat measurements skillfully done, it is possible to determine the sag to an accuracy of 0.000050 in. Since only the sag value is measured, the radius of curvature must still be determined from a table, or it must be independently calculated. A typical series of ball circles is listed in Table 6.7. The polished balls and the very light pressure excerted by the LDT point will not damage even the most sensitive surface if properly used.

Another refinement [78] is the addition of an electronic calculator. The light emitting diode (LED) number display can either show the measured sag value and the calculated radius of curvature in either millimeters or inches provided that the diameter of the ball circle used has been entered first in the corresponding dimension. The vertical measurement plunger is combined with a 2.5-$\mu$m (0.0001-in.) resolution linear position encoder.

A modern precision ball spherometer [79] provides very high precision to a theoretical accuracy of 0.1 $\mu$m (4 $\mu$in.) over the entire range of 25 mm (1 in.). Due to systematic errors in the instrument, the practical measurement accuracy to determine the sag value is claimed to be $\pm0.2$ $\mu$m or ($\pm8$ $\mu$in.). The dedicated computer calculates the radius values, which are displayed on

**Table 6.7.  Ball Circle Diameters.**

| Strasbaugh[77] (in) | Trioptics[81] (in) | Trioptics[81] (mm) | Möller-Wedel[83] (mm) |
|---|---|---|---|
| | 0.25 | 7.5 | |
| 0.75 | 0.75 | | 8.0 |
| 1.00 | | | 15.0 |
| | | 20.0 | |
| 1.50 | 1.50 | | 25.0 |
| 2.00 | | | 35.0 |
| | 2.50 | 40.0 | |
| 3.00 | | | 50.0 |
| | 3.50 | 60.0 | |
| 4.00 | | | 65.0 |
| | 4.75 | | 80.0 |
| 5.00 | | 90.0 | |
| | | 120.0 | 120.0 |
| 6.00 | 6.00 | 150.0 | |

the LED to an accuracy of better than 0.05%. The values can be displayed in millimeters to 1 $\mu$m (0.001 mm) or 1 mil (0.001 in.). The ball circles are defined by three high precision polished sapphire balls at 120° with diameters of 12, 24, 42, 66, 100, and 160 mm.

The operation of this type of spherometer proceeds as follows:

1. Select the next smaller ring diameter than the diameter of the lens.
2. Null the instrument with a reference quality plano test glass, and let it thermally stabilize. Reset button to display zero.
3. Carefully place lens surface on three balls, let it settle out, and read the sag value on LED display.
4. Enter ball circle diameter and whether the surface is convex or concave. Then select the desired dimension (millimeters or inches) for the radius, and enter the measured sag value.
5. The LED will then display the radius.

A modern hand-held digital ball spherometer [80] is a high-precision metrology instrument, unlike the customary hand-held spherometers which are designed specifically to provide the optician with a quick reference check of the curvature of tool and lens blanks. It gets its precision from a high-quality linear encoder with 1 $\mu$m resolution. It is rated to measure the sag on convex and concave surfaces over a range of 30 mm (1.2 in.). The systematic errors are less than 1 μm (0.00004 in.), so careful measurements should provide for an accuracy of less than 2 μm (0.00008 in.).

The series of interchangeable ball rings with three ruby balls are calibrated and certified. The results of the measurement are digitally displayed on an LCD (liquid crystal display). The dimensions are selectable in either inch or metric. Prior to each sagitta measurement the instrument must be calibrated against a plano testplate, and the display must be reset to zero. The zero reset can be done on any probe position. This means the instrument can also be calibrated against a radius reference plate such as a known radius testplate. The radius of curvature must still be calculated. It is best to use a suitable hand-held calculator to perform that task. The equation used with this instrument is essentially the same as Eq. (6.42) and Eq. (6.43):

$$R = \frac{(d/2)^2 + s^2}{2s} \pm r, \qquad (6.44)$$

where

$R$ = radius of curvature,
$d$ = ball circle diameter,
$s$ = sagitta (measured value),
$r$ = half the ball diameter,
$-$ = for convex surfaces,
$+$ = for concave surfaces,

The values $d$ and $r$ are precisely measured, calibrated, and certified. The geometric basis for Eq. (6.44) is shown in Fig. 6.36.

Another recent development is a precision instrument [81] that is also designed for the sag measurement on convex and concave surfaces. It is used for radius testplate manufacture, for determining the radii of spherical grinding tools, and for comparative measurements of the radius of testplates and tools to determine by how much they deviate from the desired radius. Calibrated radius references are needed for the comparative measurements.

The ball rings are calibrated with 1/10 wave radius reference plates. The calibrated values for the ring diameter and the ball diameter are stored in the microprocessor memory. Each ring has a unique ID number that must be entered into the computer prior to the radius calculation. The high-precision linear encoder has a ±15 mm (±0.6 in.) measuring range with ±1 μm (±40 μin.) resolution over that range.

The control unit has a digital liquid crystal display (LCD) which displays the measured and calculated values to a 1 μm (0.00004 in.) accuracy, a programmed microprocessor, and an integral digital printer. The preferred dimension (millimeter or inch) must be entered prior to calculating the ra-

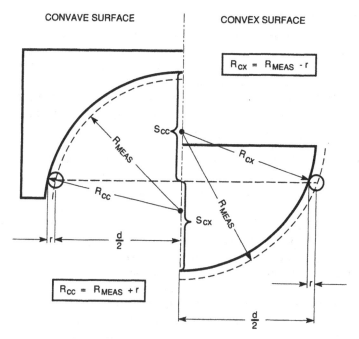

CONVAVE SURFACE | CONVEX SURFACE

$$R_{cx} = R_{MEAS} - r$$

$$R_{cc} = R_{MEAS} + r$$

**Figure 6.36.** Geometric basis for ball spherometer equation.

dius. The printer records the sag value, the calculated radius, the mean value of several independent readings (zeroed each time), and the standard deviation of these values. The operation is fast. An experienced operator can calibrate, null, measure, calculate, and print out three to four consecutive readings in one minute.

The radius range of this spherometer is from 3 mm (0.120 in.) to plano for convex and from 5 mm (0.200 in.) to plano for concave surfaces. Ball circles are available in metric or in inches. They are listed in Table 6.7.

Another advanced instrument [82] is a very accurate automatic three-ball spherometer. It has been designed for the highest testplate calibration requirement with a radius of curvature accuracy of ±0.01%. The concentricity of the ball circle to the central probe axis is within 1 μm (40 μin.). The accuracy of the ball position on the ball circle is held to 1 μm. The balls are made from extremely hard tungsten carbide. The probe point is a precision ruby ball. The axial position of the probe is measured by a high-precision linear encoder that has total systematic and random errors of less than 0.2 μm (8 μin.).

The calibration radius glasses have a spherical accuracy of better than 1/10 wave. The ball rings are calibrated against these master radius plates. The ball rings are made of high-quality stainless steel. They are first precision ground and then manually lapped to the highest precision. The three tungsten carbide balls are then precisely located on the ring and fixed in position. After calibration each ring is uniquely identified by its own ID number, and a certificate is issued that lists the precise dimensions of the ring. If a ring should be dropped or otherwise damaged, it must not be used but returned for recertification and possible repair.

The control module consists of the microprocessor to make the necessary calculations, a keyboard for data entry, a digital LCD to display the measured and calculated values, and a digital printer that provides a hard copy of the pertinent data. The microprocessor calculates the radius from the instrument constants and the measured sag value. It also provides statistical analysis, such as the average value and the standard deviation for repeated measurements. The software is capable of detecting whether the measured surface is convex or concave, and it makes the necessary correction automatically during the radius calculation. It can also record automatically the radius difference between matching pairs of testplates.

Yet another recent product [83] is a precision spherometer with digital display and programmable calculator. The operation is based on the principle of comparative measurement using one parameter (sag) to determine the value of another parameter (radius). The key elements in this relationship that determines to a great extent the accuracy with which the radius value can be calculated are the instrument characteristics such as ring diameter, ball diameter, ball location, and concentricity of the ball circle. Since the sag values that are typically measured are quite small compared to the value of the associated radii of curvature, a very precise instrument with an accuracy of $\pm 1$ $\mu$m is required.

The uncertainty in the determination of $R$ depends on random residual errors such as $\Delta d$ (ring diameter), $\Delta s$ (sag), and $\Delta r$ (ball radius):

$$\Delta R = \pm \left(\frac{d}{2s}\right) \frac{\Delta d}{2} \pm \left(\frac{R}{s-1}\right) \Delta s \pm \Delta r, \qquad (6.45)$$

where

$\Delta R$ = uncertainty in determining radius $R$,

$d$ = nominal ball ring diameter,

$s$ = measured sag value,

$\Delta d/2$ = error in the ball ring diameter,

$R$ = calculated radius of curvature,

$\Delta s$ = error in measuring the sag value,

$\Delta r$ = error in the measurement of the ball radius.

The error equation, Eq. (6.45), shows that the uncertainty ($\Delta R$) in calculating the value of $R$ is inversely proportional to the measured sag. The smaller the sag, the greater is the uncertainty. The equation also shows that the uncertainty will be least when the largest possible ring diameter is chosen. The real value of Eq. (6.45), however, is its use to predict the magnitude of the error in the value of $R$ resulting from system errors, reading errors, and calculation errors.

All ball spherometer rings are calibrated and certified. The measuring range of the spherometer, as determined by the rings (see Table 6.7), is from 4 mm convex to infinity and from 7.5 mm concave to infinity. Lenses of up to 500 mm in diameter can be measured. The maximum indicator travel is ±15 mm. (0.6 in.).

Calibration plano plates are 50 mm (2.0 in.) in diameter with 1/10 wave flatness, 100 mm (4.0 in.) in diameter (1/5 wave), and 150 mm (6.0 in.) in diameter (1/3 wave). Radius calibration sets composed of matching convex and concave plates are listed in Table 6.8. The examples for the ring spherometer (Fig. 6.33) and the ball spherometer (Fig. 6.34) use the same instrument constants. The ball circle radius $r_b$ is the arithmetic mean of the ring semidiameters $r_i$ and $r_o$ expressed by Eq. (6.46):

$$r_b = \frac{r_i + r_o}{2}. \tag{6.46}$$

These values were chosen to show that the calculated radii of curvature differ significantly despite the fact that the instrument constant (circle radius) and the sagitta measurements were identical. When converting a ring spherometer to a ball spherometer, the correct equations must be used.

### 6.3.3. Calibration of the Spherometer

The spherometer is a precision instrument and should be treated accordingly. Sag measurements should be taken with great care since even rela-

Table 6.8. Typical Radius Calibration Sets [83].

| Radius (mm) | Diameter (mm) |
|---|---|
| 7.5 | 15.0 |
| 12.5 | 22.0 |
| 17.9 | 35.0 |
| 27.9 | 45.0 |
| 39.9 | 60.0 |
| 50.0 | 77.0 |
| 63.8 | 93.0 |
| 141.8 | 132.0 |

tively small errors can result in large errors in the calculated value of the radius of curvature. The first step in using a spherometer is to calibrate it for the intended measurement against a known reference. As with any precision measuring device, this step is essential to ensure that the measurement is accurate. References used for calibrating spherometers are optical flats or standard radius plates that come supplied with high-precision instruments. Once the spherometer has been carefully calibrated for the measurement range of interest, the resulting sagitta measurements are accurate and reliable within the accuracy limits of the instrument. The following sections describe typical procedures for calibration and measurement for the basic types of spherometers.

**Using the Ring Spherometer.** The first step in using the ring spherometer is to choose the largest diameter ring for the part that must be measured to achieve the highest possible accuracy. If, for example, the diameter of the test part is 54 mm (2.16 in.), the chosen ring must be smaller, though it should of course be the next largest diameter that meets that requirement. In this example the nearest standard ring has a 50-mm (2.0-in.) diameter. It should be used for the sag measurement.

The ring must then be properly installed on the spherometer by taking reasonable care, as should be routinely done with all precision instruments. The exact nature of the installation method varies from instrument to instrument, and the operating instructions supplied with the spherometer should be consulted when in doubt.

Before proceeding further, the instrument may have to be leveled before calibration and use. This is particularly important for the Abbé type because the counterweighted plunger may not move freely and thus lead to reading errors. The leveling is usually done by adjusting two of the three leveling screws at the base of the instrument and observing the orientation of a bubble level. When the bubble is centered in the vial, the spherometer is ready for the calibration step.

The step that follows is the all-important calibration or referencing of the instrument. As already mentioned, these references are qualified plano testplates and standard radius plates that are used for calibration purposes only. Of course plano testplates of 1/8 wave or better will be adequate for most measurements, and radius testplates that have been previously measured to a degree of accuracy that exceeds that of the spherometer can also be used. However, the radius of such testplates should have been independently verified by interferometric or by optical means as described later in this chapter.

The plano reference is used for calibration prior to the measurement of relatively large radii for which the sag is only a few millimeters. The instrument can be calibrated for both convex or concave surfaces with such a plano reference. There is no rule of thumb that can be used to decide the best limits of calibrating with a plano reference. Much depends on the instrument used, and when in doubt, it is always good practice to consult the operating instructions. In general, it can be stated that if a high degree of precision is required of the measurement, then the sag difference to the plano reference should be small. When less precision is acceptable, then a greater range of sag readings can be covered by a plano calibration.

For all other radii the instrument must be calibrated against radius plates, or radius testplates, with accurately known radius of curvature. The best radius reference is the one that has a radius closest to the one that must be measured. The basic idea is to have a sagittal difference that is small compared to the actual range of sag measurement of the instrument. The reason for this is the fact that few measuring devices are very accurate over long measuring ranges, but most of them yield highly precise and repeatable measurements over a short range. The stipulation that the sag difference between the reference and the test part measurement must be small of course requires that convex surfaces be set up with convex references and concave surfaces with concave references.

When using plano or spherical reference plates for calibrating a ring spherometer, care must be taken that there is no damage to the rings or that no particles or lint can cause measurement errors. It is always good practice to repeat the calibration step several times and average the results to improve the quality of the calibration. If the instrument cannot be zeroed at the calibration point, then the average value of all readings taken must be noted and used as reference for all subsequent measurements.

Once the instrument has been properly calibrated, the part to be measured is placed on the ring and an accurate reading is taken and recorded. How to read out the sagittal displacement varies from instrument to instrument and the appropriate operating instruction should be consulted. For instance, the Abbé type has a precision engraved glass scale that is supplemented by a vernier scale. It can be read out directly to an accuracy of 1 $\mu$m (0.00004 in.) when viewed by a measuring microscope. Three to four readings should be taken and the results averaged. If the calibration was done correctly and the sag measurements made with care, then the sag readings should only differ by very little. Using the resulting sag value $s$ and the known radii $r_i$ and $r_o$ of the measuring circles, the values of the radius $R$ can be calcualted by using either Eq. (6.40) or Eq. (6.41). For details refer to examples in Fig. 6.33.

**Using the Ball Spherometer.** The ball spherometer is calibrated and used essentially in the same manner as the ring spherometer. The only differences are the comparative ease of calibration and use and the evaluation of the results. The hardened steel balls or the sapphire balls of a good-quality ball spheromenter are far less susceptible to damage and wear or to dirt contamination than the ring of even the best ring spherometer.

To illustrate the use of a typical modern ball spherometer, we will use the type schematically shown in Fig. 6.35. Such a spherometer typically utilizes interchangeable ball circles, a high-resolution micrometer, and a sensitive electronic readout. As with all sensitive electronic devices, a warm-up period of no less than 15 minutes should precede any critical measurement in order to allow the circuit to thermally stabilize.

The calibration proceeds in identical fashion as described for the ring spherometer, except that it is not as critical to ensure freedom from dust and lint since there are only three contact points between the balls and the surface. Common pratice will dictate, however, that cleanliness and care should be of utmost concern when using any precision instrument.

For near zero calibration a plano reference plate is placed on the balls. The micrometer spindle is rotated in the appropriate direction, usually to raise the plunger, until the needle of the electronic readout begins to move. The sensitivity selector switch should be switched to the least sensitive range for the initial reading. When the needle has been adjusted to be in the zero position in the center of the scale, the micrometer should be read. The reading is most likely somewhat off from zero. The micrometer should be carefully zeroed, and this causes the needle to move to either side of zero on the scale. By turning the electronic zero control in the appropriate direction, the needle can be zeroed again. When this is done, the selector switch is switched to the next higher sensitivity, and the needle is zeroed again. This procedure is repeated until the desired sensitivity range has been reached and the meter needle has been carefully zeroed for that range. The zero on the micrometer should be checked again and corrected if necessary; this must be followed by a final zero of the needle. At this state a reference plane has been established that passes through the points where the plano plate contacts the balls. This plane corresponds to the zero position of the micrometer. With some experience this process usually takes no more than one minute. Any sag reading, whether on long radius convex or on concave surfaces, can be directly read off on the micrometer after the needle has been zeroed. It is always a good practice to recalibrate both zero positions prior to every critical measurement and to take several measurements in order to improve the accuracy of the measurement by averaging the results.

The calibration and use for shorter radii is done in identical fashion, except that standard radius reference plates should be used for the calibration. Once the value of the sag $s$ has been established and the values of the ball circle radius $r$ and the ball radius $b$ are precisely known, then the radius of curvature $R$ can be calculated using Eq. (6.42) or Eq. (6.43) according to the examples in Fig. 6.34.

**Calculating the Value of $R$.** Not too long ago it was not uncommon to have to calculate the radius longhand from the measured sag value and the known radius of the ring or ball circle. This was a slow process and subject to errors. Manufacturers of spherometers attempted to simplify this process by making available books of tables that listed radius values for specific combinations of circle diameter and sag readings. This made interpolation necessary for any accurate determination. The actual value of the radius most likely fell somewhere between two readings in the book. But interpolating was no easy task either. Nomograms were constructed for a quick reference good enough for noncritical applications, that permitted the approximate radius to be read off from the known values of circle diameter and sag reading.

But with the introduction of inexpensive, powerful hand-held calculators, longhand calculations, complicated tables, difficult interpolations, and inaccurate nomograms have become obsolete. A simple scientific calculator with a square root key and one storage location can calculate the radius to an accuracy to the sixth or seventh decimal in a few seconds. A radius accuracy to the fourth decimal is all that is typically required. Programmable calculators speed up this process even further because only the values of sag $s$ and the ring diameter $d$ have to be entered to instantaneously calculate the radius value $R$.

The discussion so far has been for ring spherometers where the appropriate ring diameter must be used. For ball spherometers the ball radius must be subtracted from this result for convex surfaces and added to it for concave surfaces. But with the modern computerized spherometers even this is no longer necessary. After entering the ring diameter used for the sag measurement, the computer calculates the radius automatically and displays the result on a digital display.

### 6.3.4. Large Spherometers

There are a number of variations on the previously discussed spherometers. These can be divided into two groups as radius gauges and large spherometers. Radius gauges are small hand-held spherometers of limited accuracy that are often calibrated in diopters for a direct determination of the approximate value of the radius of curvature. Large spherometers are used for the radius determination of large optical elements such as large lenses or astronomical mirrors. Large means larger than 200 mm (8 in.) in diameter. Specially designed large spherometers are sometimes used as a quick reference for flat or near-flat ground surfaces.

Since the accuracy of the sag measurement requires a measuring circle as large as the surface will allow, the previously discussed ring or ball spherometers are unsuitable for large elements because of the limited range of their rings or ball circles which rarely exceed 150 mm (6.0 in.). The size and weight of large optical elements precludes the usual measurement approach for which the part is placed on top of the instrument. The spherometer for large surfaces is used in an inverted position. Although the theory requires that the ball circle of these instruments should be as large as the diameter will allow, it is obvious that this requirement cannot be fulfilled for large surfaces. Practicality demands an upper limit. A rule of thumb that applies here is to set a lower limit of 150-mm (6-in.) diameter for moderate sizes and an upper limit of 20% of the part diameter for the largest part diameters.

These constraints would put large spherometers in the 150 to 460 mm (6 to 18 in.) range for most large optical work.

Since the cost of making large ball circles is quite high, such instruments are usually designed with removable balls that allow for a change of the measuring circle in discrete intervals. These intervals are typically 12.5 mm (0.5 in.) over a range that best utilizes the size of the instrument. This permits the use of one instrument over a wide range of surfaces. This practice, however, leads to a lack of accuracy in the locations of the balls with the associated uncertainty in defining the ball circle. Errors in the calculated value of $R$ can result, but fortunately large radii often have sizable radius tolerances that prevent this problem.

**Tripod Flatness Tester.** One common application for a large spherometer type is the tripod flatness tester (Fig. 6.37) with typical ball circle diameters ranging from 100 mm (4.0 in.) up to 300 mm (12.0 in.). The ball circle diameters are variable in 12.5-mm (0.5-in.) increments by changing the position of the points.

Tripod flatness testers are designed to check for departure from true flatness. Since this is a relative measurement on near-plano surfaces, the exact definition of the ball, circle is not important. The instrument is first zeroed (calibrated) on a perfectly flat reference plate (see Fig. 6.37a) and then placed on the surface that needs to be checked. The departure from flatness is indicated by the position of the pointer with respect to the zero position (see Fig. 6.37b). A negative reading usually indicates a concave condition, and a positive reading indicates that the surface is convex. The value of the reading shows the magnitude of the departure from true flat. The indicator is calibrated either in metric to an accuracy of 1 $\mu$m or in English to $10^{-5}$ in.

Because of its size this type of instrument must be made from a light material, typically aluminum. It must also utilize a lightweight construction to further minimize its weight. However, such lightweighting must provide sufficient stiffness so that the reading will not be influenced by flexure of the instrument under its own weight.

The tripod flatness tester is of value only for flat surfaces or for surfaces for which the departure from true flat is regularly spherical, either convex or concave, within the typically narrow range of measurement. This condition is shown in Fig. 6.38a, b, and c. Many flat and near-flat surfaces, however, may exhibit zonal errors that can lead to contradictions in the measurements, as shown in Fig. 6.38d, e, and f.

Therefore, to use the measurement for the correction of tools or machine settings, there has to be some assurance that any departure from true flat is spherical. If zonal errors are suspected, then that should be verified by independent means such as a straightedge test or waxed surface interferometry for ground surfaces. But despite these potential limitations, in the hands of an experienced optician the tripod spherometer can be a very valuable

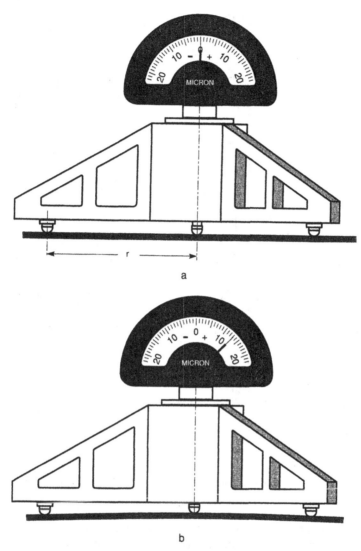

**Figure 6.37.** *(a)* Tripod base spherometer with equilateral base (third leg not shown). Tester is on a flat reference surface (pointer on zero). *(b)* Tripod base spherometer. Tester is on a nonflat convex surface (pointer is on positive).

tool for monitoring the flatness of plano surfaces of parts and tools during the various grinding stages.

Some commercial units [84, 85] are designed with adjustable legs which are interchangeable and come in two ranges (200- to 300-mm diameter and 300- to 400-mm diameter). Since the tripod flatness tester uses a Mahr gauge [86] with a 1 $\mu$m resolution, it has a limited range. The tripod gauge is used

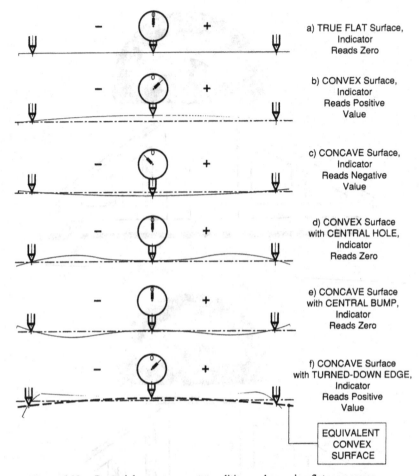

**Figure 6.38.** Potential measurement conditions when using flatness gauge.

primarily for checking plano surfaces for spherical out-of-flat condition, but it can also be used as a comparative radius measurement device as long as it can be nulled against a spherical reference such as a radius testplate. The deviation of the measured radius relative the testplate can then be easily determined. This is very useful for the correction of grinding tools and machine settings. The Mahr gauge can be replaced by a 0.01-mm resolution dial indicator when high precision is not required. The reduced sensitivity will greatly increase the range of measurement.

**Other Variants.** Two variations of the equilateral three-ball spherometer are the isoceles three-ball spherometer and the linear two-ball, or bar, spherometer. Schematic examples of these variants are shown in Fig. 6.39. The definition of the ball circles for the three types are represented by Fig. 6.40.

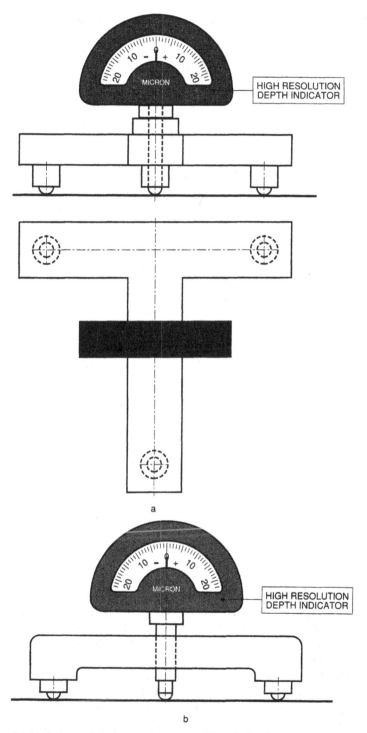

**Figure 6.39.** (*a*) Variations of the large spherometer: Three-ball spherometer with isosceles base. (*b*) Variations of the large spherometer: Two-ball bar spherometer.

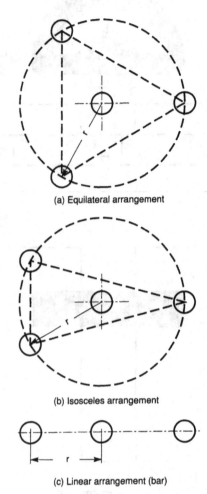

(a) Equilateral arrangement

(b) Isosceles arrangement

(c) Linear arrangement (bar)

**Figure 6.40.** Variations in the definition of the ball circle for large spherometers. *(a)* Equilateral arrangement. *(b)* Isosceles arrangement. *(c)* Linear arrangement (bar).

The isoceles three-ball spherometer allows for the construction of rather large instruments. Some of these have been built with ball circles as large as 610 mm (24 in.) in diameter. An instrument with the same ball circle using the equilateral configuration would be rather large and difficult to use. The isoceles spherometer has also removable ball feet, and it is generally used in identical fashion as the equilateral variety.

The second variant is the linear or bar spherometer. It can be thought of as a special case of the isoceles arrangement for which the two balls have been brought so close together to become only one ball. It consists of a stiff metal bar with a dial gauge or other suitable indicator mounted at its center and two spherical balls fastened to it equidistant from the centerline. This type of spherometer is handier than either the equilateral or the isoceles version, but its design introduces another constraint which is of no importance for all other spherometers discussed so far. The bar spherometer must be held perpendicular to the measured surface to yield a reliable measurement. Only if the two balls and the contact point at the center have the same radius and their centers are exactly colinear, then this constraint does not apply. Most designs, however, incorporate dial gauges with interchangeable points, so the requirement for the same radius cannot be met. With that fairly common bar spherometer arrangement, unacceptable measurement errors can result when making measurements with the instrument tilted from the normal. Bar spherometers are used for measuring the radius of curvature of cylindrical surfaces.

The calibration and use of the isoceles and the bar spherometers follow the same basic procedure as described for the equilateral design. Whether the instrument is used to make a relative measurement of departure from flatness, or the departure is measured as actual sag for the subsequent calculation of the radius of curvature $R$, a careful calibration against a plano reference surface must precede any critical measurement.

Since the large spherometers discussed in this section are designed to measure radii that are many times larger than their diameters, the sag values are usually very small. This requires the use of very sensitive gauges that have a limited range. This limits the radius ranges that these spherometers can measure.

Because the departures from plano in terms of sag values are very small, often in the micron range, the varying points of contact are of no consequence and the ball radius, which is such a small percentage of the total radius, usually does not need be considered. Also because the sag is so small, the value $s^2$ is practically zero. This permits the use of the simplified radius Eq. (6.38), which will yield sufficiently accurate values for the radius of curvature $R$.

### 6.3.5. Optical Methods

There are two optical methods for determining the radius of curvature. One is known as the *autostigmatic method*, which is usually done on an optical bench or a height gauge arrangement, the other is an *interferometric method*, which is performed with a modern shop interferometer and a radius rail. Unlike the spherometry described so far, both methods, which are based on the same principles, measure the radius directly without having to determine another value first. As a result these methods are potentially more accurate, and most of the error sources do not exist for this approach. There are, however, other problems that limit the accuracy of these measurements.

One of the limitations is that the methods require a polished spherical surface. This limits their use to the verification of the radii of curvature as a final quality buy-off. This means that they cannot replace ring or ball spherometer for making testplates because most radius measurements on these are made during the fine-grinding stage. Another limitation is the precision with which the best focus can be judged. While this is only a small problem with short radii, it can become a source of considerable error when longer radii have to be measured. Unless the parts are centered in a self-aligning mount, which usually is not the case, an off-center condition can contribute additional errors to the radius measurements.

The autostigmatic method uses an illuminated microscope objective from which a cone of light emerges. This microscope is mounted on a stage of an optical bench whose position along the rail of the bench can be determined to a high degree of accuracy. If a reflecting curved surface is mounted axially in line with the microscope objective, then the cone of light can be focused on the surface to produce an image of the target reticle on the eyepiece reticle similar to that of an autocollimator. If the stage that supports the low-power microscope is now moved until the light rays that emerge from the objective strike the surface of the spherical surface normally, another image of the reticle comes into focus. The distance between the two focus positions is the radius of curvature of the spherical surface. Figure 6.41 shows this method. With the right setup a 1 $\mu$m accuracy is possible.

The autostigmatic method can also be done vertically. For relatively short radii of 12 in. or less, a standard height gauge has proved to be very useful. The advantage of this approach is the small size of the instrument as compared to an optical bench. This permits its use right in the shop where it is needed. Figure 6.42 shows such an arrangement.

The measurement of the radius of curvature with an interferometer and a radius rail has already been discussed in detail in the section on interferometry. This has become the preferred method for in-process testing in many shops because it measures not only the radius of a spherical surface but shows the surface irregularity at the same time. An interesting arrangement combines the elements of interferometry with the autostigmatic method and

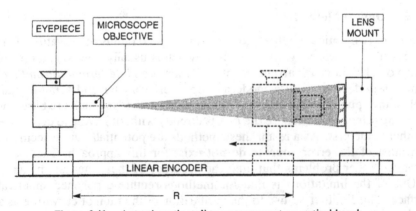

**Figure 6.41.**  Autostigmatic radius measurement on optical bench.

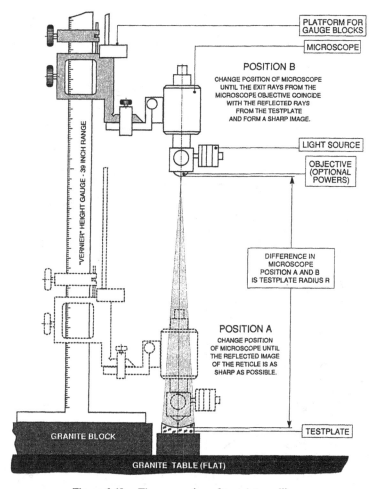

**Figure 6.42.** The measuring of testplate radii.

an $x$, $y$, $z$, $\theta$ positioning table for testing lenses on the block for radius and surface irregularity. Figure 6.43 shows this arrangement [87]. The same setup is shown in Fig. 6.20 for interferometric on-the-block testing of convex surfaces.

The vertical column of the instrument supports the interferometer head, which can be tilted at an angle with respect to the vertical ($z$) axis. An $x$ and $y$ stage provides horizontal positioning. A rotary table that accepts the lens block is mounted on this stage. In this configuration the instrument has all the elements of a standard vertical mill. When only the center of the block or lens is tested, the interferometer head is locked in a vertical position. The converging output beam of the interferometer is first focused on the surface by moving the head vertically up or down until straight fringes are seen on

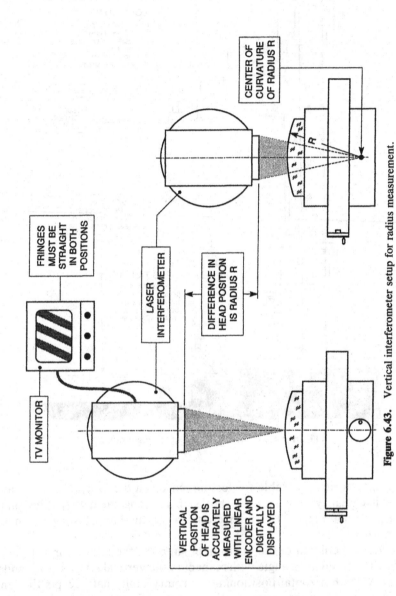

**Figure 6.43.** Vertical interferometer setup for radius measurement.

CENTER OF CURVATURE OF RADIUS R

R

FRINGES MUST BE STRAIGHT IN BOTH POSITIONS

LASER INTERFEROMETER

DIFFERENCE IN HEAD POSITION IS RADIUS R

TV MONITOR

VERTICAL POSITION OF HEAD IS ACCURATELY MEASURED WITH LINEAR ENCODER AND DIGITALLY DISPLAYED the TV monitor. This is the zero position, which requires that the digital readout for the z-axis be nulled. The interferometer head is then either raised for concave surfaces or lowered for convex surfaces until another set of interference fringes can be seen on the monitor and adjusted until they are straight. The distance traveled by the head as measured by the z-axis readout is the radius of curvature.

One advantage of this instrument lies in the ability to tilt the head for measuring other parts of the lens surface or of the lens block. However, this is for measuring the surface irregularity only because the radius cannot be measured directly in an inclined position of the head. Surfaces or blocks up to 150-mm (6-in.) diameter can be measured on this device. The range for measuring the radius of convex surfaces is from 10 to 1200 mm (0.400 to 48 in.) and from 10 to 800 mm (0.400 to 32 in.) for concave surfaces.

**References for Section 6.3 (Spherometers)**

[71] Askania, Germany.

[72] Rogers and Clarke, USA, electronic column spherometer, 1978.

[73] Ta Liang Co., Taiwan.

[74] Bothner, Germany.

[75] Loh Wetzlar, Germany.

[76] RH Strasbaugh, USA, model 18-P.

[77] RH Strasbaugh, USA, model 18-N.

[78] RH Strasbaugh, USA, model 18-AS.

[79] OTI, Ltd., England, digital spherometer.

[80] Trioptics, Germany, model Spherocompact, 1988.

[81] Trioptics, Germany, model Spherotronic, 1986.

[82] Trioptics, Germany, model Super Spherotronic, 1987.

[83] Möller-Wedel, Germany, model Spheromatic, 1988.

[84] Loh Wetzlar, Germany, tripod measuring gauge.

[85] Bothner, Germany, tripod measuring gauge.

[86] Mahr, Germany, micron gauge.

[87] Möller-Wedel, Germany, model Spheroscope.

## 6.4. AUTOCOLLIMATOR

The autocollimator has played an important role in optical shops for a long time. This instrument is used for measuring small angular deviations in plane-parallel windows and first surface mirrors. Prism angles can be measured when the autocollimator is set up with an angle reference. Two other instruments that are subcomponents of the autocollimator are sometimes used in the shop as well.

### 6.4.1. Collimator

A collimator is a simple optical instrument that is used to measure the focal length of telescope lenses and the magnification of telescopes and similar optical systems. A small lamp illuminates a reticle that lies in the focal plane of the collimator objective. The reticle is made up of cross hairs or of a set of single cross lines that are etched into a highly polished glass substrate. The design of this instrument causes the light to emerge from the objective as a collimated beam. In other words, the light travels parallel to the optical axis of the instrument and, at least in theory, never converges or diverges. Another way to express the function of a collimator is to say that it projects the image of the reticle at infinity. The basic layout of a collimator is shown in Fig. 6.44.

Collimators are an integral part of any optical bench, as are telescopes or microscopes, nodal slides, and other accessories that are required for the testing of finished optical components and lens systems. Collimators with long focal lengths are easier to set at infinity than those with shorter focal lengths. Furthermore the image distortion is more noticeable at short focal lengths and the associated expansion of the image field. Collimators with about 150-mm (6-in.) focal length are sufficient for simple adjustments for which alignment errors in the arc minute range must be resolved. With objective focal lengths of more than 300 mm (12 in.), alignment errors in the arc second regime can still be resolved.

### 6.4.2. Telescope

A telescope is constructed much like a collimator, however, it does not emit light but rather collects it from a distant source. The light source of the collimator has been replaced by an eyepiece that permits the observation of the image of a distant source as it is focussed on the reticle. This is shown in Fig. 6.45. A collimator and a telescope are often used together, where the collimator is a source of the distant target that is observed by the telescope. Telescopes are usually set at infinity but most are also focusable. When an optical component or system is introduced into the test setup, the amount of focusing the telescope must undergo is a measure of the effect that the optic has on the collimated beam. This allows for the testing of focal lengths and of other basic imaging properties.

As already mentioned when discussing collimators, telescopes are used on optical benches that are designed to test the optical performance of components and systems. Optical benches continue to be used during the assembly and final testing phase, but they do no longer play an important part in the in-process testing of optical components in the average optical shop. The more versatile interferometric test methods that are available today take care of most in-process testing requirements.

Some instruments have been built that attempt to be both collimator and telescope. When the objective is focused at infinity, the instrument is useful

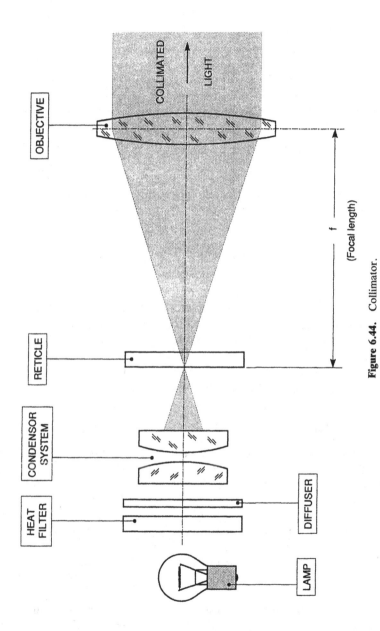

**Figure 6.44.** Collimator.

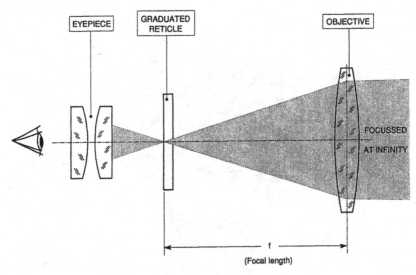

**Figure 6.45.**  Telescope.

as a collimator; when the lens is focused at a finite distance, it is turned into an alignment telescope. To serve both functions, the instrument must have an illuminated target reticle for the collimator option and a reference reticle on which the image of the target is observed through an eyepiece. In this configuration the instrument becomes an autocollimator, which is an essential test instrument found in every optical shop.

### 6.4.3.  Autocollimator

The autocollimator combines the properties of both the collimator and the telescope by projecting an image of the target reticle to infinity. The image is, upon reflection off an optical surface under test, returned through the objective which focuses it onto the eyepiece reticle where it is observed through the eyepiece. The configuration with a graduated reticle was already described more than 100 years ago [88]. Figure 6.46 shows this common arrangement. The modern autocollimator is most often used with a cube beam splitter that separates the light inside the instrument along two orthogonal beam paths. This arrangement simplifies the viewing of the projected image on the eyepiece reticle.

While most autocollimators are primarily used to test the parallelism or the wedge errors of plane surfaces, the instrument can also be adapted for testing spherical surfaces, as is customarily done for lens centering. Small angular deviations can be accurately measured with the autocollimator. When a plane surface that is inclined by a small angle $\alpha$ is placed in front of an autocollimator, the light will be reflected back into the instrument under

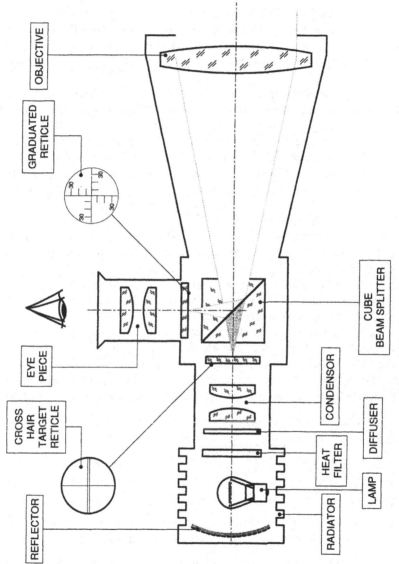

**Figure 6.46.** Schematic of an autocollimator.

the angle δ = 2α (see Fig. 6.47). In other words, the autocollimator doubles the angular errors that are detected as a displacement of the return image in both x and y relative to the orthogonal axes of the eyepiece reticle. The graduations are already corrected for this doubling effect so that the angular error can be directly read off without any further compensation. This feature makes the autocollimator a very important instrument in the optical shop because it permits the measurement of parallelism, specific wedge angles, prism angles, squareness, and perpendicularity of many optical and nonoptical components.

With a high-resolution autocollimator even very small angular deviations in the subarc second range can be detected and accurately measured. The angular resolution is a function of the focal length of the objective lens. The longer the focal length, the smaller is the angular deviation that can be resolved but the longer is the autocollimator tube and the larger the instrument. For instance, a 500-mm (20-in.) focal length objective has a theoretical angular resolution of 0.2 arc second, whereas the corresponding value for a 50-mm objective is 2 arc seconds [89]. The theoretical resolution is rarely attained even under the best of conditions. This fact is recognized by the manufacturers who advertise an angular resolution of <0.5 arc seconds for autocollimators with focal lengths exceeding 500 mm. This is summarized in Table 6.9. The claimed "sensitivities" are around 0.1 to 0.2 arc sec. This is another way of saying *theoretical resolution*. This should not be confused with attainable accuracy in the typical optical shop environment because for practical reasons it is several times lower.

The field of view is also a function of focal length with the higher resolution autocollimators having a proportionally smaller field than the less sensitive instruments. While a 50-mm focal length autocollimator has a range of at least ±20 arc minutes, a more sensitive 500-mm focal length autocollimator will have a total range of only one arc minute or perhaps two. These general

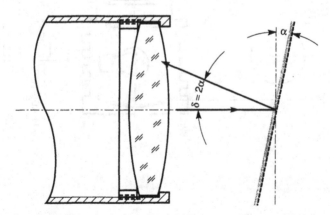

**Figure 6.47.**   Deviation of the retroreflected beam by a slightly tilted plano mirror.

**Table 6.9. Angular Resolution Limit for Long Focal Length Autocollimators**

| Maker | Ref. | F. L. | Resolution limit |
|---|---|---|---|
| Nikon | (90) | 700 mm | 0.5 arc sec. |
| Contraves | (91) | 560 mm | <0.5 arc. sec. |
| Davidson | (92) | 508 mm | 0.5 arc sec. |
| Möller | (93) | 500 mm | <0.5 arc. sec. |

relationships apply only to visual autocollimators. Modern electronic autocollimators have greatly enhanced sensitivities, so high angular resolution can be achieved over a wide range even with instruments of moderate size.

The maximum angular deviation that defines the angular range of the instrument is not only dependent on focal length but also on the distance between the autocollimator and the reflecting surface. The angular range decreases as the distance increases. This is due to the finite size of the objective aperture, which determines an acceptance limit beyond which a reflected target can no longer be intercepted by the instrument. The problem usually occurs when the reflecting surface is many feet away from the autocollimator, as is the case when large machines are aligned from a distance. This limitation, however, is typically not a problem in optics fabrication because working distances rarely exceed a few feet.

Optical autocollimators with very long focal lengths in the 1000 to 2000 mm range (1 to 2 m or 40 to 80 in.) have a theoretical angular resolution between 0.1 and 0.05 arc seconds. It is very doubtful that this level of resolution is attainable in practice. It is certain that such instruments would be not only very expensive but also quite impractical to use. Fortunately there are now electronic autocollimators that can attain this level of resolution quite reliably with objective focal lengths of only a fraction of these cumbersome instruments.

Most autocollimators have fixed objectives that are prefocused at infinity, such as a very versatile autocollimator that has been used in many optical shops for years [94]. Another popular autocollimator [95] permits a small adjustment of the focus over a narrow range to compensate for return image distortion caused by slightly convex or concave reflecting surfaces. The same feature can be used to deduce the flatness of plano surfaces to a sufficient degree of accuracy. It can be used to determine if a lens or a prism system has residual refractive powers. The light sent from the autocollimator into the system is reflected back into the instrument by the tested surface. If that surface is not flat, the eyepiece must be adjusted until the reflected image is seen sharp and free of parallax on the reticle plane. The amount of eyepiece adjustment required is a measure of errors in refractive power.

### 6.4.4. Theoretical Considerations

The displacement of the cross hair image in the $x$ and $y$ directions relative to the eyepiece reticle provides a measure of the angular displacement of the tested surface. The image displacement in $x$ (azimuth) and $y$ (elevation) is measured on the reticle. With the result of this measurement and the known focal length $f$ of the autocollimator objective, the angular error of the reflecting surface can be calculated, using the following set of equations [96].

$$\beta_x = \frac{x}{2f},\tag{6.47}$$

$$\beta_y = \frac{y}{2f},\tag{6.48}$$

$$\beta_{(x,y)} = (\beta_x^2 + \beta_y^2)^{1/2}\tag{6.49}$$

$$\beta_{(x,y)} = \frac{1}{2f}(x^2 + y^2)^{1/2}\tag{6.50}$$

where

$\beta_x$ = angular error along $x$-axis in radians,
$\beta_y$ = angular error along $y$-axis in radians,
$x$ = image displacement in horizontal ($x$) axis,
$y$ = image displacement in vertical ($y$) axis,
$f$ = nominal focal length of the objective,
$\beta_{(x,y)}$ = compound angle in radians.

These relationships, graphically represented by Fig. 6.48 are best explained by examples.

### Example 1

A 500-mm f autocollimator is used to measure the angular error of a first surface mirror (FSM). Assuming that the eyepiece reticle is calibrated in millimeters, a horizontal image displacement of 2 mm and a vertical image displacement of 1 mm are measured. What are the $x$ and $y$ errors in radians, and what is the compound angle in radians?

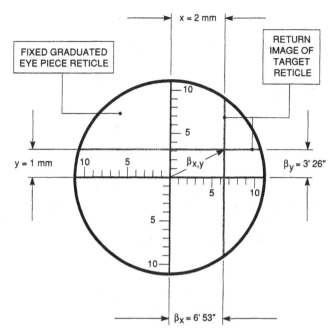

**Figure 6.48.** Definition of the compound angle $\beta_{(x,y)}$ for a $f = 500$ mm autocollimator.

*Note:*

1 radian = $180°/\pi$ = 57.296° = 3437.75 arc min = 202 264.8 arc sec.
1 mr (milliradian) = 0.001 radian = 3.438 arc min.
1′ (arc minute) = 1/3437.75 = 0.0002909 radians = 0.2909 mr.
1″ (arc sec) = 1/202 264.8 = 0.000004848 radians = 0.00485 mr.
1″ (arc sec) = 4.85 $\mu$in./in. = 4.85 $\mu$m/m.
0.001 milliradian = 0.2 arc sec.

From Eqs. (6.47), (6.48), and (6.49),

$$\beta_x = \frac{2}{2 \times 500} = \frac{1}{500} = 0.002 = 2 \text{ mr} = 2 \times 3.438' = 6.876',$$

$$\beta_y = \frac{1}{2 \times 500} = \frac{1}{1000} = 0.001 = 1 \text{ mr} = 3.438',$$

$$\beta_{(x,y)} = \sqrt{[(0.002)^2 + (0.001)^2]} = 0.002236 = 2.236 \text{ mr} = 7.687'.$$

*Result:* The $x$ error is 6.88 arc min, the $y$-error is 3.44 arc min, and the total error is 7.69 arc min.

**Example 2**

It is probably simpler to use Eq. (6.50) when $x$ and $y$ are known:

$$\beta_{(x,y)} = \frac{1}{2 \times 500} \sqrt{(2^2 + 1^2)} = 0.001\sqrt{5} = 0.001 \times 2.236 = 2.236 \text{ mr}.$$

**Example 3**

Since almost all autocollimator reticles are graduated for direct angular read-out, the actual values read instead of $x = 2$ mm and $y = 1$ mm are $\alpha_x = 6'53''$ and $\alpha_y = 3'26''$, where $\alpha_x$ and $\alpha_y$ are in degree measure. The compound or total angle $\alpha_{x,y}$ is then calculated using Eq. (6.51).

$$\alpha_{x,y} = \sqrt{[(\alpha_x)^2 + (\alpha_y)^2]}, \tag{6.51}$$

$$\alpha_x = 6 + 53/60 = 6.883 \text{ arc min},$$

$$\alpha_y = 3 + 26/60 = 3.433 \text{ arc min},$$

$$\alpha_{x,y} = \sqrt{[(6.883)^2 + (3,433)^2]} = 7.69 \text{ arc min, same as above}.$$

For the calibration of reticle divisions, the angular deviation can also be expressed trigonometrically:

$$x = f \tan 2\alpha_x, \tag{6.52}$$

$$y = f \tan 2\alpha_y. \tag{6.53}$$

**Example 4**

Based on the angular values of the displacement in the $x$ and $y$ axes from the previous example what is the physical distance of the displacement relative to the fixed reticle reference? Before the equations can be used, the angular values must be converted to degrees: $\alpha_x = 6.833$ arc min $= 6.833/60° = 0.1139°$ and $\alpha_y = 3.433$ arc min $= 3.433/60° = 0.0572°$. Using Eqs. (6.52) and (6.53) results in

$$x = 500 \tan 2(0.1139) = 500 \, (0.004) = 2.0 \text{ mm},$$

$$y = 500 \tan 2(0.0572) = 500 \, (0.002) = 1.0 \text{ mm}.$$

These are the original values from Example 1. Since the angular values are typically much less than 1 degree, the use of radian measure can simplify these equations to

$$x = 2f\beta_x, \qquad (6.54)$$

$$y = 2f\beta_y. \qquad (6.55)$$

## Example 5

Using the radian values from Example 1, the validity of these equations can be checked:

$$x = 2\,(500)\,(0.002) = 2.0 \text{ mm},$$

$$y = 2\,(500)\,(0.001) = 1.0 \text{ mm}.$$

These relationships illustrate the importance of focal length on angular resolution. A couple of examples can clarify this further:

## Example 6

$f = 100$ mm, offset in $x = 0.2$ mm, offset in $y = 0.2$ mm; then $\beta_{(x,y)} = 1/2(100)$ $(0.2^2 + 0.2^2)^{1/2} = 14.1$. $f = 500$ mm, same $x$ and $y$ offsets; then $\beta_{(x,y)} = 1/2(500)\,(0.08)^{1/2} = 70.7$.

The resolution of the 500-mm $f$ autocollimater is five times greater than that of the 100-mm $f$ unit for the same image displacement.

The preceding calculations are usually not necessary because the eyepiece reticle is graduated in arc minutes or multiples of arc seconds. The values of $x$ and $y$ can be read off directly in angular displacement. The total error must then be calculated using Eq. (6.51). To simplify matters even further, the position of the reflecting surface relative to the autocollimator can be adjusted by turning leveling or tilt adjustment screws until one of the image axes is superimposed on the corresponding axis of the eyepiece reticle. The same effect is achieved when the autocollimator is adjusted relative to a fixed reflecting surface. By doing this, one of the displacements is nulled out, and the total error can then be directly read off the reticle graduations on the other axis (see Fig. 6.49). This is the preferred method of measurement for those surfaces that can be adjusted in this way. Prior to each measurement, however, it is necessary to recalibrate the relationship of the autocollimator axis to the plane of the part support by resetting it with a suitable angle standard. The resolution limits of autocollimators are sometimes also expressed in linear measures such as micron per meter ($\mu$m/m) or microinch per foot ($\mu$in./ft): (*Note:* 1 $\mu$in. $= 25.4 \times 10^{-6}$ mm, 1 ft $= 304.8$ mm). Examples of the metric version are shown in Table 6.10.

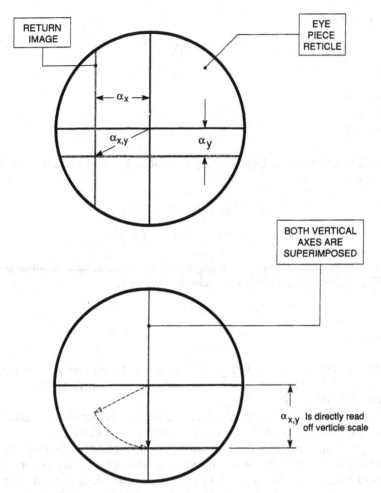

**Figure 6.49.** Nulling out one axis to read off total angular error on the other axis.

**Table 6.10.**
**Autocollimator Resolution**
**in Terms of Linear**
**Measure**

| Angular resolution | (µm/m) |
|---|---|
| 0.1 arc sec. | 0.5 |
| 0.5 arc sec. | 2.5 |
| 1.0 arc sec. | 5.0 |

### 6.4.5. Reticles

The illuminated target reticle has two fine lines etched into one side of the highly polished plane-parallel glass substrate. These lines intersect at right angles to each other to form a cross. The intersection of the cross lies at the exact center of the reticle. The image of the cross is projected to the test surface and retroreflected from there back into the instrument where it is focused by the objective on the graduated reticle surface.

There are two basic types of eyepiece reticles: One is a double-line cross without any graduations; the other is a single-line cross with graduations. The double-line cross permits the precise centering of the image of the target reticle. The centering is done with two micrometer screws that can move the eyepiece reticle independently in $x$ and $y$. The exact distance the eyepiece reticle has to be moved in one axis to position the image axis between the double line can be read off the micrometer drum which is calibrated in arc minutes or even arc seconds. The other axis is then centered, and the error is read off in identical fashion. The graduated eyepiece reticle, shown in Figs. 6.48 and 6.49, is usually calibrated in arc minutes. Some instruments with this kind of reticle are also equipped with micrometer drums to provide arc second readings between the graduations. For instance, if an $f = 500$ mm autocollimator is used with a 1-$\mu$m accuracy micrometer, then each micrometer graduation corresponds to an 0.2-arc sec angular deviation.

In both cases the compound angle must then be calculated, where $x$ and $y$ are the values that are read off the micrometer drums or the graduations of the reticle. Since both are calibrated in arc minutes (or arc seconds), the conversion factor from radian to angular measure is not used.

Reticles can be either bright field or dark field. The choice depends on the application. For instance, low-reflecting surfaces may require the use of a dark-field reticle where the dim cross line image can be detected better against a dark background. The choice is often only a matter of preference. There are three different read-out methods:

1. Double-axis graduated eyepiece reticle
2. Double cross with one micrometer per axis (vernier or digital)
3. Digital single-axis readout to 0.1 arc second accuracy

The reticle can be graduated in such a way that the angular deviation $\alpha$ can be directly read off in arc minutes. The physical distance between graduations for such a calibration can be calculated using Eq. (6.56):

$$t = 0.00029f \qquad (6.56)$$

where

$t$ = graduation in millimeters for an angular deviation of one arc minute,
$f$ = focal length of the autocollimator objective in millimeters, 0.00029 = radian measure for 1 arc minute.

The same equation applies when both graduation $t$ and focal length $f$ are expressed in inches.

## Example

The graduation $t$ for one arc minute is to be calculated for an autocollimator objective with a 300-mm focal length:

$$t = 0.00029\ (300) = 0.087\ \text{mm}.$$

### 6.4.6.  Tooling Autocollimator

There are many applications in the optical shop where arc second precision is not required. Many angular dimensions and tolerances are specified in arc minutes. The high-resolution autocollimators would not be practical for these because of their inherently narrow range of angles that can be measured. There are several short focal length autocollimators specifically designed for such applications. They typically have an angular resolution of 0.5 arc min and a range of ±30 arc min or even more. Not only are they easy to use in the shop environment because of their compact size, but they also offer a graduated reticle that is easy to read.

Although these tooling autocollimators are useful for the direct measurement of angles, they are best suited for comparative measurements after the instrument has been set up against a known angle, wedge, or parallelism standard. The angles of optical parts can then be compared to the standard to determine conformance to specifications or the angles of prisms, and the parallelism of windows can be corrected in by fine-grinding one side. With a suitable mount or stand, such a tooling autocollimator [97] can be positioned at any angle over a 120° range (Fig. 6.50). A modular measurement system [98] produced in Germany is adaptable to perform a variety of measurements. The instrument has a side-mounted eyepiece so that the technician can use it while seated.

### 6.4.7.  Pinhole Autocollimator

A simpler version of the tooling autocollimator is the pinhole autocollimator for which the two reticles have been replaced by small circular pinholes. They are typically used only for pass/fail inspection during manufacture and final quality assurance. The collimator pinhole is illuminated from the back, projected at infinity, and reflected back into the instrument where its image is focused on the pinhole reticle in the focal plane of the eyepiece. Two pinholes will be seen through the eyepiece, namely, the image of the collimator pinhole and the pinhole in the eyepiece reticle. If they are superimposed, then there is no angular error, but if they are seen as two separate pinholes,

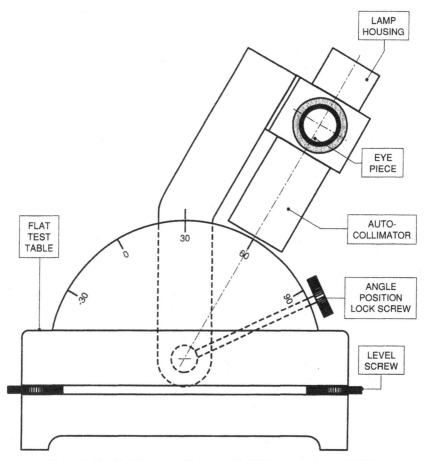

**Figure 6.50.** Tooling autocollimator with 120° range angle stand [95].

then the angular error exceeds the tolerance limit. This is so because the size of the pinholes can be chosen so that the tolerance limit of the angular error is at that point where the two pinholes just barely touch. When they are separately seen, the limit is exceeded and the part is not acceptable; when they overlap, the part is good. These conditions are shown on Fig. 6.51 [93]. The pinhole sizes correspond to angular errors ranging from several arc seconds to a few arc minutes. With this type of autocollimator, even lower-skilled inspectors can quickly and accurately determine the angular accuracy of optical parts.

The parallelism of transparent plane parallel parts can be quickly measured with such aperture images. The images that are reflected off the entrance and exit faces are needed for the parallelism measurement of plane parallel parts. It is also possible to use them to determine if the angular

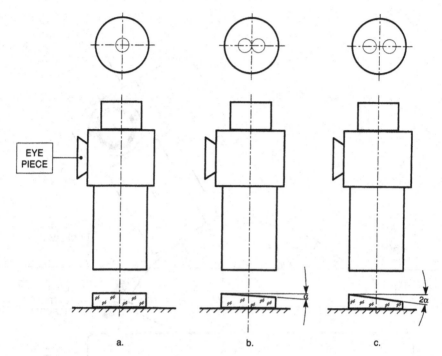

**Figure 6.51.** Measuring parallelism with a pinhole autocollimator [93]. *(a)* Pinhole images are superimposed; surfaces are parallel. *(b)* Pinhole images are just touching; limit of wedge tolerance. *(c)* Pinhole images are separated; wedge error exceeds tolerance.

tolerances of prisms have been met. Even tightly toleranced angles of roof prisms can be tested in this way.

### 6.4.8.  Electronic Autocollimator

The most accurate instruments to measure angles and angular deviations are the modern electronic autocollimators. For these, the eyepiece has been replaced by a photosensor array such as a CCD, and the position of the return image relative to a fixed reference is automatically measured to a very high degree of accuracy. These instruments are either single-axis or dual-axis devices with a resolution of less than 0.2 arc sec over a wide range of up to ±20 arc min.

The data output is compatible with most commonly used computers for analysis and graphic representation. The digitized data are processed by the computer, which uses this information to calculate statistical parameters and then display them as graphs, plots, or maps on the associated TV monitor. The angular deviations in $x$ and $y$ are displayed independently, and the compound error is calculated and displayed as well.

These modern autocollimators are equipped with highly precise rotary encoders that provide the digital output that the computer can manipulate and display. The typical resolution of these electronic autocollimators is 0.001 milliradian, which corresponds to about 0.2 arc sec. The resolution is a function of the range of angles that have to be measured. Table 6.11 lists this dependence for one type of electronic autocollimator [99]:

A particularly accurate electronic autocollimator [100] is built on a Zerodur zero-expansion glass baseplate. It has Zerodur [19] objective lenses, and the mirror mounts are made from Zerodur as well. The critical metal components such as the CCD mount are made from Invar, a very low expansion steel. The theoretical resolution is 0.005 arc sec over a 300-arc sec range. However, the absolute accuracy is claimed to be less than 0.02 arc sec over a 10 arc sec range, and less than 0.06 arc sec over the entire 300 arc sec (5 arc min) range. Like all modern electronic autocollimators, this instrument has also an electronic control unit with a computer interface.

It is doubtful that this kind of angular accuracy is needed for even the most precise production optics. The use of the electronic autocollimator is probably restricted to the critical alignment and control of complex experiments. The fact, however, that this level of precision is attainable today points to a time when shop instrumentation with reliable subarc second resolution will become the accepted standard.

### 6.4.9. Special Autocollimators

Some electronic autocollimators are not designed for shop or lab use. They are specifically designed to sense and monitor angular changes as a function of time. Like for all other electronic autocollimators, the eyepiece has been replaced by a silicon photodetector array. But instead of the usual method of detecting the reflected image of a target reticle, this type of autocollimator [101] depends on an entirely different approach.

Light emitted from two light-emitting diodes (LED) is reflected by a mirror back into the autocollimator where the return light is imaged onto a slit. The photodetector is located just behind this slit. The two LED sources are

**Table 6.11. Angular Range as Function of Resolution**

| Resolution | Angular range |
| --- | --- |
| 0.5 arc sec. | 1.0 arc sec. |
| 1.0 arc sec. | 5.0 arc sec. |
| 1.5 arc sec. | 10.0 arc sec. |
| 2.0 arc sec. | 20.0 arc sec. |

modulated 180° out of phase. The detector transmits the amplitude and phase data of the reflected light to the electronic control unit where the information is processed. The direction and amplitude of the resulting signal are a function of the angular deviation of the mirror.

Instead of measuring the displacement of the return image relative to a fixed reticle or doing the same by moving the reticle to a null position and then reading off the angular displacement on a micrometer drum, the electronic autocollimator of this type utilizes the linear repositioning of an optical wedge to redirect the deviated light onto the slit. The exact position is found where the maximum signal occurs. The amount of wedge displacement is then read off a micrometer drum which is calibrated in arc seconds. The instrument has a claimed 0.1 arc sec resolution over a ±100 arc sec range (±1.67 arc min).

The spectrometer is a precision instrument that combines the advantages of collimator, telescope, and autocollimator in one device. It is specifically designed for precision measurement of angular deviation of beams for the determination of the spectra of light sources. The spectrometer, also called *goniometer*, is typically found only in the test lab because it is too expensive for use on the production floor. The basic feature of this instrument is an autocollimating telescope that swings around a prism table. The angular swing is measured on a dividing head and read off with a microscope to arc second accuracy. In most cases the goniometer is equipped with a second collimator which is illuminated by the light of spectral lamps.

### 6.4.10. Applications

Autocollimators can check directly the parallelism of beam splitters and windows made from visually transparent materials. It must first be set up, however, with a parallel reference part to test the parallelism of parts made from opaque materials. They are also used to test the alignment of cube beam splitters during their assembly, and they are indispensible for testing the accuracy of prism angles. These are typically comparative measurements for which the autocollimator is set up against a precise angle standard. Figure 6.52 shows some of these test arrangements. Table 6.12 lists the accuracy limits of autocollimators as a function of objective focal length [93].

Other typical applications for autocollimators that are occasionally needed in an optical shop are flatness mapping of surface plates and other large plano surfaces, checking the angular accuracy of precision rotary tables, and verifying the straightness of rails and machine ways. Optical instruments, such as cameras, can also be adjusted with an autocollimator. To do that, a plane mirror is placed in the image plane of the camera. When the mirror is exactly in the focal plane of the camera objective, the reflected reticle image is seen in focus in the eyepiece.

When measuring polished plane-parallel parts made from a transparent material, any parallelism error of a few arc seconds or more will produce two

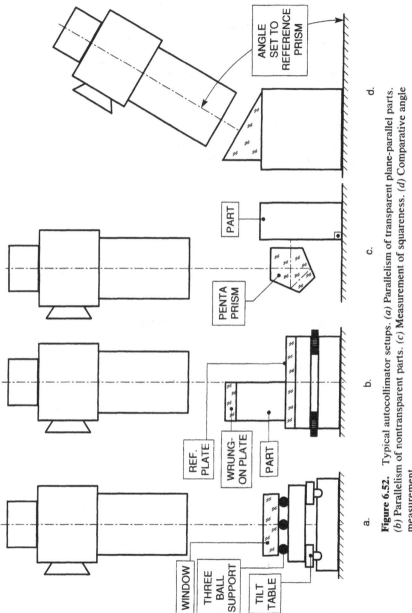

**Figure 6.52.** Typical autocollimator setups. (*a*) Parallelism of transparent plane-parallel parts. (*b*) Parallelism of nontransparent parts. (*c*) Measurement of squareness. (*d*) Comparative angle measurement.

WINDOW
THREE BALL SUPPORT
TILT TABLE

a.

REF. PLATE
WRUNG-ON PLATE
PART

b.

PENTA PRISM
PART

c.

ANGLE SET TO REFERENCE PRISM

d.

**Table 6.12.  Typical Autocollimator Accuracy Limits**

| Category | f = 140 mm | f = 300 mm | f = 500 mm |
|---|---|---|---|
| Parallelism | 8 arc sec. | 3 arc sec. | < 1 arc sec. |
| Right angles | 4 arc sec. | 2 arc sec. | < 1 arc sec. |
| Roof angles | 3 arc sec. | 2 arc sec. | 1 arc sec. |

separate images of the target cross (Fig. 6.53). One of the images is from the front surface of the window and the other from the back surface. The separation of these two images corresponds to the angular error $\delta$. This deviation is calculated using Eq. (6.57):

$$\delta = 2\alpha n, \tag{6.57}$$

where

$\delta$ = angular deviation through the reflection off the backside of the window,

$\alpha$ = parallelism error between the surfaces of the window,

$n$ = the refractive index of the window material.

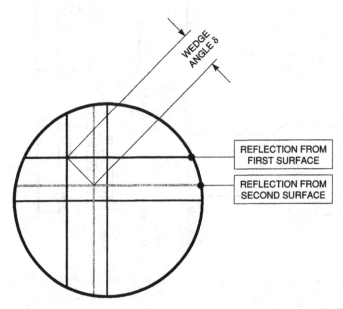

**Figure 6.53.**  Double image of test target when measuring parallelism of wedged parts.

When the deviation $\delta$ is read off during such an autocollimator measurement, the parallelism error $\alpha$ is calculated by

$$\alpha = \frac{\delta}{2n}. \tag{6.58}$$

Multiple images can be very bothersome when testing the angles of prisms. But when the prisms are placed at a slight angle with respect to the autocollimator, these undesirable images can be reflected out of the field of view.

### References for Section 6.4 (Autocollimator)

[88] Ernst Abbe, 1840–1905, director of Zeiss-Jena, Germany.

[89] Vermont Photonics (Möller-Wedel), "Optics and metrology newsletter," vol. 1, no. 2, 1987.

[90] Nikon, Japan.

[91] Contraves Goerz, USA.

[92] Davidson Optronics, USA.

[93] Möller-Wedel, Germany.

[94] Rank Hilger-Watts, England, model TA 51.

[95] Nikon, model 6B or 6D.

[96] Möller-Wedel, "Optical Test Equipment," 1987.

[97] Davidson Optronics, model D-266 with D-657 autocollimator.

[98] Möller-Wedel, modular measurement system Melos 500.

[99] Nikon, Photoelectric Autocollimator.

[100] Möller-Wedel, model Elomat HR, (Vermont Photonics) "Optics and metrology newsletter," vol. 3, no. 3, 1989.

[101] Micro-Radian Instruments, USA, 1983.

## 6.5.  SURFACE ANALYSIS

The customary visual surface inspection methods are still more than adequate for most optical surfaces. Visual methods include surface evaluation with the unaided eye by reflection or backlight transmission, eye loupe inspection, and  inspection with a binocular microscope both with low-level and high-intensity illumination. Visual inspection only yields qualitative data that are subject to variations in interpretation. The more critical requirements for low-scatter or super-smooth surfaces, however, go far beyond the level of surface quality that can be detected by visual means. Standard microscopic inspection at commonly used magnification and with high intensity illumination cannot usually resolve the fine surface detail that must be discerned for classification. Much more sensitive, and above all quantitative, methods are needed.

Very high resolution microscopy has shown promise, but since the observed conditions cannot be properly documented and quantified, their use has remained limited. Some good success was achieved with a high-quality Nomarski microscope [102] for which the critical lenses were carefully selected and matched. It was possible to resolve a great deal of fine surface detail at 1000× magnification to permit reliably grading the surfaces according to their apparent surface roughness. The visual grading corresponded very well with scatter measurements made of the same surfaces. Although the microscope was very useful as an in-process monitoring instrument for on-the-block surface analysis, it had to be backed up by a surface analyzer because the microscope image could not be quantified.

Diamond stylus profilometers have been in use for quite some time to provide surface roughness traces on a variety of surfaces on a wide range of materials. This method is inherently very sensitive, but its resolution is limited by the necessarily finite radius of the diamond stylus point. The fact that it is a contact measurement method that can damage sensitive optical surfaces has limited the use of stylus profilometry in optics.

Scatterometers became useful lab instruments with the development of the laser. Their use has not been significantly extended into the optical shop environment because of their inherently sensitive nature. Modern devices, however, have been designed for less than ideal operating conditions so that their use in the shop has become much more practical.

The most notable development in the analysis and characterization of low-scatter surfaces has occurred during the 1980s. The successful development of the ring laser gyro (RLG) created the need for a reliable and easy to use metrology to measure and quantify super-smooth optical surfaces. The efforts to meet that challenge resulted in a number of very sensitive and versatile instruments that are rugged enough to be used right on the production floor. They are collectively known as *optical profilometers* or *surface roughness profilers*.

More recently interest in the measurement and control of subsurface defects has lead to the extension of scatterometry to detect and quantify conditions that lie just below the polished surface. The instruments that are capable of doing this [103] are limited to date to the evaluation of certain single-crystal materials used by the semiconductor and the optical industries. There are also continuing efforts to develop improved comparative surface quality inspection instruments [104] for defect measurement and mapping. Concurrent reviews of international standards address the same problems from another angle.

### 6.5.1. Stylus Profilometer

Stylus profilometers measure the texture of a surface by dragging a very fine diamond stylus under a feather-light load across the surface. The minute variations in vertical position of the stylus point as it slowly traverses the surface are detected by sensitive transducers and greatly amplified to serve as input signals for driving the pen of a chart recorder. The amplification produces greatly enhanced vertical excursions of the pen, while the chart speed is synchronized with the horizontal motion of the stylus in such a way to produce a long trace on the chart for a relatively short travel of the stylus. Consequently a greatly enlarged representation of the surface roughness is obtained that can then be evaluated in one of several analytic methods.

The point of a diamond stylus has a radius of typically 12.5 $\mu$m (0.0005 in.), while the load on the stylus can be·varied from 1 up to 25 mg. The vertical resolution of most common profilometers start at about 5 Å and extend to over a million angstroms (or 100 $\mu$m) with detectable details of 1 in 40. The magnification of the vertical motion resulting from electronic amplifaction of the faint signal from the stylus ranges from 1000 to 1 million times. Such high magnification is necessary to make these fine excursions visible and traceable. The typical scan length is on the order of 1 or 2 cm with a horizontal resolution of about 400 to 500 Å. The corresponding lateral magnification ranges from 2× up to 5000×. *Note*: 1 Å = 10-4 $\mu$m = 10-7 mm= 1/254 $\mu$-in.

Stylus profilometers [105] are primarily designed to measure the texture or microtopography of the surface. The more common term used is *surface roughness*. But stylus profilometers can also measure the shape of the test piece across the length of the scan, such as radius of curvature and the angle of surface inclination. Measurements are not limited to plano surfaces because good scans can be obtained from most curved surfaces as well.

Most modern instruments [106] are programmable for automatic scanning and data acquisition. The scan data are automatically digitized for computer analysis and display of statistical parameters that can be selected from a program menu. Arithmetic averages and standard deviations of a number of parameters are generated and graphically displayed.

Simpler versions of these instruments are also available [107, 108] but they are not a suitable substitute for measuring the surface roughness of optical surfaces. They are specifically designed to measure the step height in patterned thin film and thick film coatings, and as such, they lack the vertical sensitivity needed for optical surface roughness measurement.

### 6.5.2. Scatterometer

Scatterometers measure surface roughness indirectly by detecting and analyzing light scattered by irregularities on the surface. The scatter is often not just the result of surface roughness alone. Surface contamination and bulk scatter can distort the results appreciably. There have been some recent developments to separate out the scatter caused by contaminants on the surface and that generated by scatter points in the material. It is not clear at this time if these efforts have been successful.

Scatterometers can be used on reflective, transmissive, and diffuse plano or curved surfaces. They operate on a basic principle but the measurement and interpretation of the light scatter is anything but simple. A narrow, collimated beam of light, usually a HeNe laser beam, strikes the test surface at an oblique angle and is reflected by the surface according to the laws of reflection. The reflected beam, also called *specular beam*, is then trapped and terminated to avoid any interference with the scatter measurement. Any surface irregularity will scatter a tiny fraction of the light in many directions according to scatter theory. The sum of all the light scattered by numerous surface irregularities is the light that must be detected and measured by suitably sensitive photodetectors. Depending on the type and distribution of the scattering points on the surface caused by roughness or contamination or from the bulk material, the scattered light may not be uniformly distributed but may be greater at specific angles from the specular beam. Since the scatter density can also vary radially around the specular beam for the same reasons, it is necessary to measure the scatter in that direction as well.

Most scatterometers measure the scatter along a concentric path around the specular beam at one specific angle of incidence. The detector angle can be changed for successive readings, but the proper alignment and recalibration after each angle change makes this a very tedious process. The more sophisticated computerized instruments measure the scatter distribution automatically at several angles by moving the detector a preprogrammed angle after completion of each radial scan. A three-dimensional scatter map can be generated with this information. One such instrument [109] is shown schematically in Fig. 6.54. There are several other scatterometers commercially available that have been designed for specific applications [110, 111, 112, 113].

### 6.5.3. Optical Profilometers

The development of the ring laser gyro (RLG) required the detection and measurement of very low surface roughness and resulting surface scatter. Neither the stylus profilometer nor the scatterometers could provide that level of resolution. This requirement encouraged the development of several competing optical profilometers. These instruments detect surface roughness by measuring phase changes in the light reflected and scattered off the test surface. The phase changes are directly related to the microtopography of the surface. There are to date three instruments that operate on this principle. One is the optical heterodyne profilometer, another utilizes split beam interferometry, and the third depends on the use of a Mireau interference objective. All these methods are noncontact with claimed resolutions in the subangstrom rms surface roughness range. Each of these instruments will be discussed separately.

**Heterodyne Profilometer.** This instrument was originally designed for a research effort at a U.S. national laboratory [114], and it is now commercially

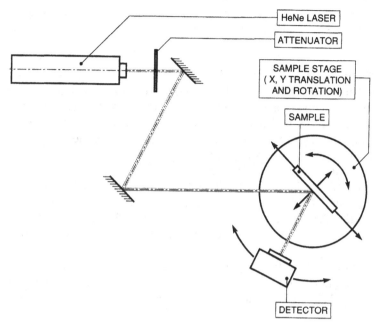

**Figure 6.54.**   Typical scatterometer [109].

available [115]. The beam of the HeNe laser used in this instrument is split, and one of the beams is polarized at a right angle to the first. Both beams are focused by the same microscope objective on the surface under test which is placed face down on a rotating stage. The two beams focus at the same plane, but they are separated by a small distance. The reference beam focuses precisely on the center of rotation of the slowly rotating table, while the other traces a circular path around it. Even though both beams originate from the same source and travel along the same path, they will not interfere with one another because they are orthogonally polarized.

The splitting of the beam also introduces a small frequency difference between them. The phase differences of the beat frequencies between the two reflected beams are detected. This phase difference is directly proportional to the height variations in the microtopography of the surface. The phase information is analyzed by computer to calculate the relative height differences between the reference beam and the measurement beam. Since the measurement beam traces a circle with a very small radius, the trace represents a tiny sample of a comparatively large surface and several readings may have to be taken at different locations on the surface to get a representative picture of the overall surface topography.

The computer analysis permits the graphic representation of surface characterization parameters such as surface roughness profile which is similar to that of a stylus profilometer, the autocovariance function which is a measure of any periodicity in the surface structure, the spectral density function, height and slope distribution, and root mean square (rms) surface roughness values.

**Split Beam Interferometric Profilometer.** This instrument [116] is also a noncontact optical profilometer with angstrom resolution. It has the advantage of long scan lengths of up to 100 mm (4.0 in.) at relatively high scan rates. This feature gives it the sample capacity of a stylus profilometer without requiring contact with the surface. Figure 6.55 shows the optical layout.

The detector measures the relative phase shift between two beams split from the output of the HeNe laser. Because it is a phase measuring instrument, it is insensitive to vibrations, so it can be used safely in the shop environment. The detected phase differences are analyzed by computer, and the resulting data are graphically displayed as surface profile, surface slope, surface height histogram, autocovariance spectrum, and power spectrum.

**Surface Texture Interferometric Microscope.** The most common type of optical profilometer goes by several names, although only two companies make them [117, 118]. It is basically an interferometric, noncontact, nonvisual

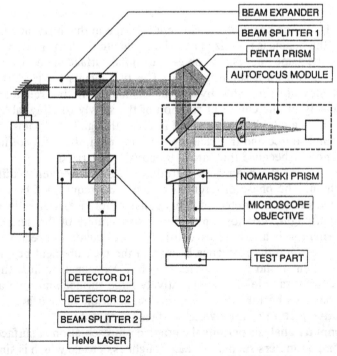

**Figure 6.55.** Optical surface profiler [116].

microscopic method of surface evaluation that detects phase differences relative to a reference. These phase differences are directly related to height variations in the surface texture. Figure 6.56 shows one such profilometer [119].

These instruments, like the others already discussed, will measure the microstructure or microtopography of reflecting optical surfaces and the overall profile of the surface over the scan length or the scanned area. One type uses white light illumination [119], while the other uses HeNe laser source [120]. The claimed advantage of a white light source is the absence of spurious fringes that can be a problem with HeNe lasers.

At the heart of each system is either a Fizeau or a Mireau interference microscope objective (see Fig. 6.57). The Mireau objective uses an equal path interferometer design that limits its use for very fine details and thin films because interference takes place only over a very narrow range. The Fizeau objective functions as an unequal path configuration that allows surfaces with appreciable topographical height variations to be evaluated. The quality of the reference surfaces of these microinterferometers limits the resolvable surface detail on the test surface. Although the Mireau uses a very small reflective spot as its reference, the reference surface of the Fizeau objective is much larger. Consequently the Mireau objective is potentially more sensitive to very fine surface detail than the Fizeau configuration. This is because it is much easier to produce a very fine surface over a small area than over one that is much larger.

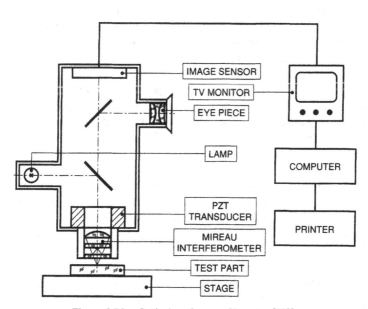

**Figure 6.56.** Optical surface profilometer [119].

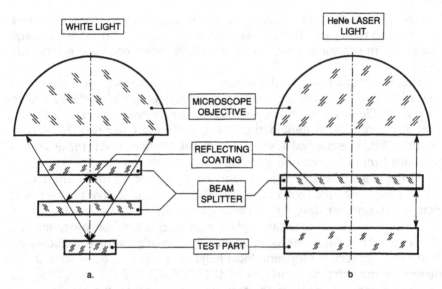

**Figure 6.57.** Basic optical profilometer heads [120]. *(a)* Mireau. *(b)* Fizeau.

The test part is placed under the microscope objective and its position is adjusted for the desired region while the operator views the interference fringes in through the eyepiece. The light from the source is partially reflected by the reference surface of the interference objective, but most of the light passes through to the test surface. From there it is reflected back into the objective where the return beam interferes with the reference beam. The entire objective head is piezoelectrically driven to modulate the reference beam. This modulates the interfering reflected beam, and the phase differences are directly proportional to surface height variations in the surface structure. The phase information becomes part of the interfering wave front.

A photodiode array detects the variations in the wave front as the reference surface is modulated. The resulting phase information is analyzed, quantified, and used by the computer to calculate surface height differences and all other surface characterization data. The measurement and data analysis cycle is quite fast.

There are two basic ways to detect, compute and display the information. The surface roughness data can be represented as a two-dimensional (2-D) line scan as height variations along a line on the surface. This is very similar to the line scan of a stylus profilometer. It can also be represented as a three-dimensional (3-D) area scan when the surface height variations are measured over a square or rectangular area. The 3-D plot results in a topographical map of the scanned area. The 2-D detector is a linear array with 1024 elements. The 3-D system uses a 256 × 256 detector array.

After computer analysis, the 2-D data is displayed as a surface profile with calculated peak-to-valley (P–V) and root mean square (rms) values. The P–V to rms ratio ranges from about 3 for smooth surfaces to 6 or more for rough surfaces. The vertical scale of the scan is in angstroms, nanometers, or micrometers, whereas the scan length is in micrometers with a scan length of about 0.65 mm (0.026 in.). Representing the data in two dimensions is best for very smooth surfaces that have little or no structure. Superpolished surfaces for low-scatter mirrors and lenses fall into this category. The surface roughness resolution limits of these instruments is 0.1 nm, or 1 Å (.004 $\mu$in.) rms. With sophisticated software treatment, surfaces with 0.5 Å rms surface roughness or even less have supposedly been measured.

The 3-D surface roughness data are displayed as a three-dimensional topographical map with greatly exaggerated vertical height. The area that can be covered is a function of the magnification of the interference objective. At low magnification the area can be several millimeters per side, while the scan area is reduced to a few tens of microns at the highest magnification. In addition electronic cursors can be used to outline a specific area of interest and only the information within the limits defined by the cursors are analyzed by the computer. This permits the close examination of unique surface details. The position resolution in $x$ and $y$ is about 1/4 $\mu$m. The vertical resolution that defines the relative height differences is 6 Å P–V or 0.02 $\mu$m. This corresponds to a surface roughness of 1 to 2 Å rms.

The 3-D contour maps are best for structural surfaces such as those resulting from single-point diamond turning (SPDT) operations, diffractive optical elements (DOE), gratings, optical data storage disks, and compact disks.

**References for Section 6.5 (Surface Analysis)**

[102] Leitz Metalloplan, 6 × 6.

[103] VTI, Inc., USA, "PBS Subsurface Measurement", 1988.

[104] SIRA Ltd., England, "Microscope Image Comparator", 1987.

[105] Rank Taylor Hobson, England, model Talysurf.

[106] Rank Taylor Hobson, model FORM Talysurf, 1990.

[107] Sloan, USA, model Dektak.

[108] Tencor, USA, model Alphastep.

[109] Breault, USA, scatterometer, 1987

[110] TMA, USA, scatterometer, 1989.

[111] Tencor, model Surfscan 4000, for Si wafers.

[112] Talandic, USA, scatterometer for RLG, 1985.

[113] Rodenstock, Germany, model RM-400.

[114] LLNL, USA (Gary Sommergren).

[115] Zygo, USA, 1987.

[116] Chapman Instruments, USA.

[117] Zygo, USA.

[118] Wyko, USA.

[119] Wyko, model TOPO.

[120] Zygo, model Maxim.

# Index